ACS SYMPOSIUM SERIES 917

Modern Aspects of Main Group Chemistry

Michael Lattman, Editor
Southern Methodist University

Richard A. Kemp, Editor
University of New Mexico
Sandia National Laboratories

Sponsored by the
ACS Division of Inorganic Chemistry, Inc

American Chemical Society, Washington, DC

Library of Congress Cataloging-in-Publication Data

Modern aspects of main group chemistry / Michael Lattman, editor ; Richard A. Kemp, editor ; sponsored by the ACS Division of Inorganic Chemistry, Inc.

    p. cm. — (ACS symposium series ; 917)

    "Developed from a symposium sponsored by the Division of Inorganic Chemistry, Inc. at the 227th National Meeting of the American Chemical Society, Anaheim, California, March 28–April 1, 2004"—Pref.

    Includes bibliographical references and index.

    ISBN-13: 978-0-8412-3926-5 (alk. paper)

    1. Chemistry, Inorganic—Congresses.

    I. American Chemical Society. Division of Inorganic Chemistry, Inc. II. American Chemical Society. Meeting (227th : 2004 : Anaheim, Calif.). III. Series.

QD151.3.M63   2005
546—dc22

                                              2005048310

The paper used in this publication meets the minimum requirements of American National Standard for Information Sciences—Permanence of Paper for Printed Library Materials, ANSI Z39.48–1984.

Copyright © 2006 American Chemical Society

Distributed by Oxford University Press

ISBN 10: 0-8412-3926-6

All Rights Reserved. Reprographic copying beyond that permitted by Sections 107 or 108 of the U.S. Copyright Act is allowed for internal use only, provided that a per-chapter fee of $30.00 plus $0.75 per page is paid to the Copyright Clearance Center, Inc., 222 Rosewood Drive, Danvers, MA 01923, USA. Republication or reproduction for sale of pages in this book is permitted only under license from ACS. Direct these and other permission requests to ACS Copyright Office, Publications Division, 1155 16th Street, N.W., Washington, DC 20036.

The citation of trade names and/or names of manufacturers in this publication is not to be construed as an endorsement or as approval by ACS of the commercial products or services referenced herein; nor should the mere reference herein to any drawing, specification, chemical process, or other data be regarded as a license or as a conveyance of any right or permission to the holder, reader, or any other person or corporation, to manufacture, reproduce, use, or sell any patented invention or copyrighted work that may in any way be related thereto. Registered names, trademarks, etc., used in this publication, even without specific indication thereof, are not to be considered unprotected by law.

PRINTED IN THE UNITED STATES OF AMERICA

Dedication

**Professor Alan H. Cowley,
The University of Texas, Austin**

The organizers and participants wish to dedicate this volume to Professor Alan H. Cowley on the occasion of his 70th birthday and in honor of his many contributions to main group chemistry.

Foreword

The ACS Symposium Series was first published in 1974 to provide a mechanism for publishing symposia quickly in book form. The purpose of the series is to publish timely, comprehensive books developed from ACS sponsored symposia based on current scientific research. Occasionally, books are developed from symposia sponsored by other organizations when the topic is of keen interest to the chemistry audience.

Before agreeing to publish a book, the proposed table of contents is reviewed for appropriate and comprehensive coverage and for interest to the audience. Some papers may be excluded to better focus the book; others may be added to provide comprehensiveness. When appropriate, overview or introductory chapters are added. Drafts of chapters are peer-reviewed prior to final acceptance or rejection, and manuscripts are prepared in camera-ready format.

As a rule, only original research papers and original review papers are included in the volumes. Verbatim reproductions of previously published papers are not accepted.

ACS Books Department

Contents

Preface..xi

Overview of This Volume..xiii
Michael Lattman and Richard A. Kemp

Unusual Oxidation States and Bonding

1. Some Retrospectives and Perspectives in Main Group Chemistry.........2
   Alan H. Cowley

2. Al-H-C Chemistry..20
   Herbert W. Roesky

3. From Gallium Hydride Chlorides to Molecular Gallium Sulfides........32
   Hubert Schmidbaur and Stefan Nogai

4. Four-Valence Electron Reactive Intermediates and the
   Philicity of Charged Carbene Analogs......................................52
   Peter P. Gaspar, Xinping Liu, Diana Ivanova, David Read,
   James S. Prell, and Michael L. Gross

5. The Diversity of Stable and Persistent Phosphorus-
   Containing Radicals..66
   A. Armstrong, T. Chivers, and R. T. Boeré

6. Stable Singlet Diradicals Based on Boron and Phosphorus...............81
   Amor Rodriguez, Carsten Präsang, Vincent Gandon,
   Jean-Baptiste Bourg, and Guy Bertrand

7. Easily Broken Strong Bonds: A New Law of Thermodynamics...........94
   Konstantin B. Borisenko, Sarah L. Hinchley, and
   David W. H. Rankin

8. Cationic Low Oxidation State Phosphorus and Arsenic Compounds 108
   Bobby D. Ellis and Charles L. B. Macdonald

9. A General Route to M_4N_4 Heterocubanes: Synthesis and Crystal Structure of $[M(\mu_3-NSiMe_3)]_4$ (M=Ge, Pb) 122
   Jack F. Eichler, Oliver Just, and William S. Rees, Jr.

10. Hydrolysis of Diborane(4) Compounds 137
    R. Angharad Baber, Jonathan P. H. Charmant, M. J. Gerald Lesley, Nicholas C. Norman, A. Guy Orpen, and Jean Rossi

11. Recent Developments in Boron–Phosphorus Ring and Cage Chemistry 152
    R. T. Paine, H. Nöth, T. Habereder, J. F. Janik, E. N. Duesler, and D. Dreissig

12. History of a Paradigm Shift: Multiple Bonds to Silicon 166
    Robert West

13. Stable Derivatives of New Isomeric Forms of Heavier Group 14 Element Alkene Analogues 179
    Philip P. Power

Coordination Chemistry

14. β-Diketiminates of Some Main Group Elements: New Structural Motifs 192
    Laurence Bourget-Merle, Yanxiang Cheng, David J. Doyle, Peter B. Hitchcock, Alexei V. Khvostov, Michael F. Lappert, Andrey V. Protchenko, and Xue-hong Wei

15. Fluoride Ion Complexation by Chelating 1,8-Diborylnaphthalene Lewis Acids and Their Isoelectronic Dicarbocationic Analogs 208
    Huadong Wang, Stéphane Solé, and François P. Gabbaï

16. New Complexes of Lanthanides with Unusual Main Group Ligands 221
    Richard A. Jones, Xiaoping Yang, Abdul Waheed, Michael Wiester, and Lilu Zhang

17. Main Group Element Calixarenes: Molecular Constraint and Flexibility...237
   Michael Lattman

18. *N*-Heterocyclic Carbene Adducts of High-Oxidation-State Metal Halides: An Unexpected Instance of Lewis Acidic $C_{carbene}$ Character..252
   Mark D. Spicer, Christopher A. Dodds, John P. Culver, and Colin D. Abernethy

19. Imidazol-2-ylidenes and Their Reactions with Small Reagents.........266
   Taramatee Ramnial and Jason A. C. Clyburne

20. Coordination Chemistry of Phosphorus(III) as a Lewis Acceptor...280
   Neil Burford and Paul J. Ragogna

21. Metallacarboranes of Main Group, Transition, and Lanthanide Elements: Syntheses, Structures, and Reactivities..............................293
   Jianhui Wang, John A. Maguire, and Narayan S. Hosmane

Materials, Polymers, and Other Applications

22. Polyhedral Boranes in the Nanoworld...312
   M. Frederick Hawthorne, Omar K. Farha, Richard Julius, Ling Ma, Satish S. Jalisatgi, Tiejun Li, and Michael J. Bayer

23. Synthesis and Reactivity of (Silylamino)- and (Silylanilino)phosphines..325
   Robert H. Neilson, Pradeep Devulapalli, Bethany K. Jackson, Andrew R. Neilson, Sahrah Parveen, and Bin Wang

24. Cyclic and Polymeric Alkyl/Aryl Phosphazenes.................................335
   Patty Wisian-Neilson, June-Ho Jung, and Srinagesh K. Potluri

25. Giant Dendrimer Construction: Hydroboration versus Hydrosilylation as a Growth Strategy...347
   Jaime Ruiz, Gustavo Lafuente, Sylvia Marcen, Catia Ornelas, Sylvain Lazare, Eric Cloutet, Jean-Claude Blais, and Didier Astruc

26. **Electrical Properties of Boron Nanowires**............362
    Carolyn Jones Otten, Dawei Wang, Jia G. Lu, and
    William E. Buhro

27. **Molecular Design of Precursors for the Chemical Vapor Deposition of Group 13 Chalcogenides**............376
    Claire J. Carmalt, Emily S. Peters, Simon J. King,
    John D. Mileham, and Derek A. Tocher

28. **Phosphate Ester Cleavage with Binuclear Boron Chelates**............390
    Amitabha Mitra and David A. Atwood

29. **Synthesis and Characterization of Divalent Main Group Diamides and Reactions with CO_2**............410
    Yongjun Tang, Ana M. Felix, Virginia W. Manner,
    Lev N. Zakharov, Arnold L. Rheingold, Bahram Moasser,
    and Richard A. Kemp

30. **Organotin–Sulfur Intramolecular Interactions: An Overview of Current and Past Compounds and the Biological Implications of Sn---S Interactions**............422
    Teresita Munguia, Francisco Cervantes-Lee, László Párkányi,
    and Keith H. Pannell

Author Index............437

Subject Index............439

Preface

This volume is based on a symposium entitled *Modern Aspects of Main Group Chemistry,* held at the 227th National Meeting of the American Chemical Society (ACS) in Anaheim, California, March 28–April 1, 2004. Our goal was to bring together many of the well-established leaders in the field of main group chemistry with several of the "up-and-coming" younger scientists. As the invitation to participate in the symposium was based only on the very broad topic of main group chemistry, this book reflects the diversity and richness in creativity that currently defines this branch in inorganic chemistry.

In addition to the scientific reasons for hosting a symposium on main group chemistry, the organizers had another motive in mind. One of the leading practitioners of main group chemistry during the past 40 years has been Professor Alan Herbert Cowley, F.R.S., who is the Robert A. Welch Professor of Chemistry at the University of Texas at Austin. Because Professor Cowley was about to celebrate his 70th birthday, we decided that there would be no more appropriate manner to recognize his contributions to inorganic chemistry than to organize this symposium in his honor. Dubbed "CowleyFest 2004" by the participants, the symposium successfully brought together scientists from around the world, from small universities to the world's leading institutions, with the common themes being our interests in main group chemistry and our affection and respect for Professor Cowley.

As the end product of the meeting, this book is targeted to academic and industrial chemists and materials scientists working in areas related to main group chemistry, including nanotechnology and polymer chemistry, as well as to industrial chemists who want to become familiar with main group chemistry in a minimum of time. In addition, we believe this book will be a valuable resource for postdoctoral students, graduate students, and advanced undergraduates who are conducting research in main group chemistry. This book is designed to be used as a textbook or reference book in a graduate level main group topics course as well.

We are very grateful to all of the participants in this symposium, many of whom took part while paying their own travel expenses. Additionally, we are particularly grateful to those participants who contributed chapters to this book. The assortment of topics presented in this volume is a testament to the importance of main group chemistry in the areas of science and technology.

Financial support for this symposium came from a variety of sources. A type SE Grant from the Petroleum Research Fund, administered by the ACS, was instrumental in allowing several of the foreign speakers to attend. Significant financial contributions from the following sources are gratefully acknowledged as well: The Welch Foundation; Shell Chemical Company; ExxonMobil Corporation; Prosortium, Inc.; Slusser, Wilson, and Partridge LLP; The Seaborg Institute at Los Alamos National Laboratory; Dr. Stephen G. Baxter; The University of Texas at Austin; The University of New Mexico (UNM); and Southern Methodist University (SMU). The editors thank their own institutions—SMU, UNM, and Sandia National Laboratories—for participating in the symposium and for allowing time for us to prepare this book. The ACS Books Department is also noted for its help in preparing this book for publication. We also gratefully acknowledge Dr. Stephen Ritter and the ACS for their coverage of this symposium in *Chemical and Engineering News* (May 10, 2004).

Michael Lattman
Department of Chemistry
Southern Methodist University
Dallas, TX 75275

Richard A. Kemp
Department of Chemistry
MSC03 2060
University of New Mexico
Albuquerque, NM 87131

and

Advanced Materials Laboratory
Sandia National Laboratories
1001 University Boulevard, SE
Albuquerque, NM 87106

Overview of this Volume

Michael Lattman[1] and Richard A. Kemp[2,3]

[1]Department of Chemistry, Southern Methodist University, Dallas, TX 75275
[2]Department of Chemistry, University of New Mexico, MSC03 2060, Albuquerque, NM 87131
[3]Sandia National Laboratories, Advanced Materials Laboratory, 1001 University Blvd., SE, Albuquerque, NM 87106

The last twenty-five years has seen a resurgence in main group chemistry, beginning with the fundamental breakthroughs involving multiple bonding in silicon, phosphorus, and heavier elements to the present-day applications in materials and nanotechnology. Simply stated, main group compounds are ubiquitous. In materials research, main group elements are used in the synthesis of nano- and mesoporous materials via sol-gel methods (particularly in silicate chemistry), as well as in the preparation of materials such as boron nitrides and chemically modified single-walled nanotubes. Ceramics and dense materials are almost all made using oxides, carbides, nitrides, or borides. Interestingly, a main group element – hydrogen – is viewed by many as the "fuel of the future." Main group elements such as magnesium (photosynthesis) and phosphorus (ATP and ADP) are important in biological processes. In transition metal chemistry many useful ligands contain main group elements as donor atoms. Inorganic polymers such as polyphosphazenes, semiconductor group 13-15 and 14-16 compounds, coatings, organometallic catalysts – all have main group elements as key components. The chapters of this book cover a broad spectrum of main group chemistry, ranging from the fundamental themes of synthesis and bonding to the more applied aspects. Many of these uses for main group elements are highlighted below and discussed in more detail in the individual chapters.

What defines main group chemistry? As with almost all areas currently in chemistry, the definition seems to be rapidly changing and expanding to areas previously unforeseen. For the purpose of this symposium, we view main group chemistry as the study of elements of the s and p block, particularly those in Groups 1, 2, and 13-17. While not strictly in these groups, the elements of Group 12 are often considered to be main group due to their similarities in reaction chemistry. All of these elements generally use the outermost s and p

orbitals in bonding, yet there are no restrictions against expansion of the valence shell, *e.g.*, in hypervalent compounds.

The book is organized into three broad sections in an effort to categorize the contributions. This organization is somewhat artificial – there is obvious overlap, and certain chapters could easily be in other sections. The first section, *Unusual Oxidation States and Bonding*, highlights some of the most traditional aspects of main group chemistry and updates the current state-of-the-art in this area. The book opens with a chapter by the symposium honoree (Alan H. Cowley) describing recent developments from his laboratory, particularly with respect to carbene analogs. This is followed by chapters on aluminum hydroxides, thiols, selenols, and amides (Herbert W. Roesky) and the preparation of large gallium sulfide arrays from gallium hydride chlorides (Hubert Schmidbaur). The next chapter describes electron-deficient intermediates focusing on four-electron species but also includes six-electron structures (Peter P. Gaspar). Two chapters on stable phosphorus radicals follow: the first describes a variety of neutral radical species (Tristram Chivers), while the second depicts diradical species containing one bond less than usual valence rules predict (Guy Bertrand). The next chapter describes a striking example of solid-state vs. gas phase structural changes and a detailed analysis of the underlying reasons for the differences (David W. H. Rankin). Chapters on structure and reactivity of phosphorus and arsenic cations in a formal oxidation state of +1 (Charles L. B. Macdonald) and M_4N_4 (M = Ge, Pb) cubane-like structures showing increased distortions with the heavier congeners (William S. Rees, Jr.) follow. Two boron-related chapters describe the hydrolysis reactivity of diborane(4) $[B_2R_4]$ compounds (Nicholas C. Norman) and boron-phosphorus rings and cages with potential applications to boron phosphide (Robert T. Paine). The last two chapters of this section are devoted to arguably the most significant fundamental advances in main group chemistry in the last 25 years – multiple bonds to elements below the first row of the periodic table. A chapter on the history of multiple bonding to silicon (Robert West) is followed by one on recent advances to prepare isomeric alkene analogs of elements heavier than silicon (Philip P. Power).

The second section, *Coordination Chemistry*, represents the broad definition that includes most types of donor/acceptor bonds. This section opens with a chapter on β-diketiminates, ligands which stabilize low oxidation states as well as low molecular aggregation (Michael F. Lappert). This is followed by chapters on bidentate Lewis acids with specific fluoride ion binding capabilities (François P. Gabbaï), new "salen"-type main group ligands for lanthanide metal ion binding (Richard A. Jones), and the use of calixarenes to control geometries of main group atoms and ligand-metal interactions (Michael Lattman). Two chapters follow on *N*-heterocyclic carbenes. The first describes carbene/high-oxidation state metal interactions where the carbene displays Lewis acidic behavior (Colin D. Abernethy), and the second illustrates redox chemistry of the carbenes (Jason A. C. Clyburne). This section is rounded out by chapters on phosphorus (III) coordination as a Lewis acid (Neil Burford) and the rich and varied chemistry of metallacarboranes (Narayan S. Hosmane).

The third section, *Materials, Polymers, and Other Applications*, emphasizes how the fundamental aspects of main group chemistry are being transformed into applications. The section opens with a chapter on polyhedral boranes with applications to nanotechnology, including molecular delivery (M. Frederick Hawthorne). This is followed by chapters on precursors to (Robert H. Neilson) and properties of cyclic- and polyphosphazenes (Patty Wisian-Neilson). Chapters on giant dendrimers (Didier Astruc), boron nanowires (William E. Buhro), and molecular precursors for CVD of Group 13 chalcogenides (Claire J. Carmalt) follow. The final three chapters describe the ability of certain boron chelates to cleave phosphate esters with applications to decontamination of nerve agents and pesticides (David A. Atwood), the reactions of main group diamides with carbon dioxide to eventually prepare ^{11}C-labelled radiopharmaceuticals (Richard A. Kemp), and the biological implications of tin/sulfur interactions (Keith H. Pannell).

Unusual Oxidation States
and Bonding

Chapter 1

Some Retrospectives and Perspectives in Main Group Chemistry

Alan H. Cowley

Department of Chemistry and Biochemistry, The University of Texas at Austin, Austin, TX 78712

Introduction

First of all, I offer sincere thanks to the many graduate students and postdoctoral associates it has been my privilege and pleasure to work with over the years, two of whom, Professors Richard A. Kemp and Michael Lattman, organized this symposium so effectively. Equally, I am grateful both to the undergraduates who have participated in our research program and also to the several scientists with whom we have had fruitful collaborations.

For the present symposium, a theme was chosen that represents a more-or-less continuous thread through our research program over a three-decade period. The particular theme is the chemistry of main group analogues of carbenes - species that bear a resemblance to carbenes in terms of the number of valence electrons, frontier orbitals, or coordination number. The most obvious carbene analogues are silylenes, germylenes, stannylenes, and plumbylenes (Table 1).

Table I. Carbenoids

Four-Electron	Six-Electron	Six-Electron	
RB	R_2C	RN	R_2N^+
RAl	R_2Si	RP	R_2P^+
RGa	R_2Ge	RAs	R_2As^+
RTn	R_2Sn	RSb	R_2Sb^+
RTi	R_2Pb	RBi	R_2Bi^+

In regard to group 15 chemistry, nitrenes, phosphinidenes, and their heavier congeners are related to carbenes in the sense that they feature six valence electrons. Nitrenium, phosphenium, and heavier congeneric cations are also six-valence electron species; however, like carbenes they are two-coordinate. Clearly, the one-coordinate borylenes and the related heavier group 13 fragments only possess four valence electrons. Nevertheless, like carbenes, they can, in principle, exist in a singlet or triplet ground state.

My interest in carbene-analogous compounds started in 1965 when I was on a research leave at the University of Southern California in the laboratory of the late Professor Anton B. Burg. One of the outcomes of this sojourn was the synthesis of $(CF_3As)_4$, (**1**), the first example of a cyclotetraarsine (*1*). The first

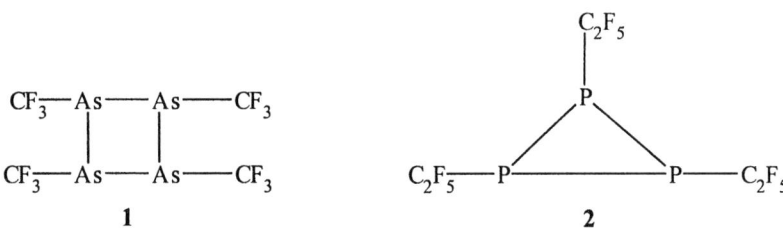

example of a cyclotriphosphine (**2**) was synthesized somewhat later (*2*). Compounds **1** and **2** can be regarded as cyclic oligomers of arsenidenes and phosphinidenes, respectively. As such, it was surmised that these compounds might behave as sources of RP and RAs moieties. The first indication of the validity of this idea came from the reaction of the cyclotetraphosphine, $(CF_3P)_4$, with $Me_2AsAsMe_2$ (**3**) which resulted in the insertion of a trifluoromethylphosphinidene moiety into the As-As bond to form **4** (*3*) (Equation 1). A similar CF_3P insertion reaction took place with MeSSMe; however, no such reaction was observed with $Me_3SiSiMe_3$ (*3*). Since the disilane does not possess lone pair electrons, it could imply that the initial step

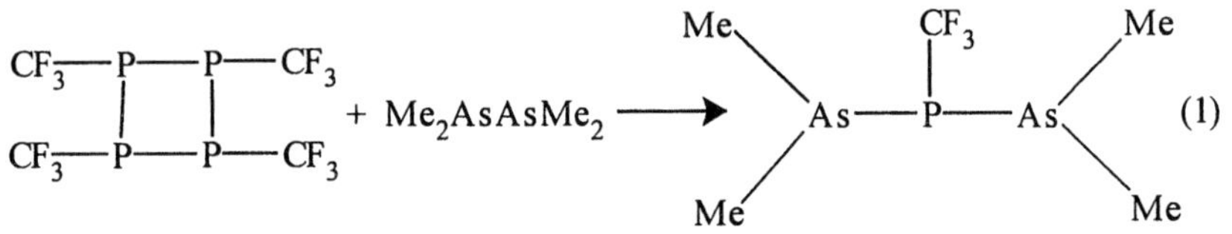

$$CF_3-P-P-CF_3 \atop CF_3-P-P-CF_3 \quad + \text{ Me}_2\text{AsAsMe}_2 \longrightarrow \begin{array}{c} \text{Me} \quad \text{CF}_3 \quad \text{Me} \\ \diagdown \quad | \quad \diagup \\ \text{As}-\text{P}-\text{As} \\ \diagup \quad \quad \diagdown \\ \text{Me} \quad \quad \text{Me} \end{array} \quad (1)$$

                                            3   4

in the insertion reactions is the nucleophilic attack of a diarsine or disulfide on $(CF_3P)_4$. Indeed, some years earlier it was found that phosphines will disrupt this cyclotetraphosphine to form phosphine-phosphinidenes **5** (4,5) (Equation 2),

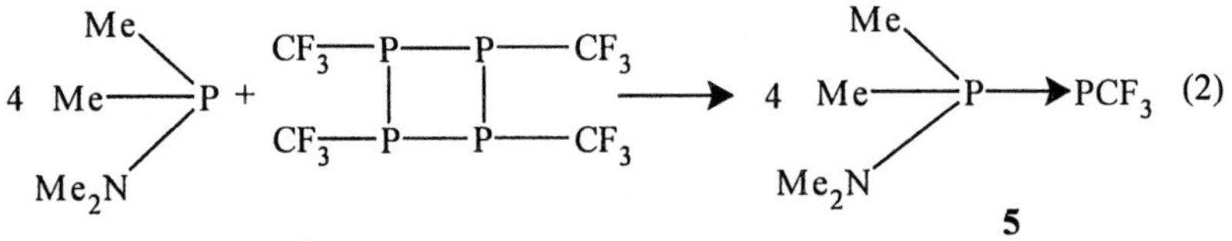

$$4 \text{ Me} \underset{Me_2N}{\overset{Me}{\diagdown}} P \quad + \quad {CF_3-P-P-CF_3 \atop CF_3-P-P-CF_3} \longrightarrow 4 \text{ Me} \underset{Me_2N}{\overset{Me}{\diagdown}} P \rightarrow PCF_3 \quad (2)$$

                                                                                                            5

$$R = Me \text{ (4), } Me_2N \text{ (5)}$$

which can be regarded as phosphorus analogues of Wittig reagents. In principle, such compounds should be capable of serving as phosphinidene transfer agents. To gain more insight into this possibility, we studied the variable temperature ^{19}F NMR spectrum of $Me_3P \rightarrow PCF_3$ in the presence of Me_3P (**6**). The four-line ^{19}F NMR spectrum collapsed to two peaks between 65 and 80 °C. Of the three mechanisms that were considered (Scheme 1), mechanism 1 was found to be the

Scheme 1

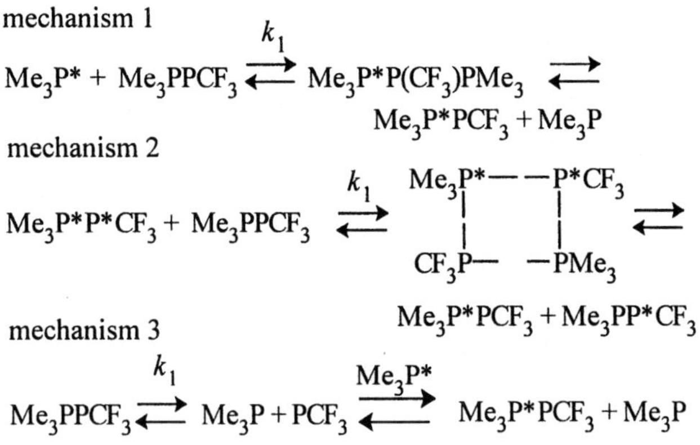

mechanism 1
$$Me_3P^* + Me_3PPCF_3 \underset{}{\overset{k_1}{\rightleftarrows}} Me_3P^*P(CF_3)PMe_3 \rightleftarrows Me_3P^*PCF_3 + Me_3P$$

mechanism 2
$$Me_3P^*P^*CF_3 + Me_3PPCF_3 \overset{k_1}{\rightleftarrows} \begin{array}{c} Me_3P^*---P^*CF_3 \\ | \quad\quad\quad | \\ CF_3P---PMe_3 \end{array} \rightleftarrows Me_3P^*PCF_3 + Me_3PP^*CF_3$$

mechanism 3
$$Me_3PPCF_3 \overset{k_1}{\rightleftarrows} Me_3P + PCF_3 \overset{Me_3P^*}{\rightleftarrows} Me_3P^*PCF_3 + Me_3P$$

most appropriate because of the dependence of the rate on the concentration of Me$_3$P hence there was no evidence for the formation of free trifluoromethylphosphinidene in solution. Other methods were therefore developed for the production of free phosphinidenes. Three of the potential candidates for phosphinidene precursors are shown below. Upon photolysis with 254 nm light, the bis(azide) **6** produces the cyclic secondary phosphine, **9** (7).

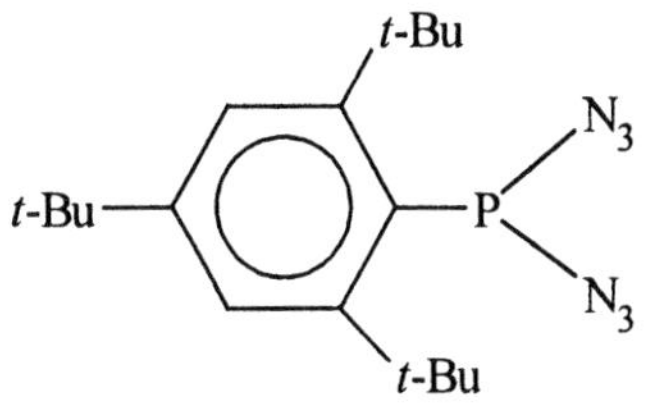

6 (7) 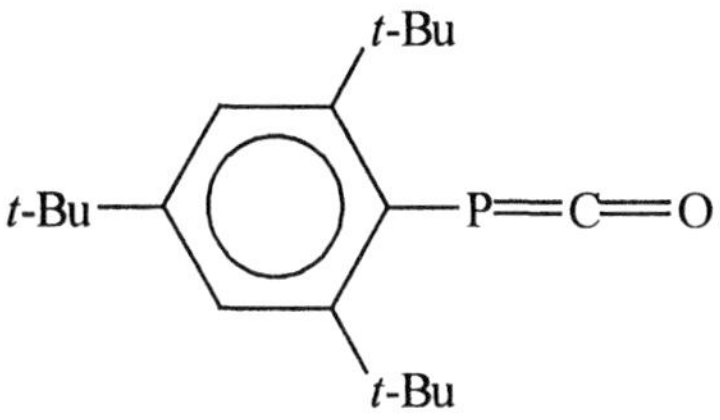 **7 (8)**

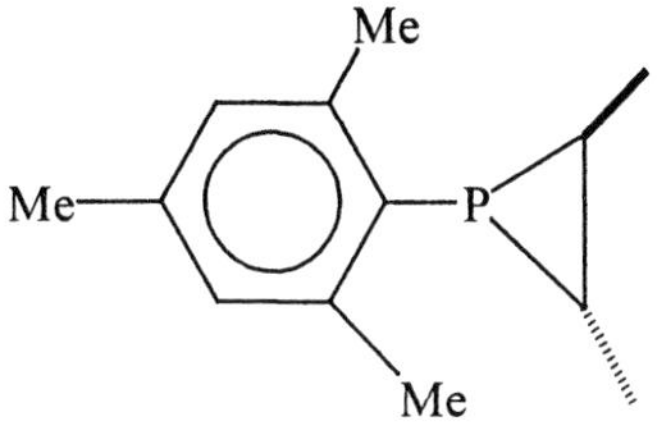

8 (9)

Although the formation of **9** is consistent with the intermediacy of a phosphinidene, alternative approaches to phospinidene production became

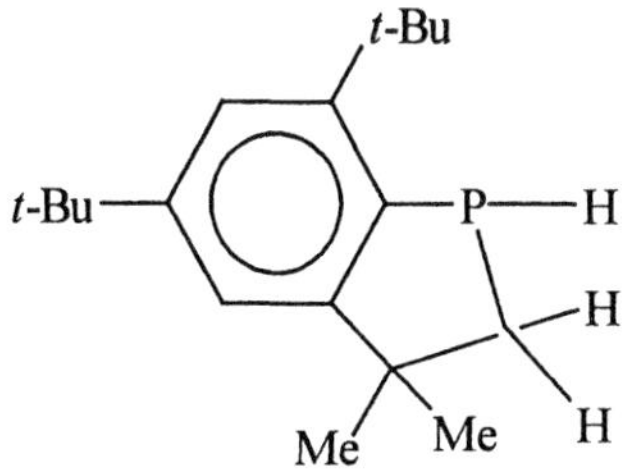

9

necessary. Photolysis of the phosphaketene **7** was carried out with a 320 nm pulsed Na/YAG laser (7). Species were detected with half lives of 8.3 x 10^{-7}s and 5.2 x 10^{-6}s. The longer-lived species resulted from the decay of the shorter-lived species. These observations are consistent with the initial formation of a singlet phosphinidene via photolytic cleavage of the P=C bond of **7**, followed by spin crossover to the more stable triplet state. In collaboration with Professor Gaspar's group (9), photolysis of the phosphirane **8** in methylcyclohexane glass at 77 K resulted in the isolation of an individual phosphinidene. The detection of an ESR signal at 11,492 G led to the conclusion that mesitylphosphinidene possesses a triplet ground state.

The next question to arise was whether dimers of carbene-like fragments could be isolated and, if so, what type(s) of bonding exist between such fragments? The first structurally authenticated example of a compound of this type is the ditin derivative, $\{(Me_3Si)_2CH\}_2SnSn\{CH(SiMe_3)_2\}_2$ **10** (10) which was shown to possess a *trans*-bent solid-state geometry and a relatively long Sn-Sn bond distance, *i.e.* a significantly different structure than that of a typical alkene. Moreover, **10** undergoes facile monomerization in solution. This general area received another significant boost in 1981 by the disclosure of West et al. of (mesityl)$_4$Si$_2$ (**11**), the first example of a stable disilene (11). The

structure of **11** exhibits only slight *trans* bending and a Si-Si bond distance consistent with a bond order of two. Within a short period of time, the first structurally characterized examples of stable compounds with phosphorus-

phosphorus, **12** (12), arsenic-arsenic, **13** (13), and phosphorus-arsenic, **14** (14), double bonds were reported. Although the identity of the phosphastibene, **14** (E = Sb), was clear from spectroscopic data (14), the first X-ray crystal structure of such a compound was reported later by Power et al. (15). The present situation is such that many of the multiple bonding possibilities for dimers of the general type $R_nE = ER_n$ for the group 13, 14, and 15 elements have now been realized (16). Nevertheless, several significant challenges exist, such as the isolation of the first examples of structurally authenticated diborenes (RBBR), dialuminenes (RAlAlR), disylynes (RSiSiR) and heteronuclear analogues of the latter in which the silicon is bonded to a different group 14 element.

The question was raised earlier regarding the nature of the bonding between carbene-like dimers and, of course, the same question is pertinent to carbyne-analogous group 14 dimers. A considerable body of useful structural data has been acquired for such compounds and several theoretical calculations has been performed on actual and model systems (16). As an alternative approach, we performed an electron deformation density (EDD) study of the diphosphenes **12** (12) and $(Me_3Si)_3CP=PC(SiMe_3)_3$ (17) in collaboration with Professor C. Krüger and his colleagues (18). As shown in Figure 1a, highly positive EDD values are

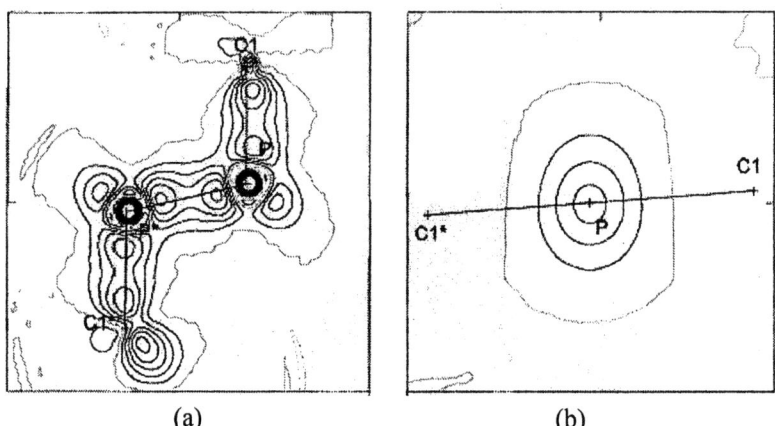

(a)   (b)

Figure 1 (Reproduced from reference 18. Copyright 1997 American Chemical Society.)

evident on the phosphorus atoms in a region corresponding to that anticipated for lone pairs. Figure 1b shows that the EDD is distinctly noncylindrical at the midpoint of the P-P vector in the plane perpendicular to the C-P-P-C plane. A cut through a pure σ-bond would result in circular contours. The elongation of the EDD in the direction expected for the π-bond implies that a phosphorus-

phosphorus double bond description is appropriate. Moreover, the experimental electron deformation densities were found to be in good agreement with those obtained by DFT calculations.

Like carbenes, phosphinidenes are able to bond in a terminal fashion to suitable organometallic fragments. Unlike carbenes, however, phosphinidenes possess four electrons that are potentially available for bonding to the metal rather than two as is the case for R_2C fragments. As a consequence, both linear and bent terminal phosphinidene complexes were anticipated. The first example of a terminal phosphinidene complex to be isolated and structurally characterized was **15** (19). The Mo-P-C bond angle of 115.8(2)° is consistent

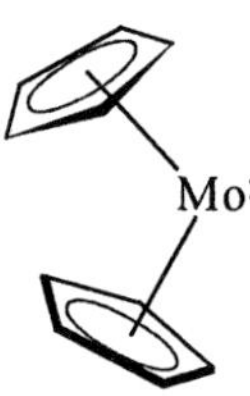

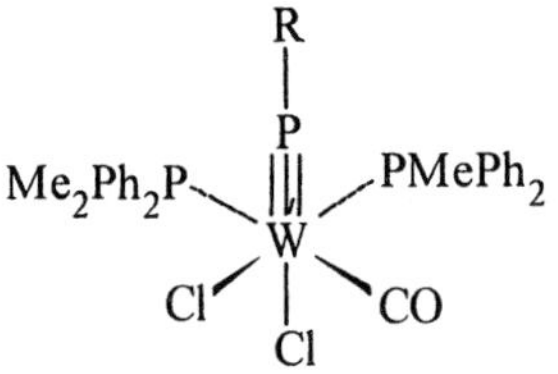

15

R = 2,4,6-t-Bu$_3$C$_6$H$_2$

16

R = 2,4,6-t-Bu$_3$C$_6$H$_2$

with the presence of a molybdenum-phosphorus double bond and a stereochemically active lone pair at phosphorus. The first linear terminal phosphinidene complex, **16**, which was prepared by oxidative addition of phosphaketene **7** to WCl$_2$(PMePh$_2$)$_4$, has a short tungsten-phosphorus bond distance (2.169(1) Å) and a relatively upfield ^{31}P chemical shift indicative of a triple bonding description (20). It is also possible to prepare "terminal phosphinidene complexes" of N-heterocyclic carbenes, **17** (21). Note, however, that in **17** the role of the phosphinidene is that of an acceptor rather than a donor,

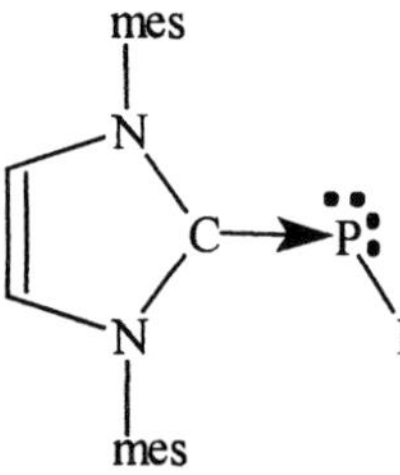

17 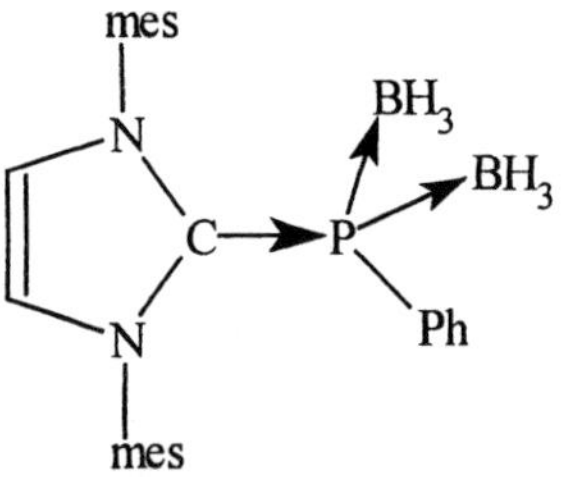 **18**

as in the case of terminal phosphinidene transition metal complexes. The appropriateness of the structural formula depicted for **17** was confirmed by the fact that treatment of this compound with BH_3·thf resulted in the exclusive formation of the bis(borane) adduct, **18**.

Bridging pnictinidene complexes have also attracted attention and, in some cases, presented bonding dichotomies. For example, the reaction of $(Me_3Si)_2CHSbCl_2$ with $K_2[W(CO)_5]$ resulted in the "open" stibinidene complex, **19** (22). The geometry at antimony is trigonal planar and there is evidence for a modicum of multiple bonding between antimony and tungsten. However, the tungsten-tungsten separation is such as to preclude a W· · ·W interaction. Interestingly, the reaction of $(Me_3Si)_2CHSbCl_2$ with $Na_2[Fe(CO)_4]$, followed by treatment with $Fe_2(CO)_9$, resulted in a "closed" stibinidene complex, **20** (23). In terms of isolobal relationships, **19** and **20** are related to the allyl anion and

$$\begin{array}{cc}
CH(SiMe_3)_2 & (Me_3Si)_2CH \\
| & \diagdown \\
Sb & Sb: \\
\diagup \diagdown & \diagup \diagdown \\
(OC)_5W \quad W(CO)_5 & (OC)_4Fe \text{———} Fe(CO)_4 \\
\mathbf{19} & \mathbf{20}
\end{array}$$

cyclopropane, respectively. Clearly, the coordination number of the transition metal plays an important role because metal-metal bond formation in **19** would require seven-coordinate tungsten centers.

Phosphenium ions, $[R_2P]^+$, resemble singlet carbenes in that they feature a lone pair and a formally vacant valence p-orbital on the central atom (24). Although various synthetic methods are available, phosphenium ions are typically prepared by halide ion abstraction from precursor halophosphines (24). The structures of several phosphenium ion salts have been established by X-ray crystallography. An early example (25) is shown in Figure 2. The skeletal geometry of $[i\text{-}Pr_2N)_2P]^+$ is consistent with the view that the phosphorus atom is approximately sp^2 hybridized (N(1)-P-N(2) = 114.8(2)°) and the subunit C(1)C(2)N(1)PNC(2)C(3)C(4) is essentially planar, thus maximizing dative π-bonding between the nitrogen lone pairs and the formally vacant P(3p) orbital.

Being analogs of carbenes, phosphenium ions react readily with unsaturated hydrocarbons. For example, phosphenium ions react with 1,3-dienes to form phospholenium ions (**21**), which are converted to phospholenium oxides upon hydrolysis. These phosphenium ion reactions are considerably faster than the corresponding reactions with dihalophosphines (the so-called McCormack reaction (26)). Phosphenium ions also react readily with 1,4-dienes to afford

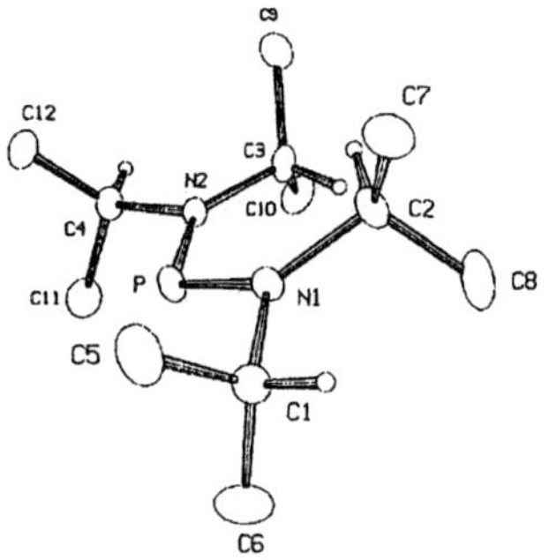

Figure 2. Structure of the [(i-Pr$_2$N)$_2$P]$^+$ cation (25). (Reproduced from reference 25. Copyright 1978 American Chemical Society.)

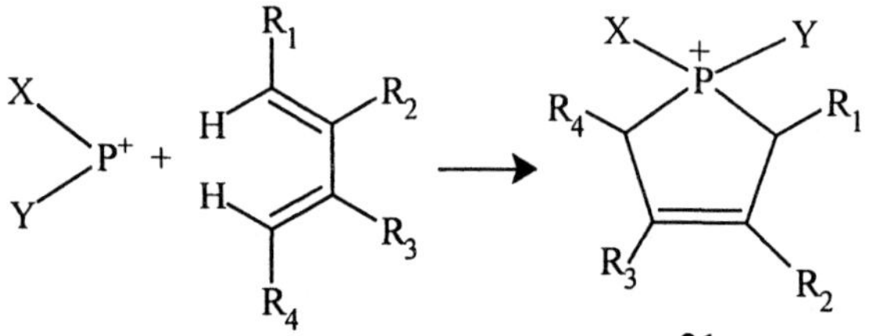

bicyclic phosphenium cations such as **22** (27). The reaction of the

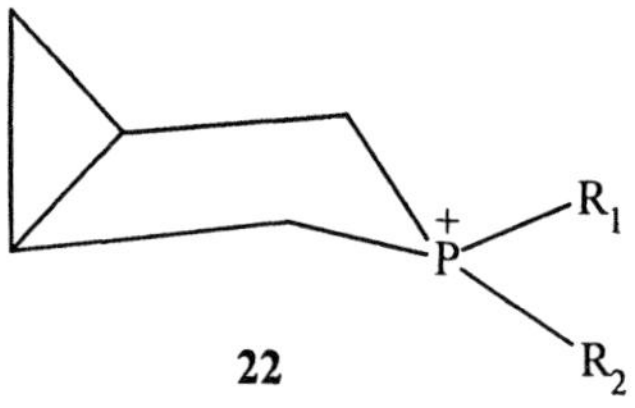

[(Me$_2$N)(Cl)P]$^+$ cation with cyclooctatetraene was of particular interest because the metrical parameters of the resulting phosphenium ion (**23**) indicate an impressively close approach to the transition state for the Cope rearrangement (28). Given that phosphenium ions possess (i) an electropositive center, (ii) a formally vacant P(3p) orbital at phosphorus, and (iii) they are coordinatively unsaturated, they are expected to cause C-H activation. The first example of

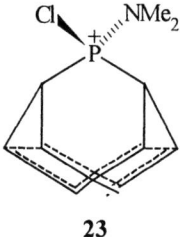

23

such a reaction was the insertion of [(*i*-Pr$_2$N)$_2$P]$^+$ into a C-H bond of stannocene to form **24** (29).

24

The presence of a lone pair of electrons, a vacant π-orbital and a formal positive charge at phosphorus renders phosphenium cations excellent π-acceptor ligands. Early examples of charged phosphenium complexes such as [(R$_2$N)$_2$PFe(CO)$_4$]$^+$ were formed by halide ion abstraction from precursor bis(amido)halophosphine complexes (30, 31) while neutral phosphenium complexes were prepared via the metathetical reaction of e.g. cyclic diamminohalophosphines with [(η^5-C$_5$H$_5$)Mo(CO)$_3$]$^-$ (32).

DFT calculations on the singlet and lowest-lying triplet states of the four-valence-electron group 13 fragments (η^5-C$_5$H$_5$)M, (η^5-C$_5$Me$_5$)M, (H$_3$Si)$_2$NM (M

= B, Al, Ga, In) reveal that, regardless of the substituents R, the ground state is a singlet in each case and that the singlet-triplet energy gap generally increases with atomic number (33). As a consequence of the singlet ground state, the HOMO's of group 13 RM fragments typically exhibit pronounced lone pair character (Figure 3).

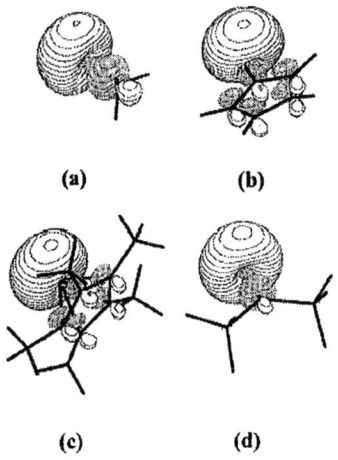

Figure 3. Boron lone pair orbitals for (a) CH_3B, (b) $(\eta^5\text{-}C_5H_5)B$, (c) $(\eta^5C_5Me_5)B$, and (d) $(H_3Si)_2NB$. (Reproduced from reference 33. Copyright 1999 American Chemical Society.)

Depending on the steric demands, electronic characteristics and crystal packing, group 13 RM fragments can assemble into clusters rather than the dimers that were discussed above. Such clusters are known for all the group 13 elements and are exemplified by tetrahedral (t-BuB)$_4$ (34), [(η^5-C$_5$Me$_5$)Al]$_4$ (35) (Fig. 4), [(Me$_3$Si)$_3$CIn]$_4$ (36) [(Me$_3$Si)$_3$CTl]$_4$ (37), and octahedral (η^5-C$_5$Me$_5$Ga)$_6$ (38). Several of these clusters undergo facile dissociation in solution and can therefore serve as useful sources of ligands for the synthesis of complexes in which the RM fragments bind in a primarily σ-donor fashion to transition metal entities (39). Unfortunately, however, (t-BuB)$_4$ is quite unreactive and [(η^5-C$_5$Me$_5$)B]$_4$ is not known, probably because of the large size of the [C$_5$Me$_5$]$^-$ ligand in relation to the covalent radius of boron. Accordingly, until recently terminal borylene (boranediyl) complexes were unknown. The first examples of such complexes were prepared by metathetical reactions of (η^1-C$_5$Me$_5$)BCl$_2$ or (Me$_3$Si)$_2$NBCl$_2$ with organometallic dianions as exemplified by Equation 3 (40,

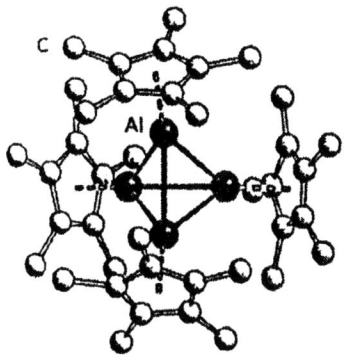

Figure 4. Structure of $[(\eta^5\text{-}C_5Me_5)Al]_4$ (35). (Reproduced from reference 35. Copyright 2000 Wiley-VCH.)

$$(\eta^1\text{-}C_5Me_5)BCl_2 \xrightarrow[\text{toluene, -78 °C}]{K_2[Fe(CO)_4]} (\eta^5\text{-}C_5Me_5)BFe(CO)_4 \quad (3)$$
$$25$$

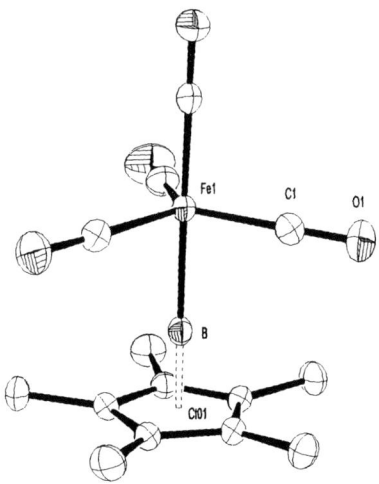

Figure 5. Structure of $(\eta^5\text{-}C_5Me_5)B \longrightarrow Fe(CO)_4$ (40). (Reproduced from reference 40. Copyright 1998 American Chemical Society.)

41). The structure of $(\eta^5\text{-}C_5Me_5)B{\to}Fe(CO)_4$ (**25**) is shown in Figure 5. The implication that the bonding in **25** involves primarily σ-donation on the part of the RB ligand suggested that borylenes and heaver congeners should be able to function as Lewis bases. This anticipated Lewis basicity is evidenced by the observation that e.g. $[(\eta^5\text{-}C_5Me_5)Al]_4$ reacts with $B(C_6F_5)_3$ and $Al(C_6F_5)_3$ to form 1:1 complexes with aluminum(I)-boron(III) and aluminum(I)-aluminum(III) donor-acceptor bonds, respectively (42, 43). The structure of $(\eta^5\text{-}C_5Me_5)Al{\to}B(C_6F_5)_3$ (**26**) is illustrated in Figure 6. An assessment of the Lewis

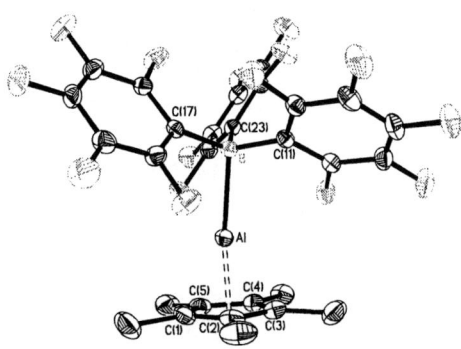

Figure 6. Structure of $(\eta^5\text{-}C_5Me_5)Al{\longrightarrow}B(C_6F_5)_3$ (42). (Reproduced from reference 42. Copyright 2000 American Chemical Society.)

basicity of $(\eta^5\text{-}C_5Me_5)Al$ can be made from the extent of distortion of the acceptor molecule $B(C_6F_5)_3$ as complex formation takes place. On this basis, since the sum of bond angles at boron in e.g. $(\eta^5\text{-}C_5Me_5)Al{\to}B(C_6F_5)_3$ is 339.8(2)° (42) and virtually identical to that in $Ph_3P{\to}B(C_6F_5)_3$ (44), it can be implied that $(\eta^5\text{-}C_5Me_5)Al$ and Ph_3P are of comparable Lewis basicity with respect to $B(C_6F_5)_3$.

In attempting to prepare Lewis acid-base complexes akin to the group 13 complexes of the type $RE{\to}ER_3'$, decamethylstannocene was treated $Ga(C_6F_5)_3$. However, instead of complex formation, this reaction resulted in the first example of a triple-decker main group cation (45). Prior to this development, only two triple-decker main group anions had been reported (46, 47). In contrast to the *trans*-type geometry of the triple-decker anions, $[(\eta^5\text{-}C_5Me_5)Sn(\mu\text{-}\eta^5\text{-}C_5Me_5)Sn(\eta^5\text{-}C_5Me_5)]^+$ (**27**) adopts a *cis*-type geometry (Fig. 7). The distance from the tin atoms to the centroids of the terminal C_5Me_5 rings (av. 2.246(18) Å) is shorter than that to the bridging C_5Me_5 moiety (av. 2.644(19) Å) which explains the fluxional behavior of this cation in solution. It was surmised that the initial step of the reaction is the abstraction of a $[C_5Me_5]^-$ anion by the strong Lewis acid $Ga(C_6F_5)_3$, to form $[Ga(C_6F_5)_3(C_5Me_5)]^-$ and $[C_5Me_5Sn]^+$, the latter of

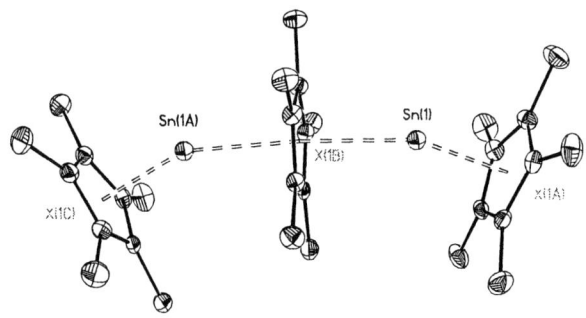

*Figure 7. Structure of the triple-decker cation,
[(η^5-C_5Me_5)Sn(μ-η^5-C_5Me_5)Sn(η^5-C_5Me_5)]$^+$ (45). (Reproduced from reference 45. Copyright 2001 The Royal Society of Chemistry.)*

which attacks (η^5-C$_5$Me$_5$)$_2$Sn electrophilically to form the triple decker cation, **27**. Clearly, at some point the [Ga(C$_6$F$_5$)$_3$(C$_5$Me$_5$)]$^-$ anion undergoes redistribution to form [Ga(C$_6$F$_5$)$_4$]$^-$. In subsequent work, a more rational approach to the synthesis of triple decker cations has been developed, namely

- Addition of positively-charged fragments to neutral metallocenes
- Maximization of the lattice energies of the resulting salts by matching the sizes of the anion and the cation
- Understanding packing interactions

The use of this approach has permitted the recent synthesis (48) of the triple-decker dilead cation (η^5-C$_5$Me$_5$)Pb(μ-η^5-C$_5$Me$_5$)Pb(η^5-C$_5$Me$_5$)]$^+$ (**28**).
The further use of this approach, coupled with the recognition that (η^6-C$_6$R$_6$)In and (η^5-C$_5$R$_5$)Sn are isoelectronic, prompted us to synthesize the cation [(η^6-toluene)In(μ-η^5-C$_5$Me$_5$)In(η^6-toluene)]$^+$ (**29**) (49) as summarized in Scheme 2.

Although **29** can be regarded as a group 13 triple decker cation, it should be pointed out that the indium to ring centroid distances are extremely large hence **29** can be viewed more appropriately as the first example of an inverse-sandwich main-group cation. Such a view is in accord with the observation (50) that the replacement of the large anion [(C$_6$F$_5$)$_3$BO(H)B(C$_6$F$_5$)$_3$]$^-$ by the smaller anion [B(C$_6$F$_5$)$_4$]$^-$ results in the removal of the capping η^6-toluene molecules.

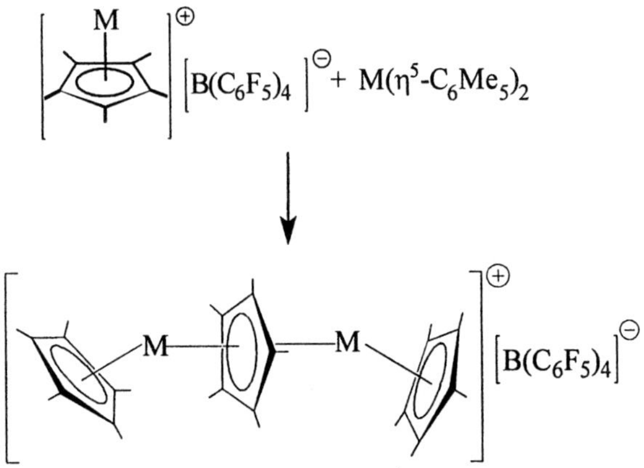

28

M = Sn, Pb

Scheme 2

$$(\eta^5\text{-}C_5Me_5)In + H_2O\cdot B(C_6F_5)_3 + B(C_6F_5)_3 \longrightarrow$$

$$In^+ + [(C_6F_5)_3BOB(C_6F_5)_3]^-$$

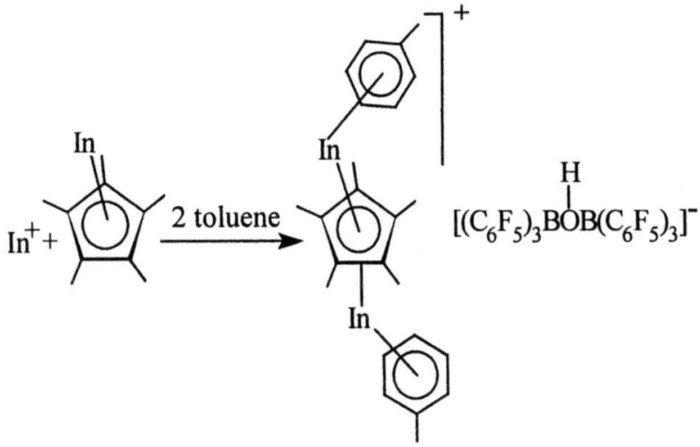

29

Acknowledgements

In addition to my sincere thanks to graduate students, postdoctoral associates, undergraduates, and collaborators for their enthusiasm, dedication, and hard work, I also wish to express my gratitude to the National Science Foundation, the ACS Petroleum Research Fund, and the Robert A. Welch Foundation (Grant F-135) for generous financial support.

References

1. Cowley, A. H.; Burg, A. B.; Cullen, W. R. *J. Am. Chem. Soc.* **1966**, *88*, 3178.
2. Cowley, A. H.; Furtsch, T. A.; Dierdorf, D. S. *Chem. Commun.* **1970**, 523.
3. Cowley, A. H. *J. Am. Chem. Soc.* **1967**, *89*, 5990.
4. Burg, A. B.; Mahler, W. *J. Am. Chem. Soc.* **1961**, *83*, 2388.
5. Cowley, A. H.; Dierdorf, D. S. *J. Am. Chem. Soc.* **1969**, *91*, 5919.
6. Cowley, A. H.; Cushner, M. C. *Inorg. Chem.* **1980**, *19*, 515.
7. Cowley, A. H.; Gabbaï, F. P.; Schluter, R.; Atwood, D. A. *J. Am. Chem. Soc.* **1992**, *114*, 3142.
8. Appel, R.; Paulen, W. *Angew. Chem., Int. Ed. Engl.* **1983**, *22*, 785.
9. Li, X.; Weissman, S. I.; Lin, T.-S.; Gaspar, P. P.; Cowley, A. H.; Smirnov, A. I. *J. Am. Chem. Soc.* **1994**, *116*, 7899.
10. Goldberg, D. E.; Harris, D. H.; Lappert, M. F.; Thomas, K. M. *J. Chem. Soc., Chem. Commun.* **1976**, *261*; Davidson, P. J.; Harris, D. H.; Lappert, M. F. *J. Chem. Soc., Dalton Trans.* **1976**, 2268.
11. West, R.; Fink, M. J.; Michl, J. *Science* **1981**, *214*, 1343.
12. Yoshifuji, M.; Shima, I.; Inamoto, N.; Hirotsu, K.; Higuchi, T. *J. Am. Chem. Soc.* **1981**, *103*, 4587.
13. Cowley, A. H.; Lasch, J. G.; Norman, N. C.; Pakulski, M. *J. Am. Chem. Soc.* **1983**, *105*, 5506.
14. Cowley, A. H.; Lasch, J. G.; Norman, N. C. Pakulski, M.; Whittlesey, B. R. *Chem. Soc., Chem. Commun.* **1983**, 5659.
15. Twamley, B.; Power, P. P. *J. Chem. Soc. Chem. Commun.* **1998**, 1979.
16. For a review, see Power, P. P. *Chem. Rev.* **1999**, *99*, 3463.
17. Cowley, A. H.; Kilduff, J. E.; Norman, N. C.; Pakulski, M.; Atwood, J. L.; Hunter, W. E. *J. Am. Chem. Soc.* **1983**, *105*, 4845.

18. Cowley, A. H.; Decken, A.; Norman, N. C.; Krüger, C.; Lutz, F.; Jacobsen, H.; Ziegler, T. *J. Am. Chem. Soc.* **1997**, *119*, 3389.
19. Hitchcock, P. B.; Lappert, M. F.; Leung, W. P. *J. Chem. Soc., Chem. Commun.* **1987**, 1282.
20. Cowley, A. H.; Pellerin, B.; Atwood, J. L.; Bott, S. G. *J. Am. Chem. Soc.* **1990**, *112*, 6734.
21. Arduengo, A. J., III; Carmalt, C. J.; Clyburne, J. A. C.; Cowley, A. H.; Pyati, R. *Chem. Commun.* **1997**, 981.
22. Arif, A. M.; Cowley, A. H.; Norman, N. C.; Orpen, A. G.; Pakulski, M. *Chem. Soc., Chem. Commun.* **1985**, 1267.
23. Arif, A. M.; Cowley, A. H.; Norman, N. C.; Pakulski, M. *Inorg. Chem.* **1986**, *25*, 4836.
24. For reviews, see Cowley, A. H.; Kemp, R. A. *Chem. Rev.* **1985**, *85*, 367; Sanchez, M. M.; Mazières, R.; Lamande, L.; Wolf, R. *Multiple Bonds and Low Coordination Chemistry in Phosphorus Chemistry*; Regitz, M.; Sherer, O, Eds.; Georg Thieme Verlag: Stuttgart, 1990; D1, p129ff; Gudat, D. *Coord. Chem. Rev.* **1997**, *173*, 71, and references cited therein.
25. Cowley, A. H.; Cushner, M. C.; Szobota, J. S. *J. Am. Chem. Soc.* **1978**, *100*, 7784.
26. McCormack, W. B. U.S. Patents 2,663,736 and 2,663,737, 1953. *Chem. Abs.* **1955**, *49*, 7601.
27. Cowley A H.; Stewart, C. A.; Whittlesey, B. R.; Wright, T. C. *Tetrahedron Lett.* **1984**, *25*, 815.
28. Weissman, S. A.; Baxter, S. G.; Arif, A. M.; Cowley, A. H. *J. Am. Chem. Soc.* **1986**, *108*, 529.
29. Cowley, A. H.; Kemp, R. A.; Stewart, C. A. *J. Am. Chem. Soc.* **1982**, *104*, 3239.
30. Montemayer, R. G.; Sauer, D. T.; Fleming, S.; Bennett, D. W.; Thomas, M. G.; Parry, R. W. *J. Am. Chem. Soc.* **1978**, *100*, 2231.
31. Cowley, A. H.; Kemp, R. A.; Wilburn, J. C. *Inorg. Chem.* **1981**, *20*, 4289.
32. Light, R. W.; Paine, R. T. *J. Am. Chem. Soc.* **1978**, *100*, 2230; Hutchins, L. D.; Paine, R. T.; Campana, C. F. *J. Am. Chem. Soc.* **1980**, *102*, 4521.
33. Macdonald, C. L. B.; Cowley, A. H. *J. Am. Chem. Soc.* **1999**, *121*, 12113.
34. Mennekes, T.; Paetzold, P.; Boese, R.; Bläser, D. *Angew. Chem., Int. Ed. Engl.* **1991**, *30*, 173.
35. Dohmeir, C.; Loos, D.; Schnöckel, H. *Angew. Chem., Int. Ed. Engl.* **1996**, *35*, 129; Schulz, S.; Roesky, H. W.; Koch, J.; Sheldrick, G. M.; Stalke, D.; Kuhn, A. *Angew. Chem., Int. Ed. Engl.* **1993**, *32*, 1729.

36. Schluter, R. D.; Cowley, A. H.; Atwood, D. A.; Jones, R. A.; Atwood, J. L. *J. Coord. Chem.* **1993**, *30*, 215; Uhl, W.; Graupner, R.; Layh, M.; Schütz, U.; *J. Organomet. Chem.* **1995**, *493*, C1.
37. Uhl, W.; Miller, W.; Layh, M.; Schwarz, W. *Angew. Chem., Int. Ed. Engl.* **1992**, *31*, 1364.
38. Tacke, M.; Plaggenborg, L.; Schnöckel, H.; *Z. Anorg. Allg. Chem.* **1991**, *604*, 35; Jutzi, P.; Neumann, B.; Reumann, G.; Stamnler, H.-G. *Organometallics* **1998**, *17*, 1305.
39. For a review, see Fischer, R. A.; Weiss, J. *Angew. Chem., Int. Ed. Engl.* **1999**, *38*, 2830.
40. Cowley, A. H.; Lomelí, V.; Voigt, A. *J. Am. Chem. Soc.* **1998**, *120*, 6401.
41. Braunsweig, H.; Kollann, C.; Englert, U. *Angew. Chem. Int. Ed. Engl.* **1998**, *37*, 3179.
42. Gorden, J. D.; Voigt, A.; Macdonald, C. L. B.; Silverman, J. S.; Cowley, A. H. *J. Am. Chem. Soc.* **2000**, *122*, 950.
43. Gorden, J. D.; Macdonald, C. L. B.; Cowley, A. H. *Chem. Commun.* **2001**, 75.
44. Jacobsen, H.; Berke, H.; Döring, S.; Kehr, G.; Erker, G.; Fröhlich, R.; Meyer, O. *Organometallics* **1999**, *18*, 1724.
45. Cowley, A. H.; Macdonald, C. L. B.; Silverman, J. S.; Gorden, J. D.; Voigt, A. *Chem. Comm.* **2001**, 175.
46. Beswick, M. A.; Palmer, J. S.; Wright, D. S.; *Chem. Soc. Rev.* **1998**, *27*, 225; Beswick, M. A.; Gornitzka H.; Kärcher, J.; Mosquera, M. E. G.; Palmer, J. S.; Russell, C. A.; Stalke, D. Steiner, A.; Wright, D. S. *Organometallics* **1999**, *18*, 1148.
47. Harder, S.; Prosenc, M. H. *Angew Chem., Int. Ed. Engl.* **1996**, *35*, 97.
48. Jones, J. N.; Cowley, A. H., The University of Texas at Austin, *unpublished*.
49. Cowley, A. H.; Macdonald, C. L. B.; Silverman, J. S.; Gorden, J. D.; Voigt, A. *Chem. Comm.* **2001**, 175.
50. Jones, J. N.; Macdonald, C. L. B.; Gorden, J. D.; Cowley, A. H.; *J. Organomet. Chem.* **2003**, *666*, 3.

Chapter 2

Al-H-C Chemistry

Herbert W. Roesky

Institut für Anorganische Chemie der Universität Göttingen,
Tammannstrasse 4, D–37077 Göttingen, Germany

The synthesis of $LAl(OH)_2$, $LAl(SH)_2$, $LAl(SeH)_2$, and $LAl(NH_2)_2$ (L = $HC(CMeNAr)_2$, Ar = $2,6\text{-iPr}_2C_6H_3$) is described. For the preparation of $LAl(SH)_2$ from $LAlH_2$ in the presence of sulfur a catalyst is required, whereas $LAl(OH)_2$ and $LAl(NH_2)_2$ are obtained in the presence of a N-heterocyclic carbene as a HCl acceptor. The starting material for the latter reaction is $LAlCl_2$. The organometallic hydroxides are environmentally friendly, in comparison to the metal halides, due to their OH functionality resembling that of water. The compounds are interesting precursor for the preparation of heterobimetallic compounds e.g. $LAl(\mu\text{-}S)_2ZrCp_2$.

The most fundamental and lasting objective of synthesis is not production of new compounds but production of properties.
George S. Hammond

Introduction

In nature there are a number of hydrated forms of alumina (Al_2O_3) known with the composition AlOOH and Al(OH)$_3$. Prominent examples of composition AlOOH are the minerals boehmite and diaspore. The natural aluminum trihydroxides are gibbsite (γ-Al(OH)$_3$) and bayerite (α-Al(OH)$_3$) (*1*). All the polymorphs of alumina can also be prepared using various synthetic methods. Gibbsite, boehmite, and bayerite are obtained from aluminum salts by precipitation from aqueous solutions depending on temperature and pH conditions. Another route for the preparation of the aluminum hydroxides is the hydrothermal method. The vapor pressure of water and the temperature are the key factors for the nature of the aluminum polymorphs prepared by the latter method.

The different polymorphs of aluminum have in common a well defined (*2*) oxygen network with interstices. The aluminum atoms are located in these interstices and are octahedrally coordinated. The arrangement of the oxygen network governs the type of product formed after dehydration. The hydrogen atoms play an important role in the cohesion of the layered structures. Moreover the positions of the hydrogen atoms also determine the space group of the aluminum hydroxides although their positions in many systems remain controversial. Characteristic of all aluminum hydroxides is the arrangement of the OH groups in a bridging rather than in a terminal position (*3*). Furthermore compounds with the valence isoelectronic functional groups such as SH, SeH, and NH$_2$ are not known in nature.

Herein we report on soluble molecular organoaluminum compounds containing the OH, SH, SeH, and NH$_2$ functionalities.

Organoaluminum Dihydroxide

Hydroxides of aluminum can be considered as reactive intermediates in the controlled hydrolysis of organoaluminum compounds for the preparation of alumoxanes with the general formula [RAlO]$_n$ and [R$_2$AlOAlR$_2$]$_n$. In recent years there have been a few reports on the preparation of organoaluminum hydroxides with bridged or capped OH groups (*4-9*) that have been isolated and structurally characterized. Although the arylboronic acids RB(OH)$_2$ have found interesting application in Suzuki (*10*) coupling, the corresponding organoaluminum dihydroxide was prepared only recently (*11*). In the meantime, however, several routes for the preparation of LAl(OH)$_2$ **1** have been developed. They are summarized in Scheme 1, with L = HC(CMeNAr)$_2$ (Ar = 2,6-iPr$_2$C$_6$H$_3$). The reaction of LAlI$_2$ with KOH required the application of the two phase system toluene and liquid ammonia at -78 °C. Moreover, KH has to be added to remove the amount of water present in the KOH (*11*).

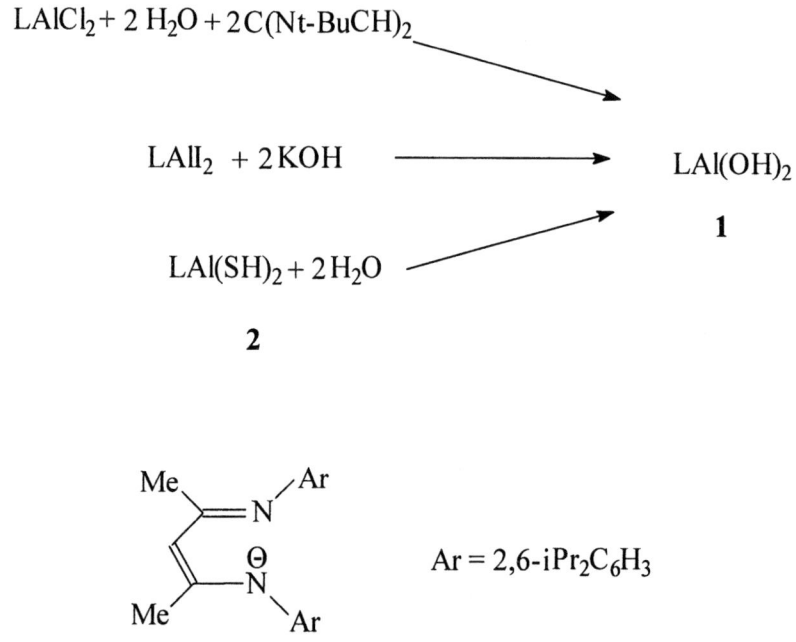

Scheme 1. Alternative routes for the preparation of LAl(OH)$_2$

The conversion of LAlI$_2$ to **1** mainly proceeds at the interface of toluene/ammonia due to the different solubility of the starting materials. The total yield of **1** is 48%. In summary, this route is rather complicated and needs expensive precursors.

The synthesis of **1** using LAlCl$_2$ and water in the presence of the N-heterocyclic carbene 1,3-di-t-butylimidazol-2-ylidene in benzene resulted in a 65% yield of the desired product (*12*). The N-heterocyclic carbene C(Nt-BuCH)$_2$ functions as an HCl acceptor to yield the imidazolium chloride, which can be easily recovered by filtration and recycled, to the free carbene using a strong base such as KOtBu and NaH, respectively (*12*). Finally, treatment of LAl(SH)$_2$

2 with water yields **1** under elimination of H_2S (*13*). The latter route to **1** is quite facile due to the easy conversion of $LAlH_2$ to $LAl(SH)_2$.

The reaction of $LAlH_2$ with KOH in a two phase system liquid ammonia/toluene with the appropriate stoichiometry and the addition of KH for the 10-15% content of H_2O in KOH resulted in the formation of an alumoxane **3** with terminal hydroxide groups (*14*) (Scheme 2).

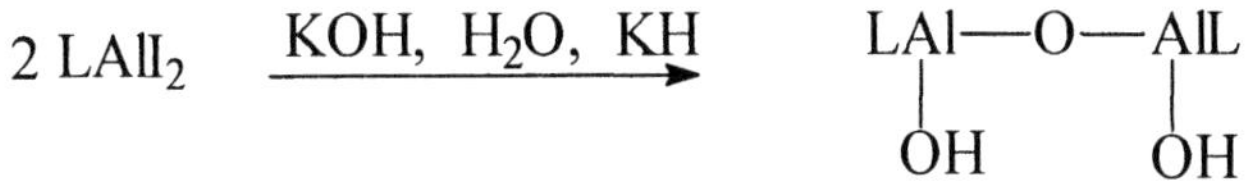

$$2\ LAlH_2 \quad \xrightarrow{KOH,\ H_2O,\ KH} \quad \underset{OH}{LAl}-O-\underset{OH}{AlL}$$

3

Scheme 2

Compound **3** is formed in a 67% yield. The reaction of **3** with Me_2AlH proceeds under elimination of hydrogen and methane to yield the trimeric alumoxane **4** (*14*) (Scheme 3).

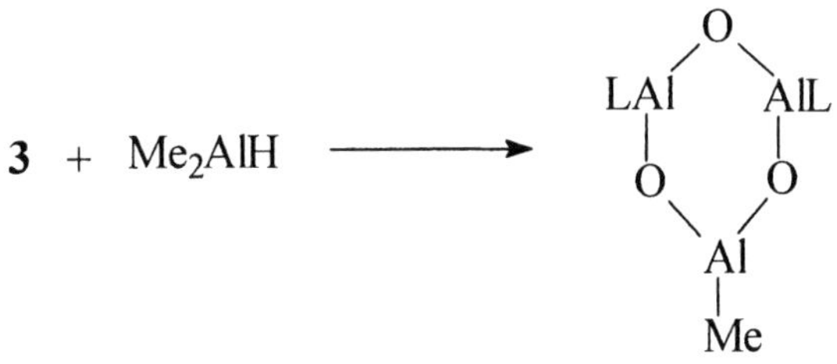

3 + Me_2AlH ⟶

[structure of compound **4**: a six-membered Al_3O_3 ring with two LAl groups bridged by O, and an Al-Me group]

4

Scheme 3

Compound **4** contains two four-coordinate Al and one three-coordinate Al linked by μ-oxo bridges to form a highly distorted six-membered Al_3O_3 ring. The coordination sphere around the four coordinated aluminum atoms consists of two nitrogen and two oxygen atoms for each aluminum. The coordination sphere of the three coordinated aluminum is completed by two oxygen and one carbon of a methyl group. Compound **4** can be considered to contain a trapped monomeric MeAlO (MAO) (*14*).

Organoaluminum Monohydroxide

In a quite different manner an aluminum monohydroxide **5** was obtained by the reaction of LAlH$_2$ **6** with tBuN=C=O in a molar ratio of 2:1 in the presence of water (Scheme 4) (*15*). tBuN=C=O and LAlH$_2$ without water yields the 1:1 insertion product **7**.

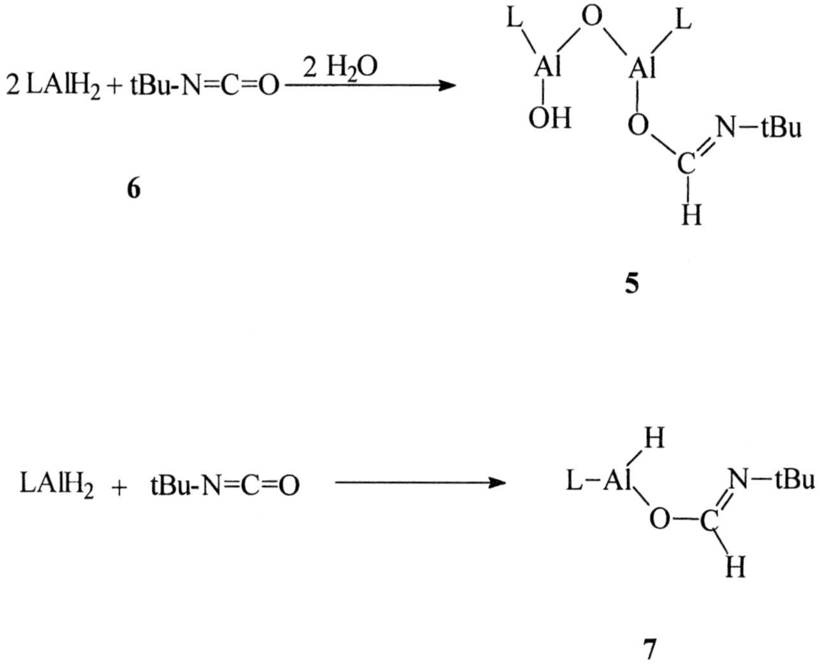

$$2\ \text{LAlH}_2 + \text{tBu-N=C=O} \xrightarrow{2\ \text{H}_2\text{O}}$$

6

[Structure of compound **5**: L\Al(OH)-O-Al/L with a second O bridging to C(H)=N-tBu]

5

$$\text{LAlH}_2 + \text{tBu-N=C=O} \longrightarrow$$

[Structure of compound **7**: L-Al(H)-O-C(H)=N-tBu]

7

Scheme 4

Compound **7** is obviously an intermediate in the formation of **5**. Finally the reaction of LAlH$_2$ and **7** in the presence of water yields **5**. However, an excess of water results in the decomposition of **5**. Compound **5** was characterized by a single X-ray crystal structural analysis (*15*). Compound **5** contains a bent Al-O-Al unit (112.35(9)°) with two tetrahedral distorted aluminum centers. The Al-OH bond length (1.727(2) Å) is comparable to those in LAl(OH)$_2$ (*11,16*) (1.6947(15) and 1.7107(16) Å).

Hydrolysis of trimesitylaluminum and –gallium was monitored by ^{1}H NMR spectroscopy in the temperature range -60 °C to room temperature using deuterated THF as a solvent. As a result of this reaction the hydrolysis of trimesitylaluminum leads to $(Me_3Al·OH_2)·nTHF$ **8** a water adduct, and the monohydroxide $(Mes_2AlOH)_2·2THF$ **9**. The structure of **9** shows a central Al_2O_2 four-membered ring. Two THF molecules are coordinated to the bridging μ-OH groups with short μ-OH···O(THF) distances of 1.912 and 1.861 Å respectively (*17*). In the case of gallium the water adduct $Mes_3Ga·OH_2·2THF$ **10** was isolated and characterized at low temperature by a single crystal structural analysis (*17*). (Figure 1) The relative stability of **10** is assigned to the influence of two THF molecules forming hydrogen bonds with the water protons

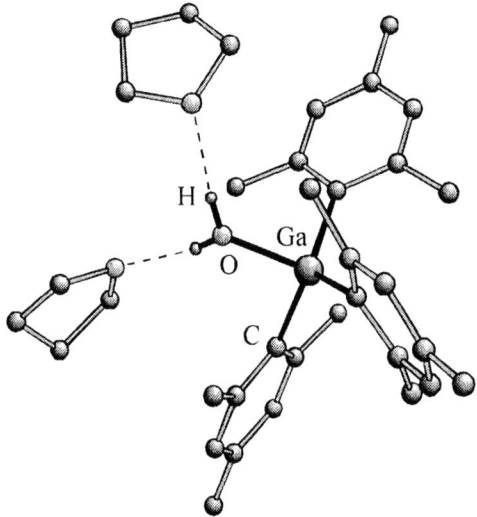

Figure 1. The molecular structure of the water adduct $Mes_3Ga·OH_2·2THF$ (**10**). Hydrogen atoms have been omitted for clarity.

At elevated temperature elimination of mesitylene leads to the dimeric hydroxide (Mes$_2$GaOH)$_2$·THF **11** whereas excess of water gives Mes$_6$Ga$_6$O$_4$(OH)$_4$·4THF **12** (*17*). The molecular structure of **12** is given in Figure 2.

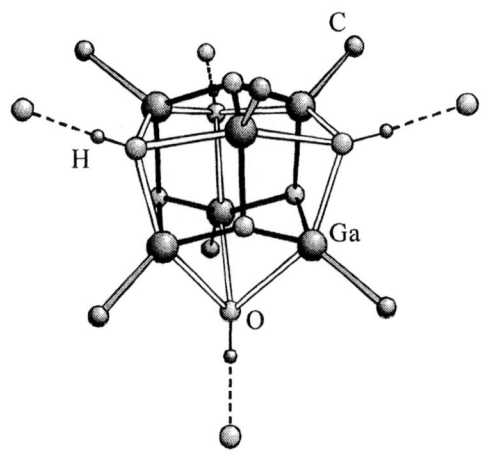

Figure 2. The X-ray single crystal structure of Mes$_6$Ga$_6$O$_4$(OH)$_4$·4THF (**12**) is shown. The mesityl substituents are replaced by six C atoms and the four THF molecules are represented by 4 oxygen atms.

Organoaluminum Dithiol

The SH functionality plays an important role as an intermediate in the formation of metal sulfides such as ZnS, PbS, and FeS$_2$. There are few examples of structurally characterized transition metal complexes known containing two terminal hydrogen sulfide groups (*18*). Among the main group metals only recently the LAl(SH)$_2$ **2** (*19*) was prepared. LAlH$_2$ **13** prepared from LH and

AlH$_3$·NMe$_3$ reacts with elemental sulfur to give a mixture of products, where **2** is formed only in small amounts. The addition of P(NMe$_2$)$_3$ as a catalyst increases the yield of **2** to 90% (Scheme 5)

$$\text{LAlH}_2 + \tfrac{1}{4} S_8 \xrightarrow{\text{P(NMe}_2)_3} \text{LAl(SH)}_2$$
$$\quad \mathbf{13} \qquad\qquad\qquad\qquad \mathbf{2}$$

Scheme 5

The proposed mechanism for the insertion of sulfur into the Al-H bonds is given in Scheme 6.

$$\text{P(NMe}_2)_3 \xrightarrow{1/8\ S_8} \text{S=P(NMe}_2)_3 \xrightarrow{1/8\ S_8} \text{S}_2\text{P(NMe}_2)_3$$

$$\text{LAlH}_2 + \text{S}_2\text{P(NMe}_2)_3 \longrightarrow \text{LAl(H)-S-P(NMe}_2)_3\text{(SH)}$$
$$\mathbf{13}$$
$$\text{LAl(H)-S-P(NMe}_2)_3\text{SH} \longrightarrow \text{LAl(H)(SH)} + \text{SP(NMe}_2)_3$$
$$\qquad\qquad\qquad\qquad\qquad\qquad \mathbf{14}$$

$$\text{LAl(H)SH} \xrightarrow[{-\text{SP(NMe}_2)_3}]{\text{S}_2\text{P(NMe}_2)_3} \text{LAl(SH)}_2$$
$$\qquad\qquad\qquad\qquad\qquad \mathbf{2}$$

Scheme 6

To investigate the role of the P(NMe$_2$)$_3$ in this reaction the ^{1}H and ^{31}P NMR spectra were monitored. From the ^{1}H and ^{31}P NMR studies it is obvious that P(NMe$_2$)$_3$ and S=P(NMe$_2$)$_3$ do not function as a catalyst. We assume that S$_2$P(NMe$_2$)$_3$ plays the important role as a reactive intermediate. The intermediate **14** was identified in the ^{1}H NMR spectrum. Compound **2** forms pale yellow crystals which were suitable for an X-ray structural analysis. The S-H bonds (1.2 Å) fall in the range of other metal complexes containing S-H groups (*18*).

The latent acidic nature of the SH protons in **2** makes them prone to react with Cp$_2$ZrMe$_2$ to generate LAl(μ-S)$_2$ZrCp$_2$ **15** (Scheme 7) (*13*).

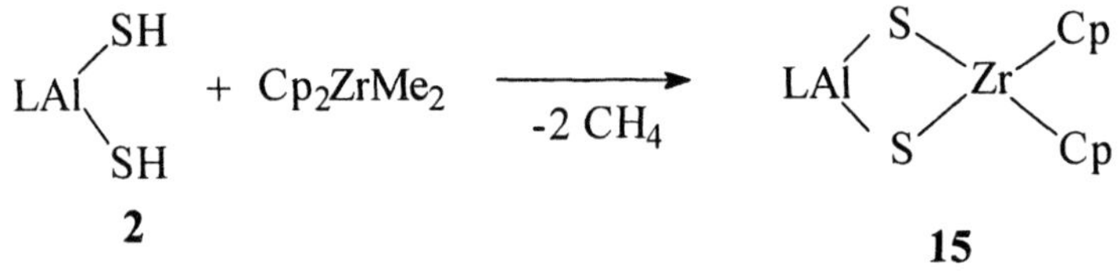

$$\text{LAl}(SH)_2 + Cp_2ZrMe_2 \xrightarrow{-2\ CH_4} \text{LAl}(\mu-S)_2ZrCp_2$$
$$\mathbf{2} \hspace{5cm} \mathbf{15}$$

Scheme 7

Compound **15** is a crystalline solid which was characterized by an X-ray structural analysis (*13*). The reaction shown in Scheme 7 opens up an interesting new field for the preparation of heterobimetallic disulfides.

Organoaluminum Diselenol

For the preparation of LAl(SeH)$_2$ **16** compound LAlH$_2$ **13** and elemental selenium were used in toluene at room temperature. The diselenol **16** was obtained in 58% yield.

Both the sulfur in the reaction of LAl(SH)$_2$ **2** as well as the selenium for preparing LAl(SeH)$_2$ **16** function as oxidizing agents and convert the hydridic hydrogens in **13** to the protonic ones in **2** and **16**. Compound **16** is stable in the solid state but undergoes condensation in solution under elimination of H$_2$Se to yield L(HSe)Al-Se-Al(SeH)L **17** (Scheme 8, (*20*)).

$$2\text{LAl(SeH)}_2 \xrightarrow{-H_2Se} \text{L(HSe)Al-Se-Al(SeH)L}$$
$$\mathbf{16} \hspace{5cm} \mathbf{17}$$

Scheme 8

The X-ray structures of both, **16** and **17**, determined from their single crystals, clearly show them to be well-separated monomers excluding the presence of any intermolecular hydrogen bonds. This observation is in contrast to the oxygen congener **1** which forms a dimer in the solid state with O··H-O bonds. Obviously the Brønsted acidity decreases in the order LAl(OH)$_2$ < LAl(SH)$_2$ < LAl(SeH)$_2$.

Organoaluminum Diamide

Aluminum compounds containing terminal NH_2 groups are rare (21). However, they could function as precursor for the preparation of aluminum nitride under mild conditions. Therefore we became interested in the synthesis of $LAl(NH_2)_2$ **18**. Indeed, the reaction of $LAlCl_2$ in the presence of the N-heterocyclic carbene (C(Nt-BuCH)$_2$) and ammonia in toluene as a solvent leads to $LAl(NH_2)_2$ in a 75% yield (Scheme 9). The presence of the N-heterocyclic carbene is essential for the preparation of **18** due to the high reactivity of this compound towards protons.

$$LAlCl_2 + 2NH_3 \xrightarrow[-2HCl]{2C(Nt\text{-}BuCH)_2} LAl(NH_2)_2$$
$$\phantom{LAlCl_2 + 2NH_3 \xrightarrow[-2HCl]{2C(Nt\text{-}BuCH)_2}} \mathbf{18}$$

Scheme 9

Therèfore, using an amine instead of the N-heterocyclic carbene causes side reactions due to the equilibrium of free base and protons within the amine system. Moreover, the resulting imidazolium chloride is sparingly soluble in hydrocarbon solvents and allows an easy separation from the reaction mixture by filtration (12).

Compound **18** was characterized by an X-ray single crystal structural analysis. **18** is monomeric in the solid state and the NH_2 groups are not involved in any hydrogen bonding.

Summary

Organometallic hydroxides of aluminum containing terminal OH groups are easily available. They are promising precursors for the preparation of heterobimetallic systems of the type Al-O-M (M = metal). So far compounds with the Al-O-M skeleton are rare. However, the reported synthetic strategies allow an access to this new class of compounds. The synthesis of $LAl(SH)_2$, $LAl(SeH)_2$, and $LAl(NH_2)_2$ are likely to change the way in which some of the "unknown" might be accessible using the reported new synthetic methods.

Acknowledgment

Financial support by the Deutsche Forschungsgemeinschaft and the Göttinger Akademie der Wissenschaften is highly acknowledged. I am very thankful to my coworkers for their contributions, the names are given in the references.

References

1. Levin, J.; Brandon, D. *J.Am.Ceram Soc.* **1998**, *81*, 1995.
2. Digne, M.; Sautet, P.; Raybaud, P.; Toulhoat, H.; Artacho, E. *J. Phys. Chem. B* **2002**, *106*, 5155.
3. F. Ullmann, *Aluminum Oxide* A1 **1985**, 557-594.
4. Feng, T.L.; Gurian, P.L.; Healy, M.D.; Barron, A.R. *Inorg. Chem.* **1990**, *29*, 408.
5. Harlan, C.J.; Mason, M.R.; Barron, A.R. *Organometallics* **1994**, *13*, 2957.
6. Mason, M.R.; Smith, J.M.; Bott, S.G.; Barron, A.R. *J. Am. Chem. Soc.* **1993**, *115*, 4971.
7. Wehmschulte, R.J.; Grigsby, W.J.; Schiemenz, B.; Bartlett, R.A.; Power, P.P. *Inorg. Chem.* **1996**, *35*, 6694.
8. Storre, J.; Klemp, A.; Roesky, H.W.; Schmidt, H.-G.; Noltemeyer, M.; Fleischer, R.; Stalke, D. *J. Am. Chem. Soc.* **1996**, *118*, 1380.
9. Storre, J.; Schnitter, C.; Roesky, H.W.; Schmidt, H.-G.; Noltemeyer, M.; Fleischer, R.; Stalke, D. *J. Am. Chem. Soc.* **1997**, *119*, 7505.
10. Duggan, P.J.; Tyndall, E.M. *J. Chem. Soc. Perkin Trans.* **2002**, 1325.
11. Bai, G.; Peng, Y.; Roesky, H.W.; Li, J.; Schmidt, H.-G.; Noltemeyer, M. *Angew. Chem.* **2003**, *115*, 1164; *Angew. Chem. Int. Ed.* **2003**, *42*, 1132.
12. Jancik, V.; Pineda, L.W.; Pinkas, J.; Roesky, H.W.; Neculai, D.; Neculai, A.M.; Herbst-Irmer, R. *Angew. Chem.* **2004**, *116*, 2194; *Angew. Chem. Int. Ed.* **2004**, *43*, 2142.
13. Jancik, V.; Roesky, H.W. unpublished results.
14. Bai, G.; Roesky, H.W.; Li, J.; Noltemeyer, M.; Schmidt, H.-G. *Angew. Chem.* **2003**, *115*, 5660; *Angew. Chem. Int. Ed.* **2003**, *42*, 5502.
15. Peng, Y.; Bai, G.; Fan, H.; Vidovic, D.; Roesky, H.W.; Magull, J. *Inorg. Chem.* **2004**, *43*, 1217.
16. Roesky, H.W.; Murugavel, R.; Walawalkar, M.G. *Chem .Eur. J.* **2004**, *10*, 324.
17. (a) Storre, J.; Klemp, A.; Roesky, H.W.; Schmidt, H.-G.; Noltemeyer, M.; Fleischer, R.; Stalke, D. *J. Am. Chem. Soc.* **1996**, *118*, 1380. (b) Storre, J.; Belgardt, Th.; Stalke, D.; Roesky, H.W. *Angew. Chem.* **1994**, *106*, 1365; *Angew. Chem. Int. Ed. Engl.* **1994**, *33*, 1244.
18. (a) Briant, C.E.; Hughes, G.R.; Minshall, P.C.; Michael, D.; Mingos, P. *J. Organomet. Chem.* **1980**, *202*, C18. (b) Ghiraldi, C.A.; Midollini, S.; Nuzzi, F.; Orlandini, A. *Transition Met. Chem.* **1983**, *8*, 73. (c) Arif, A.M.;

Hefner, J.G.; Jones, R.A.; Koschmieder, S.U. *J. Coord. Chem.* **1991**, *23*, 13. (d) Khorasani-Motlag, M.; Safari, N.; Pamplin, C.B.; Patrick, B.O.; James, B.R. *Inorg. Chim. Acta* **2001**, *320*, 184. (e) Bottomley, F.; Drummond, D.F.; Egharevba, G.O.; White, P.S. *Organometallics* **1986**, *5*, 1620. (f) Jessop, P.G.; Lee, C.-L.; Rastar, G.; James, B.R.; Lock, C.J.L.; Faggiani, R. *Inorg. Chem.* **1992**, *31*, 4601. (g) Schwarz, D.E.; Dopke, J.A.; Rauchfuss, T.B.; Wilson, S.R. *Angew. Chem.* **2001**, *113*, 2413; *Angew. Chem. Int. Ed.* **2001**, *40*, 2351. (h) Howard, W.A.; Parkin, G. *Organometallics* **1993**, *12*, 2363.
19. Jancik, V.; Peng, Y.; Roesky, H.W.; Li, J.; Neculai, D.; Neculai, A.M.; Herbst-Irmer, R. *J. Am. Chem. Soc.* **2003**, *125*, 1452.
20. Cui, C. Roesky, H.W.; Hao, H.; Schmidt, H.-G.; Noltemeyer, M. *Angew. Chem.* **2000**, *112*, 1885; *Angew. Chem. Int. Ed.* **2000**, *39*, 1815.
21. Chang, C.-C.; Li, M.-D.; Chiang, M.Y.; Peng, S.-M.; Wang, Y.; Lee, G.-H. *Inorg. Chem.* **1997**, *36*, 1955.

Chapter 3

From Gallium Hydride Chlorides to Molecular Gallium Sulfides

Hubert Schmidbaur and Stefan Nogai

Department Chemie, Technische Universität München
Lichtenbergstrasse 4, 85748 Garching, Germany

In an effort to prepare molecular components for the construction of larger aggregates – from nanostructures to multidimensional arrays – of gallium sulfide, the chemistry of the (chloro)gallium hydrides GaH_nCl_{3-n} has been developed further and employed for selective substitution reaction. The synthetic pathways to $[HGaCl_2]_2$ have been optimized and the structure of the crystalline product determined. The dinuclear compound was converted into mononuclear 1:1 complexes with series of both tertiary phosphines and substituted pyridines. The complexes are stable and soluble in common organic solvents to allow efficient purification. The crystal structures of selected examples have been determined including those of coordination compounds of ditertiary phosphines. Reference compounds with $GaCl_3$ and GaH_3 have also been prepared and fully characterized. On controlled thermal decomposition, the $HGaCl_2$ complexes undergo dehydrogenative coupling to give almost quantitative yields of $GaCl_2$ complexes which are diamagnetic dinuclear species with a Ga-Ga single bond, as demonstrated for phosphine and pyridine adducts. The 2:1 complex of GaH_3 with 2,4-dimethylpyridine undergoes an internal 1,4-hydrogallation reaction to afford the 4-hydro-pyridyl-(pyridine)gallium dihydride. $(HGaCl_2)_2$ can be readily converted into ternary compounds [GaYX] which are soluble in pyridines (L) as the trinuclear complexes [GaYX(L)]$_3$ with X = Cl, Br and Y = S, Se. Their molecular structures (six-membered rings) and conformations have been determined. With excess ligand L, these trinuclear compounds can be transformed into bicyclic tetranuclear dications of the type [Ga$_4$Y$_5$(L)$_6$]$^{2+}$ and finally into molecular species [Ga$_4$Y$_6$(L)$_4$] containing e.g. molecular gallium sulfide Ga$_4$S$_6$ as a soluble pyridine complex.

Introduction

Owing to their relevance to a variety of advanced technologies, there is currently wide-spread interest in metal and mixed-metal chalcogenides of all dimensionalities, from bulk materials to nano-sized clusters (*1*). While there has been great progress in the chemistry of 1:1 sulfur, selenium and tellurium binaries e.g. of zinc and cadmium (*2*), the research is less advanced for the 2:3 chalcogenide analogues of the neighbouring elements gallium and indium (*3*).

The modifications of bulk gallium(III) sulfide Ga$_2$S$_3$ are generally described as complicated "defect ZnS structures" ("random wurtzite or zincblende" etc.) based on arrays of tetracoordinate gallium atoms linked by two-coordinate sulfur atoms in a network of six-membered rings. In ordered modifications these six-membered rings may be fused to generate adamantane type building blocks cross-linked by sulfide bridges, again reminiscent of the ZnS-type lattices (*4*). The adamantane cages have also been found to be components of nano-sized molecular species and the Ga$_4$S$_6$ core has recently been classified as an "artificial tetrahedron atom" (T-atom) which can be used as the building block of supertetrahedral clusters T$_n$ (*3,5*).

In our own previous studies (*6,7,8*) we were able to find convenient synthetic routes to the ternary compounds GaYX (Y = chalcogen, X = halogen) (*9*), starting from dichlorogallane [HGaCl$_2$]$_2$ and elemental chalcogen or chalcogen halides, and to show that these can be readily dissolved in polar organic solvents upon addition of pyridines. Novel trinuclear complexes of the type [(L)GaYX]$_3$ with Y = S, Se; X = Cl, Br, and L = pyridine) are thus obtained which have molecular six-membered ring structures in standard chair or boat conformations, depending on the nature of the substituents X, Y and L (*6*).

The prototypes like [(pyr)GaSCl]$_3$ are unique in that they can be redissolved and crystallized in a variety of modifications (*7,8*). The halogen functionalities at their Ga$_3$S$_3$ rings allow the construction of more extended secondary products, one of which was identified as a compound with a bicyclic dication of the borax-type structure (8). The more basic ligand L' = 4-dimethylamino-pyridine proved to be the most efficient donor for such transformations.

In subsequent investigations we have now found access to a complex of the elusive Ga_4S_6 molecule with this ligand. The preparative pathway has its origin in the 1:1 complexes of dichlorogallane $[HGaCl_2]_2$ (10,11) which can be converted to a family of adducts with tertiary phosphines (11) and substituted pyridines (12).

In the present account the preparation and structural characterization of these precursor complexes are introduced here and subsequently the synthesis of the chalcogenides using bis(trimethylsilyl)sulfide $(Me_3Si)_2S$ as the sulfide source for the gallane complexes is described.

Results

Gallium trihydride, trihydride/halide and trihalide complexes of tertiary phosphines $(R_3P)GaX_nH_{3-n}$ (X = Cl, Br)

Tertiary $[R_3P]$ and ditertiary phosphines $[R_2P(CH_2)_nPR_2]$ are known to stabilize the elusive gallane unit GaH_3 (13) in 1:1 /1:2 complexes of the type $(R_3P)GaH_3$ (14,15,16) or $H_3Ga[R_2P(CH_2)_nPR_2]GaH_3$ (16), but is was only in one case $[R_3P = tri(cyclohexyl)phosphine]$ that an adduct has been converted into the corresponding chlorogallane and dichlorogallane complexes, $(Cy_3P)GaH_2Cl$ and $(Cy_3P)GaHCl_2$, respectively, using HCl (17). A selection of gallane complexes mainly of the type $(R_3P)GaH_3$ has been structurally characterized, complemented only by the structure of $(Cy_3P)GaH_2Cl$ (15-17), and very few complexes of (poly)tertiary phosphines with the gallium trihalides have been reported. These include mainly $(Me_3P)GaCl_3$ (18), $(Et_3P)GaCl_3$ (11) and $[CHPPh_2GaBr_3]_2$ (19).

With $[HGaCl_2]_2$ now readily available and structurally characterized (11), there is direct access to its complexes from the components. This was demonstrated with PEt_3, PCy_3, PPh_3, PVi_3 and $[CH_2PPh_2]_2$ (dppe), for which the products were obtained from equivalent quantities of the reactants in almost quantitative yields (Equations 1, 2).

$(HGaCl_2)_2 + 2 PR_3 \longrightarrow 2 HGaCl_2(PR_3)$ (1)

$(HGaCl_2)_2 + (Ph_2PCH_2)_2 \longrightarrow (CH_2Ph_2PGaHCl_2)_2$ (2)

The complex with PEt₃ is a colorless liquid at room temperature (m.p. 4-7°C), all others are colorless solids, sensitive to air and moisture. With the exception of the PVi₃ complex which is insoluble in all common organic solvents, the compounds are soluble i. a. in di- and trichloromethane, benzene, toluene, and diethylether. Their composition has been confirmed by elemental analysis and infrared and NMR spectroscopy. In all cases a band appearing close to 1950 cm^{-1} can be assigned to the Ga-H stretching frequency, except for the insoluble adduct of the net composition (Vi₃P)GaHCl₂ , in which the ligand probably has undergone a hydrogallation side-reaction linking the molecular units into polymers. This suggestion is supported by the observation, that all complexes except for "(Vi₃P)GaHCl₂" show a proton resonance at ca. 5.5 ppm of the hydrogen atom attached to the gallium center (*20*).

Surprisingly, the ^{31}P resonances of the adducts appear upfield from the resonances of the corresponding free ligands, but there are parallels for this phenomenon in the data published and discussed for complexes of tertiary phosphines with other metal trihalides (*19*).

The crystal structures have been determined for two crystalline modifications of (Ph₃P)GaHCl₂, for (Cy₃P)GaHCl₂ and for [CH₂Ph₂PGaHCl₂]₂ as a 1:1 solvate with benzene and as a 1:2 solvate with diethyl ether. Selected results are shown in Figures 1-3. In all cases some disorder of the GaHCl₂ part over at least two sites was observed, but this could be accounted for (Cy₃P)GaHCl₂ and for the benzene solvate of [CH₂Ph₂PGaHCl₂]₂ by the usual split models. In the latter structures the sites of the Ga-*H* hydrogen atoms could be located and refined to an acceptable accuracy. The Ga-H distances agree well with previous data for gallium hydrides. The monoclinic and triclinic modification of (Ph₃P)GaHCl₂ differ only by the rotation of the PPh₃ ligand about the Ga-P bond: The Cl-Ga-P-C torsional angles are larger by ca. 15° in the monoclinic form. The projections of the GaCl₂H units along the Ga-P axis are virtually superimposible (Figure 4).

The dinuclear dppe complex has a crystallographically imposed center of inversion midway between the ethylidene carbon atoms in both solvates implying a fully extended *trans* conformation of the staggered *PCH₂CH₂P* unit, but disrotatory *gauche* conformations for the *CH₂CH₂PPh₂Ga* and *CH₂PGaH* units (referring to the dihedral angles of the four lead atoms in *italics*).

The molecular structure of the liquid (Et₃P)GaHCl₂ has been calculated by quantum-chemical methods on the RI-MP2 level with an SV(P) basis set (TURBOMOLE) (*21*). The results are similar to the data obtained by X-ray crystallography for the other members of the series of complexes (Figure 5). The structure and conformation represent a true energy minimum for the molecule in the gas phase as demonstrated by a detailed frequency analysis. A superposition of the experimental (IR) and calculated vibrational spectra showed an excellent

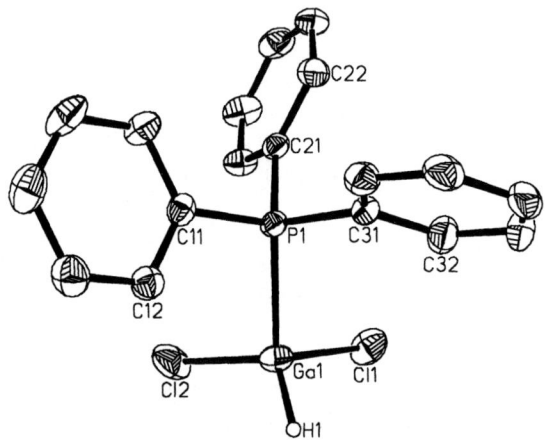

Figure 1: Molecular structure of (Ph₃P)GaHCl₂ in the monoclinic modification.

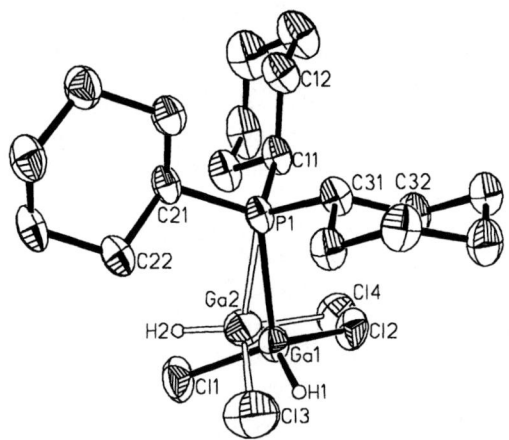

Figure 2: Molecular structure of (Cy₃P)GaHCl₂.

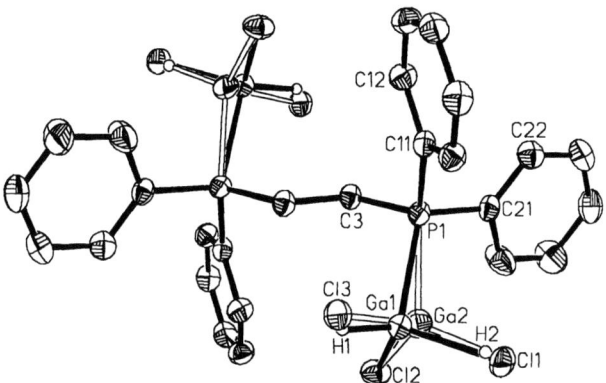

Figure 3: Molecular structure of [CH$_2$Ph$_2$PGaHCl$_2$]$_2$ in the benzene solvate.

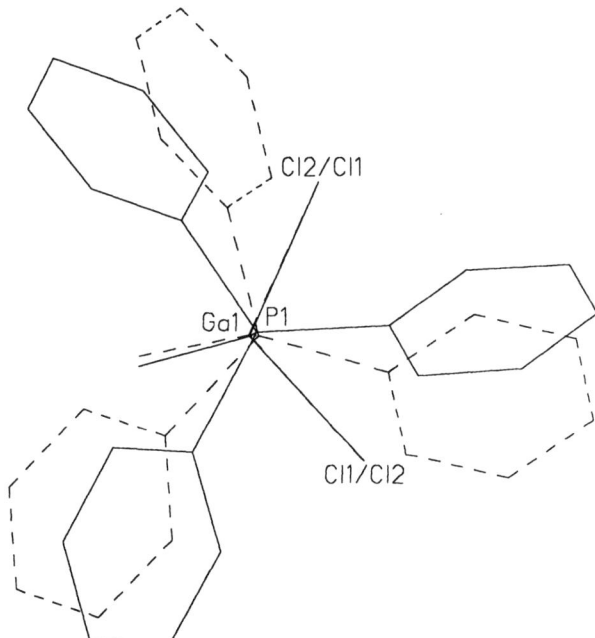

Figure 4: Superposition of the (Ph$_3$P)GaHCl$_2$ formula units in the monoclinic and the triclinic modifications.

agreement. [Calculated dimensions: Ga-H 1.5570, Ga-P 2.4355, Ga-Cl 2.1976/2.1978 Å.]

Thermal decomposition of $(Et_3P)GaHCl_2$ affords hydrogen and a dinuclear gallium(II) compound of the formula $(Et_3P)Cl_2GaGaCl_2(PEt_3)$ (Equation 3). According to a single crystal structure analysis this molecule is centrosymmetrical (point group C_i) with a Ga-Ga distance of 2.4269(5) Å (Figure 6). The same product is obtained from "gallium dichloride" in toluene (22), which has long been identified as the mixed-valent salt $Ga[GaCl_4]$, and PEt_3. In this reaction a comproportionation with formation of a Ga-Ga single bond takes place, which was also observed with other donor molecules. $Ga[GaBr_4]$ gives the same reaction (Ga-Ga 2.427(1) Å) (11) (Equation 4).

$$2\ (Et_3P)GaHCl_2 \longrightarrow (Et_3P)Cl_2Ga\text{-}GaCl_2(PEt_3) + H_2 \qquad (3)$$

$$Ga[GaX_4] + 2\ Et_3P \longrightarrow (Et_3P)X_2Ga\text{-}GaX_2(PEt_3)$$
$$X=Cl, Br \qquad (4)$$

Gallium trihydride, trihydride/halide and trihalide complexes of pyridines

The preparation and characterization of pyridine complexes of dichlorogallane have only recently been reported from this Laboratory (12). Representative examples of the 1:1 stoichiometry are shown in Figures 7 and 8. The compounds are readily obtained from the components (Equation 5) and can be crystallized as colorless solids with low melting points from mixtures of inert polar and non-polar solvents.

$$(HGaCl_2)_2 + 2\ Pyr \longrightarrow 2\ (Pyr)GaHCl_2 \qquad (5)$$

It is particularly noteworthy that with 4-dimethylamino-pyridine the Ga-N bond is not formed with the nitrogen atom of the amino group, but with the pyridine nitrogen atom. This indicates the strong preference of GaX_3 acceptors for pyridine donors which is obvious also from the broad range of stable gallium halide (12, 23) and chalcogenide (6-8) complexes with pyridines. Pyridines are definitely the favourite ligands of most gallium acceptor molecules.

With two equivalents of pyridine, the 2:1 complexes are obtained as shown for the example with 3,5-dimethyl-pyridine (Figure 9). This structure approaches C_{2v} symmetry. The ν(Ga-H) band appears at 1873 cm^{-1} as compared to 1978 cm^{-1} in the 1:1 complex.

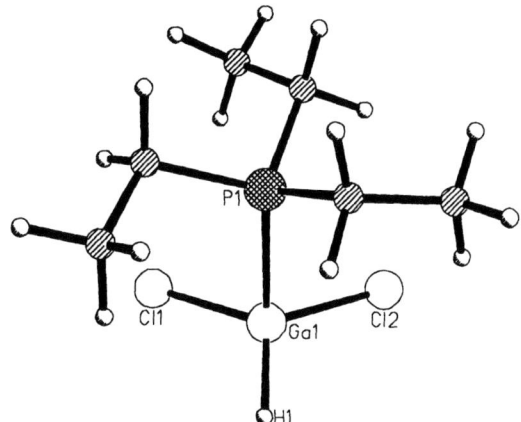

Figure 5: Calculated minimum geometry for $(Et_3P)GaHCl_2$.

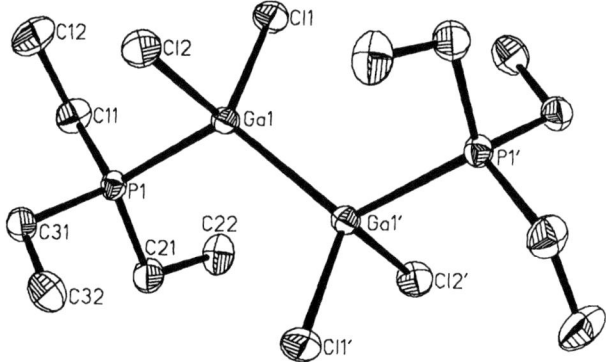

Figure 6: Molecular structure of $(Et_3P)Cl_2GaGaCl_2(PEt_3)$.

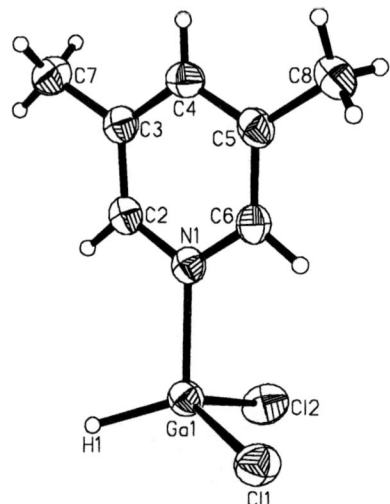

Figure 7: Molecular structure of (3,5-Me$_2$-Py)GaHCl$_2$.

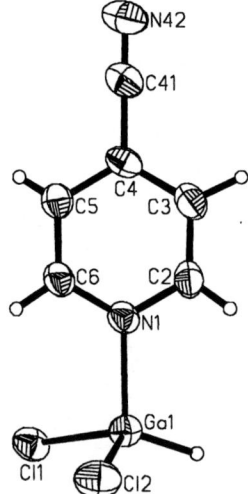

Figure 8: Molecular structure of (4-NC-Py)GaHCl$_2$.

In each case, the characterization is also based on elemental analyses, infrared and NMR spectra, and crystal structure analysis of representative examples. In the crystal, the adduct molecules show weak intermolecular interactions as demonstrated in Figure 10. The deviations of the coordination sphere of the gallium atoms from a tetrahedral environment induced by the dimerization are small.

An adduct of 3,5-dimethylpyridine (L) with gallane of the formula (L)GaH$_3$ was obtained from LiGaH$_4$ and the corresponding pyridinium chloride (12). With evolution of hydrogen gas and precipitation of LiCl the product is formed in high yield (Equation 6).

$$\text{Li[GaH}_4\text{]} + \text{[3,5-Me}_2\text{-PyH]Cl} \longrightarrow \text{H}_3\text{Ga(3,5-Me}_2\text{-Py)} + \text{H}_2 + \text{LiCl} \tag{6}$$

The complex was readily identified by standard analytical, spectroscopic and structural methods. The molecular structure (two independent molecules in the unit cell) is shown in Figure 11.

On prolonged storage or upon treatment with excess L (3 d at 20°C) the compound undergoes fundamental changes as easily recognized from a color change to red and from new IR and NMR spectral features. The analytical data show that the main component of the reaction mixture is an adduct of a second ligand to the 1:1 complex. The incoming ligand undergoes hydrogallation to give the 3,5-dimethyl-pyridine adduct of a 3,5-dimethyl-1,4-dihydro-pyrid-1-yl-gallane. In the latter, one of the three Ga-H bonds of the starting material has been added across the new pyridine ligand (in 1,4-positions). The structure has been determined by a single crystal X-ray diffraction analysis (Figure 12).

The two Ga-N bonds are significantly different indicating a much stronger bonding of the pyridyl group [Ga-N 1.898(2) Å] as compared to the pyridine ligand [Ga-N 2.072(2) Å]. The Ga-H bonds are equivalent within the limits of standard deviations [average 1.44 Å]. The pyridyl ligand has strongly alternating C-C bond lengths, while the pyridine ligand has retained its aromaticity-equilibrated dimensions.

The IR absorptions and proton resonances of the GaH$_2$ unit in the addition product differ significantly from those of the (L)GaH$_3$ substrate. In the region of the ligand absorptions there is a new set of bands which can be assigned to the 4-hydro-pyridyl group. The bands of the remaining pyridine ligand are largely unchanged. In the ^{1}H and ^{13}C NMR spectra (in benzene solution at 20°C) a new set of signals is characteristic of the 4-hydro-pyridyl group. A proton singlet resonance at 3.21 ppm, shifted strongly upfield from pyridine resonances, represents the hydrogen atoms at the sp^3-carbon atom in the 4-position. The ^{13}C resonance of the C-4 carbon atom of the pyridyl group is even shifted upfield by no less than ca. 100 ppm (δ 36.0 vs. 138.8 ppm).

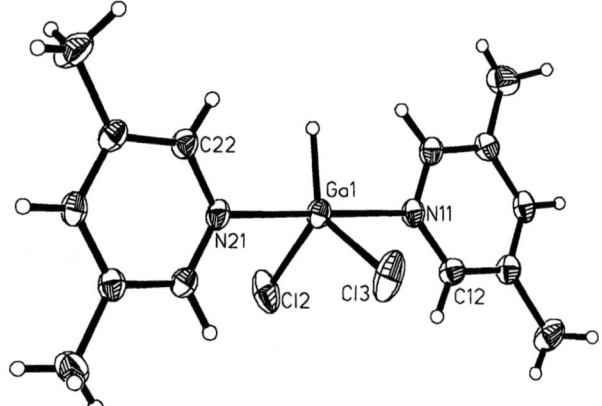

Figure 9: Molecular structure of (3,5-Me$_2$-Py)$_2$GaHCl$_2$.

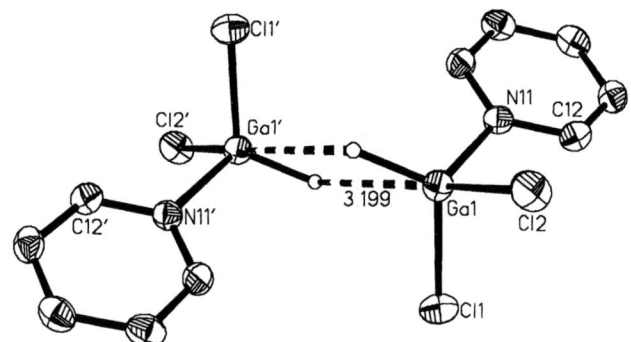

Figure 10: Intermolecular contacts in the lattice of (Py)GaHCl$_2$.

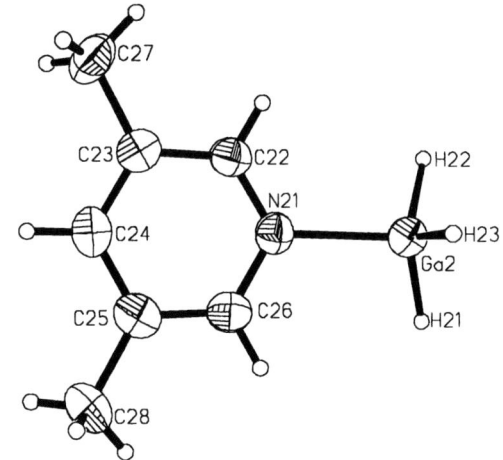

Figure 11: Molecular structure of (3,5-Me$_2$-Py)$_2$GaH$_3$.

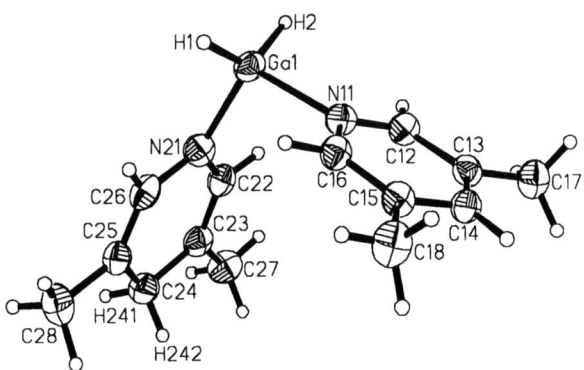

Figure 12: Molecular structure of (3,5-Me$_2$-Py)(4-H$_2$-3,5-Me$_2$-Py)GaH$_2$.

The hydrogallation of pyridine is a parallel of the hydroalumination reported previously (24). Since aluminum hydrides are more powerful reducing agents than gallium hydrides, the hydroalumination may even occur with all four Al-H functions of e.g. LiAlH$_4$. In the present case, only one Ga-H function of (L)GaH$_3$ was used even if an excess of the pyridine was available. More forcing conditions ($\geq$ 20°C, $\geq$ 3 d) have not yet been tested. Already under the mild conditions the reaction mixture was quite complex, the yield of the product was very low ($\leq$ 15%) and the formation of elemental gallium was observed.

As a product of dehydrogenative coupling of (L)GaHCl$_2$ molecules the dinuclear complexes (L)Cl$_2$Ga-GaCl$_2$(L) are formed. With L = 3,5-dimethyl-pyridine, an almost 90% yield is obtained after 5 h in refluxing toluene. This very stable compound (m.p. 228°C) is a true gallium(II) species with a standard Ga-Ga bond as proved by a crystal structure analysis (Figure 13). The two halves of the dinuclear complex (point group C_{2h}) are related by symmetry. A staggered ethane-type conformation is found with the pyridine rings coplanar with the N-Ga-Ga'-N' plane. In solution there is free rotation of the rings about the Ga-N bonds as indicated by the NMR spectra which show equivalent methyl groups at ambient temperature. The following parameters are accurate benchmark data for this type of Ga(II) compounds: Ga-Ga 2.4000(8), Ga-N 2.015(4), Ga-Cl 2.189(1) Å). Note that the Ga-Ga distance is significantly shorter than in the corresponding PEt$_3$ complex (above).

The reaction of 3,5-dimethyl-pyridine with gallium trichloride gives quantitative yields of the 1:1 adduct, which shows no anomalies (Figure 14).

Molecular Gallium Sulfides with Ring and Cage Structures

Monocyclic gallium sulfides of the type [GaSX(L)]$_3$ with a variety of substituents can be prepared by dissolving gallium sulfide halides (GaSX), with X = Cl, Br, in a solution of suitable ligands L in polar solvents. While this is generally true, it has been discovered a few years ago that pyridines are by far superior to any other ligands L (6-8). Attempts to dissolve [GaSX]$_3$ compounds with other ligands – including tertiary phosphines – have to-date been largely unsuccessful. It therefore again appears that pyridines are the ligands of choice for gallium acceptors.

A large family of these trinuclear complexes has been prepared and their structures determined, two of which are shown in Figures 15,16. The Ga$_3$S$_3$ rings are in boat or chair conformations with the halogen and pyridine substituents distributed in various motifs over axial and equatorial positions (C_1 or C_s symmetry). Each gallium atom is always bearing one halogen atom and one pyridine ligand, its tetrahedral coordination being completed by two ring sulfur atoms. Details of this extensive work have been published (6-8). The preparative

45

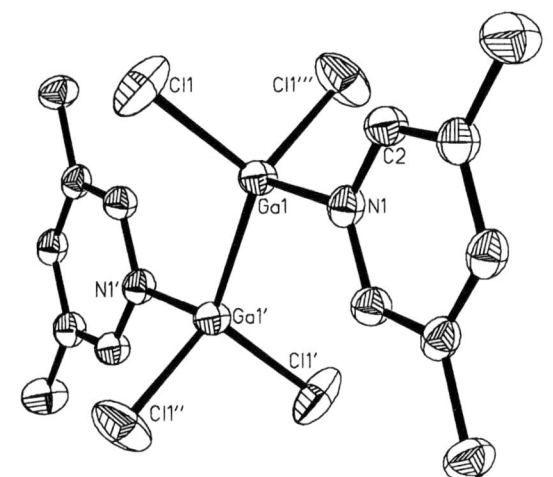

Figure 13: Molecular structure of [(3,5-Me$_2$-Py)GaCl$_2$]$_2$.

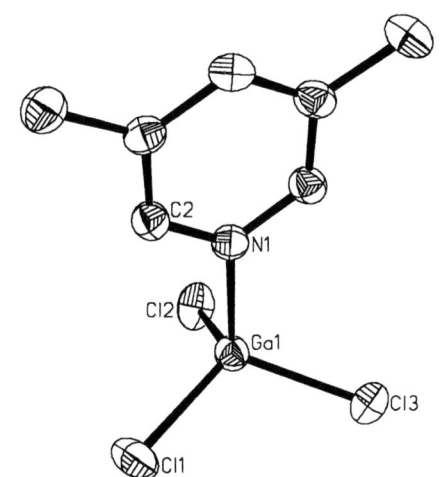

Figure 14: Molecular structure of (3,5-Me$_2$-Py)GaCl$_3$.

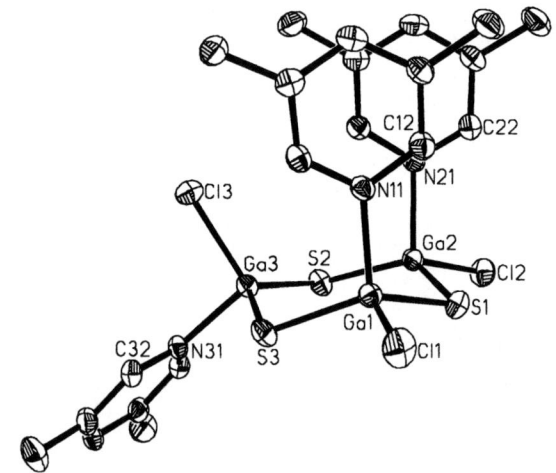

Figure 15: Molecular structure of [(3,5-Me₂Py)GaSCl]₃.

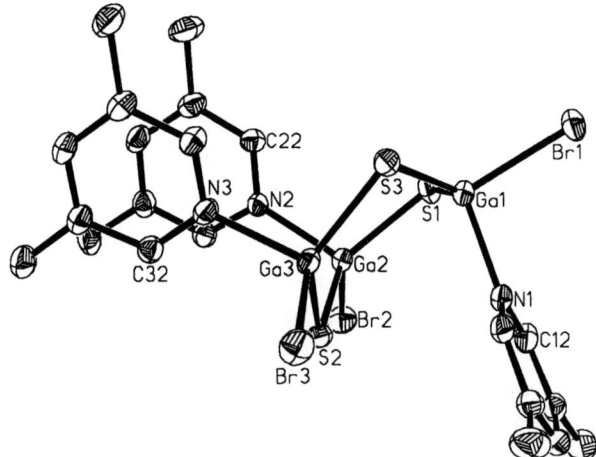

Figure 16: Molecular structure of [(3,5-Me₂Py)GaSBr]₃.

work in this area is based on chlorogallane chemistry, because the ternary phases [GaSX] are in fact best prepared from dichlorogallane [HGaCl$_2$]$_2$: Its thermolysis affords quantitative yields of Ga[GaCl$_4$] (*10*) which can be converted e.g. into [GaSCl] by treatment with sulfur or sulfur chlorides (*6,7*).

Treatment of one of these trinuclear complexes with *excess* pyridine (or substituted pyridines) leads to a ring opening and the buildup of condensed systems. A typical example is the tetranuclear dicationic species [Ga$_4$S$_5$(L)$_6$]$^{2+}$ with two fused six-membered rings and with L = 4-dimethylamino-pyridine (Figure 17), which is obtained as the dibromide salt upon reacting [GaSBr(L)]$_3$ with excess L in refluxing acetonitrile. The framework of this dication resembles the structure of the borax dianion [B$_4$O$_5$(OH)$_4$]$^{2-}$, or of a neutral trimethylamine adduct of an aluminum sulfide hydride: Al$_4$S$_5$H$_2$(NMe$_3$)$_4$ (*25a*).

An alternative method for building more complex gallium sulfide structures is the reaction of the pyridine complexes of dichlorogallane with bis(trimethylsilyl)sulfide. Thus (L)$_2$GaHCl$_2$ (with L = 4-Me$_2$N-Py) affords an insoluble mixed hydride halide on treatment with (Me$_3$Si)$_2$S in acetonitrile (Equation 7). The presence of Ga-H functions in the product has been confirmed by a prominent IR absorption at 1890 cm^{-1}. The analytical composition suggests a polymeric structure with –[(L)GaS]- repeating units, and a random distribution of hydrogen and chlorine substituents at the gallium atoms [(L)GaSH$_x$Cl$_{1-x}$]$_n$. Independent experiments with (L)GaH$_3$ and (Me$_3$Si)$_2$S confirmed that compounds with the general composition [(L)GaS(H)]$_n$ are indeed insoluble polymers.

n HGaCl$_2$(Pyr) + n (Me$_3$Si)$_2$S ⟶ [GaSH$_x$Cl$_{1-x}$(Pyr)]$_n$ + 2n Me$_3$SiCl   (7)

The polymers were found to dissolve almost completely in boiling acetonitrile upon addition of excess ligand L, and a colorless crystalline compound could be isolated in ca. 30% yield. This was identified as a complex of molecular gallium sulfide Ga$_4$S$_6$ bearing four ligands L (Figure 18). The gallium and sulfur atoms form an adamantane-type structure, and each gallium atom bears a pyridine ligand.

The new complex is a rare example in which a molecular metal sulfide, Ga$_4$S$_6$, is preserved in a stable complex. The analogous aluminum compound (*25b*) has been stabilized by trimethylamine ligands, which can not be used for Ga$_4$S$_6$. If the same reaction - successfully executed with pyridines - is carried out with tertiary amines, no analogous gallium sulfide adducts are obtained.

The new molecular gallium sulfide complexes are an excellent basis for the construction of larger frameworks.

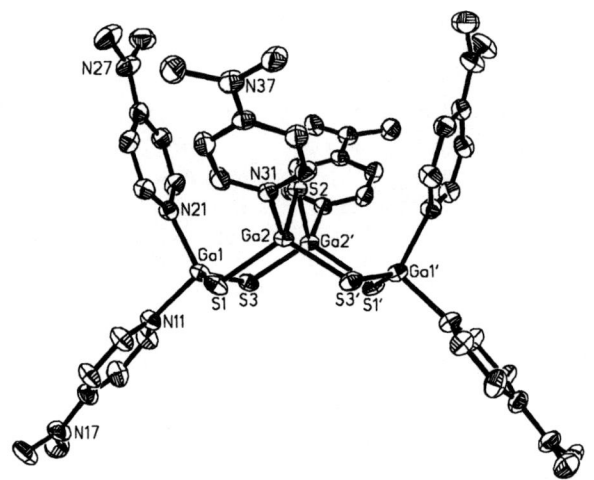

Figure 17: Structure of the dication $[(4-Me_2N-Py)_6Ga_4S_5]^{2+}$ in the bromide salt

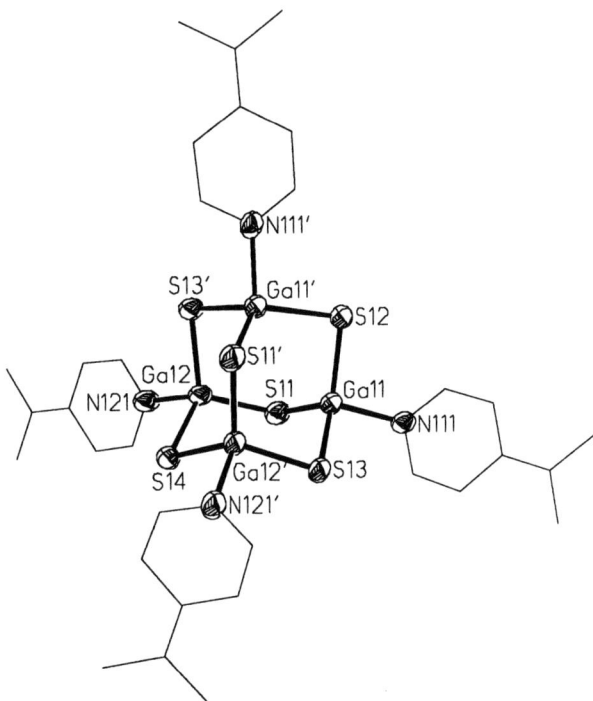

Figure 18: Molecular structure of [(4-Me$_2$N-Py)Ga$_4$S$_6$]

Acknowledgement

The work presented in this account was generously supported by Deutsche Forschungsgemeinschaft, Fonds der Chemischen Industrie, Alexander von Humboldt Foundation, Deutscher Akademischer Austauschdienst, and Volkswagenstiftung. The authors gratefully acknowledge the contributions of collaborators to previous studies in this research area (W. Findeiss, T. Zafiropoulos, J. Ohshita, and E. S. Schmidt.).

References

1. a) Collier, C. P.; Vossmeyer, T.; Heath, J. R.; *Ann. Rev. Phys. Chem.* **1998**, *49*, 371; b) Dance, I.; Fisher, K.; in *Progress in Inorganic Chemistry*, Karlin, K. D.; Ed., Vol 41, p637, 1994; c) Krebs, B.; Henkel, G.; *Angew. Chem. Int. Ed. Engl.* **1991**, *30*, 769; d) Krebs, B.; Voelker, D.; Stiller, K.-O.; *Inorg. Chim. Acta* **1982**, *65*, 101; e) Downs, A. J.; *Chemistry of Aluminium, Gallium, Indium, and Thallium*, Blackie Academic & Professional, Glasgow, 1993.
2. a) Pfistner, H.; Fenske, D.; *Z. Anorg. Allg. Chem.* **2001**, *627*, 575; b) Soloviev, V. N.; Eichhöfer, A.; Fenske, D.; Banin, U.; *J. Am. Chem. Soc.* **2001**, *123*, 2354.
3. Zheng, N.; Bu, X.; Feng, P.; *J. Am. Chem. Soc.* **2003**, *125*, 1138.
4. a) Wells, A. F.; *Structural Inorganic Chemistry*, 5. Ed., Oxford University Press, pp. 763, 1987; b) Hahn, H.; Klinger, W.; *Z. Anorg. Allg. Chem.* **1949**, *259*, 135; c) Goodyear, J.; Steigmann, G. A.; *Acta Cryst.* **1963**, *16*, 946; d) Collin, G.; Flahaut, J.; Guittard, M.; Loireau-Lozach, A.; *Mat. Res. Bull.* **1976**, *11*, 285; e) Tomas, A.; Pardo, M. P.; Guittard, M.; Guymont, M.; Famery, R.; *Mat. Res. Bull.* **1987**, *22*, 1549.
5. a) Zheng, N.; Bu, X.; Wang, B.; Feng, P.; *Science* **2002**, *298*, 2366; b) Li, H.; Laine, A.; O'Keeffe, M.; Yaghi, O. M.; *Science* **1999**, *283*, 1145; c) Li, H.; Eddaoudi, M.; Laine, A.; O'Keeffe, M.; Yaghi, O. M.; *J. Am. Chem. Soc.* **1999**, *121*, 6096; d) Li, H.; Kim, J.; Groy, T. L.; O'Keeffe, M.; Yaghi, O. M.; *J. Am. Chem. Soc.* **2001**, *123*, 4867.
6. Ohshita, J.; Schier, A.; Schmidbaur, H.; *J. Chem. Soc., Dalton Trans.* **1992**, 3561.
7. Nogai, S. D.; Schier, A.; Schmidbaur, H.; *Z. Naturforsch.* **2001**, *56b*, 711.
8. Nogai, S. D.; Schmidbaur, H.; *Dalton Trans.* **2003**, 2488.
9. a) Kniep, R.; Wilms, A.; Beister, H. J.; *Mater. Res. Bull.* **1983**, *18*, 615; b) Kniep, R.; Wetzel, W.; *Z. Naturforsch.* **1984**, *40b*, 26; c) Wilms, A.; Kniep, R.; *Z. Naturforsch.* **1981**, *36b*, 1658.

10. Schmidbaur, H.; Findeiss, W.; Gast, E.; *Angew. Chem. Int. Ed. Engl.* **1965**, *4*, 152.
11. Nogai, S.; Schmidbaur, H.; *Inorg. Chem.* **2002**, *41*, 4770.
12. Nogai, S.; Schriewer, A.; Schmidbaur, H.; *Dalton Trans.* **2003**, 3165.
13. a) Downs, A. J.; Goode, M. J.; Pulham, J. C. R.; *J. Am. Chem. Soc.* **1989**, *111*, 1936; b) Pulham, C. R.; Downs, A. J.; Goode, M. J.; Rankin, D. W. H.; Robertson, H. E. J.; *J. Am. Chem. Soc.* **1991**, *113*, 5149.
14. Greenwood, N. N.; Ross, E. J. F.; Storr, A.; *J. Chem. Soc.* **1965**, 1400.
15. Tang, C. Y.; Coxall, R. A.; Downs, A. J.; Greene, T. M.; Kettle, L.; Parsons, S.; Rankin, D. W. H.; Robertson, H. E.; Turner, A. R.; *Dalton Trans.* **2003**, 3526.
16. a) Atwood, J. L.; Robinson, K. D.; Bennett, F. R.; Elms, F. M.; Kousantonis, G. A.; Raston, C. L.; Young, D. J.; *Inorg. Chem.* **1992**, *31*, 2673; b) Elms, F. M.; Gardiner, M. G.; Koutsantonis, G. A.; Raston, C. L.; Atwood, J. L.; Robinson, K. D.; *J. Organomet. Chem.* **1993**, *449*, 45.
17. Elms, F. M.; Koutsantonis, G. A.; Raston, C. L.; *J. Chem. Soc., Chem. Commun.* **1995**, 1669.
18. Carter, J. C.; Jugie, G.; Enjalbert, R.; Galy, J.; *Inorg. Chem.* **1978**, *17*, 1248.
19. Sigl, M.; Schier, A.; Schmidbaur, H.; *Z. Naturforsch.* **1998**, *53b*, 1301.
20. Nogai, S.; Schmidbaur, H.; *Z. Anorg. Allg. Chem.* **2004** (in press).
21. a) Ahlrichs, R.; Bär, M.; Häser, M.; Horn, H.; Kömel, C.; *Chem. Phys. Letters* **1989**, *162*, 165; b) Weigend, F.; Häser, M.; *Theor. Chem. Acc.* **1997**, *97*, 331; c) Weigend, F.; Häser, M.; Patzelt, H.; Ahlrichs, R.; *Chem. Phys. Letters* **1998**, *294*, 143.
22. a) Schmidbaur, H.; Thewalt, U.; Zafiropoulos, T.; *Organometallics* **1983**, *2*, 1550; b) Schmidbaur, H.; *Angew. Chem. Int. Ed. Engl.* **1985**, *24*, 893; c) Schmidbaur, H.; Thewalt, U.; Zafiropoulos, T.; *Angew. Chem. Int. Ed. Engl.* **1984**, *23*, 76; d) Schmidbaur, H.; Nowak, R.; Huber, B.; Müller, G.; *Polyhedron* **1990**, *9*, 283.
23. a) Greenwood, N. N.; Perkins, P. G.; *Pure Appl. Chem.* **1961**, *2*, 55; b) Greenwood, N. N.; Srivastava, T. S.; Straughan, B. P.; *J. Chem. Soc. A* **1966**, 699; c) Restivo, R.; Palenik, G. J.; *J. Chem. Soc., Dalton Trans.* **1972**, 341; Sinclair, I.; Small, R. W. H.; Worrall, I. J.; *Acta Crystallogr.* **1981**, *B37*, 1290; Gordon, E. M.; Hepp, A. F.; Duraj, S. A.; Habash, T. S.; Farnwick, P. E.; Schupp, J. D.; Eckles, W. E.; Long, S.; *Inorg. Chim. Acta* **1997**, *257*, 247.
24. a) Lansbury, P. T.; Peterson, J. O.; *J. Am. Chem. Soc.* **1963**, *85*, 2236; b) Hensen, K.; Lemke, A.; Stumpf, T.; Bolte, M.; Fleischer, H.; Pulham, C. R.; Gould, R. O.; Harris, S.; *Inorg. Chem.* **1999**, *38*, 4700.
25. a) Godfrey, P. D.; Raston, C. L.; Skelton, B. W.; Tolhurst, V.-A.; White, A. H.; *Chem. Commun.* **1997**, 2235; b) Wehmschulte, R. J.; Power, P. P.; *J. Am. Chem. Soc.* **1997**, *119*, 9566.

Chapter 4

Four-Valence Electron Reactive Intermediates and the Philicity of Charged Carbene Analogs

Peter P. Gaspar, Xinping Liu, Diana Ivanova, David Read, James S. Prell, and Michael L. Gross

Department of Chemistry, Washington University, St. Louis, MO 63130

Studies of six-valence electron charged carbene analogs have been undertaken via quadrupole ion trap mass spectroscopy. Mechanistic differences between charged and neutral carbene analogs may be explained by philicity differences, with postive ions reacting as powerful electrophiles. Four-valence electron species may form as many as three bonds at a time, and some reactions of SiH^+ are reported.

Why Study Electron-Deficient Species?

Many years ago we were led to the study of electron-deficient species by the question: how can the mechanistic ideas of first-row chemistry be extended to the understanding of the covalent bond throughout the periodic table? Our approach has been to investigate heavier element analogs of first-row reactive intermediates whose reactions are so distinctive that product studies could provide mechanistic screens capable of revealing mechanistic similarities and differences between the heavier species and their first-row parents.

We chose neutral, six-valence electron species as our models because of the distinctive addition and insertion reactions of carbenes, in which the formation of two bonds can be concerted because a singlet carbene is equipped with both a nucleophilic and an electrophilic orbital, the s-weighted lone pair of the HOMO and the pure-p LUMO. These are shown in Fig. 1 for a singlet silylene along with the addition and insertion processes that silylenes share with carbenes.

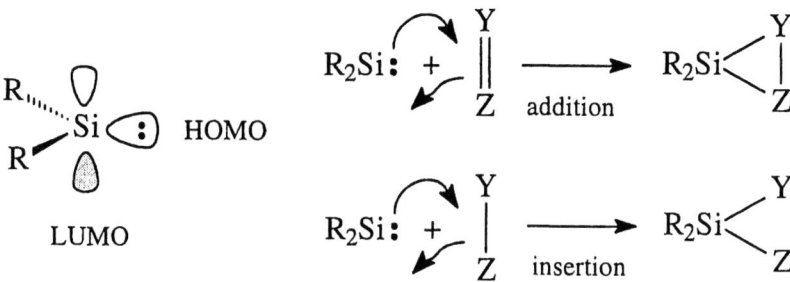

Figure 1. *FMO view of singlet silylenes and their carbene-like reactions*

The finding of both similarities and differences between the behavior of heavier carbene analogs and that of their first-row parents has been of practical as well as mechanistic interest. Several classes of organosilicon compounds became accessible through the reactions of silylenes: silirenes (*1*) thanks to the similarity of the addition reactions of silylenes and carbenes, and disilenes from the dimerization of silylenes (*2, 3*), a reaction made possible for silylenes by their reduced tendency, compared with that of carbenes, to be consumed by intramolecular processes.

Charged Six-Valence Electron Carbene Analogs

Study of charged six-valence electron carbene analogs poses questions to which we return in discussing four-valence electron species. Relatively little is known about the boryl anions $R_2B:^-$ (*4*) or the nitrenium ions $R_2N:^+$ (*5*) of the "carbene family" (Fig. 2), but phosphenium ions $R_2P:^+$ are more familiar, largely through the work of Alan Cowley, whom this symposium honors. Reaction studies on the oxenium ions RO^+ (*6*) and sulfenium ions RS^+ (*7*) of the "nitrene family" (Fig. 2) have been equivocal about whether concerted processes forming two bonds occur.

At first sight the phosphenium ion reactions shown in Fig. 3 (*8, 9*) appear carbene-like, forming two bonds by addition and insertion processes, but they look

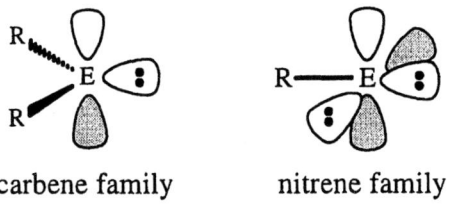

carbene family   nitrene family

Figure 2. Families of six-valence electron reactive intermediates differing in the symmetry of their HOMOs.

somewhat odd from the perspective of carbene chemistry, because concerted 1,4-addition of a carbene, silylene or germylene to a diene is generally disfavored relative to 1,2-addition, and the formal 1,4-adducts from silylenes and germylenes arise largely from ring-expansion of vinylsilirane and germirane intermediates. (*10*). Concerted addition of carbenes to the termini of dienes is limited to

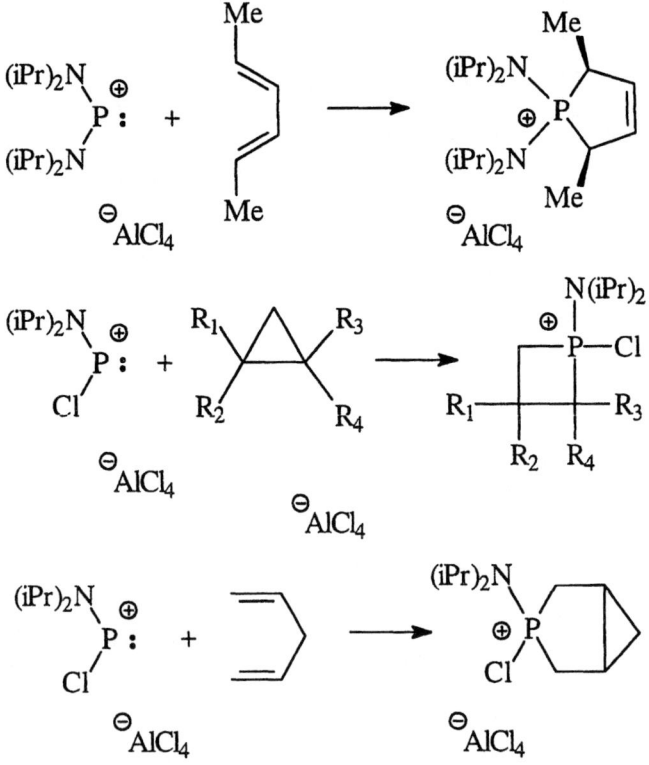

Figure 3. Some solution-phase addition and insertion reactions of phosphenium ions (references 8, 9)

norbornadienes (*11*), and insertion into C-C bonds of cyclopropanes is also nearly unknown.

The understanding of these reactions is growing, however, and has produced a fringe benefit in refocusing attention on the electronic factors that control carbene reactions. Calculations have indicated that the addition of phosphenium ions to butadiene is *stepwise*, but **not** via a carbene-like addition simultaneously forming two bonds by concerted acceptance and donation of electron pairs. Both concerted 1,4-addition forming a phospholenium ion product in a single step, and 1,2-addition forming a vinylphosphiranium ion with subsequent ring-expansion to a five-membered ring product, are predicted, admittedly by a low-level semi-empirical calculation, to have higher barriers than a Lewis acid - Lewis base interaction forming one bond. The kinetically favored pathway is predicted to be an electrophilic addition of a phosphenium ion to 1,3-butadiene, forming a 2-phosphaethyl-substituted allyl cation intermediate that can close rapidly to a phospholenium ion product, as shown it Fig. 4.

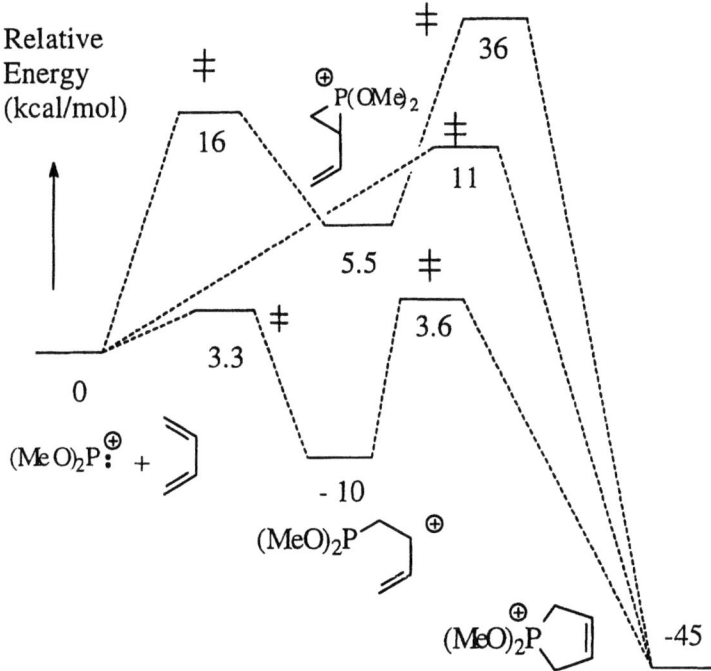

Figure 4. *Predicted (PM3) energy profile for addition of (MeO)$_2$P$^+$ to butadiene (reference 12).*

Once one thinks of phosphenium ions as Lewis acid-like, the reactions with 1,4-dienes and with cyclopropanes discovered by Cowley and Baxter no longer seem unusual. The addition of bis(methoxy)phosphenium ions to butadiene whose predicted energy profile is shown in Fig. 4 has been studied in the gas-phase. Gevrey, Taphanel and Morizur (*13*) in an early application of quadrupole ion trap mass spectrometry QITMS (*14*) to the study of ion-molecule reactions found the product of formal 1,4-addition by $(MeO)_2P^+$ to 2,3-dimethylbutadiene. The product structure was supported by collision-induced dissociation (CID) experiments.

It is hoped that light will be shed on the reactions of a variety of charged carbene-analogs via QITMS experiments. Quadrupole ion trap mass spectrometers operate at *ca.* 10^{-3} torr, pressures high enough to collisionally deactivate significant fractions of vibrationally-excited ion-molecule reaction products before they undergo unimolecular decompostion. Hence it is possible to generate charged carbene analogs in the ion source of the QITMS and study their simple ion-molecule association products like the addition and insertion products from charged carbene analogs. Long ion residence times translate into long reaction times, and the capability for increasing ion kinetic energy permits MS/MS experiments up to $(MS)^5$. Thus the structural characterization of ion-molecule reaction products is facilitated by several stages of collision-induced dissociation.

Time constraints permit only two results to be presented from our QITMS studies of six-valence electron charged carbene analogs. The formation of oxenium ions has been problematic, but electron impact on anisole has been shown by Zagorevskii, Holmes, and Stone to be a clean source of phenyloxenium ions Ph-O$^+$ (*15*). As indicated in Fig. 5, adducts with ethylene are formed at *m/z* 121 whose collision-induced dissociation gives both phenyl cations (*m/z* 77) and tropylium or benzyl cations (*m/z* 91) (*16*). This raises the hope that a carbene-like addition occurred to give the O-phenyloxiranium ion, the likely source of *m/z* 77 upon CID. The alternative structure is more likely to dissociate to *m/z* 91, making it likely that the *m/z* 121 adduct ions consist of a mixture of the structures shown.

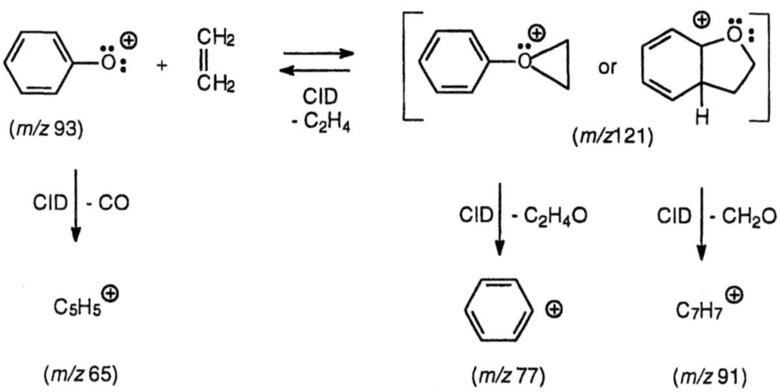

Figure 5. *QITMS observations on the gas-phase reaaction of Ph-O$^+$ with ethylene (reference 16).*

It is well-known that π-electron delocalization in cyclopropenylidenes reduces the electron-deficiency of the carbene center to the point that 2,3-diphenylcyclopropenylidene reacts as a nucleophile (17). It was of interest to examine the reactivity of the charged cyclopropenylidene analog phosphorenylium ion $cyclo$-$(CH)_2P^+$, a substituted version of which was made in solution by Laali and Regitz (18) and recently in the gas-phase by Eberlin and Laali (19).

Liu has implemented a general approach to the preparation in the gas-phase of heterocyclopropenium ions to generate the unsubstituted phosphirenylium ion (20). In the addition of the radical-cation •PBr^+ to acetylene the bromine atom acts as a sacrificial group, homolysis of the P-Br bond carrying away ca. half the exothermicity of additon. Even at the millitorr pressures at which a QITMS is operated, stabilization of small, rigid ions formed in highly exothermic reactions is difficult. The energy profile predicted for this reaction is displayed in Figure 6.

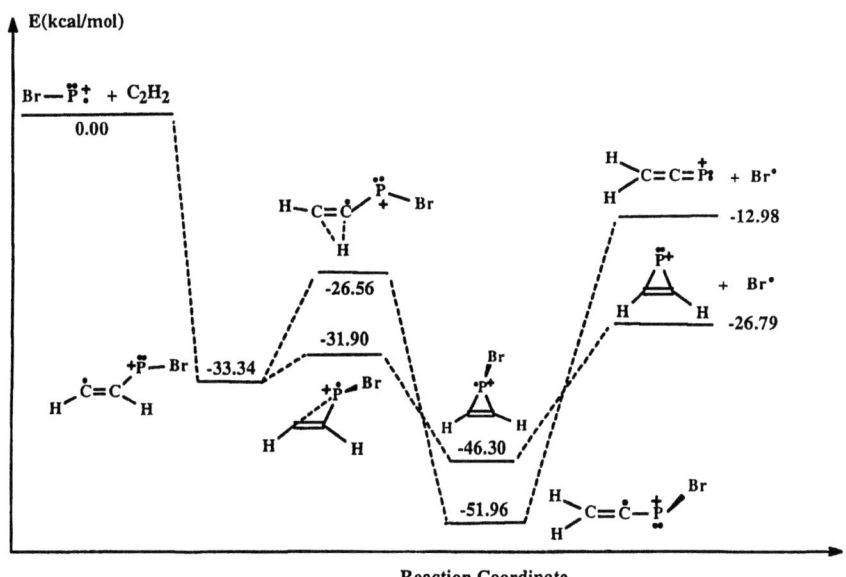

Figure 6. Energy profile for the exothermic reactions of PBr^+ and acetylene predicted by B3LYP/6-31G* DFT calculations (reference 20)

Ions of m/z 57 constitute >90% of the product ions from reaction of PBr^+ with acetylene. The absence of any observed reactivity of the m/z 57 product ions toward acetylene is in accord with the aromatic stabilization expected for $cyclo$-$(CH)_2P^+$, and this phosphirenylium ion is predicted to be the lowest energy singlet state $C_2H_2P^+$ isomer (21). However, formation of the isomeric vinylidenephosphenium ion $H_2C=C=P:^+$ (22) is also predicted to be exothermic.

The *m/z* 57 product ions from PBr⁺ and acetylene were characterized as *cyclo*-$(CH)_2P^+$ with the aid of B3LYP/6-31G˙ density functional theory electronic structure calculations. As shown in Figure 6, formation of the cyclic product is favored both thermodynamically and kinetically. The intitial interaction is predicted to form an open-chain adduct whose cyclization has a lower barrier than a hydrogen-migration, and loss of a bromine atom from the cyclic intermediate is favored over bromine atom-loss from the open-chain rearranged species.

Theoretical predictions also suggest that the lack of reactivity observed for the $C_2H_2P^+$ ions toward acetylene is compatible with the phosphirenylium ion cyclic structure but not with vinylidenephosphenium ion alternate structure. As indicated by the energy profile of Figure 7, both the $C_2H_2P^+$ isomeric phosphenium

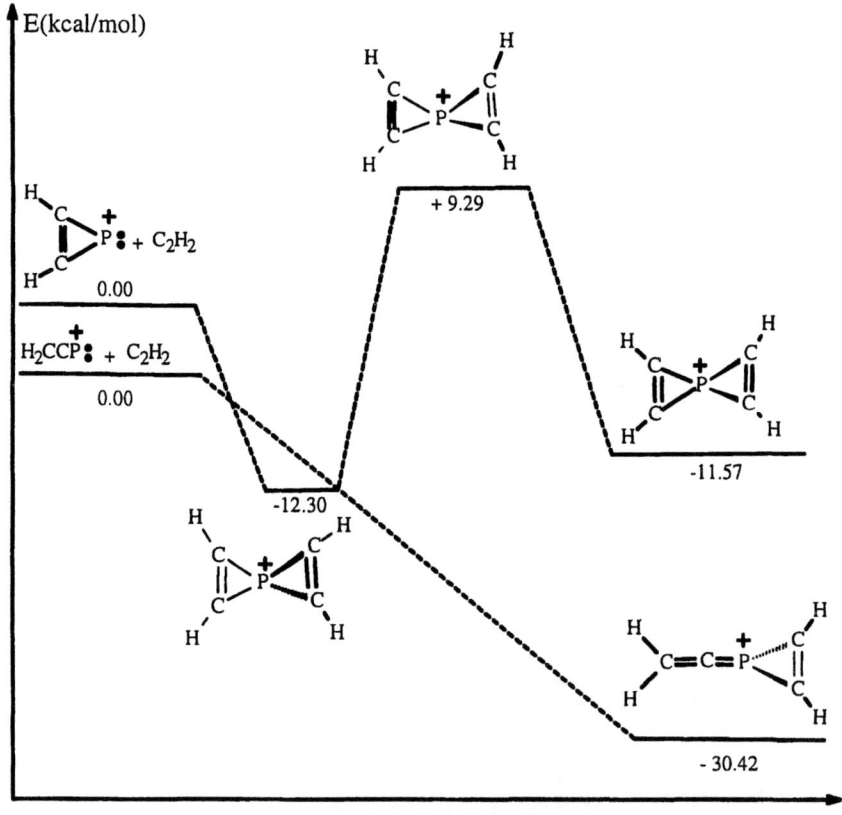

Figure 7. Energy profiles for the reactions of cyclo-$(CH)_2P:^+$ *and* $H_2C=C=P:^+$ *with acetylene*

ions, the aromatically stabilized cyclic phosphirenylium ion and the open-chain vinylidene phosphenium ion are predicted to form cycloadducts with acetylene in exothermic processes, but there is a significant kinetic barrier predicted for the formation of the spiro-adduct from $cyclo$-$(CH)_2P^+$, and this could account for the slowness of this process. The formation of the cycloadduct from $H_2C=C=P:^+$ is not only predicted to be more favored thermodynamically, but is predicted to occur without a barrier. Thus if the observed $C_2H_2P^+$ product ions had the $H_2C=C=P:^+$ structure, they would be expected to undergo rapid addition to acetylene.

Four-Valence Electron Species

Turning at last to the generation and reactivity of four-valence electron species, one can take as examples monovalent silicon cations, whose electronic structure and representative reactions are schematically presented in Fig. 8. Equipped with a *pair* of degenerate LUMO's, a filled HOMO, and a positive charge, such species first attracted our attention because their frontier orbital symmetry resembles that of the "carbene family" of six-valence electron reactive intermediates, but their geometries are those of the "nitrene family."

Thus equipped, a silyne cation has the potential for forming one, two, or even three bonds in concert. It is the possibility that three bonds might be formed simultaneously that led us to optimistically refer to them as "supersilylenes." It must be admitted that evidence has not been found that demands such supersilylene reactivity, but Fig. 9b depicts a reaction found in our QITMS experiments that does seem to indicate the formation of three bonds within the lifetime of a single ion-molecule collision complex between $HSi:^+$ and diethylamine..

Given the clear indication of Lewis acid-like attack of phenylsilicon cations Ph-Si:$^+$ on pyridine shown in Fig. 9a, the stepwise path to the cyclic ammonium ion seems likely. Lewis acid-like addition to the amino group of diethylamine would form a distonic silylene whose insertion into a C-H bond closes the four-membered ring.

Benzene has proven to be a most interesting substrate for $HSi:^+$, whose major product may be the long-sought silatropylium ion: It was an embarrassment of riches a decade ago when, as seen in Fig. 10, Beauchamp found that electron impact on phenylsilane $PhSiH_3$ yielded *two* forms of $C_6H_7Si^+$ ions, one of which reacts further with its precursor, and the other one does not (23). The structure assigned to the reactive form of $C_6H_7Si^+$ was the 1-silabenzyl cation, also known as the phenylsilyl cation $PhSiH_2^+$, and this assignment continues to be accepted.

To the unreactive form of $C_6H_7Si^+$ Beauchamp assigned the silatropylium ion structure, a silacycloheptatrienyl cation. To explain the failure of this ion to

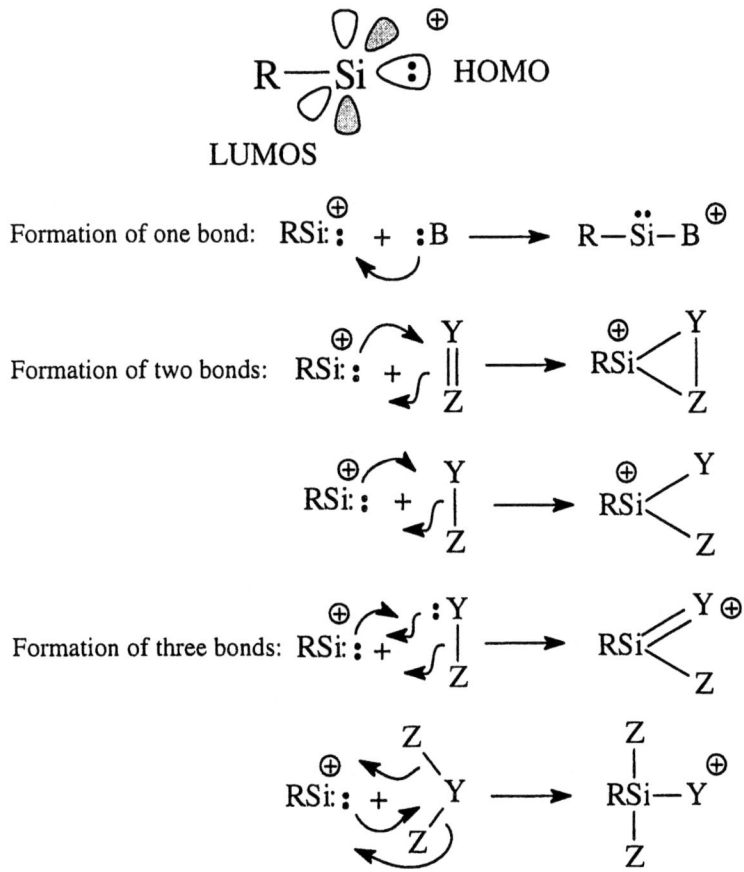

Figure 8. Four-valence electron chemistry: FMO view of silanetriyl (silyne) cations and their reactions

accept a hydride ion from cycloheptatriene, Beauchamp suggested that the silatropylium ion may have a lower hydride affinity than does the parent tropylium ion. But electronic structure calculations by Radom (24) and Shin (25) predicted that the silatropylium ion has a greater hydride affinity than the all-carbon tropylium ion. It was therefore felt that the silatropylium ion structure for unreactive $C_6H_7Si^+$ was ruled out by the observed inertness of the unreactive form of $C_6H_7Si^+$ toward cycloheptatriene.

Jarek and Shin proposed a π-complex of $HSi:^+$ and benzene as the structure of unreactive $C_6H_7Si^+$, and in support of this suggestion mentioned (but did not

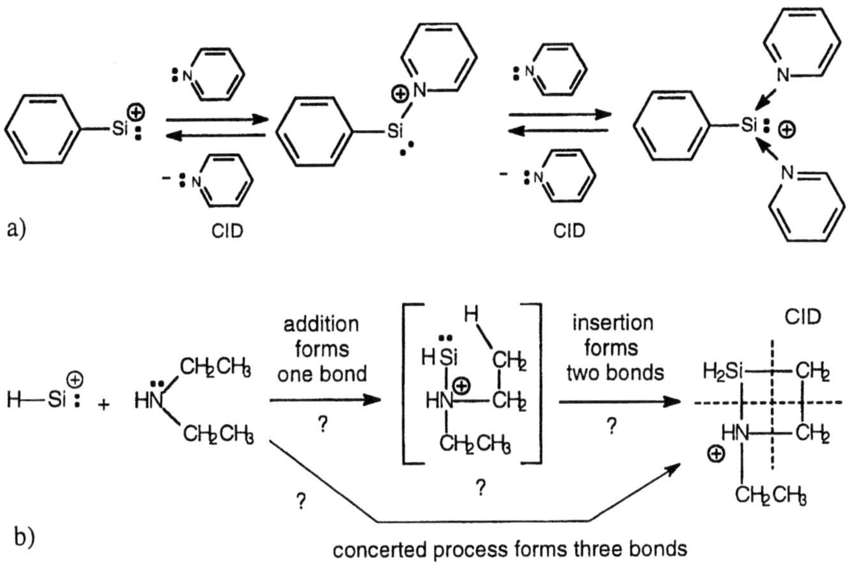

Figure 9. Reactions of: a) PhSi+ with pyridine, and b) HSi:+ with diethylamine in a QITMS (reference 20).

publish) CID experiments that indicated formation of SiH+ and Si+ as major fragmentation processes.

In recent QITMS experiments (26) it has been found that the reactions of HSi:+ and benzene form only the unreactive form of $C_6H_7Si^+$ in addition to PhSi:+ + H_2. Allen and Lampe were the first to react HSi:+ with benzene, but did not suggest a structure for the $C_6H_7Si^+$ product of their high-pressure mass spectrometric experiments (27). In agreement with the observations of the Beauchamp and Shin laboratories, we found that this form of $C_6H_7Si^+$ does *not* react with phenylsilane, cycloheptatriene, or benzene on the 100 ms timescale of our QITMS experiments.

As shown in Table I, however, while the results from low-energy CID experiments confirm that the $C_6H_7Si^+$ from HSi:+ and benzene and the unreactive form of $C_6H_7Si^+$ undergo identical fragmentation, loss of benzene to form HSi+ or Si+ did not contribute significantly to the MS/MS spectra, and only minute quantities of HSi+ and Si+ are formed in high energy CID experiments.

We feel that the π-complex structure suggested by Jarek and Shin and the 7-silanorbornadienyl cation structure considered by Nicolaides and Radom are ruled out by the CID spectra of Table I - both would be expected to eliminate benzene with the formation of HSi$^+$ and Si$^+$ fragment ions - and by the slowness of the ligand exchange reaction with benzene-d$_6$.

Table I. CID Spectra of Unreactive C$_6$H$_7$Si$^+$ From Two Sources

m/z:	107 (M$^+$)	106	105	104	103	81	79	65	55	53
HSi$^+$ + C$_6$H$_6$	100	18	75	12	85	33	22	8	8	11
PhSiH$_3$, EI	100	20	75	11	86	35	19	7	9	11

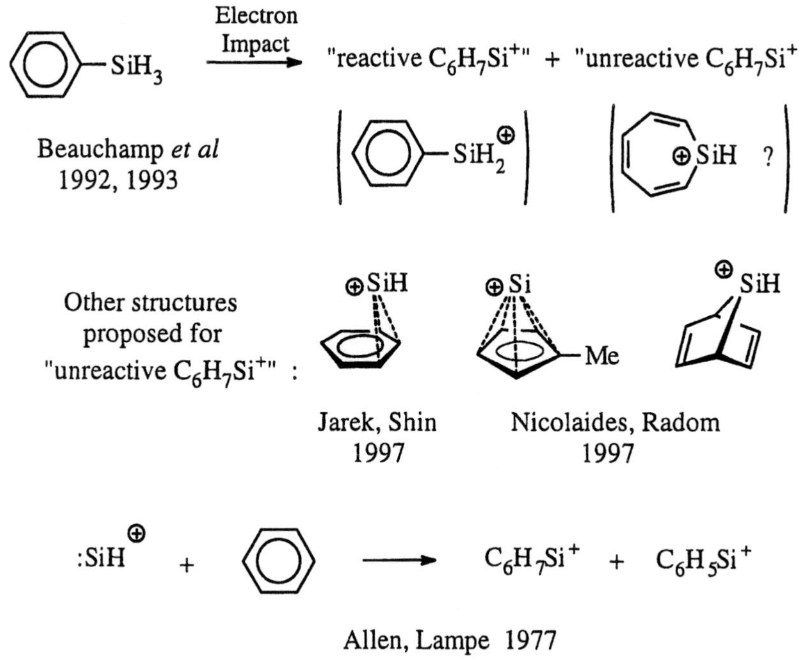

Ph–SiH$_3$ →(Electron Impact) "reactive C$_6$H$_7$Si^{+}" + "unreactive C$_6$H$_7$Si$^+$"

Beauchamp et al
1992, 1993

[Ph–SiH$_2^\oplus$]   [cycloheptatrienyl–SiH ?]

Other structures proposed for "unreactive C$_6$H$_7$Si^{+}":

⊕SiH (on benzene)   ⊕Si–Me (on cyclopentadienyl)   ⊕SiH (on norbornadienyl)

Jarek, Shin   Nicolaides, Radom
1997   1997

:SiH$^\oplus$ + C$_6$H$_6$ ⟶ C$_6$H$_7$Si$^+$ + C$_6$H$_5$Si$^+$

Allen, Lampe 1977

Figure 10. Ion-molecule reactions forming C$_6$H$_7$Si$^+$ and suggested structures (references 23-25)

The π-complex form of C$_6$H$_7$Si$^+$ and the 7-silanorbornadienyl cation, which is a σ-complex of the same components, HSi:$^+$ and benzene, are expected to undergo facile interconversion, and, in our calculations, the latter is not found as a local minimum on the potential surface. B3LYP/6-31G* calculations predict that the a C$_6$H$_7$Si$^+$ π-complex of HSi:$^+$ and C$_6$H$_6$ will react with C$_6$D$_6$ to form the corresponding C$_6$HD$_6$Si$^+$ π-complex with no barrier beyond the recovery as potential energy of the predicted 14.7 kcal/mol exothermicity of formation of a symmetric η^3, η'^3-dibenzene-sandwich intermediate stored as internal energy.

We believe that the silatropylium ion structure is the one most likely for the unreactive form of $C_6H_7Si^+$, and the energy profile for its formation predicted by B3LYP/6-31G* DFT calculations is shown in Fig. 11. The first step is predicted

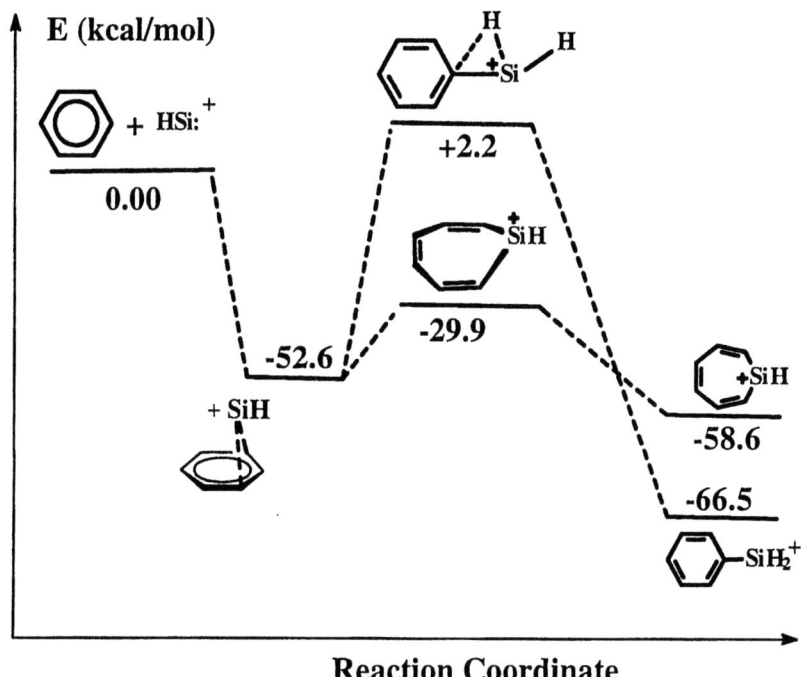

Figure 11. Energy profile for the reaction of $HSi:^+$ and benzene

to be the formation of a π-complex with a structure and binding energy similar to those predicted by Jarek and Shin (25). There is a reasonably low energy barrier for rearrangment to the silatropylium ion which is 6 kcal/mol more stable than the π-complex. The barrier for formation of the even more stable 1-silabenzyl cation is predicted to lie ca. 30 kcal/mole higher than the barrier for rearrangement of the π-complex to the silatropylium ion, and this is in accord with the exclusive formation of the less reactive form of $C_6H_7Si^+$ from $HSi:^+$ and benzene.

The failure of unreactive $C_6H_7Si^+$ to undergo reaction with cycloheptatriene can be explained by a large energy barrier hindering kinetically the thermodynamically favored hydride transfer. In short, objections to the silatropylium ion structure for unreactive $C_6H_7Si^+$ do not stand up to close scrutiny, and stronger arguments can be made against the alternative structures. This does not establish the silatropylium ion structure, but does render it quite plausible.

The nature of the product from $HSi:^+$ and benzene is important for the study of four-valence electron reactive intermediates, because this reaction could be regarded as an example of a carbene-like reaction process. Studies of both neutral and charged species in both the gas-phase and in solution are being enthusiastically continued. The effect of charge on the philicity of both six-valence electron carbene analogs and four-valence electron "supercarbenes" deserves careful consideration. Results presented here suggest that single-bond formation via Lewis acid-Lewis base interactions may be preferred over concerted cycloaddition pathways for positively charged species.

Acknowledgments. This chapter is dedicated to Professor Alan Cowley. Financial support was from the National Science Foundation (CHE-0316124) and the National Institutes of Health (P41RR00954). X. L. thanks Dr. Dong Zhou for preparative assistance and Dr. Daryl Giblin for help with high energy CID experiments.

References

1. Conlin, R. T.; Gaspar, P. P. *J. Am. Chem. Soc.*, **1976**, *98*, 3715-3716.
2. Conlin, R. T.; Gaspar, P. P. *J. Am. Chem. Soc.*, **1976**, *98*, 868-870.
3. West, R.; Fink, M. J.; Michl, J. *Science (Washington, DC)*, **1981**, *214*, 1343-1344..
4. Wagner, M.; van Eikema Hommes, N. J. R.; Nőth, H.; Schleyer, P. v. R. *Inorg. Chem.*, **1995**, *34*, 607-614.
5. Falvey, D. E. *Reactive Intermediate Chemistry*, Moss, R.A.; Platz, M. S.; Jones, M., Jr. Eds., Wiley-Interscience, Hoboken, 2004, pp. 593-650.
6. Li, Y.; Abramovitch, R. A.; Houk, K. N. *J. Org. Chem.*, **1989**, *54*, 2911-2914 and earlier references contained therein.
7. Bortolini, O.; Guerrini, A.; Lucchini, V.; Modena, G.; Pasquato, L. *Tetrahedron Lett.*, **1999**, *40*, 6073-6076.
8. Cowley, A. H.; Kemp, R. A.; Lasch, J. G.; Norman, N. C.; Stewart, C. A.; Whittlesey, B. R.; Wright, T. C. *Inorg. Chem.*, **1986**, *25*, 740-749.
9. Weissman, S. A.; Baxter, S. G. *Tetrahedron Lett.*, **1988**, *29*, 1219-1222.
10. Bobbitt, K. L.; Lei, D.; Maloney, V. M.; Parker, P. S.; Raible, J. M.; Gaspar, P. P. In *Frontiers in Organogermanium, -tin, and -lead Chemistry*, Lukevics, E; Ignatovich, L. Eds.; Latvian Institute of Organic Synthesis; Riga, 1993; pp. 29-40.

11. Moss, R. A.; Jones, M., Jr. In *Reactive Intermediates (Wiley)*, Jones, M. Jr., Moss, R. A. Eds., **1981**, *2*, 59-133.
12. Ivanova, D. unpublished results.
13. Gevrey, S.; Taphanel, M.-H.; Morizur, J.-P. *J. Mass Spectrom.*, **1998**, *33*, 399-402.
14. March, R. E. *J. Mass Spectrom.*, **1997**, *32*, 351-369.
15. Zagorevskii, D. V.; Holmes, J. L.; Stone, J. A. *Eur. Mass Spectrom.*, **1996**, *2*, 341-345.
16. Prell, J. S. unpublished results.
17. Jones, W. M.; Stowe, M. E.; Well, E. E., Jr.; Lester, E. W. *J. Am. Chem. Soc.*, **1968**, *90*, 1849-1859.
18. Laali, K. K.; Geissler, B.; Wagner, O.; Hoffmann, J.; Armbrust, R.; Eisfeld, W.; Regitz, M. *J. Am. Chem. Soc.*, **1994**, *116*, 9407-9408.
19. Sabino, A. A.; Eberlin, M. N.; Moraes, L. A. B.; Laali, K. K. *Org. Biomol. Chem.*, **2003**, *1*, 395-400.
20. Liu, X.; Ivanova, D.; Giblin, D.; Gross, M. L.; Gaspar, P. P. submitted for publication.
21. MacLagan, R. G. A. R. *Chem. Phys. Lett.*, **1989**, *163*, 349-353.
22. Largo, A.; Barrientos, C.; Lopez, X.; Cossio, F. P.; Ugalde, J. M. *J. Phys. Chem.*, **1995**, *99*, 6432-6440.
23. Murthy, S.; Nagano, Y.; Beauchamp, J. L. *J. Am. Chem. Soc.*, **1992**, *114*, 3573-357.
24. Nicolaides, A.; Radom, L. *J. Am. Chem. Soc.*, **1997**, *119*, 11933-11937 and earlier references contained therein.
25. Jarek, R. L.' Shin, S. K. *J. Am. Chem. Soc.*, **1997**, *119*, 6376-6383.
26. Liu, X.; Gross, M. L.; Gaspar, P. P. submitted for publication.

Chapter 5

The Diversity of Stable and Persistent Phosphorus-Containing Radicals

A. Armstrong[1], T. Chivers[1], and R. T. Boeré[2]

[1]Department of Chemistry, University of Calgary, Calgary, Alberta, T2N 1N4, Canada
[2]Department of Chemistry and Biochemistry, University of Lethbridge, Lethbridge, Alberta, T1K 3M4, Canada

> Stable and persistent phosphorus-containing radicals can be divided into five general categories: phosphinyl, diphosphanyl, 1,3-diphosphaallyl, lithiated tetrakisimidophosphates, and phosphaverdazyl systems. The syntheses, structures, EPR spectroscopic characterization, and sources of stability of these neutral radicals are presented and compared.

Introduction

Though the history of stable radicals dates back to the mid-19th century (*1, 2*), significant advances in the synthesis of inorganic main group radicals have only been made in the past twenty-five years (*3*). The recent rapid development of this field coincides with increased interest in radical species as potential magnetic materials (*4*), polymerization catalysts (*5*), and spin-labels for biomolecules (*6*). Stable radicals are also of interest from a more fundamental standpoint, in that they challenge one of the basic tenets of chemical theory: that electrons are always present as electron pairs in compounds of the main group elements.

While most radicals are highly reactive transient species due to their open-shell electron configurations, numerous examples of both persistent and stable (7) radicals are now known. Such radicals are generally stabilized by delocalization of the unpaired electron over several electronegative atoms such as oxygen or nitrogen, or by delocalization over a π manifold as is often seen in sulfur-nitrogen ring systems (8). Kinetic stabilization of these systems is also possible; the introduction of bulky substituents can impede the reactivity of a radical considerably, enabling its isolation. Despite these efforts to render them stable, numerous radicals dimerize either in solution or in the solid state in order to remove the instability caused by the presence of unpaired electrons.

Electron Paramagnetic Resonance (EPR) spectroscopy is the single most important technique for characterizing radicals, as hyperfine coupling of the unpaired electron to neighboring spin-active nuclei can be used to deduce its chemical environment. The magnitude of the hyperfine coupling (hfc) constant gives information regarding the location of the unpaired electron both within a molecule and within a particular atomic or molecular orbital. The hfc constant for a particular spin-active nucleus is dependent upon the amount of interaction which occurs between the unpaired electron and that nucleus. The type of orbital in which the electron is located can thus be deduced, as an electron in an s orbital will have a significantly larger nuclear interaction than an electron located in a p or d orbital. In addition, the spin density on a particular atom in a delocalized system can be estimated, as the magnitude of the hfc constant will be directly proportional to the amount of time the unpaired electron resides on that atom. Finally, the value of a hfc constant is also dependent upon the gyromagnetic ratio of the nucleus to which it is coupled.

More recently, researchers have used theoretical calculations in order to confirm and even predict the orbital contributions of various atoms to the singly occupied molecular orbital (SOMO), and to verify experimentally observed hyperfine coupling constants. At the present time, persistent or stable radicals are usually identified through a combination of theoretical calculations and experimental techniques such as EPR spectroscopy.

Among the main group elements, there is particular interest in phosphorus-based radicals since phosphorus has only one isotope (^{31}P, $I = \frac{1}{2}$) and gives rise to much larger hyperfine couplings than those observed for other abundant spin-active nuclei such as ^{14}N ($I = 1$, 99%). These properties render phosphorus an ideal nucleus for spin-labeling experiments, as the anisotropy or orientation-dependence of the hyperfine coupling to ^{31}P can be used to glean information about rapidly moving molecules (6).

The past decade has witnessed the isolation and characterization of a handful of stable and persistent phosphorus-containing radicals. What is perhaps most interesting about this group of radicals as a whole is their chemical and structural diversity. Persistent radicals containing both P(III) and P(V) are known; species with one, two, or even three phosphorus atoms have been characterized. In some cases the unpaired electron is localized on one atom, while in others it is delocalized over an extensive π manifold. This article examines the similarities and differences amongst the known persistent and stable phosphorus-containing radicals, including aspects of their synthesis, EPR characterization, and the underlying reasons for their stabilities. Beginning with the simplest [R$_2$P]$^{\bullet}$ systems, the discussion progresses to related P(III) systems, then on to more complex P(V) radicals, including the heterocyclic phosphaverdazyl radicals. The coverage is limited to *neutral* radicals.

Phosphinyl Radicals [R$_2$P] $^{\bullet}$

A two-coordinate phosphinyl radical is, from a structural standpoint, the simplest of the phosphorus-centered radicals; it is also one of the most thoroughly understood radicals of the main group elements. Numerous radicals of this type can be produced via the photolytic reduction of the appropriate chloro(dialkyl)phosphine in the presence of an electron-rich alkene (*9, 10*). The most persistent of the phosphinyl radicals that has been prepared since their initial discovery (*11*) nearly forty years ago is the bis(trimethylsilyl)methyl derivative $^{\bullet}$P[CH(SiMe$_3$)$_2$]$_2$ **1** which is stable indefinitely both in solution and in the gas phase (*9, 10*).

EPR Characterization

The EPR spectrum (*9*) of a solution of reduced ClP[CH(SiMe$_3$)$_2$]$_2$ consists of two 1:2:1 triplets, characteristic of a single unpaired electron coupling to one ^{31}P nucleus and to two equivalent ^{1}H (I = ½, 100%) nuclei. This confirms the identity of the radical species as **1**, with coupling to the two methine protons observed. A simulation (*12*) of this spectrum is shown in Figure 1. The value of the phosphorus hyperfine coupling constant (a_{31P} = 96.3 G) suggests that the unpaired electron is located in an orbital of predominantly *p* character, presumably the 3$p\pi$ orbital of the phosphorus atom.

Interestingly, the intensity of the EPR signal does not decrease over time, indicating that the radical does not dimerize in solution, as most phosphinyl radicals do, to yield the diamagnetic diphosphine R$_2$P-PR$_2$ [R = CH(SiMe$_3$)$_2$] **2**. This is somewhat surprising for two reasons: first, the unpaired electron is essentially localized on one atom, which should render **1** highly reactive. Secondly, the phosphorus atom is only two-coordinate and thus should not be

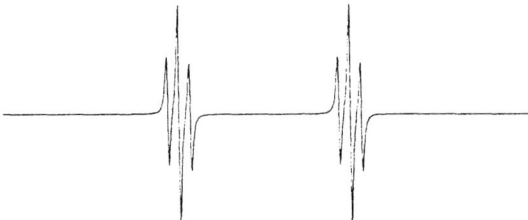

Figure 1. Simulated EPR spectrum of 1.

sufficiently kinetically stabilized to prevent dimerization, its bulky bis(trimethylsilyl)methyl substituents notwithstanding. An investigation of the intensity of the EPR signal as a function of concentration may help to clarify the solution behavior of **1**.

Structure and Stability

In order to probe the reasons for the unexpectedly high stability of this radical, a structural study of **1** and its parent diphosphine **2** was undertaken (*13, 14*) using a combination of X-ray crystallography, gas-phase electron diffraction (GED) and theoretical calculations. The results of this investigation elucidated for the first time the considerable rôle that thermodynamics plays in the stabilization of **1**. The structure of **1** was determined by GED measurements, which showed that it exists as a V-shaped species ($\angle$CPC = 104.0°) with the CH(SiMe$_3$)$_2$ groups oriented exclusively in the *syn,syn* conformation. Density Functional Theory (DFT) calculations were used to confirm this structure. An X-ray crystal structure of **2** showed that, contrary to what is observed in **1**, the diphosphine bis(trimethylsilyl)methyl groups adopt a *syn,anti* conformation in the solid state. While this allows for more efficient packing of the CH(SiMe$_3$)$_2$ substituents, it also results in significant crowding of the SiMe$_3$ groups, both between groups attached to the same phosphorus, and between those bonded to different P atoms.

The corollary of these structural data is simply this: the dissociation of the dimer **2** to yield two units of **1** involves not only homolytic cleavage of the phosphorus-phosphorus bond, but also isomerization of the bis(trimethylsilyl)methyl groups. According to theoretical calculations, the first step in the dissociation process of 2 is homolysis of the P-P bond, which is an endothermic process (95 kJ mol^{-1}). This bond cleavage yields two phosphinyl radicals, both of which are in the *syn,anti* conformation: that is, they are geometrical isomers of the established structure of **1**. Relaxation of the CH(SiMe$_3$)$_2$ groups to relieve steric strain, followed by the rotation of one such group about the P-C bond to yield a *syn,syn* radical releases an estimated 67.5 kJ mol^{-1} for each unit of **1**. Overall, the conversion of the dimer **2** to two fragments of **1** is an exothermic process, releasing approximately 40 kJ for each mole of **2**. Thus the fact that **1** does not dimerize in solution is attributed not to the kinetic stabilization of the bulky ligands, but to the energetic input required for the rearrangement prior to dimerization.

Recent studies (*15*) have shown that the unsymmetrical phosphinyl radical $^{\bullet}$P[N(SiMe$_3$)$_2$](N^{i}Pr$_2$) is produced by the thermally reversible dissociation of its diphosphine dimer. As was seen for **1**, the persistence of $^{\bullet}$P[N(SiMe$_3$)$_2$](N^{i}Pr$_2$) is attributed to the reorganization energy necessary for dimerization to occur, rather than to any intrinsic stability associated with this radical.

The Diphosphanyl Radical [R$_2$PPR]$^{\bullet}$

Related to the phosphinyl radical **1** is the diphosphanyl radical R$_2$PP$^{\bullet}$R. While both these species contain a two-coordinate phosphorus atom, the diphosphanyl radical is somewhat more complex in that, as its name implies, it contains a second phosphorus atom. Unlike the phosphinyl radicals, the first stable diphosphanyl radical [Mes*MeP-PMes*]$^{\bullet}$ **3** (Mes* = 2,4,6-tri(*tert*-butyl)phenyl) was synthesized only recently (*16*), and at present, **3** remains the only such radical known.

The radical **3** was first observed in a cyclic voltammetry experiment with the diphosphene salt [Mes*MeP=PMes*][O$_2$SCF$_3$] **4**, which was found to undergo a chemically reversible one-electron reduction. This result prompted an attempt to synthesize the reduction product using chemical means, specifically via an electron transfer reaction from an electron-rich alkene. Reminiscent of the facile reduction of R$_2$PCl to produce the phosphinyl radical **1** (*9*) reduction of **4** is effected in an acetonitrile solution containing tetrakis(dimethylamino)ethene at room temperature, yielding yellow crystals of **3**.

Unlike the phosphinyl radicals, **3** is sufficiently stable to be isolated as a solid material; however, magnetic measurements indicate that a small amount of dimerization (~10%) occurs during the liquid to solid transition. Slow decomposition of **3** is observed in the solid state, while in solution, the half-life of this persistent radical is approximately 90 minutes.

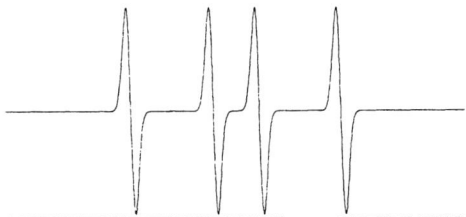

3

EPR Characterization

A solution EPR spectrum of **3** consists of four lines of equal intensities, indicating that the unpaired electron is interacting with two inequivalent ^{31}P nuclei and confirming the identity of **3** as a diphosphanyl radical. The larger of the two coupling constants (a_{31P} = 139.3 G) was attributed to the two-coordinate phosphorus atom, where the larger part of the spin density is thought to be located; the smaller coupling constant (a_{31P} = 89.3 G) was assigned to the three-coordinate phosphorus atom. A simulation (*17*) of this spectrum is shown in Figure 2.

Figure 2. Simulated EPR spectrum of 3.

In connection with the growing interest in using ^{31}P-centered radicals as spin-labels, the anisotropy or direction-dependence of the ^{31}P coupling of **3** was examined. By doping **3** into a single crystal of the related diphosphane Mes*MeP-PMes*Me, solid-state EPR spectra of **3** were obtained, varying the angle of the crystal to the magnetic field. A difference of 420 G between the high- and low-field transitions was observed when the magnetic field is perpendicular to the P-P bond, while a much smaller separation of 130 G was observed with the magnetic field parallel to the P-P bond, i.e. perpendicular to the *p* orbital in which the unpaired electron resides. The large anisotropies of these phosphorus couplings indicate that diphosphanyl radicals such as **3** may well find use in the future as spin-labels.

Structure and Stability

As is seen for phosphinyl radicals, the phosphorus couplings observed in the EPR spectra of **3** are relatively large, indicating that the unpaired electron resides almost exclusively on the two phosphorus atoms. DFT calculations revealed the expected small s contribution (~10%) to the SOMO, while confirming that the largest part (~74%) of the spin density resides in a p orbital of the two-coordinate phosphorus atom, while the remaining 15% is located in a p orbital of the three-coordinate phosphorus. Consistently, the calculated P-P bond distance (2.18 Å) is only slightly shorter than a typical P-P single bond (2.22 Å); a considerably shorter bond would be expected if the electron were delocalized equally over the two phosphorus atoms, as this would result in a P-P bond order of 1.5. Since minimal delocalization of the unpaired electron occurs in **3**, the stability of this radical is attributed to the steric protection afforded by the Mes* substituents, which reduces the reactivity of **3** making it possible to characterize this species.

A 1,3-Diphosphaallyl Radical [R$_2$NP(CNR$_2$)PNR$_2$]$^\bullet$

In addition to the diphosphanyl radical, several other species are known which contain a single unpaired electron delocalized over two phosphorus (III) atoms, including the diphosphiranyl **5**, diphosphirenyl **6**, and 1,3-diphosphaallyl **7** radicals. Of these, **5** and **6** are transient species that have been observed in EPR spin-trapping experiments (*18*) and postulated as reaction intermediates (*19*). Similar behavior has been noted for the related 1,3-diphosphaallyl radicals, which have widely been considered to be unstable; however, a recent study (*20*) has shown that the introduction of amino groups on the three ring atoms has a powerful stabilizing effect.

As in the case of the diphosphanyl radical, the stable radical [(^{i}Pr$_2$NP)$_2$CN^{i}Pr$_2$]$^\bullet$ **8** was first observed in a cyclic voltammetry experiment as the product of the one electron reduction of the diphosphacyclopropenium salt **9**. Preparative scale reductions involving either electrolysis or the reduction of **9** with lithium metal led to the isolation of **8** in the form of paramagnetic red crystals. Although the crystals were not suitable for X-ray analysis, a FAB mass spectrum showed the expected molecular ion. Unlike the phosphinyl and diphosphanyl radicals discussed earlier, **8** can be isolated and shows no sign of decomposition in the solid state over a period of weeks.

EPR Characterization

Solution EPR spectra of **8** were obtained in an attempt to identify the radical species. Each spectrum displayed five broad lines of relative intensities 1:3:4:3:1 as shown in Figure 3; additional hyperfine coupling could not be

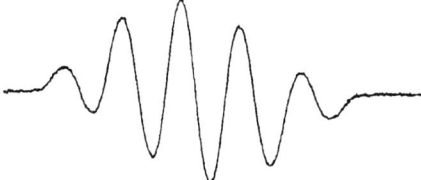

*Figure 3. EPR spectrum of **8**. (Reproduced from reference 20. Copyright 1997 American Chemical Society.)*

resolved even at low temperature. The increased complexity of this spectrum compared to those of the phosphinyl and diphosphanyl radicals can be attributed to delocalization of the unpaired electron over more spin-active nuclei, specifically ^{14}N. Indeed, the best simulation of this spectrum is obtained by including hyperfine coupling to two equivalent ^{31}P nuclei (a_{31P} = 9.4 G), two equivalent ^{14}N nuclei (a_{14N} = 1.5 G), and a unique nitrogen atom (a_{14N} = 9.9 G). The relatively small value of the ^{31}P coupling constant in **8** compared to those observed in the two P(III) radicals discussed earlier reflects a decrease in the phosphorus character of the SOMO, which results from delocalization of the unpaired electron over the carbon and nitrogen centers.

Structure and Stability

Based on the EPR data, two structures **8'** and **8''** can be proposed for the radical **8**. In order to determine the true identity of **8**, *ab initio* calculations were carried out on these two isomers. Structure **8''** was found to be significantly higher in energy (114.6 kJ mol^{-1}) than **8'**, essentially precluding the possibility that **8''** is the correct formulation of this radical. In addition, the optimized

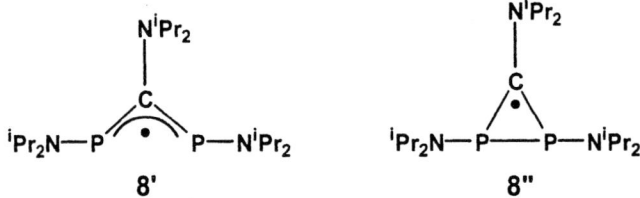

geometry of **8''** has the CN^{i}Pr$_2$ unit bent out of the plane of the three-membered ring, destroying the symmetry of the molecule; hence, its geometry is not consistent with the observed EPR hyperfine couplings. The stable radical **8** was therefore identified as the 1,3-diphosphaallyl species **8'**; the mechanism of the rearrangement of **9** to yield **8'** has yet to be elucidated. The increased stability of **8'**, compared to that of **1** or **3**, can be attributed to the delocalization of the unpaired electron over a total of five atoms, three of which are highly electronegative.

Dilithiated Tetrakisimidophosphate Radicals

In moving from **1**, **3**, and **8** to a tetrakisimidophosphate radical, the oxidation state of the phosphorus atom increases from those P(III)-based systems to a P(V) species. However, the basic principles of stabilizing the radical through the use of bulky substituents and delocalization of the unpaired electron over several atoms remain unchanged. Like the diphosphanyl radical, the dilithiated tetrakisimidophosphate radical "Li$_2$[P(N^{t}Bu)$_3$(NSiMe$_3$)]$^{•}$" **10** was discovered only recently (*21*). The P(V) radical **10** was prepared via the oxidation of the trilithiated tetrakisimidophosphate Li$_3$[P(N^{t}Bu)$_3$(NSiMe$_3$)] (*22*) with iodine or bromine in tetrahydrofuran (THF), producing blue solutions that persist for several weeks. From these reactions, paramagnetic blue solids can be isolated that are stable indefinitely in the solid state in the absence of air.

Structure and Stability

In order to identify the radical **10**, the blue solid was crystallized and an X-ray structure was obtained. The crystallographic data showed that, in the solid state, the oxidation product **10** acts to trap a monomeric unit of lithium iodide. The resultant adduct {Li$_2$[P(N^{t}Bu)$_3$(NSiMe$_3$)](LiI)•3THF}$^{•}$ **11** exists as a highly distorted PN$_3$Li$_3$I cube; an X-ray structure of the isostructural LiO^{t}Bu adduct **12** has also been obtained (*23*). At present, **11** and **12** are the only inorganic phosphorus-containing radicals to have been characterized by X-ray crystallography.

In order to probe the source of the remarkable stability of these radicals, DFT calculations were carried out (24) to determine the extent of delocalization of the unpaired electron. As expected, nearly all (98%) of the spin density was found to reside on the three cluster nitrogen atoms. Thus, the stability of **11** and **12** can in large part be attributed to the delocalization of spin density over multiple nitrogen centers.

In order to examine the extent of kinetic stabilization in **11** and **12**, the related tetrakisimidophosphates $Li_3[P(NAd)_3(NSiMe_3)]$ **13** and $Li_3[P(NCy)_3(NSiMe_3)]$ **14** were also reacted with halogens (25). Oxidation of **13**, in which the *tert*-butyl-imido groups have been replaced by sterically demanding adamantyl groups, was found to produce stable radicals; however, oxidation reactions of **14**, which contains smaller cyclohexyl-imido groups, result in highly transient radicals. These results indicate that the steric bulk of the alkyl-imido groups plays a critical role in stabilizing the tetrakisimidophosphate radicals.

EPR Characterization

The EPR spectrum of **10** in THF solution is highly dependent upon both temperature and concentration, suggesting that more than one radical may be present in solution. At extreme dilution a limiting spectrum is obtained which contains approximately fifty lines, all of which can be attributed to a single species. The best simulation of the experimental spectrum shown in Figure 4

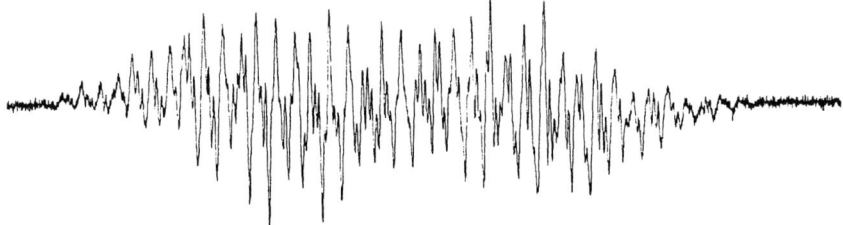

Figure 4. EPR spectrum of 10. (Reproduced with permission from reference 21. Copyright Wiley-VCH 2004.)

was obtained by including hyperfine couplings to one phosphorus atom (a_{31P} = 23.1 G), two equivalent nitrogen centers (a_{14N} = 5.38 G), two unique nitrogen atoms (a_{14N} = 7.38G and 1.93 G), and a single lithium nucleus (a_{7Li} = 0.30 G). These EPR parameters are consistent with the formation of the radical monoanion **15** and a $[Li(THF)_4]^+$ counterion via solvation of the cube **11**. The proposed structure of the solvent-separated ion pair {Li[P(N^{t}Bu)$_3$(NSiMe$_3$)][Li(THF)$_4$]}$^\bullet$ 15 contains two equivalent bridging nitrogen atoms and a single lithium cation, which is consistent with the experimentally obtained EPR spectrum. The ^{31}P coupling constant in this EPR spectrum is small compared to the values observed for radicals 1 and 3 in which the spin density is primarily on the phosphorus atom(s). This is in agreement with the DFT calculations, which indicated that the SOMO is predominantly Np in character.

The difference between the solution and solid-state structures of the radical 10 can be explained by the fact that the crystals of 11 were grown from a concentrated THF syrup, while the EPR spectrum corresponding to 15 was recorded in a very dilute THF solution. In dilute solutions, solvation of the lithium cations of 11 occurs, which ruptures the cubic structure forming the solvent-separated ion pair 15 and a molecule of (THF)$_3$LiI (*21*).

The Phosphaverdazyl Radicals

The phosphaverdazyls represent another example of phosphorus(V)-nitrogen radicals. These species are six-membered heterocycles composed of four nitrogen atoms, one phosphorus atom, and one carbon atom. Since their discovery five years ago, four persistent phosphaverdazyls radicals 16 – 19 have been prepared and characterized (*26*). Like the tetrakisimidophosphate radicals, phosphaverdazyls can be considered nitrogen-centered radicals which exhibit spin delocalization over a neighboring phosphorus atom.

The radicals 16 – 19 are produced via the homolytic cleavage of an N-H bond in the corresponding diamagnetic tetraazaphosphorines using common oxidizing agents such as iodine, periodate, or benzoquinone (*27-29*). While they can be isolated as red-brown semi-solids, the phosphaverdazyls decompose both in solution and in the solid state over a period of days.

EPR Characterization

The phosphaverdazyl radicals exhibit multiple hyperfine couplings due to interactions of the unpaired electron with neighboring protons, nitrogen atoms, and phosphorus nuclei. This extensive coupling results in numerous overlapping peaks, and frequently produces poorly resolved EPR spectra with quite broad lines. The EPR spectrum of the most complex of these radicals, the spirocyclic cyclophosphazene-phosphaverdazyl hybrid **19**, could not be interpreted without the aid of a double-resonance technique. An electron nuclear double resonance ENDOR) experiment (*29*) produced the simplified spectrum shown in (Figure 5. The reduced number of lines in the ENDOR spectrum, combined

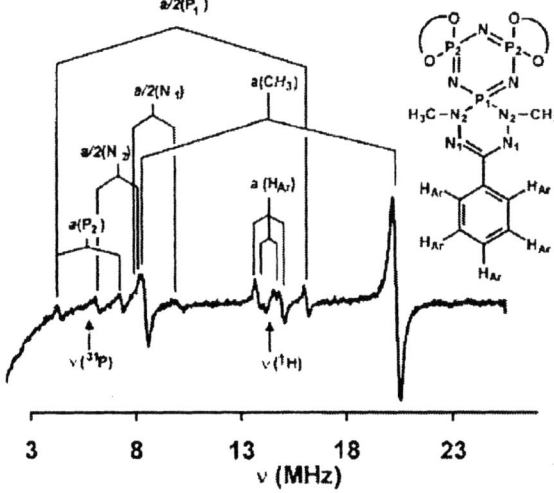

Figure 5. ENDOR spectrum of 19. (Reproduced with permission from reference 29. Copyright NRC Press 2002.)

with its superior resolution compared to a traditional EPR spectrum, made it possible to unequivocally assign hyperfine coupling constants to the two pairs of equivalent nitrogen atoms in the verdazyl ring (a_{14N} = 5.18 G and 6.45 G), the six equivalent protons from the methyl groups attached to the verdazyl nitrogen atoms (a_{1H} = 4.24 G), and two types of protons on the phenyl ring (a_{1H} = 0.40 G and 0.20 G).

Interestingly, two different ^{31}P couplings were observed: one due to an interaction with the phosphorus in the phosphaverdazyl ring (7.29 G) and one caused by coupling to the two remote phosphorus atoms in the cyclophosphazene ring (1.06 G). Though this second hfc constant is quite small, it indicates delocalization of the unpaired electron not only over the phosphaverdazyl ring, but throughout the cyclophosphazene ring as well. While two other radicals with cyclophosphazene substituents are known (*30, 31*), coupling in those systems is only observed to the phosphorus nucleus closest to the radical center. The evidence of spin-spin communication between the two rings in **19** suggests that spirocyclic phosphaverdazyls radicals of this type may find applications as magnetic materials.

Structure and Stability

The stability of the persistent phosphaverdazyls radicals **16 – 19** can be attributed to the delocalization of the unpaired electron over at least the four nitrogen atoms of the heterocycle. Detailed theoretical calculations (*28*) on **16 – 18** confirmed that the SOMOs of these three radicals are fairly similar, with large contributions from the four heterocycle nitrogens and smaller contributions from the phosphorus and other atoms.

The lower stability of **16 – 19** as compared to that of the tetrakisimidophosphate radical in **11** and **12** can be attributed to the lack of kinetic stabilization in the phosphaverdazyls systems. Despite their extensive delocalization, there are no bulky substituents on the heterocycle to hinder the reactions of these species.

Conclusion

Recent investigations have led to the the synthesis and identification of a variety of structurally diverse, stable and persistent phosphorus-conatining radicals. To date only one class of these systems, the dilithiated tetraimidophosphates, has been characterized in the solid state by X-ray crystallography. For the other radicals, EPR hyperfine coupling parameters, in conjunction with theoretical calculations, have provided detailed insights into the spin distribution in these paramagnetic systems. Possible applications of these materials as spin labels and in magnetic materials are being considered (26). Although this article has been limited to neutral phosphorus-containing radicals, recent progress in the generation and characterization of the radical cations $[R_3P]^{+\bullet}$ by the one-electron oxidation of triarylphosphines is noteworthy (32, 33). Persistent radicals of this type are obtained by using bulky aryl groups with two isopropyl groups in the *ortho* positions. Isotropic EPR hfcs indicate substantial *s* orbital character for the SOMO (33).

References

1. Fremy, E. *Ann. Chim. Phys.* **1845**, *15*, 459.
2. Atkins, P. W.; Symons, M. C. R. *The Structure of Inorganic Radicals: An Application of Electron Spin Resonances to the Study of Molecular Structure*, Elsevier, Amsterdam, 1967.
3. Power, P. *Chem. Rev.* **2003**, *103*, 78.
4. Turnbull, M. M.; Sugimoto, T.; Thompson, L. K.; *Molecule-Based Magnetic Materials: Theory, Techniques, and Applications*; American Chemical Society: Washington DC, 1996.
5. See, for example, Hawker, C. *Acc. Chem. Res.* **1997**, *30*, 373.
6. Berliner, L. J. *Spin Labelling: Theory and Applications*; Academic Press: New York, 1979; Vol. 2.
7. As defined in Reference 2, a "stable" radical is one which is inherently stable as an isolated species and does not decompose under an inert atmosphere at room temperature. A "persistent" radical is one which has a relatively long lifetime under the conditions which are used to generate it.
8. Cordes, A. W.; Haddon, R. C.; Oakley, R. T. In *Chemistry of Inorganic Ring Systems*; Steudel, R., Ed.; Elsevier: Amsterdam, 1992, Chapter 16, pp. 295-322.
9. Gynane, M. J. S.; Hudson, A.; Lappert, M. F.; Power, P. P. *J. C. S. Chem. Comm.* **1976**, 623.
10. Gynane, M. J. S.; Hudson, A.; Lappert, M. F.; Power, P. P.; Goldwhite, H. *J. C. S. Dalton*, **1980**, 2428.
11. Schmidt, U.; Kabitzke, K.; Markau, K.; Müller, A. *Chem. Ber.* **1966**, *99*, 1497.
12. Simulation created by the authors with the aid of the WINEPR SimFonia software provided by Bruker, using the experimental hfc constants from reference 9 and a linewidth of 1.5 G.

13. Hinchley, S. L.; Morrison, C. A.; Rankin, D. W. H.; Macdonald, C. L. B.; Wiacek, R. J.; Cowley, A. H. Lappert, M. F.; Gundersen, G.; Clyburne, J. A. C.; Power, P. P. *Chem. Commun.* **2000**, 2045.
14. Hinchley, S. L.; Morrison, C. A.; Rankin, D. W. H.; Macdonald, C. L. B.; Wiacek, R. J.; Voigt, A.; Cowley, A. H. Lappert, M. F.; Gundersen, G.; Clyburne, J. A. C.; Power, P. P. *J. Am. Chem. Soc.* **2001**, *123*, 9045.
15. Bezombes, J. P.; Borisenko, K. B.; Hitchcock, P. B.; Lappert, M. F.; Nycz, J. E.; Rankin, D. W. H.; Robertson, H. E. *Dalton. Trans.* **2004**, 1980.
16. Loss, S.; Magistrato, A.; Cataldo, L.; Hoffmann, S.; Geoffroy, M.; Rothlisberger, U.; Grützmacher, H. *Angew. Chem. Int. Ed.* **2001**, *40*, 723.
17. Simulation created by the authors with the aid of the WINEPR SimFonia software provided by Bruker, using the experimental hfc constants from reference 15 and a linewidth of 10 G.
18. Gouygou, M.; Tachon, C.; Koenig, M.; Dubourg, Al. Declercq, J.-P.; Jaud, J.; Etemad-Moghadam, G. *J. Org. Chem.* **1990**, *55*, 5750.
19. Canac, Y.; Bourissou, D.; Baceiredo, A.; Gornitzka, H.; Schoeller, W. W.; Bertrand, G. *Science*, **1998**, *279*, 2080.
20. Canac, Y.; Baceiredo, A.; Schoeller, W. W.; Gigmes, D.; Bertrand, G. *J. Am. Chem. Soc.* **1997**, *119*, 7579.
21. Armstrong, A.; Chivers, T.; Parvez, M.; Boeré, R. T. *Angew. Chem. Int. Ed.* **2004**, *43*, 502.
22. Armstrong, A.; Chivers, T.; Krahn, M.; Parvez, M.; Schatte, G. *Chem. Commun.* **2002**, 2332.
23. Armstrong, A.; Chivers, T.; Parvez, M.; Schatte, G.; Boeré, R. T. *Inorg. Chem.* **2004**, *43*, 3453.
24. Armstrong, A.; Chivers, T.; Szabo, M.; Ziegler, T. University of Calgary, *Unpublished Results*, **2003**
25. Armstrong, A.; Chivers, T. *Unpublished Results*, **2004**
26. Hicks, R. G. *Can. J. Chem.* **2004**, *82*, in press.
27. Hicks, R. G.; Hooper, R. *Inorg. Chem.* **1999**, *38*, 284.
28. Hicks, R. G.; Ohrstrom, L.; Patenaude, G. W. *Inorg. Chem.* **2001**, *40*, 1865.
29. Barclay, T.; Hicks, R. G.; Ichimura, A. S.; Patenaude, G. W. *Can. J. Chem.* **2002**, *80*, 1501.
30. Carriedo, G. A.; Garcia-Alonso, F. J.; Gomez-Elipe, P.; Brillas, E.; Julia, L. *Org. Lett.* **2001**, *3*, 1625.
31. Haddon, R. C.; Mayo, S. L.; Chichester, S. V.; Marshall, J. H. *J. Am. Chem. Soc.* **1985**, *107*, 7585.
32. Sasaki, S.; Sutoh, K.; Murakami, F.; Yoshifuji, M.; *J. Am. Chem. Soc.* **2002**, *124*, 14830.
33. Boeré, R. T.; Masuda, J. D.; Tran.P. University of Lethbridge, *Unpublished Results*, **2004**.

Chapter 6

Stable Singlet Diradicals Based on Boron and Phosphorus

Amor Rodriguez, Carsten Präsang, Vincent Gandon,
Jean-Baptiste Bourg, and Guy Bertrand*

UCR-CNRS Joint Research Chemistry Laboratory (UMR 2282),
Department of Chemistry, University of California,
Riverside, CA 92521-0403
*Corresponding author: gbertran@mail.ucr.edu

Using the specific properties of boron and phosphorus, a room temperature stable localized singlet diradical featuring a PBPB-four membered ring skeleton has been isolated. Its structure and reactivity are discussed. Moreover, it is shown that depending of the nature of substituents at phosphorus and boron, this type of diradical undergoes a ring closure that is thermally allowed.

Intoduction

Diradicals are even-electron molecules that have one bond less than the number permitted by the standard rules of valence (*1-5*). The two remaining electrons occupy two orbitals that have the same or nearly the same energy and can have antiparallel (singlet state) or parallel spins (triplet state).

As emphasized by Borden (*2*), there are several types of organic diradicals but this work mainly focuses on singlet localized diradicals. In these diradicals the partially filled orbitals reside on two different atoms that are connected by one or more atoms that are saturated. These diradicals are also called "nonconjugated", however the partially filled atomic orbitals that are formally localized on the two radical centers can interact either through space or through the σ bonds of the atoms that connect them. Trimethylene ($CH_2CH_2CH_2$) is the archetypal representative.

Singlet diradicals are involved as intermediates in many reactions such as the ring opening and closure of strained cycloalkanes (6-10), Cope rearrangement (11-16), bicyclo-butane or -pentane inversion (17,18)...

Catenation of singlet diradicals, via appropriate linkers, are predicted to lead to antiferromagnetic low-spin polymers, in which the half-filled electron bands would confer the capability for metallic conduction without doping (3,19). However, as noted by Berson (3), a vast and largely unmapped terrain still must be explored before these practical objectives can be reached. Not only are we deficient in knowledge about the specifically solid state intermolecular interactions between chains, which can be decisive in the ultimate bulk ferromagnetism or conductivity of a polymer, but we also need to understand more about the spin interactions among weakly bound electrons within individual diradicals. Nevertheless, many laboratories already have made strides toward the actual synthesis of organic electrical conductors and ferromagnets by ligation of non-Kekulé units (3,4,20-25).

One of the major problems is the inherent instability of diradicals. Indeed, diradicals are even more ephemeral than monoradicals since their bifunctionality permits intermolecular as well as intramolecular coupling reactions. Even the triplet 1,3-cyclobutanediyls **A(T)** can only be observed in matrices at very low temperature, although the combination of the two unpaired electrons, leading to bicyclo[1.1.0]butanes **B**, is impeded by a huge ring strain and a spin barrier (26). The corresponding singlet diradicals **A(S)**, formally resulting from the homolytic cleavage of the bridging σ-bond of **B**, are predicted only as transition states for the inversion of **B** (17). Optimizing the substituent effects (vide infra), a few localized singlet 1,3-diradicals have been observed (27-29), but the half-life in solution at room temperature of the most persistent, **A1** (27), is only in the microsecond range (Figure 1).

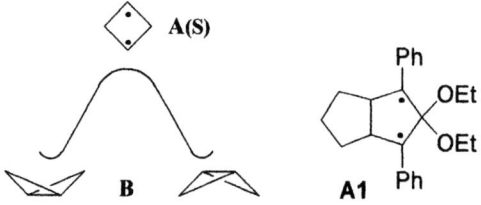

*Figure 1. Energy profile for the inversion of **B** and the most stable singlet diradical based on carbon(**A1**)*

Synthesis and Structure of the first stable diradical

In 2002, we succeeded in synthesizing the localized singlet 1,3-diradical **1a**, which is indefinitely stable at room temperature both in solution and in the solid-state (*30*) (Figure 2).

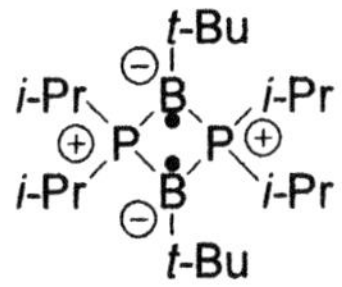

Figure 2. The first stable singlet diradical

The choice of our target molecule of type **1** was dictated by several factors. (i) As already mentioned, the ring strain in the bicyclo[1.1.0] system **2** should weaken the central BB bond (Figure 3). (ii) The spin multiplicity and relative stability of the carbon-base diradicals such as **A1** were explained by the presence of low-lying σ*(C-X) orbitals allowing through-bond interaction of the two radical sites (*28,31,32*). In the same way, diradicals of type **1** should benefit from the presence of low-lying σ* orbitals at the phosphorus atoms. (iii) The σ(B-B) bond should be weakened by the electrostatic repulsion between the two adjacent negative charges.

Figure 3. Diradical form 1 and bicyclic isomer 2

The synthetic strategy was based on previous findings from our group. We had already shown that due to the reluctance of phosphorus to become planar and to form π-bonds, di(phosphino)carbocations **C** adopt a cyclic form **D** rather than an amidinium-type structure **E** (*33*). By analogy, it was reasonable to believe that 1,2-diphosphino-diboranes **3** should adopt a bicyclo[1.1.0]butane form **2** and hopefully the diradical structure **1** rather than a butadiene structure **4** (*34*) (Figure 4). The valence isomerization of butadienes into bicyclo[1.1.0]butane derivatives has already been predicted computationally and postulated experimentally in the case of heavier main-group element containing derivatives (*35-37*).

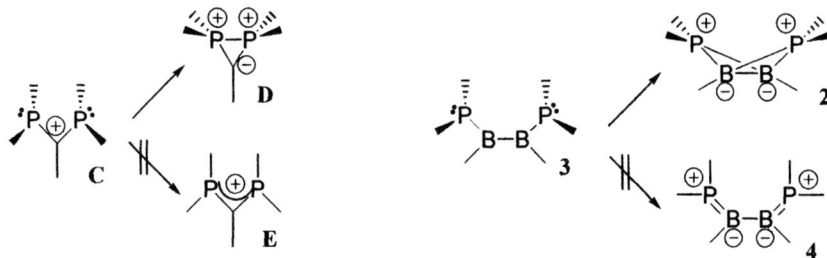

Figure 4. Rational of our synthetic approach

The obvious route to the desired 1,2-diphosphinodiboranes was to react two equivalents of a lithium phosphide with a 1,2-dichlorodiborane. To assure a kinetic protection of the desired diradical moiety, sterically demanding substituents, namely *tert*-butyl and isopropyl groups, were used at boron and phosphorus, respectively. Compound **1a** was isolated in 68% yield as extremely air-sensitive, but highly thermally stable yellow crystals.

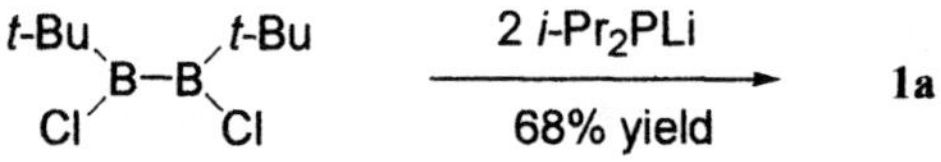

$$\underset{\text{Cl}}{\text{t-Bu}}\text{B-B}\underset{\text{Cl}}{\text{t-Bu}} \xrightarrow[\text{68\% yield}]{2\ i\text{-Pr}_2\text{PLi}} \textbf{1a}$$

Scheme 1. Synthesis of diradical 1a

According to a single crystal X-ray diffraction study, the PBPB four-membered ring of **1a** is perfectly planar, the boron atoms are in a planar environment and the P-B bond lengths are equal, but a little shorter than expected for single bonds (Figure 5). Note that a planar environment has also been observed by Power for the only structurally characterized boron-centered monoradical (*38*). The most striking feature of **1a** is the very large B-B distance of 2.57 Å, which clearly indicates the cleavage of the B-B bond.

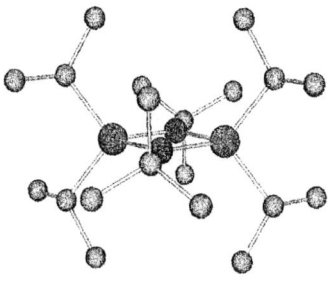

Figure 5. Molecular view of diradical 1a

Calculations carried out on the experimentally obtained molecule **1a** found that the singlet state is lower in energy than the triplet state (17.2 kcal/mol at UB3LYP/6-311++g** level) demonstrating an interaction between the two radical sites. UV spectroscopy [**1a** (toluene): λ_{max} = 446 nm, absorption coefficient ε = 2200] indicates a molecular orbital splitting, that is also observed for the carbon-based diradicals **A1** (*27,28*). The nature of the coupling of the two radical sites (positive p-orbital overlap) was apparent from the highest occupied molecular orbital (HOMO). It involves the participation of the σ^*(P-C) orbitals which indicates some through-bond interaction, and an overlap of the 2p(B) orbitals, despite the long boron-boron distance that suggests some through-space interaction. The latter has been confirmed by an electron localization function (ELF) study (*39*). Both the through-bond and through space interaction explains the unusual stability of this species, although it cannot be adequately described by two-center two-electron bonds (*40*).

Reactivity of stable singlet diradical 1a

As soon as they were discovered, radicals found applications for the large scale production of halogenoalkanes and for the polymerization of unsaturated monomers. However, for a long time, they were avoided in organic synthesis because of their high instability and reactivity. Nowadays, their behavior is sufficiently understood to allow their use in complex reaction steps, which can involve the formation of several bonds.

The intermolecular reactivity of localized singlet diradicals is virtually unknown; their lifetime is too short to allow intermolecular chemistry to compete efficiently. In a recent paper Abe *et al.* wrote (*27*): "All our extensive efforts to trap the localized singlet diradical **A1** (the most stable known so far), by external additives and, thus, to explore the intermolecular reactivity of a localized singlet diradical met with failure.... Not even in the time-resolved spectroscopic experiments, in which the exceedingly long lifetime (*e.g.* in chloroform *ca.* 3.7 µs) should facilitate the observation of subtle effects, did we obtain clear-cut evidence for trapping by external additives". However, note that some intermolecular reactions of π-conjugated non-Kekule derivatives have been reported when the diradicals were generated in the presence of alkenes, or dioxygen (*3*). Berson mentioned that their dimerization must be among the fastest bimolecular reactions known; the rates of these dimerizations are essentially at the diffusion-limited value (*3*).

The unusual stability of diradical **1a** gave us the opportunity to study its reactivity (*41*).

During NMR experiments we observed that compound **1a** slowly reacted with deuterated chloroform. The reaction was complete after three days at room temperature. The 2,4-dichloro adducts **5** were obtained in an approximate 3/1 cis/trans ratio (Scheme 2). The two isomers were separated by crystallization and characterized by multi-nuclear NMR spectroscopy and X-ray analyses. Derivatives ***cis*-5** and ***trans*-5** do not interconvert in solution, however the mechanism for the reaction of **1a** with chloroform is not clearcut.

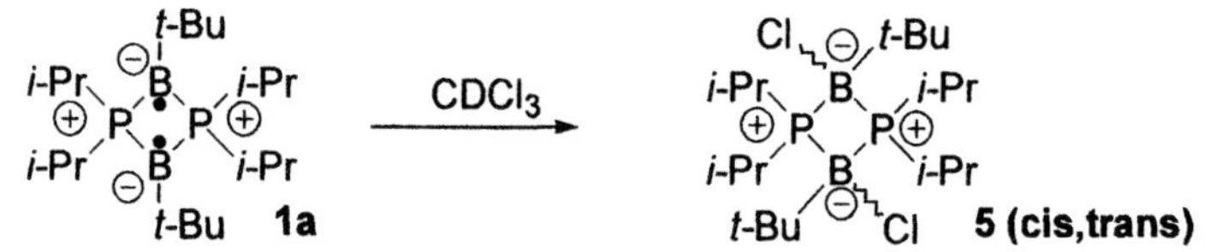

Scheme 2. Reaction of 1a with CDCl$_3$

Since **1a** readily reacts with mild oxidizing agents such as chloroform, we subsequently studied its behavior towards elemental selenium. Complete conversion was observed after 8 hours at room temperature in toluene solution. The [1.1.1]bicyclic structure (asterane) of the resulting derivative **6** (70% isolated yield) was unambiguously established by an X-ray diffraction analysis. Interestingly, compound **6** was also obtained in high yield by reacting **1** with diphenyl diselenide (*42*) (Scheme 3).

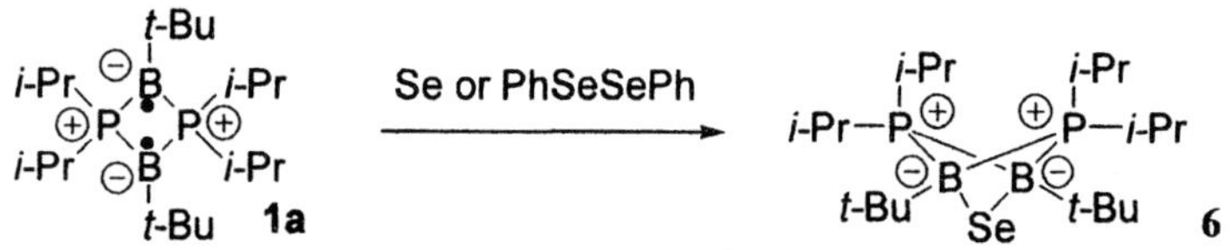

Scheme 3. Reaction of 1a with elemental selenium and PhSeSePh

This result prompted us to investigate the reactivity of **1** towards typical reagents for radical-type reactions.

A spontaneous and clean reaction was observed with trimethyltin hydride at room temperature, the *trans* 1,3-adduct **7** being isolated as colorless crystals in 73% yield. The *trans* geometry of **7** was unambiguously deduced from the X-ray analysis (Figure 6), and is likely to result from a stepwise, rather than a concerted reaction.

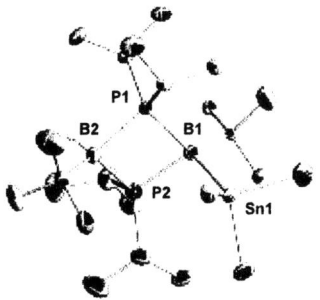

Figure 6. Molecular structure of the tin hydride adduct **7**

The reaction with bromotrichloromethane afforded further evidence for the radical-type behavior of **1a**. The reaction did not require any radical initiator and was complete in few minutes at room temperature in toluene solution. Compound **8** was obtained in 56% yield along with a small amount of the *trans* 1,3-dibromo adduct **9** (Scheme 4). Crystallization from a saturated dichloromethane solution at –30°C afforded single crystals of **8**. The X-ray diffraction study revealed a novel B-spiro structure. The BPBP four-membered ring is retained; one of the boron centers bears a bromine atom while the other is engaged in a BCC three-membered ring (*43*).

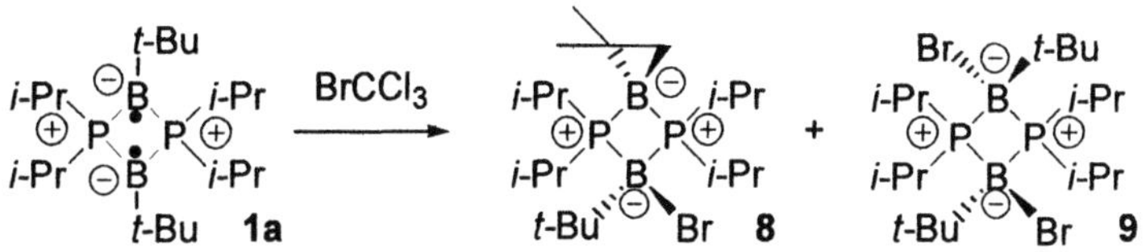

Scheme 4. Reaction of **1a** *with bromotrichloromethane*

The formation of **8** most probably results from a stepwise reaction: i) **1a** abstracts a bromine atom from the bromotrichloromethane, ii) the resulting radical pair disproportionates to give **8** and chloroform. Note that an alternative pathway for the second step of this reaction can be envisaged: the radical CCl_3 would act as an oxidizing agent leading to a boron center with an empty orbital, which would insert into the C-H bond via a three-center two-electron BHC bond, the latter intermediate would finally transfer a proton to the CCl_3 anion. Although the exact structure of the radical intermediate (opened or bridged) could not be determined to date, the postulated mechanism is supported by the formation of an increasing amount of the 1,3-dibromo adduct **9** when increasing amounts of bromotrichloromethane were used.

These results as a whole demonstrate that although the heteroatom-containing 1,3-diradical **1a** benefits from an increased thermal stability compared to its transient congeneers, it does feature some radical-type behavior.

Diradical versus bicyclo[1.1.0] structure

2,4-Diphosphacyclobutane-1,3-diyls **F**, recently reported by Niecke (*44-48*) and Yoshifuji (*49*) feature a trans-annular *anti-bonding* π-overlap, which makes the thermal ring closure into **G** forbidden. In marked contrast, 1,3-dibora-2,4-diphosphoniocyclobutane-1,3-diyls **H** (such as **1a**) feature a trans-annular *bonding* π-overlap, which allows for the thermal ring closure into the bicyclo[1.1.0] isomer **I** (Figure 7). Therefore, variation of the phosphorus and boron substituents was expected to strongly influence the ground-state structure of compounds **H/I** and thus offered an opportunity not only to isolate the structural extremes, as reported for **F** and **G** (*44*), but also to mimic the whole reaction profile for the inversion of **I** (*50*).

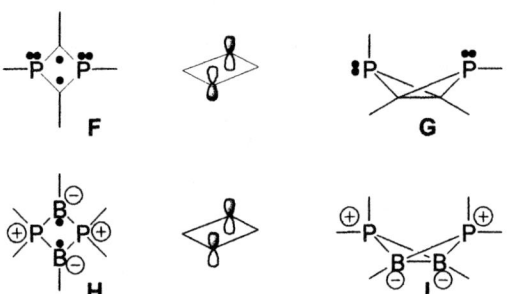

*Figure 7. Schematic representations of bicyclic compounds **G** and **I**, diradicals **F** and **H**, and of their HOMO.*

Compounds **10-12** were obtained by the synthetic route used previously for the preparation of **1a**. However, since the 1,2-diphenyl-1,2-dichloro-diborane is known to be highly unstable (*51*), derivative **13** was synthesized by reduction of the corresponding 1,3-dichloro-1,3-diborata-2,4-diphosphoniocyclobutane (*52*) with two equivalents of lithium naphthalenide in toluene solution. All of the compounds **10-13** were isolated in moderate to good yields as very air-sensitive, but thermally highly stable crystalline materials (Scheme 5).

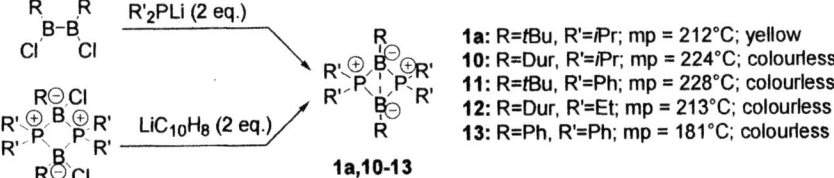

Scheme 5. Synthesis of derivatives 1a and 10-13

The impact of the substituents at boron upon structure was first investigated by replacing the *tert*-butyl groups of **1a** by duryl rings (duryl = 2,3,5,6-tetramethylphenyl), while keeping the *iso*-propyl groups at phosphorus. X-ray diffraction analysis revealed that **10** adopts a very different structure to that observed for **1a**. The BPBP core deviates from planarity (interflap angle between the two PBB units 130°), the B-B distance is significantly shortened (2.24 Å). Steric hindrance probably does not favor a coplanar arrangement of the duryl rings and BPBP core in **10**, thereby preventing efficient stabilization of the radical centers via π-delocalization.

The influence of the substituents at phosphorus was then studied by replacing the *iso*-propyl substituents of **1a** by phenyl rings. The B-B distance of **11** is noticeably shortened again (1.99 Å), while the BPBP core deviates further from planarity (interflap angle 118°). This result suggests that the less sterically demanding substituents at phosphorus favor the folded structure by decreasing the 1,3-diaxial interactions. This hypothesis was confirmed by comparing the solid-state structures of compounds **12** (BB: 1.89 Å; interflap angle 115°) and **10**, both featuring duryl groups at boron, but ethyl and *iso*-propyl substituents at phosphorus, respectively.

Lastly, the most folded structure (interflap angle 114°) was obtained for the perphenylated derivative **13** (Figure 8) for which the B-B distance (1.83 Å) is in the range typical for B-B single bonds, and about 40% shorter than in **1a**.

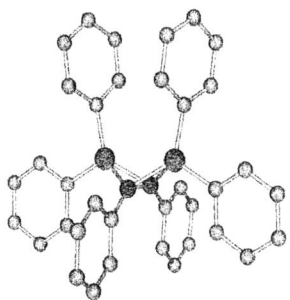

Figure 8. Molecular structure of butterfly 13

The geometric parameters observed for **13** are very similar to those calculated for the parent bicyclic compound **H** (H at boron and phosphorus) while the data for **1a** are very close to those predicted for the parent diradical **I**, which is the transition state for the inversion of **H**. Therefore, in the solid state, derivatives **10-12** adopt structures intermediate between those for **H** and **I**.

These results illustrate to some extent the conceptually most simple reactions: the stretching and eventual rupture of a σ-bond to afford two single-electron species, together with the reverse bond-forming process.

Conclusions and Perspectives

Aside from any fundamental curiosity concerning the synthesis and isolation of these species, the availability of diradicals or diradicaloids (*40,53-56*), which can be handled under standard laboratory conditions, opens the way for new developments in various fields. Their chemical behavior will define their potential use as radical scavengers and initiators, while the study of their physical properties will allow for a better understanding of the factors that control singlet-triplet energy spacing and spin interactions in diradicals, which is decisive for the rational development of new molecular organic materials such as electrical conductors and ferromagnets. The catenation of our singlet diradicals via appropriate linkers is one of our priorities.

Interestingly, the concept of "bond-stretch isomerism" has been introduced by Stohrer and Hoffmann using strained tricyclic hydrocarbons: "In the 2,2,2-system the optimum alignment for through–bond coupling of radical lobes creates the conditions for a new type of isomerism – two stable conformations related by a simple bond stretching. These are the normal tricyclic form **K** and the stabilized diradical **J**" (*57*) (Figure 9). The early attempts to characterize bond-stretch isomers either failed or were eventually rejected as crystallographic artifacts, and therefore the existence of bond-stretch isomers became questionable (*58*).

Figure 9. Hoffmann predictions of the existence of two bond-stretch isomers

According to the most recent review on this topic (*59*), the 1,3-diphosphacyclobutane-2,4-diyl **F** and 1,3-diphosphabicyclo-[1.1.0]butane **G** are the first and only known stretch isomers that have been isolated and independently characterized. As mentioned before, because of a trans-annular *anti-bonding* π-overlap (Figure 7) the thermal ring closure of **F** into **G** is forbidden. One of the exciting challenges is to bring evidence for the existence of two bond-stretch isomers featuring a trans-annular *bonding* π-overlap, which allows for the thermal ring closure and opening processes, as for **H/I**. We are currently investigating this problem.

References

(1) Salem, L.; Rowland, C. *Angew. Chem., Int. Ed. Engl.* **1972**, *11*, 92-111.
(2) Borden, W. T. In *Encyclopedia of Computational Chemistry*; Schleyer, P. v. R., Ed.; Wiley: New York, 1998, pp 708-722.
(3) Berson, J. A. *Acc. Chem. Res.* **1997**, *30*, 238-244.
(4) Dougherty, D. A. *Acc. Chem. Res.* **1991**, *24*, 88-94.
(5) Johnston, L. J.; Scaiano J. C. *Chem. Rev.* **1989**, *89*, 521-547.
(6) Zewail, A. H. *Angew. Chem., Int. Ed.* **2000**, *39*, 2587-2631.
(7) Pedersen, S.; Herek, J. L.; Zewail, A. H. *Science* **1994**, *266*, 1359-1364.
(8) Berson, J. A. *Science* **1994**, *266*, 1338-1339.
(9) Marcus, R. A. *J. Am. Chem. Soc.* **1995**, *117*, 4683-4690.
(10) Cramer, C. J.; Kormos, B. L.; Seierstad, M.; Sherer, E. C.; Winget, P. *Org. Lett.* **2001**, *3*, 1881-1884.
(11) Hrovat D. A.; Chen, J. G.; Houk, K. N.; Borden, W. T. *J. Am. Chem. Soc.* **2000**, *122*, 7456-7460.
(12) Hrovat, D. A.; Beno, B. R.; Lange, H.; Borden, W. T. *J. Am. Chem. Soc.* **1999**, *121*, 10529-10537.
(13) Doering, W. v. E.; Wang, Y. H. *J. Am. Chem. Soc.* **1999**, *121*, 10112-10118.
(14) Staroverov, V. N.; Davidson, E. R. *J. Mol. Struct-Theochem* **2001**, *573*, 81-89.
(15) Roth, W. R.; Gleiter, R.; Paschmann, V.; Hackler, U. E.; Fritzsche, G.; Lange, H. *Eur. J. Org. Chem.* **1998**, 961-967.
(16) Staroverov, V. N.; Davidson, E. R. *J. Am. Chem. Soc.* **2000**, *122*, 186-187.
(17) Nguyen, K. A.; Gordon, M. S.; Boatz, J. A. *J. Am. Chem. Soc.* **1994**, *116*, 9241-9249.
(18) Johnson, W. T. G.; Hrovat, D. A.; Skancke, A.; Borden, W. T. *Theor. Chem. Acc.* **1999**, *102*, 207-225.
(19) *Conjugated Polymers and Related Material;* Salaneck, W. R.; Lundstrom, L.; Ranby, B. Eds; Oxford: New York **1993**.
(20) Crayston, J. E.; Devine, J. N.; Walton, J. C. *Tetrahedron* **2000**, *56*, 7829-7857.

(21) Miller, J. S.; Epstein, A. J. *MRS Bulletin* **2000**, *25*, 21-28
(22) Veciana, J.; Iwamura, H. *MRS Bulletin* **2000**, *25*, 41-51.
(23) Adam, W.; Baumgarten, M.; Maas, W. *J. Am. Chem. Soc.* **2000**, *122*, 6735-6738.
(24) Rajca, A. *Chem. Rev.* **1994**, *94*, 871-893.
(25) Iwamura, H.; Koga, N. *Acc. Chem. Res.* **1993**, *26*, 346-351.
(26) Jain, R.; Sponsler, M. B.; Coms, F. D.; Dougherty, D. A. *J. Am. Chem. Soc.* **1988**, *110*, 1356-1366.
(27) Abe, M.; Adam, W.; Heidenfelder, T.; Nau, W. M.; Zhang, X. Y. *J. Am. Chem. Soc.* **2000**, *122*, 2019-2026.
(28) Adam, W.; Borden, W. T.; Burda, C.; Foster, H.; Heidenfelder, T.; Heubes, M.; Hrovat, D. A.; Kita, F.; Lewis, S. B.; Scheutzow, D.; Wirz, J. *J. Am. Chem. Soc.* **1998**, *120*, 593-594.
(29) Abe, M.; Adam, W.; Nau, W. M. *J. Am. Chem. Soc.* **1998**, *120*, 11304-11310.
(30) Scheschkewitz, D.; Amii, H.; Gornitzka, H.; Schoeller, W. W.; Bourissou, D.; Bertrand, G. *Science* **2002**, *295*, 1880-1881.
(31) Skancke, A.; Hrovat, D. A.; Borden, W. T. *J. Am. Chem. Soc.* **1998**, *120*, 7079-7084.
(32) Xu, J. D.; Hrovat, D. A.; Borden, W. T. *J. Am. Chem. Soc.* **1994**, *116*, 5425-5427.
(33) Canac, Y.; Bourissou, D.; Baceiredo, A.; Gornitzka, H.; Schoeller, W. W.; Bertrand, G. *Science* **1998**, *279*, 2080-2082.
(34) Kaufmann, B.; Jetzfellner, R.; Leissring, E.; Issleib, K.; Noth, H.; Schmidt, M. *Chem. Ber.* **1997**, *130*, 1677-1692.
(35) Kira, M.; Iwamoto, T.; Kabuto, C. *J. Am. Chem. Soc.* **1996**, *118*, 10303-10304.
(36) Driess, M.; Pritzkow, H.; Rell, S.; Janoschek, R. *Inorg. Chem.* **1997**, *36*, 5212-5217.
(37) Weidenbruch, M. *Eur. J. Inorg. Chem.* **1999**, 373-381.
(38) Olmstead, M. M.; Power, P. P. *J. Am. Chem. Soc.* **1986**, *108*, 4235-4236.
(39) Savin, A.; Nesper, R.; Wengert, S.; Fassler, T. F. *Angew. Chem., Int. Ed. Engl.* **1997**, *36*, 1809-1832.
(40) Schoeller, W. W.; Rozhenko, A.; Bourissou, D.; Bertrand, G. *Chem. Eur. J.* **2003**, *9*, 3611-3617.
(41) Amii, H.; Vranicar, L.; Gornitzka, H.; Bourissou, D.; Bertrand, G. *J. Am. Chem. Soc.* **2004**, *126*, 1344.
(42) *The Chemistry of Organoselenium and Tellurium Compounds;* Pataï, S.; Rappaport, Z., Eds.; Wiley: New-York, **1986**.
(43) Kropp, M. A.; Baillargeon, M.; Park, K. M.; Bhamidapaty, K.; Schuster, G. B. *J. Am. Chem. Soc.* **1991**, *113*, 2155-2163.

(44) Niecke, E.; Fuchs, A.; Nieger, M. *Angew. Chem. Int. Ed.* **1999**, *38*, 3028-3030.
(45) Niecke, E.; Fuchs, A.; Baumeister, F.; Nieger, M.; Schoeller, W. W. *Angew. Chem. Int. Ed. Engl* **1995**, *34*, 555-557.
(46) Schmidt, O.; Fuchs, A.; Gudat, D.; Nieger, M.; Hoffbauer, W.; Niecke, E.; Schoeller, W. W. *Angew. Chem. Int. Ed. Engl* **1998**, *37*, 949-952.
(47) Schoeller, W. W.; Begemann, C.; Niecke, E.; Gudat, D. *J. Phys. Chem.* **2001**, *105*, 10731-10738.
(48) Sebastian, M.; Nieger, M.; Szieberth, D.; Nyulaszi, L.; Niecke, E. *Angew. Chem. Int. Ed.* **2004**, *43*, 637-641.
(49) Sugiyama, H.; Ito, S.; Yoshifuji, M. *Angew. Chem. Int. Ed.* **2003**, *42*, 3802-3804.
(50) Scheschkewitz, D.; Amii, H.; Gornitzka, H.; Schoeller, W. W.; Bourissou, D.; Bertrand, G. *Angew. Chem. Int. Ed.* **2004**, *43*, 585-587.
(51) Hommer, H.; Nöth, H.; Knizek, J.; Ponikwar, W.; Schwenk-Kircher, H. *Eur. J. Inorg. Chem.* **1998**, 1519-1527.
(52) Lube, M. S.; Wells, R. L.; White, P. S. *Inorg. Chem.* **1996**, *35*, 5007-5014.
(53) Seierstad, M.; Kinsinger, C. R.; Cramer, C. J. *Angew. Chem. Int. Ed.* **2002**, *41*, 3894-3896.
(54) Cheng, M. J.; Hu, C. H. *Mol. Phys.* **2003**, *101*, 1319-1323.
(55) Jung, Y.; Head-Gordon, M. *Chem. Phys. Chem.* **2003**, *4*, 522-525.
(56) Jung, Y.; Head-Gordon, M. *J. Phys. Chem. A* **2003**, *107*, 7475-7481.
(57) Stohrer, W. D.; Hoffmann, R. *J. Am. Chem. Soc.* **1972**, *94*, 779-786.
(58) Parkin, G. *Chem. Rev.* **1993**, *93*, 887-891.
(59) Rohmer, M. M.; Benard, M. *Chem. Soc. Rev.* **2001**, *30*, 340-354.

Chapter 7

Easily Broken Strong Bonds: a New Law of Thermodynamics

Konstantin B. Borisenko, Sarah L. Hinchley, and David W. H. Rankin

School of Chemistry, University of Edinburgh, West Mains Road, Edinburgh, EH9 3JJ, United Kingdom

Determination of the molecular structures of tetrakis[bis-(trimethylsilyl)methyl]diphosphine and its arsenic analog in the crystalline phase showed that the central P-P and As-As bonds were little longer than in other diphosphines and diarsines. Nevertheless, their vapors consisted entirely of the bis[bis(trimethylsilyl)methyl]phosphido and bis[bis(trimethylsilyl)methyl]arsenido free radicals. This thermodynamic conundrum has been solved by series of *ab initio* calculations, which have shown that the energy needed to break the strong central bonds is stored in deformed ligands. Detailed analysis of the dissociation process, aided by a simple ball and spring model, has allowed the intrinsic energy of the central bond to be separated from the overall dissociation energy. The intrinsic bond energies are consistent for a wide range of disphosphines, even though their dissociation energies are scattered over a range of hundreds of kJ mol^{-1}.

At first sight, the dissociation of a sterically crowded tetraalkyldiphosphine into two dialkyl phosphido radicals (Figure 1) was not unexpected, although unusual. Large substituents would naturally cause lengthening, and eventually breaking, of bonds. So the observation[1] that tetrakis[bis(trimethylsilyl)-methyl]diphosphine had an EPR spectrum, consistent with the formation of bis[bis(trimethylsilyl)-methyl]phosphido radicals in the liquid phase, was seen as a natural consequence of the size of the bis(trimethylsilyl)methyl (also known as disyl) ligands. The transition from crystalline to liquid and gaseous phases was associated with a change in colour from pale yellow to intense purple,[1] so it appeared that the radical was predominant in these phases, but absent from the crystal. It was only when the structures, in crystalline and gaseous phases, were determined[2] that it became clear that the situation is considerably more complex.

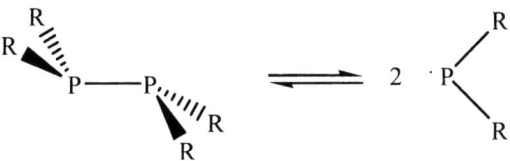

Figure 1. Dissociation of a tetra-alkyldiphosphine to give two dialkylphosphido radicals

Structures of solid and gaseous tetrakis(disyl)diphosphine and tetrakis(disyl)diarsine

The structure of tetrakis(disyl)diphosphine, $\{P[CH(SiMe_3)_2]_2\}_2$, in the crystalline phase was straightforward to determine. There was one molecule per asymmetric unit, in which the four asymmetric groups were arranged so that they packed in an efficient way (Figure 2).[2,3] The steric crowding was demonstrated by large deviations of bond lengths and inter-bond angles from the standard values (Table I). Thus, for example, the PCSi angles ranged from 110.9 to 125.2°, while CSiC angles covered the range from 103 to 117°. However, the central P-P bond length was 231.0(7) pm, only about 8 pm longer than a typical value for an unstrained diphosphine. Thus it was clear that the packing of the four disyl groups in each molecule was enabled by their asymmetry, and that this allowed the central P-P bond to be relatively unstrained, although it did not avoid the need for considerable distortions of the packing ligands.

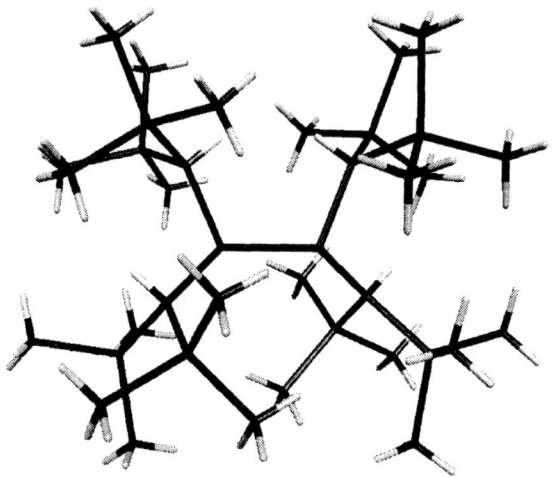

Figure 2. The structure of {P[CH(SiMe₃)₂]₂}₂ in the crystalline phase

Table I. Selected geometrical parameters for crystalline {P[CH(SiMe$_3$)$_2$]$_2$}$_2$

Parameter	Number of occurrences	Range
r P-C	4	189.2 to 189.6 pm
r (P)C-Si	8	189.2 to 192.6 pm
<PPC	4	104.8 to 107.9°
<PCSi	8	110.9 to 125.2°
<CPC	2	103.0 to 103.6°

Determination of the structure of this compound in the gas phase thus became particularly important, but at the time this was a formidable undertaking. In gas electron diffraction (GED) studies, resolution of similar interatomic distances is difficult or impossible. In the case of the bis(disyl)phosphido radical, there are P-C and inner and outer Si-C distances, which are all expected to be similar, and distortions due to steric crowding, as well as the symmetry of the radical, lead to a further increase in the number of unresolvable distances. There is a similar problem with non-bonded distances; for example, two-bond P...Si, Si...Si and C...C distances are all similar, and as they therefore also overlap in the radial distribution curve (Figure 3), resolution of them is impossible. Thus, despite heroic efforts, all attempts to determine the structure of the radical proved to be fruitless. However, what was clear was that the compound in the gas phase consisted entirely, or almost entirely, of radicals. The short, strong P-P bond was being broken on vaporization.

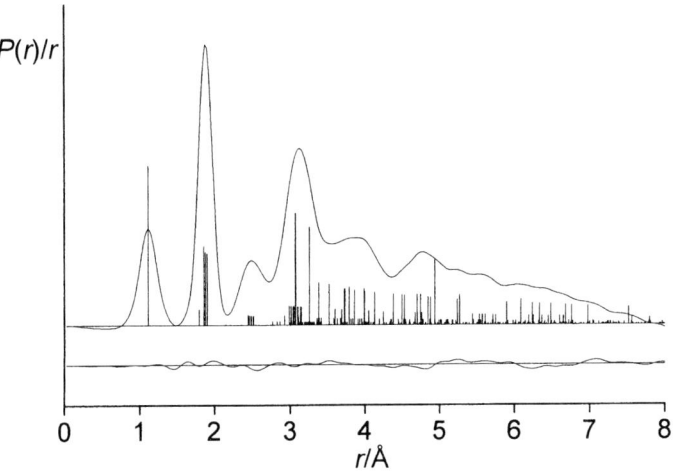

Figure 3. Radial distribution curve for the bis(disyl)phosphido radical

At the same time as the group in Edinburgh was struggling to interpret the GED data for the bis(disilyl)phosphido radical, the Oslo GED group were having parallel difficulties with the arsenic analog, As[CH(SiMe$_3$)$_2$]$_2$}, derived from {As[CH(SiMe$_3$)$_2$]$_2$}$_2$. Arsenic is only a little larger than phosphorus, so the problems were essentially the same, but with the As-C bonds slightly longer than Si-C, whereas the P-C bonds are somewhat shorter. It seemed that the problems were insoluble.

Solving gas-phase structures with the SARACEN method

Eventually, developments in the methodology of gas-phase structure determination provided a way out. The continual increases in speed and capacity of computing hardware, with parallel advances in computational chemistry software, eventually allowed experimental and theoretical data to be brought together in combined analyses. The SARACEN (Structure Analysis Restrained by Ab initio Calculations for Electron diffractioN) method[4] uses values of geometrical parameters that cannot be determined reliably from experimental data to be taken from computed structures, and used as flexible restraints in least-squares refinements. The weights assigned to the restraints are estimated by consideration of a series of structures computed at increasing levels of theory and size of basis set. In the refinements all geometrical parameters are normally refined, and the structure that is obtained should be the best that is possible, using all information, both experimental and theoretical, that is available.

When the SARACEN method was applied to the bis(disyl)phosphido and bis(disyl)arsenido radicals, their structures could be refined without any great difficulty.[2,3] It was found that in both cases the conformation of one of the disyl groups had changed from the position adopted in the corresponding fragment in the crystalline phase, and that this change had reduced the steric crowding very substantially (Figure 4). The bond lengths, and particularly the inter-bond angles, covered much narrower ranges than their counterparts in the crystal structures, and average values were much closer to those expected for unstrained molecules (Table II, cf. Table I for crystal data).

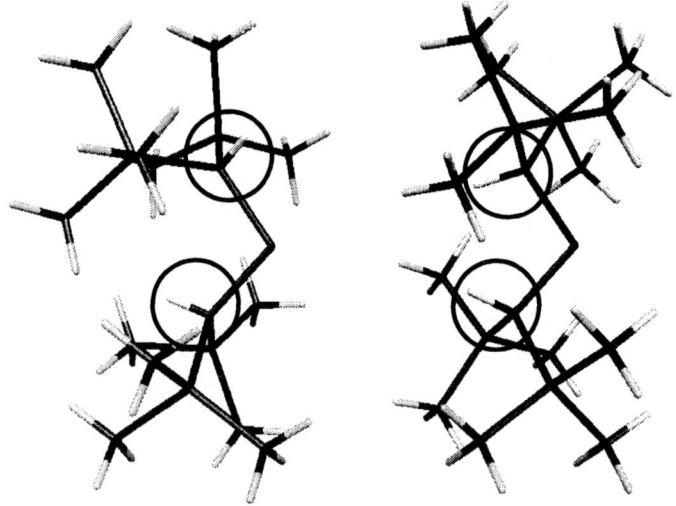

Figure 4. Structures of half of tetrakis(disyl)diphosphine (left) and the bis(disyl)phosphido radical (right). The ringed C-H bonds of the disyl groups show the conformational change on dissociation.

Table II. Selected geometrical parameters for gaseous P[CH(SiMe$_3$)$_2$]$_2$

Parameter	Number of occurrences	Range
r P-C	2	185.6 pm
r (P)C-Si	4	190.2 to 190.5 pm
<PPC	-	
<PCSi	4	109.1 to 109.8°
<CPC	1	104.0°

The crystal structure of tetrakis(disyl)diarsine

Shortly after the gas-phase structures were at last determined, the crystal structure of tetrakis(disyl)diarsine was also solved.[3] The origin of the difficulties that this had presented were revealed: there were no less than eight independent molecules in the asymmetric unit. In such circumstances one should always look for any symmetry element that has been overlooked, but the correctness of the space group assignment was unequivocally confirmed when it was realised that seven of the independent molecules had the same conformation, but that the eighth one was different. In fact, the unique molecule had almost exactly the same conformation as the single independent molecule of tetraks(disyl)diphosphine in the crystalline phase, but the other seven were different. Although the large number of atomic coordinates to be refined made refining the structure a little more difficult, it presented a rare opportunity to analyse the effects of the crystal environment on the measured geometrical parameters, and thus to see the significance of the structure and conformation that is adopted.

The method used in this analysis involved comparison of equivalent parameters for the seven molecules. For example, four of the eight different AsCSi angles for each molecule are listed in Table III. It is immediately obvious that adjacent angles in the same disyl group may differ enormously, but that the equivalent angles in different molecules differ very little.

In Table IV the data for all eight AsCSi angles are given, in the form of means and standard deviations. It can be seen that the standard deviation for each set of corresponding angles is approximately the same as, or a very little larger than, the estimated standard deviation for a single refined angle. In other words, the scatter of values caused by the variation in the environments of the seven independent molecules is extremely small.

Table III. Variations of some AsCSi angles in $\{As[CH(SiMe_3)_2]_2\}_2$ molecules in the crystalline phase

Molecule number	AsCSi(n1)	AsCSi(n2)	AsCSi(n3)	AsCSi(n4)
#1	120.3	113.6	104.5	132.0
#3	120.9	114.6	105.6	129.9
#4	121.3	114.0	104.9	130.7
#5	121.1	113.7	104.9	130.8
#6	121.0	113.6	105.9	131.0
#7	120.5	113.6	105.3	130.4
#8	120.3	113.6	105.6	129.7
mean	120.8	113.8	105.2	130.6
rms deviation	0.3	0.3	0.5	0.7
typical esd	0.4	0.3	0.4	0.3

The differences between these geometrical parameters in these seven molecules (parameters for the eighth one are also given in Table IV, along with those for the single bis(disyl)phosphine molecule) and those for the radical are therefore highly significant, reflecting the internal energies of the molecules and radicals. We therefore have reliable stuctural data for both the diarsine and diphosphine and for the arsenido and phosphido radicals, and should be able to use them to interpret these changes, in particular accounting for the easy cleavage of the short, strong P-P and As-As bonds.

The thermodynamics of dissociating diphosphines and related compounds

To understand the relationships between structural and energetic changes that occur when tetrakis(disyl)diphosphine and diarsine, and other related molecules, dissociate we have used computational methods to dissect the dissociation process. As the molecules with disyl substituents are large for high-level calculations, we experimented with various methods of calculation and basis set size, to find the simplest calculation that would give reasonably reliable energies. For these molecules we eventually selected B3LYP/3-21G*, although this is not a method we would normally employ. For somewhat smaller molecules we used MP2/6-311+G* calculations, but otherwise with the same methodology.

Table IV. Comparison of AsCSi angles in {As|CH(SiMe$_3$)$_2$|$_2$}$_2$ conformers and {P|CH(SiMe$_3$)$_2$|$_2$}$_2$

AsCSi angle	mean, molecules #1 and #3-#8	rms deviation	molecule #2	P analog
AsCSi(n1)	120.8	0.3	121.8	123.8
AsCSi(n2)	113.8	0.3	109.7	110.9
AsCSi(n3)	105.2	0.5	110.2	112.8
AsCSi(n4)	130.6	0.7	125.1	125.1
AsCSi(n5)	121.0	0.4	123.2	123.3
AsCSi(n6)	113.9	0.8	109.0	111.9
AsCSi(n7)	106.4	0.6	110.0	112.4
AsCSi(n8)	117.7	1.1	123.4	125.2

In the analysis of the processes for tetrakis(disyl)diphosphine, we first calculated the energy of a diphosphine molecule, starting with coordinates taken from the crystal structure determination, allowing all atomic positions to optimize. For a general dimer X-X, this energy is E_{opt}(X-X), as shown in Figure 5. This molecule was then split into two halves, which were identical, because the diphosphine had C$_2$ symmetry. A single-point calculation then gave the energy of one of these radicals, E_{sp}(X), and the difference between E_{opt}(X-X) and $2E_{sp}$(X) is then the instantaneous dissociation energy, ΔE_{inst}. (We include a correction for basis set superposition error.)

Allowing the structure of the radical to relax would, for most molecules, give the true dissociation energy, ΔE_{diss}. However, in the case of the bis(disyl)phosphido radical simply allowing the fragment formed by splitting the diphosphine into two to relax does not yield the observed structure of the radical. To reach this state one of the two disyl groups must be rotated. A twist of 180° is more than is needed, but from this position it will optimize to the potential minimum, where the energy is E_{opt}(X). Care must therefore be taken to ensure that the true value of ΔE_{diss} is obtained. The relaxation energy for one radical is defined as ΔE_{reorg}, and for the system as a whole this must be doubled (Figure 5).

The Morse curve for a sterically crowded diphosphine differs from that for an ideal, unstrained diphosphine in two major ways. First, the potential minimum is higher, and secondly, the interatomic distance at the equilibrium position is increased. The lower potential energy curve in Figure 5 represents such an idealized system. We have modeled the relationship between strained and unstrained systems using a very simple ball and spring model,[5,6] which allows us to extract the intrinsic energy of the central bond, which does not necessarily

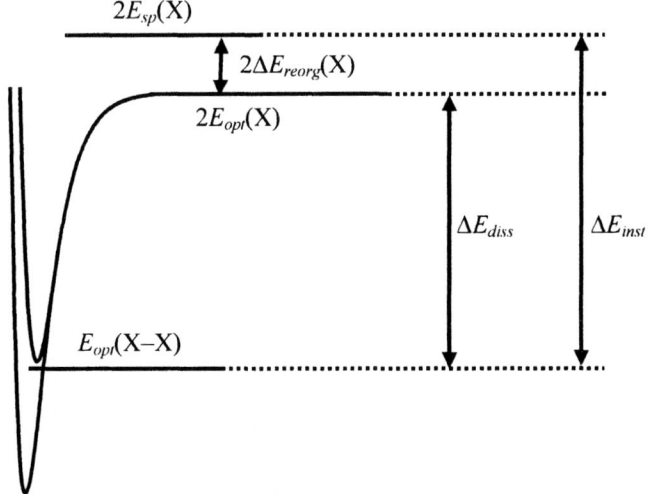

Figure 5. Energetics of dissociation of a molecule X-X to give two X radicals

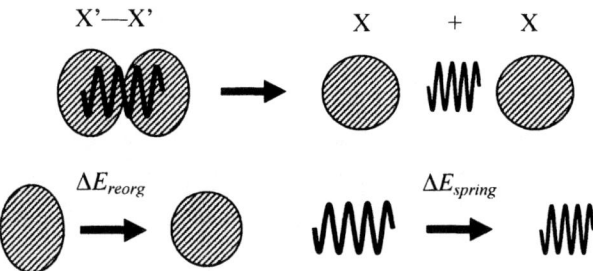

Figure 6. Ball and spring model of dissociation

resemble the actual dissociation energy of the bond, because of the reorganisation of the radicals.[7] This model is illustrated in Figure 6.

In this model the two half-molecules (PR_2 in the case of a diphosphine) are regarded as flexible balls, linked by a spring. We define a single force constant, f_b, which represents deformation of the ball along the direction of the connecting spring, which has a force constant f_s. At equilibrium (i.e. as in the structure of the dimer) the spring is strained and the balls are also distorted, and the total potential energy of the system is thus equal to the sum of the potential energies stored in the balls and the spring.

Values of these two force constants were computed by calculating the derivative of the total energy with respect to the P-P internuclear distance, first keeping the structures of the two half-molecules unchanged, and secondly allowing them to relax. Using the method described earlier,[5,6] the energies stored in the spring and the distorted balls (half-molecules) were calculated, and the energy relationships shown in Figure 7 then gave the intrinsic energy of the P-P bond (Equation 1).

$$D_0 = \Delta E_{diss} + 2\Delta E_{reorg} + \Delta E_{spring} \tag{1}$$

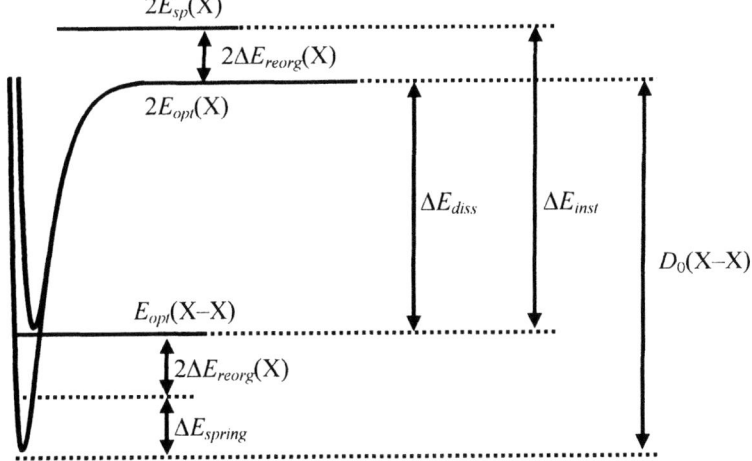

Figure 7. Relationships between energies of dissociation of a sterically crowded X-X molecule and a hypothetical unstrained molecule.

The results of these calculations for a series of diphosphines and disilanes are shown in Figure 8, and data are given in Table V. The heavy straight line represents equality of the intrinsic bond energy and the actual dissociation energy. The data for two diphosphines lie very close to this line. These are tetramethyldiphosphine and tetrasilyldiphosphine, both compounds with small substituents, for which steric effects should be minimal. The remaining diphosphines compounds all have larger substituents, and as a consequence of the interactions between the groups the dissociation energies are smaller, and in the case of tetrakis(disyl)diphosphine, even negative. Nevertheless, the intrinsic dissociation energies of the P-P bonds are remarkably consistent, particularly so given the approximations of the methodology. The smallest intrinsic energy is about 140 kJ mol^{-1}, for tetrakis(disyl)diphosphine, but the calculations for this system were only done with the B3LYP/3-21G* method and basis set. Using the conditions that were applied to the other systems, we would expect to obtain a rather larger value, as indicated by the arrow in Figure 8.

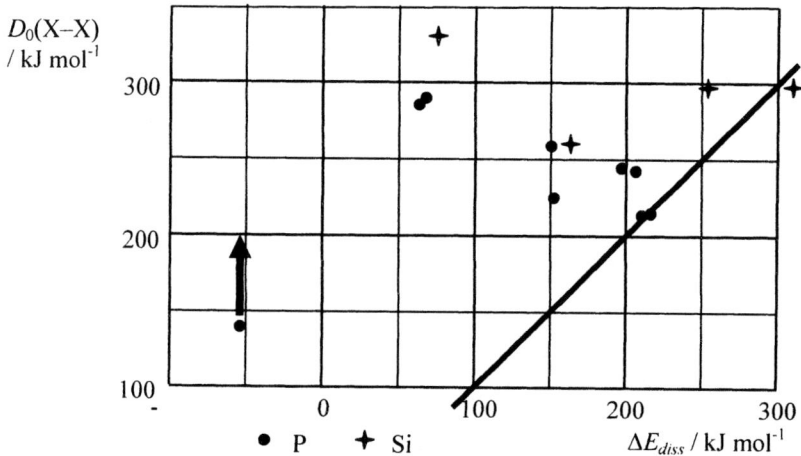

Figure 8. Comparison of intrinsic bond dissociation energies, $D_0(X-X)$, with dissociation energies, ΔE_{diss}, for a series of diphosphines (circles) and disilanes (crosses).

The results resolve most of the mysteries about tetrakis(disyl)diphosphine. Although the central P-P bond is not much longer than normal, the molecule can dissociate freely in the gas phase, because the relaxation of the substituents when the bis(disyl)phosphido radicals are formed releases sufficient energy to make the bond-fission process energetically favorable overall. However, there is one further feature of this system that distinguishes it from other diphosphines that can dissociate to give free radicals, which we discuss below.

The data for the disilanes follow a similar pattern. There is one compound, unsubstituted disilane, for which the intrinsic Si-Si bond energy is close to the molecular dissociation energy, and the more crowded molecules have similar intrinsic bond energies, despite having a very wide range of dissociation energies.

Table V. Dissociation energies and intrinsic bond energies for some diphosphine and disilanes.

System	Dissociation energy	Intrinsic bond energy
	MP2/6-311+G*	
$P_2(CH_3)_4 \to 2P(CH_3)_2$	212	214
$P_2(SiH_3)_4 \to 2P(SiH_3)_2$	213	216
	MP2/6-31+G*	
$P_2[C(CH_3)_3]_4 \to 2P[C(CH_3)_3]_2$	151	258
$P_2[CH(SiH_3)_2]_4 \to 2P[CH(SiH_3)_2]_2$	153	226
$P_2[C(SiH_3)_3]_4 \to 2P[C(SiH_3)_3]_2$	68	290
$P_2[SiH(CH_3)_2]_4 \to 2P[SiH(CH_3)_2]_2$	216	242
$P_2[Si(CH_3)_3]_4 \to 2P[Si(CH_3)_3]_2$	207	245
$\{P[N(SiMe_3)_2](NPr^i_2)\}_2$ $\to 2P[N(SiMe_3)_2](NPr^i_2)$	65	286
	B3LYP/3-21G*	
$P_2[CH(SiMe_3)_2]_4$ $\to 2P[CH(SiMe_3)_2]_2$	-54	140
$Si_2H_6 \to 2SiH_3$	306	309
$Si_2H_2Bu^t_4 \to 2SiHBu^t_2$	254	306
$Si_2H_2[C(SiH_3)_3]_4$ $\to SiH[C(SiH_3)_3]_2$	163	261
$Si_2Bu^t_6 \to 2SiBu^t_3$	~75	~330

Dissociation energies (for vibrationless states at 0K) and intrinsic bond energy terms (corrected for BSSE, kJ mol^{-1}) are from theoretical calculations.

Ligand shape is as important as ligand size

Although the large size of the disyl substituents in tetrakis(disyl)diphosphine is a critically important factor in the facile cleavage of the P-P bond, the shape of these groups is also important. The structure of the bis(trimethylsilylamido)(di-*iso*-propylamido)phosphido radical, P[N(SiMe$_3$)$_2$](N^{i}Pr$_2$), in gaseous and solid phases has recently been determined.[8] The crystal structure suffers from severe disorder, and the conclusions that can be drawn are therefore limited, although it is clear that the crystal consists of diphosphine molecules. However, the reorganization energy is substantially smaller than for the bis(disyl)diphosphido radical. This difference can be attributed to the fact that the -N(SiMe$_3$)$_2$ group is planar at nitrogen, and so rotation about the P-N bond makes no difference to the crowding of the phosphido substituents (Figure 9). In contrast, rotation of a disyl group of P[CH(SiMe$_3$)$_2$]$_2$ around the P-C bond makes a great deal of difference to the crowding of the groups (Figure 4), and it is therefore the asymmetry of this group that provides the key to the apparent anomaly of an easily broken but strong P-P bond.

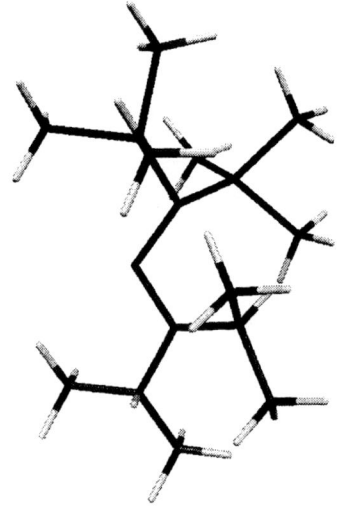

Figure 9. The bis(trimethylsilylamido)(di-iso-propylamido)phosphido radical

Conclusions

Tetrakis(disyl)diphosphine and the analogous diarsine exist as the molecular diphosphine and diarsine in the crystalline phase, but dissociate completely to free radicals in the gas phase. The central P-P and As-As bonds are not exceptionally long. The energy required to break these bonds on vaporization comes from the substituent groups, which are highly distorted in the dimer molecules.

On dissociation, strain caused by interactions between the two ends of the molecule is removed. A special feature of the disyl compounds is that rotation of one disyl substituent of each radical removes interactions between the two groups in each radical, and thus permits a further substantial reduction in strain energy. It is the total energy released in this way that allows the strong central bonds of the diphosphine and diarsine to be broken.

The energetics of the processes have been analysed with the aid of computational methods, and a simple ball and spring model. These enable the intrinsic dissociation energies of the central bonds to be derived. Applications to series of diphosphines and disilanes show that these intrinsic energies are remarkably constant, for compounds with extremely wide ranges of dissociation energies.

The unexpected behaviour of tetrakis(disyl)diphosphine and diarsine does not therefore depend on a new law of thermodynamics. However, study of this behaviour has lead to a new understanding of the thermodynamics of homolytic bond dissociation.

Acknowledgements

We thank Professor Alan H. Cowley (University of Texas at Austin) and Professor Michael F. Lappert (University of Sussex) for provision of samples, ideas, support and patience.

References

1. (a) Gynane, M. J. S.; Hudson, A.; Lappert, M. F.; Power, P. P.; Goldwhite, H. *J. Chem. Soc., Chem. Commun.*, **1976**, 623; (b) Gynane, M. J. S.; Hudson, A.; Lappert, M. F.; Power, P. P.; Goldwhite, H. *J. Chem. Soc., Dalton Trans*. **1980**, 2428.
2. Hinchley, S. L.; Morrison, C. A.; Rankin, D. W. H.; Macdonald, C. L. B.; Wiacek, R. J.; Cowley, A. H.; Lappert, M. F.; Gundersen, G.; Clyburne, J. A. C.; Power, P. P. *Chem. Commun.* **2000**, 2045.
3. Hinchley, S. L.; Morrison, C. A.; Rankin, D. W. H.; Macdonald, C. L. B.; Wiacek, R. J.; Voigt, A.; Cowley, A. H.; Lappert, M. F.; Gundersen, G.; Clyburne, J. A. C.; Power, P. P. *J. Am. Chem. Soc.*, **2001**, *123*, 9045.
4. (a) Blake, A. J.; Brain, P. T.; McNab, H.; Miller, J.; Morrison, C. A.; Parsons, S.; Rankin, D. W. H.; Robertson, H. E.; Smart, B. A. *J. Phys. Chem.* **1996**, *100*, 12280. (b) Brain, P. T.; Morrison, C. A.; Parsons, S.; Rankin, D. W. H. *J. Chem. Soc., Dalton Trans.* **1996**, 4589.
5. Borisenko, K. B.; Rankin, D. W. H. *J. Chem. Soc., Dalton Trans.*, **2002**, 3135.
6. Borisenko, K. B.; Rankin, D. W. H. *Inorg. Chem.*, **2003**, *42*, 7129.
7. (a) Morokuma, K. *J. Chem. Phys.*, **1971**, *55*, 1236. (b) Sanderson, R. T. *J. Org. Chem.*, **1982**, *47*, 3835.
8. Bezombes, J.-P.; Borisenko, K. B.; Hitchcock, P. B.; Lappert, M. F.; Nycz, J. E.; Rankin, D. W. H.; Robertson, H. E. *Dalton Trans.*, in press.

Chapter 8

Cationic Low Oxidation State Phosphorus and Arsenic Compounds

Bobby D. Ellis and Charles L. B. Macdonald*

Department of Chemistry and Biochemistry, University of Windsor, Windsor, Ontario, Canada

Compounds containing main group elements in unusually low oxidation states exhibit structural features and reactivities that are significantly different from those of analogous compounds containing the elements in their more typical oxidation states. This work summarizes the investigations of cationic univalent group 15 compounds, with a particular focus on the research derived from the seminal work of Schmidpeter concerning P(I) cations. These unusually stable "triphosphenium" cations consist of a P^+ center stabilized by two phosphine donors and have the general form $[R_3P–P–PR_3]^+$. Recently, interest in this type of compound has been re-kindled because of the unique modes of reactivity they may display. Improved synthetic strategies to such compounds, and their arsenic analogues, have been developed and current research exploits the unique chemistry of Pn(I) cations to produce unprecedented chemicals and materials.

Introduction

Heavier group 15 elements (pnictogens; Pn = P, As, Sb, Bi) are generally found in either of their typical oxidation states: +3 or +5. There exists only a handful of types of compounds containing pnictogen atoms in the +1 oxidation state. Neutral examples include transient pnictenidenes (Pn-R), which must be stabilized by either a Lewis base or transition metal complex to prevent oligomerization to $(Pn-R)_x$ rings or double-bonded dimers (R–Pn=Pn–R) (1-3). Examples of charged species containing +1 oxidation state pnictogen atoms are even more rare, typically consisting of a Pn^+ ion stabilized by two Lewis bases, which are usually phosphines (4).

The electronic structure of these cations is described by the various canonical forms illustrated in Figure 1. The nature of the bonding implied by the different models in Figure 1 range from a base-stabilized Pn^+ ion having formal single Pn-P bonds (a) to a situation in which there is a double bond between the Pn^+ ion and each Lewis base (b); in theory, the actual bond order would depend on the degree of back-bonding from Pn to the stabilizing ligands. Structural and computational evidence both indicate that there is a significant amount of Pn to ligand back-bonding in these types of molecules and that canonical form (c) is the most adequate description of the electronic structure. Such a description suggests that back-bonding effectively "oxidizes" the Pn(I) center (5) by removing excess electron density and helps to explain the relative stability of such cations as to the point that some of these Pn(I) salts are even air stable.

In contrast to +3 oxidation state pnictogen cations, pnictogenium cations, R_2Pn^+, in which a significant amount of the positive charge is located on the pnictogen center (6), the positive charge in the +1 oxidation state cations is localized on the substituents. The various canonical structures depicted in Figure 1 suggest many modes of reactivity for these P(I) cations, several of which are distinct from those of the higher oxidation state analogues. Some examples of unique chemistry that has been investigated recently are described herein.

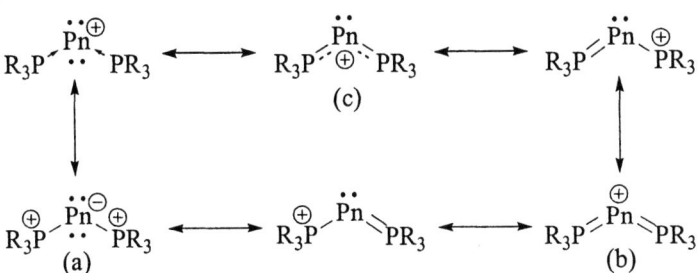

Figure 1. Canonical Forms of Pn^+ Cations Stabilized by Phosphines.

The following account outlines research stemming from the seminal work of Schmidpeter in the development of synthetic strategies for the generation of cations containing phosphorus atoms in the +1 oxidation state and the initial investigation of their reactivity. Recently, interest in these types of compounds has been re-ignited, both in terms of their synthesis and potential uses as reagents and as sources of P^+ ions. Research in this area has since been extended to include arsenic analogues, which are also discussed.

Early Work

The initial work of Schmidpeter and co-workers on +1 oxidation state pnictogen cations was done exclusively for Pn = P. The first example of such a cation was synthesized by the reduction of PCl_3 by $SnCl_2$ in the presence of an equimolar amount of a chelating diphosphine, bis(diphenylphosphino)ethane (dppe), as shown in Scheme 1 (7). The structure of this cation was confirmed by X-ray crystallography and the P–P bond distances, 2.122(1) Å and 2.128(2) Å, are found to fall between those of P–P single bonds, Ph_2P–PPh_2: P–P = 2.217 Å (8), and P–P double bonds, Mes*P=PMes*: P=P = 2.046 Å (9). Such intermediate bond lengths suggest a degree of multiple bonding in this structure. The investigators later discovered that an additional equivalent of dppe improves the reaction by sequestering the $SnCl_4$ as a dppe·$SnCl_4$ by-product (4).

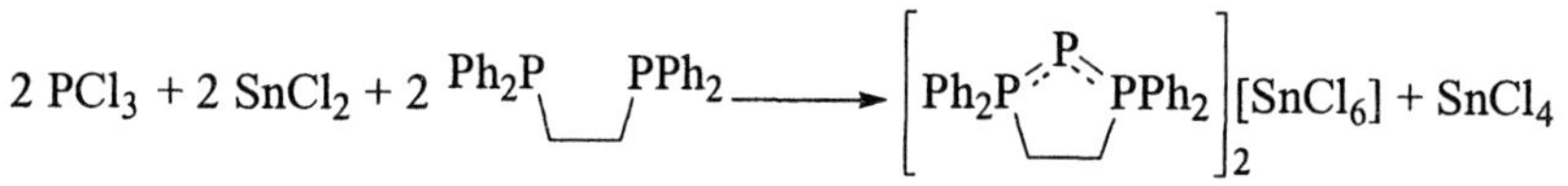

$$2\ PCl_3 + 2\ SnCl_2 + 2\ Ph_2P\frown PPh_2 \longrightarrow \left[Ph_2P\overset{P}{\frown}PPh_2\right]_2 [SnCl_6] + SnCl_4$$

Scheme 1

Schmidpeter and co-workers also demonstrated that, in certain cases, one additional equivalent of phosphine may be used to reduce phosphorus trichloride, instead of tin(II) chloride, as shown in Scheme 2. The stability of the salt is enhanced by either the presence of a chloride acceptor, such as $AlCl_3$, or by a concurrent metathesis anion exchange with, for example, [Na][BPh_4] (10,11). The tetrachloroaluminate salt of the [Ph_3P–P–PPh_3]$^+$ cation was structurally characterized and again the average P–P bond distance of 2.132 Å is intermediate between those of single and double P–P bonds (10).

Cations stabilized by PPh_3 may be converted to other cations by displacement of PPh_3 by more basic phosphines to produce new symmetric and asymmetric cations (*10*). Schmidpeter and Lochschmidt also showed that cations such as these can undergo substitution of the phosphines by more basic anions, X^-, such as CN^-, $SnPh_3^-$, PPh_2^- and $POPh_2^-$, which generate either neutral $R_3P=PX$ or anionic PX_2^- molecules with P(I) centers (*4,12*).

The symmetric P(I) stabilized cations contain a characteristic AX_2 spin system, which allows for easy identification of their structure. The dicoordinate P atom exhibits a triplet splitting pattern that has a chemical shift range of approximately $\delta = -156$ to -261, depending on the identity of the stabilizing phosphines. The tetracoordinate P atoms give rise to a doublet splitting pattern and range from $\delta = +12$ to $+104$. The $^1J_{P-P}$ coupling constants range from 347 to 566 Hz for both symmetrical and unsymmetrical examples, and the $^2J_{P-P}$ coupling constants range from 15 to 41 Hz in the unsymmetrical cases (*10,13*).

$$PCl_3 + 3\ PR_3 + 2\ AlCl_3 \longrightarrow [R_3P-P-PR_3][AlCl_4] + [R_3PCl][AlCl_4]$$

$$R = Ph, NMe_2$$

$$PCl_3 + 3\ PR_3 + 2\ [Na][BPh_4] \xrightarrow[-2\ NaCl]{} [R_3P-P-PR_3][BPh_4] + [R_3PCl][BPh_4]$$

Scheme 2

Acyclic "triphosphenium" cations may be oxidized by the addition of an organic chloride (or HCl) in the presence of an equivalent of $AlCl_3$ to produce dications as tetrachloroaluminate salts, in which the dicoordinate P(I) atom is oxidized from +1 to +3, as shown in Scheme 3A (*14*). With the exception of oxidation by HCl, cyclic cations were thought to be resistant to oxidization. The analogous cyclic dications may be produced, however, through chloride abstraction by two equivalents of $AlCl_3$ of a dichlorophosphine in the presence of a chelating phosphine, as shown in Scheme 3B (*14*). The oxidation of the dicoordinate P atom results in a deshielding of about 100 ppm for the chemical shift of the central P(I) atom, ranging form $\delta = -23$ to -157, and a shielding of 5 to 10 ppm of the tetracoordinate P atoms in the stabilizing ligands. The $^1J_{P-P}$ coupling constants decreased to range from 239 to 358 Hz.

It should be noted that the salt $[Ph_3P-P(H)-PPh_3][AlCl_4]_2$ was characterized by X-ray crystallography and the most important change in the metrical parameters upon oxidation is the increase in the P–P bond lengths to 2.205(1) Å and 2.224(1) Å, which are indicative of single bonds (*14*).

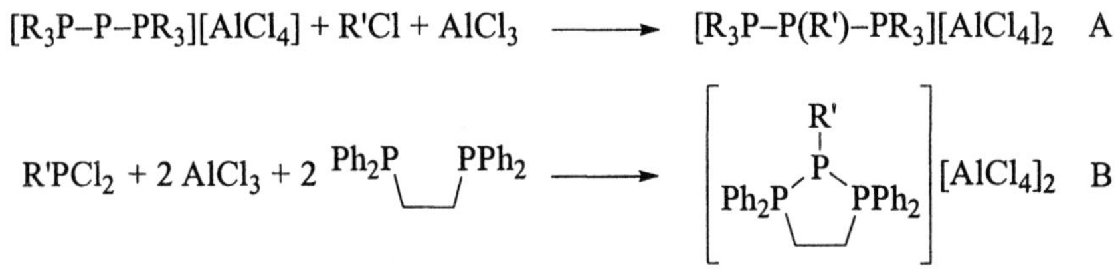

$[R_3P-P-PR_3][AlCl_4] + R'Cl + AlCl_3 \longrightarrow [R_3P-P(R')-PR_3][AlCl_4]_2$ A

$R'PCl_2 + 2\,AlCl_3 + 2\,Ph_2P\smile PPh_2 \longrightarrow \left[\begin{array}{c} R' \\ | \\ Ph_2P\diagup\overset{P}{}\diagdown PPh_2 \\ \smile \end{array}\right][AlCl_4]_2$ B

Scheme 3

Schmidpeter and co-workers also demonstrated the nucleophilicity of the P(I) cations through coordination of the cation to $AlCl_3$ (*11,15*). Coordination of the Lewis acid results in a deshielding of about 50 ppm for the resonance of the P(I) atom, and broadening of the signal. The chemical shift of the P(III) atoms were also shielded by an additional 15 ppm.

One of these P(I) cations has been suggested to act as a source of P^+, as shown by the insertion of P^+ into a C=C double bond, to generate 2-phosphaallylic cations (*16*). It was speculated that the P(I) cation undergoes electrophilic attack by an electron-rich olefin, followed by sequential loss of the two phosphines to result in a phosphaallylic cation, as shown in Scheme 4 (R = NMe_2, R' = Me, Et).

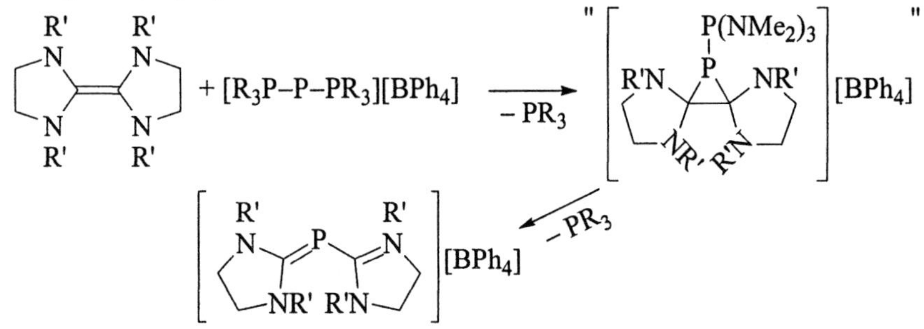

Scheme 4

Recent Developments

Almost 10 years following Schmidpeter's initial synthesis of the first P(I) cation, in 1993 Gamper and Schmidbaur extended the $SnCl_2$ reduction reaction to arsenic. By mixing equimolar amounts of a chelating diphosphine, tin(II) chloride and either arsenic or phosphorus trichloride, they synthesized both the

As(I) and P(I) cations (17). The arsenic compound was structurally characterized by single crystal X-ray diffraction. As with the phosphorus analogues, the As–P bond lengths of 2.250(1) Å and 2.244(1) Å are intermediate between those of As–P single bonds, P(As(C(O)^{t}Bu)$_2$)$_3$: As–P = 2.305 Å (18), and double bonds, Mes*P=AsCH(SiMe$_3$)$_2$: As=P = 2.124 Å (19). The phosphorus analogue may be deprotonated to form a zwitterionic species analogous to cyclic carbodiphosphoranes, as illustrated in Scheme 5.

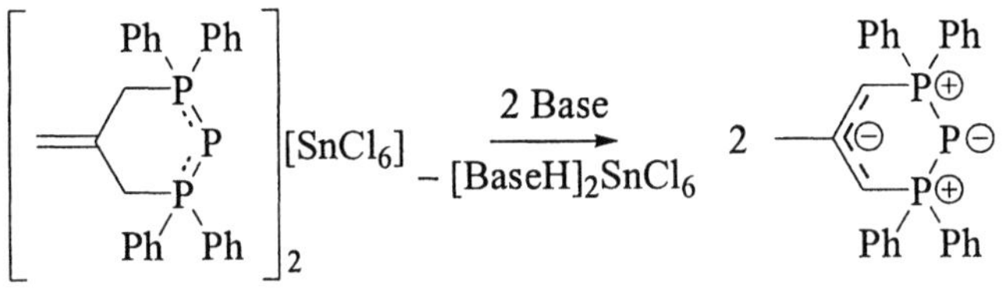

Scheme 5

Karsch and co-workers synthesized related zwitterions containing Pn(I) atoms by reduction of pnictogen trichlorides using LiC(PR$_2$)$_2$(SiMe$_3$) (LiCPRSi), R = Me or Ph, to generate four-, six- and eight-membered rings. The size of the ring depended upon the substituents on phosphorus and whether the equivalents of the lithium salt were added at once or stepwise, as shown in Scheme 6 (20). Both lithium salts reduced SbCl$_3$ and BiCl$_3$ to elemental Sb and Bi. When R = Me, AsCl$_3$ was reduced to a mixed valent species with an average oxidation state of 0.5 (Scheme 7).

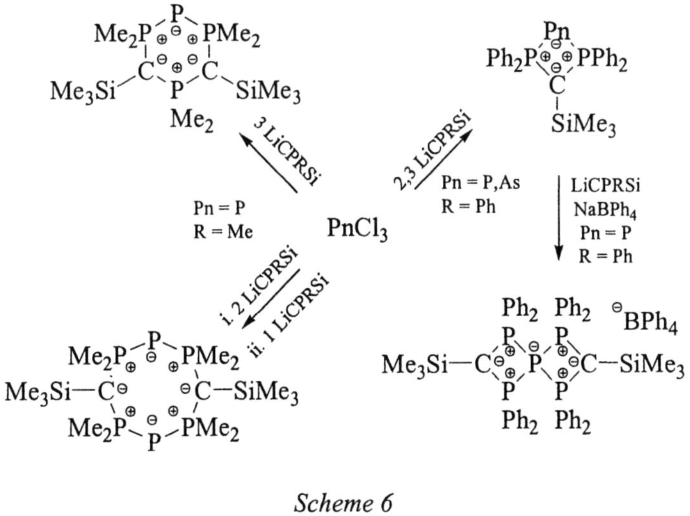

Scheme 6

$4 \text{ AsCl}_3 + 12 \text{ LiC(PMe}_2)_2(\text{SiMe}_3) \longrightarrow$

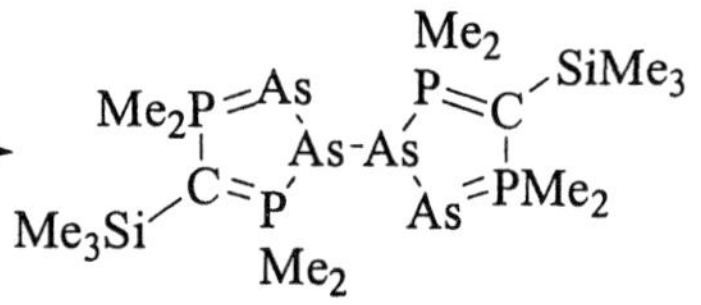

[structure diagram]

Scheme 7

Ellermann and co-workers concurrently produced similar salts using an analogous nitrogen-based on the lithium reagent, LiN(PPh$_2$)$_2$ *(21)*. The reactions were performed using pnictogen triiodides and resulted in the successful synthesis of an eight-membered zwitterionic ring and a seven-membered cationic ring for arsenic. The cation was structurally characterized as the iodide salt and the distances between As(I) and P(III) were again consistent with partial multiple-bond character. Complete reduction to the element was also observed for Sb and Bi.

In 2000, Dillon and co-workers continued to build on the seminal work of Schmidpeter. The group synthesized many new cyclic triphosphenium cations from PX$_3$ (X = Cl, Br, I) based on ^{31}P NMR data, and they structurally characterized the six-membered ring analogue of Schmidpeter's original five-membered ring, also as the hexachlorostannate salt *(22)*. They noted that the formation of the P(I) salt was observed whether or not a tin(II) halide was present in the reaction mixture. It was suggested that the reaction occurring involved the reduction of PX$_3$ and the oxidation of dppe to either [dppeX][X] (Scheme 8A) or [dppeX$_2$][X]$_2$ (Scheme 8B) depending on the stoichiometry. The oxidized diphosphine is evident by peaks in ^{31}P NMR spectra between δ = 66 to 30 ppm (for the chloro-systems) either as two doublets for [dppeX][X] or a singlet for [dppeX$_2$][X]$_2$.

$\text{PX}_3 + 2 \text{ dppe} \longrightarrow [(\text{dppe})\text{P}]\text{X} + [\text{dppeX}][\text{X}]$ A

$2 \text{ PX}_3 + 3 \text{ dppe} \longrightarrow 2 [(\text{dppe})\text{P}]\text{X} + [\text{dppeX}_2][\text{X}]_2$ B

Scheme 8

The following year, Dillon and co-workers reported additional P(I) cyclic cations and new cyclic As(I) cations *(23)*. They again showed that the halide salts could be made through stirring of PnX$_3$ (Pn = P, As) with diphosphines based on ^{31}P NMR data. However, only the hexachlorostannate salts of [(dppben)Pn]$^+$, Pn = P, As, dppben = 1,2-bis(diphenylphosphino)benzene, and the unusual stannate C$_2$H$_2$(Ph$_2$POSnCl$_5$)$_2^{-2}$ salt of [(dppE)As]$^+$, dppE = *cis*-1,2-bis(diphenylphosphino)ethene were structurally characterized. The Pn(I)–P(III) bonds lengths all lie in the expected region intermediate between single and double Pn–P bonds.

We have also recently reported a hexachlorostannate salt of a cyclic As(I), [(dppe)As]$_2$SnCl$_6$, the arsenic analogue to Schmidpter's first cyclic P(I) cation, prepared using his method (*24*). The X-ray crystal structure showed bond lengths or angles in good agreement with similar structures reported. The molecular structure is depicted in Figure 2.

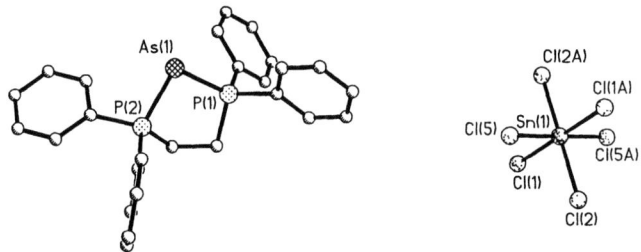

Figure 2. Molecular Structure of [(dppe)As]$_2$[SnCl$_6$]. Hydrogen Atoms and One Molecule of CH$_2$Cl$_2$ Have Been Removed for Clarity. (Adapted from Reference (24). Copyright 2004 Taylor & Francis.)

In our investigation of the synthesis of and chemistry iodide salts of P(I) and As(I) cyclic cations, however we have noticed that the oxidation of excess diphosphine is not part of the redox couple with the reduction of PnI$_3$. Rather the interaction of PnI$_3$ with diphosphines, such as dppe, helps to promote the reduction of P(III) to P(I) and the oxidation of iodide to iodine, as shown in Scheme 9. ^{31}P NMR spectra of the reaction mixtures exhibit no indication of either [dppeI][I] or [dppeI$_2$][I]$_2$ during the reaction (*5*). The iodine can be washed out of the system and colorless crystals of [(dppe)P][I] suitable for single crystal X-ray crystallography can be obtained by the slow evaporation of dichloromethane. The structure of one of the two independent molecules in the asymmetric unit is depicted in Figure 3. If the reaction is performed in donor solvents, such as THF or MeCN, there is evidence of the formation of the oxidized iodo-phosphonium iodide salt contaminants, presumably generated from the reactive iodine by-product.

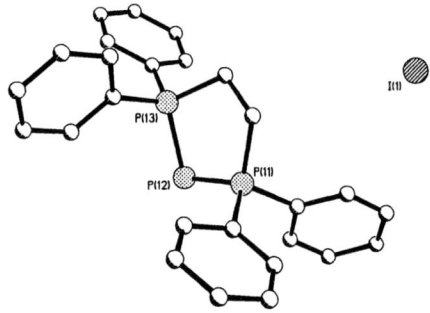

Figure 3. Molecular Structure of [(dppe)P][I]. Hydrogen Atoms Have Been Removed for Clarity. (Adapted from Reference (5). Copyright 2003 Royal Society of Chemistry.)

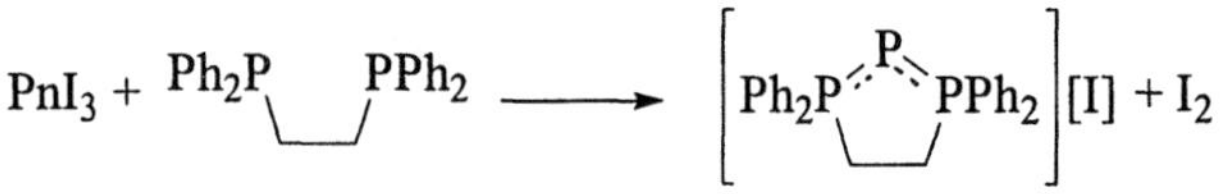

$$PnI_3 + Ph_2P\underset{}{}PPh_2 \longrightarrow \left[Ph_2P\overset{P}{\underset{}{}}PPh_2\right][I] + I_2$$

Scheme 9

The iodide salt of [(dppe)P]$^+$ is remarkably stable, even in the presence of H$_2$O, however reactivity studies of the P(I) cations can be complicated through side-reactions involving the I$^-$ anion. Thus, it is necessary to convert the iodide to a more robust, less reactive anion. This is conveniently achieved through simple metathesis reactions, as shown in Scheme 10. We have structurally characterized the tetraphenylborate salt of the [(dppe)P] cation, in which the cation has the expected metrical parameters in terms of bond lengths and angles (5). The molecular structure of [(dppe)P][BPh$_4$] in depicted in Figure 4.

$$[(dppe)Pn][I] + MX \longrightarrow [(dppe)Pn][X] + [M][I]\downarrow$$

Scheme 10

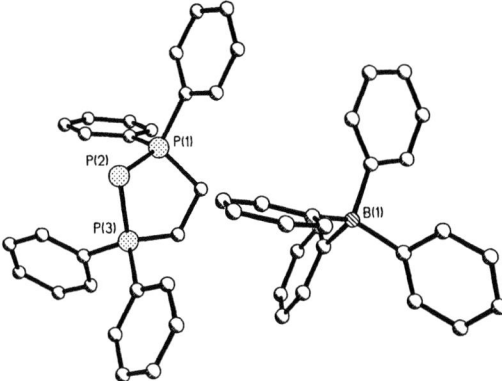

Figure 4. Molecular Structure of [(dppe)P][BPh$_4$]. Hydrogen Atoms Have Been Removed for Clarity. (Adapted from Reference (5). Copyright 2003 Royal Society of Chemistry.)

The [(dppe)P][BPh$_4$] salt is incredibly stable, even in the presence of air and moisture; we performed a computational investigation in an attempt to understand the origin of this surprising stability. Density functional theory (DFT) calculations on the model cation [(dmpe)P]$^+$ (dmpe = bis(dimethylphosphino)ethane) reveal backing-bonding of π-electrons from the dicoordinate P(I) to anti-bonding orbitals on the tetracoordinate P(III) centers. This increases the bond order, which is exhibited in the shortening of the bonds, relative to single bonds, noted during structural investigations. Furthermore, the removal of electron density effectively increases the oxidation state of the central phosphorus atom and explains the stability of the cation (5).

Just as Schmidpeter showed [R$_3$P–P–PR$_3$][BPh$_4$] (R = NMe$_2$) to be a source of P$^+$, in 1999 Driess and co-workers utilized the same salt with Schwartz's reagent, Cp$_2$ZrHCl, in the presence of base, to generate the unprecedented cationic four-coordinate, planar P(I) atom bound to four zirconicenes bridged by four hydrides Scheme 11 (25). Subsequently, Driess' group synthesized the As(I) salt [R$_3$P–As–PR$_3$][BPh$_4$] (R = NMe$_2$) using Schmidpeter's method, and showed that under similar conditions, but without any base, both the planar P(I) and As(I) salts could be synthesized from the As(I) salt, suggesting that it is a simultaneous source of both P$^+$ and As$^+$ (26,27).

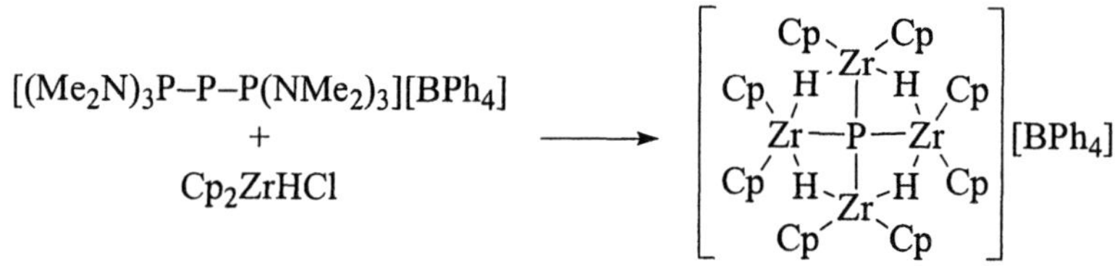

[(Me₂N)₃P–P–P(NMe₂)₃][BPh₄]
+
Cp₂ZrHCl

⟶

$\begin{bmatrix} Cp & Cp \\ & Zr & \\ Cp & H & H & Cp \\ & Zr-P-Zr & \\ Cp & H & H & Cp \\ & Zr & \\ Cp & Cp \end{bmatrix}$ [BPh₄]

Scheme 11

In a similar vein, we have shown that [(dppe)As]I can be used as a source of "As–I". Oxidation of the stabilizing diphosphine ligand appears to liberate As–I fragments, which oligomerize to form six-membered As rings in the chair conformation with I atoms in the equatorial positions. The ring is capped by two iodide ions (*28*) and the resulting structure is a thus a distorted As_6I_2 cube, as depicted in Figure 5.

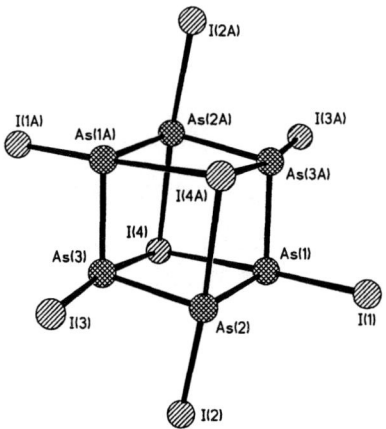

Figure 5. Molecular Structure of the $As_6I_8^{-2}$ Anion. (Adapted from Reference (28). Copyright 2004 American Chemical Society.)

Dillon and Olivey have recently extended the number cyclic dications analogous to Schimpeter's early dications. Whereas Schmidpeter and co-workers were only able to oxidize the cyclic P(I) cations with hydrochloric acid, methylation was also found to be successful. Although reaction with methyl iodide proved unsuccessful, the stronger methylating reagent methyl triflate resulted in successful oxidation of the P(I) atom based on ^{31}P NMR chemical shifts to higher frequency (*29*). Four equivalents of methyl triflate (MeOTf) were necessary to generate a significant amount of the oxidized dication, relative to the starting monocation, as shown in Scheme 12. Unfortunately, the only product that could be structurally characterized was the salt [dppeH$_2$][OTf]$_2$.

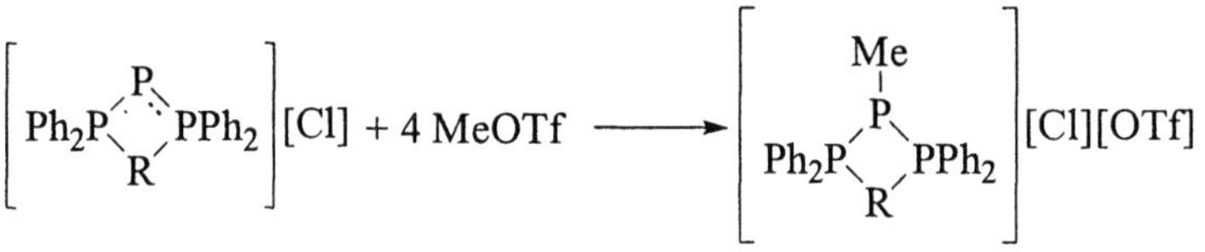

$$\left[Ph_2P \overset{P}{\underset{R}{\diagup\hspace{-0.5em}\diagdown}} PPh_2 \right][Cl] + 4\ MeOTf \longrightarrow \left[Ph_2P \overset{\overset{Me}{|}}{\underset{R}{\diagup\hspace{-0.5em}\diagdown}} PPh_2 \right][Cl][OTf]$$

Scheme 12

Shah and Protasiewicz have shown the utility of the closely related phosphine stabilized phosphinidenes, RP← PR$_3$, as phospha-Wittig reagents, as shown in Scheme 13. Reactions with aryl aldehydes successfully resulted in the generation of phosphaalkenes of the type ArHC=PR (*30*).

$$Ar-\overset{O}{\underset{H}{C}} + R_3P=PAr' \longrightarrow Ar-\overset{PAr'}{\underset{H}{C}} + O=PR_3$$

Scheme 13

We have investigated the utility of P(I) cations as possible phosha-Wittig reagents, using computational methods (DFT). Calculations indicate that phospha-Wittig reactions should be favorable for reactions of P(I) cations with aldehydes and even more favorable (by ~100 kJ mol^{-1}) with epoxides (*24*). Some potential types of products from these reactions are depicted in Scheme 14. The reactivity of these cations should be enhanced by changing the alkyl phosphines to either aminophosphines or phosphites, based on the computational results. Following the first metathesis reaction there is still another P–P bond available for a further phospha-Wittig reaction, which makes these cations potential diphospha-Wittig reagents. We are currently undertaking experimental investigations to verify the computational predictions.

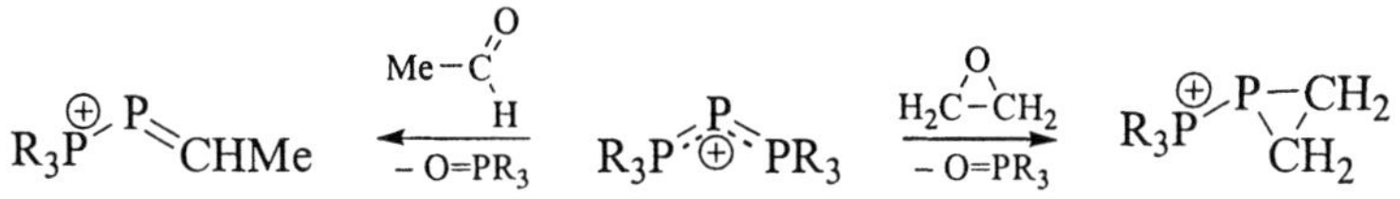

$$R_3\overset{\oplus}{P}\diagdown_{CHMe}^{P} \xleftarrow[-O=PR_3]{Me-\overset{O}{\underset{H}{C}}} R_3P\overset{P}{\underset{\oplus}{\diagup\hspace{-0.5em}\diagdown}}PR_3 \xrightarrow[-O=PR_3]{H_2C-\overset{O}{\diagdown}CH_2} R_3\overset{\oplus}{P}\overset{P-CH_2}{\underset{CH_2}{\diagdown\diagup}}$$

Scheme 14

Conclusions

Recent progress in the chemistry of low oxidation state Pn(I) cations provides new and convenient synthetic methods for the production of such stable low oxidation state compounds. The facile production of such compounds has allowed for the development of these species as useful reagents; of particular note is their unprecedented use as a source of "Pn$^+$" fragments. The exploitation of the unique reactivity of these Pn(I) cations has only begun to be investigated and continuing research should prove these cations to be versatile reagents for the synthesis of new compounds and valuable materials.

References

1. Macdonald, C. L. B.; Ellis, B. D. In *Encyclopedia of Inorganic Chemsitry*; 2nd ed.; King, R. B., Ed.; John Wiley & Sons Ltd., 2004.
2. Lammertsma, K. *Top. Curr. Chem.* **2003**, *229*, 95-119.
3. Cowley, A. H. *Acc. Chem. Res.* **1997**, *30*, 445-451.
4. Schmidpeter, A.; Lochschmidt, S. *Inorg. Synth.* **1990**, *27*, 253-8.
5. Ellis, B. D.; Carlesimo, M.; Macdonald, C. L. B. *Chem. Commun.* **2003**, 1946-1947.
6. Gudat, D. *Coord. Chem. Rev.* **1997**, *163*, 71-106.
7. Schmidpeter, A.; Lochschmidt, S.; Sheldrick, W. S. *Angew. Chem., Int. Ed. Engl.* **1982**, *21*, 63-64.
8. Dashti-Mommertz, A.; Neumuller, B. *Z. Anorg. Allg. Chem.* **1999**, *625*, 954-960.
9. Cowley, A. H.; Decken, A.; Norman, N. C.; Krueger, C.; Lutz, F.; Jacobsen, H.; Ziegler, T. *J. Am. Chem. Soc.* **1997**, *119*, 3389-3390.
10. Schmidpeter, A.; Lochschmidt, S.; Sheldrick, W. S. *Angew. Chem., Int. Ed. Engl.* **1985**, *24*, 226-227.
11. Schmidpeter, A.; Lochschmidt, S. *Angew. Chem., Int. Ed. Engl.* **1986**, *25*, 253-4.
12. Lochschmidt, S.; Schmidpeter, A. *Z. Naturforsch.* **1985**, *40B*, 765-73.
13. Schmidpeter, A. In *Multiple Bonds and Low Coordination in Phosphorus Chemistry*; Regitz, M., Scherer, O. J., Eds.; Thieme Medical Publishers, Inc.: New York, 1990.
14. Schmidpeter, A.; Lochschmidt, S.; Karaghiosoff, K.; Sheldrick, W. S. *J. Chem. Soc., Chem. Commun.* **1985**, 1447-8.
15. Lochschmidt, S.; Mueller, G.; Huber, B.; Schmidpeter, A. *Z. Naturforsch.* **1986**, *41B*, 444-54.
16. Schmidpeter, A.; Lochschmidt, S.; Willhalm, A. *Angew. Chem., Int. Ed. Engl.* **1983**, *22*, 545-6.

17. Gamper, S. F.; Schmidbaur, H. *Chem. Ber.* **1993**, *126*, 601-4.
18. Jones, C.; Junk, P. C.; Williams, T. C. *J. Chem. Soc., Dalton Trans.* **2002**, 2417-2418.
19. Cowley, A. H.; Kilduff, J. E.; Lasch, J. G.; Mehrotra, S. K.; Norman, N. C.; Pakulski, M.; Whittlesey, B. R.; Atwood, J. L.; Hunter, W. E. *Inorg. Chem.* **1984**, *23*, 2582-93.
20. Karsch, H. H.; Witt, E. *J. Organomet. Chem.* **1997**, *529*, 151-169.
21. Dotzler, M.; Schmidt, A.; Ellermann, J.; Knoch, F. A.; Moll, M.; Bauer, W. *Polyhedron* **1996**, *15*, 4425-4433.
22. Boon, J. A.; Byers, H. L.; Dillon, K. B.; Goeta, A. E.; Longbottom, D. A. *Heteroat. Chem.* **2000**, *11*, 226-231.
23. Barnham, R. J.; Deng, R. M. K.; Dillon, K. B.; Goeta, A. E.; Howard, J. A. K.; Puschmann, H. *Heteroat. Chem.* **2001**, *12*, 501-510.
24. Ellis, B. D.; Macdonald, C. L. B. *Phosphorus, Sulfur Silicon Relat. Elem.* **2004**, *179*, 775-778.
25. Driess, M.; Aust, J.; Merz, K.; Van Wullen, C. *Angew. Chem. Int. Ed.* **1999**, *38*, 3677-3680.
26. Driess, M.; Ackermann, H.; Aust, J.; Merz, K.; Von Wullen, C. *Angew. Chem. Int. Ed.* **2002**, *41*, 450-453.
27. Ackermann, H.; Aust, J.; Driess, M.; Merz, K.; Monse, C.; Van Wullen, C. *Phosphorus, Sulfur Silicon Relat. Elem.* **2002**, *177*, 1613-1616.
28. Ellis, B. D.; Macdonald, C. L. B. *Inorg. Chem.* **2004**, *in press*.
29. Dillon, K. B.; Olivey, R. J. *Heteroat. Chem.* **2004**, *15*, 150-154.
30. Shah, S.; Protasiewicz, J. D. *Coord. Chem. Rev.* **2000**, *210*, 181-201.

Chapter 9

A General Route to M_4N_4 Heterocubanes: Synthesis and Crystal Structure of $[M(\mu_3\ NSiMe_3)]_4$ (M=Ge, Pb)

Jack F. Eichler, Oliver Just, and William S. Rees, Jr.*

School of Chemistry and Biochemistry, and School of Materials Science and Engineering, and Molecular Design Institute, Georgia Institute of Technology, Atlanta, GA 30332–0400

Novel germanium- and lead-nitrogen heterocubanes containing the trimethylsilyl moiety, $[M(\mu_3\text{-NSiMe}_3)]_4$ (M = Ge, Pb), obtained in the reaction of N-lithio(trimethylstannyl)(trimethylsilyl)amine with MCl_2 (M = Ge, Pb), have been structurally characterized by single crystal X-ray diffraction studies. Structural comparison to the previously reported $[Sn(\mu_3\text{-NSiMe}_3)]_4$ reveals a trend of increased distortion in the M_4N_4 tetrameric core with the heavier congeners of the Group 14 series.

The Sn_4N_4 heterocubanes, described previously in the literature, have been prepared by a variety of synthetic pathways (1). However, the heavier analogues containing germanium or lead are not commonly found, with structurally characterized examples being limited to two lead, $[PbN(C_6H_{11})]_4$ (2) and $[PbN(2,6-i-Pr_2C_6H_3)]_4$, (3) and one germanium species, $[GeN(C_6H_5)]_4$ (4). It has recently been demonstrated that reaction of lithiated stannyl amines with $SnCl_2$ affords the tin-nitrogen cubane structural motif (5). Thus, it is of interest to determine if reactions of lithiated stannyl amines with $PbCl_2$ and $GeCl_2$ result in the corresponding lead and germanium heterocubanes. This report discusses the synthesis and structural characterization of the rare germanium-nitrogen, $[Ge(\mu_3\text{-}NSiMe_3)]_4$ (2) and lead-nitrogen cubanes $[Pb(\mu_3\text{-}NSiMe_3)]_4$ (3), as well as the trimethylstannyl-trimethylsilyl lithium amide species $[(Me_3Sn)(Me_3Si)NLi\cdot Et_2O]_2$ (1).

Results and Discussion

The preparation of compound **1** is depicted in Scheme 1. The synthesis of the THF adduct of **1** was previously reported (5); however, a satisfactory X-ray structure was not obtained. The alternate synthetic route reported here was explored in an attempt to obtain higher quality crystals. Me_3SiCl was added dropwise to a diethyl ether solution of $[(Me_3Sn)_2NLi\cdot THF]_2$ (1/2 equivalent) at 0°C, and subsequently stirred overnight at room temperature. The colorless solution was filtered to remove the LiCl precipitate, and one equivalent of MeLi was added drop-wise *in situ* to this filtrate at −30°C. The solution was then allowed to attain ambient temperature and stirred overnight. After reducing the volume of the clear, colorless solution, **1** was recrystallized directly from diethyl ether at −40°C.[‡] Scheme 2 illustrates the synthesis of compounds **2** and **3**. Both species were prepared by adding a diethyl ether solution of **1** (1/2 equivalent) to a diethyl ether slurry of the appropriate metal dichloride at 0°C. After stirring overnight at room temperature, the orange (**2**) or yellow (**3**) reaction mixture was filtered to remove LiCl and Me_3SnCl. Recrystallization from approximately 15 mL of CH_2Cl_2 at -80°C produced orange (**2**)[§] or yellow (**3**)[¥] crystals. Elemental analyses of compounds **1-3** confirmed the formulae given in Schemes 1 and 2 and the single crystal X-ray diffraction studies of **1-3** revealed the structures depicted therein.

The crystal structures of **2** and **3** are shown in Fig. 1[§] and 2[¥] respectively. The incorporation of the trimethylsilyl moiety into germanium- or lead-nitrogen cubanes has to our knowledge not been previously reported. However, the configuration of the central M_4N_4 skeleton does not differ drastically from other

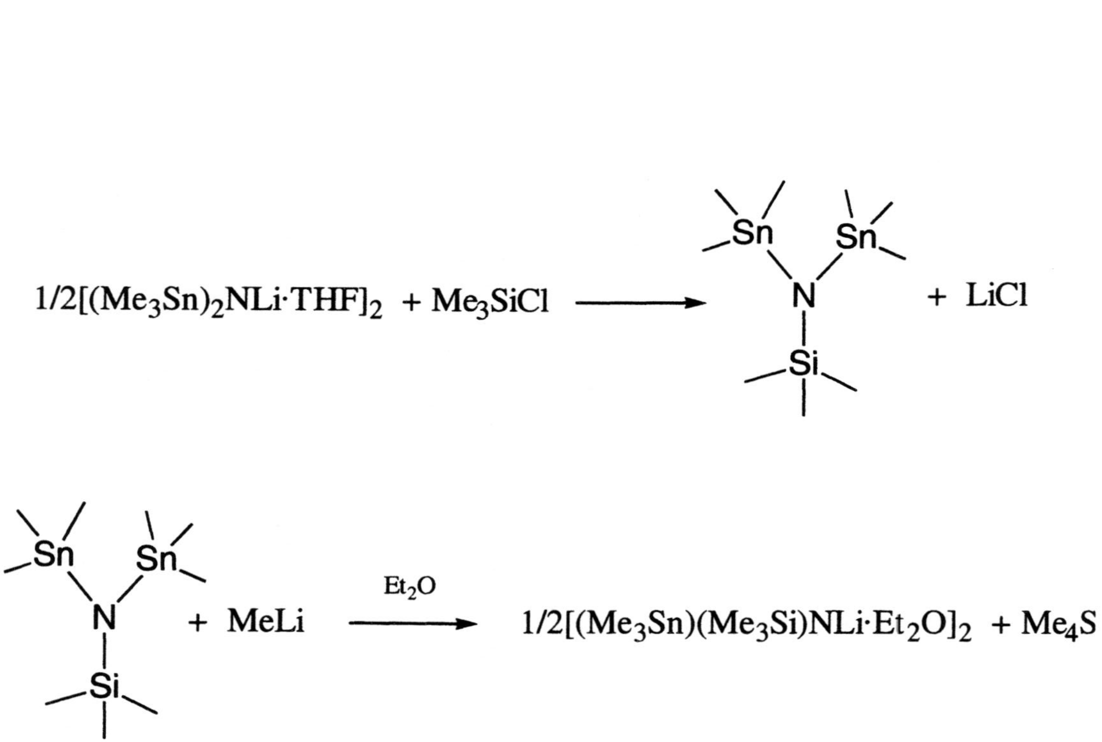

$1/2[(Me_3Sn)_2NLi \cdot THF]_2 + Me_3SiCl \longrightarrow$ (Me_3Sn)_2N-SiMe_3 + LiCl

(Me_3Sn)_2N-SiMe_3 + MeLi $\xrightarrow{Et_2O}$ $1/2[(Me_3Sn)(Me_3Si)NLi \cdot Et_2O]_2 + Me_4Sn$

Scheme 1: Synthesis of [(Me_3Sn)(Me_3Si)NLi·Et_2O]_2 (1)

1/2[(Me$_3$Sn)(Me$_3$Si)NLi·Et$_2$O]$_2$ + MCl$_2$ $\xrightarrow{}$ 1/4 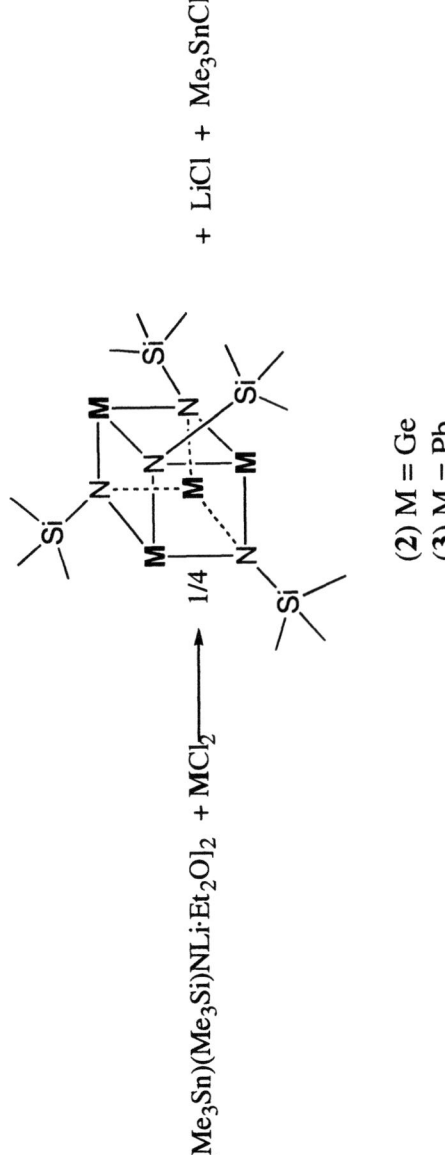 + LiCl + Me$_3$SnCl

(2) M = Ge
(3) M = Pb

Scheme 2: Synthesis of [Ge(μ_3-NSiMe$_3$)]$_4$ (2) and [Pb(μ_3-NSiMe$_3$)]$_4$ (3)

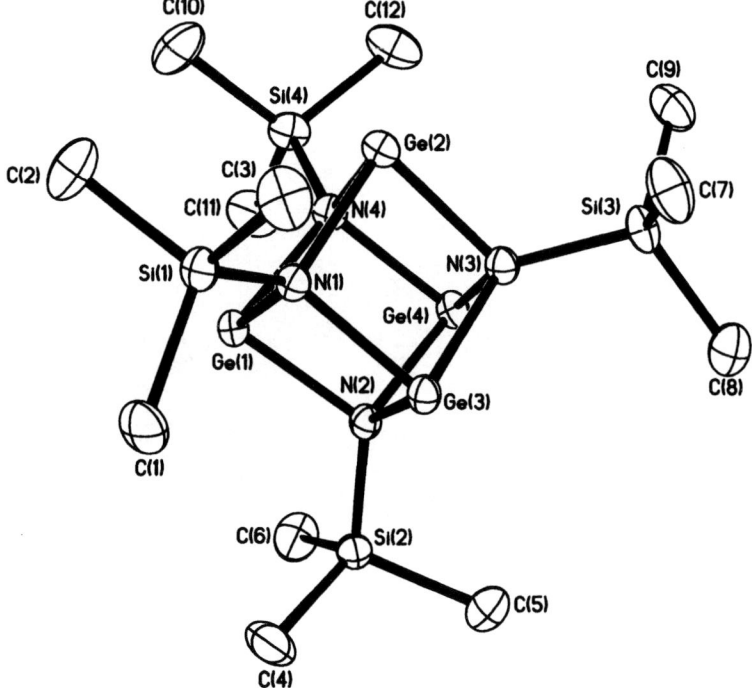

Fig. 1 ORTEP representation (30% probability) of [Ge(μ_3-NSiMe$_3$)]$_4$ (**2**); hydrogen atoms omitted for clarity.

127

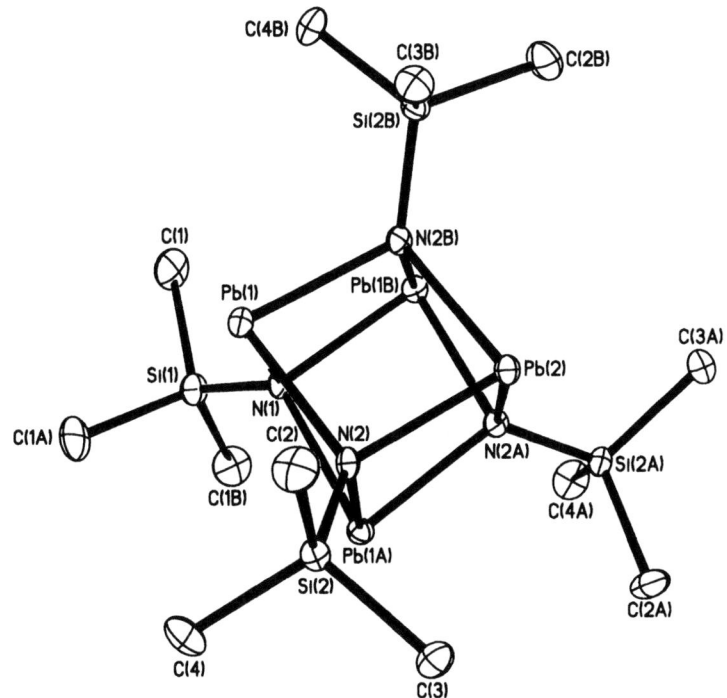

Fig. 2 ORTEP representation (30% probability) of [Pb(μ_3-NSiMe$_3$)]$_4$ (**3**); hydrogen atoms omitted for clarity.

Group 14-nitrogen cubane species formerly described (*1-4*). The M_4N_4 core for both **2** and **3** exhibits a distortion from a perfect cube, as evidenced by the average angles of 84.06° (N-Ge-N) and 96.65° (Ge-N-Ge) in **2**, as well as the average angles of 81.8° (N-Pb-N) and 97.4° (Pb-N-Pb) in **3**. This type of distortion is common for similar Group 14-nitrogen cubanes. Comparison to the previously reported, structurally analogous tin-nitrogen cubane, [Sn(μ_3-NSiMe$_3$)]$_4$ (*5,6*) (average N-Sn-N = 82.27° and average Sn-N-Sn = 97.26°) indicates an increasing trend of distortion from perfect cube geometry from germanium to lead. This can be attributed to the decrease in sp-hybridization in the heavier tin and lead atoms. A summary of the angles found in the M_4N_4 tetrameric core is given in Table 1. The average Ge-N and Pb-N interatomic distances of 2.013 Å in **2** and 2.304 Å in **3** are comparable to those found in previous reports: [GeNPh]$_4$, 2.019 Å (*4*); [PbNCy]$_4$, 2.303 Å (*2*); and [PbN(2,6-*i*-Pr$_2$C$_6$H$_3$)]$_4$, 2.337 Å (*3*). The distorted tetrahedral geometry about the silicon atoms in **2** and **3**, with average Si-N interatomic distances of 1.742 and 1.704 Å, respectively, is in agreement with previous structures containing the trimethylsilyl moiety in a similar chemical environment: (Me$_3$C)Al$_2$Li$_2$[μ_3-N(SiMe$_3$)]$_4$, Si-N$_{avg}$ = 1.691 Å (*7*); [MeGa(μ_3-NSiMe$_3$)]$_4$, Si-N$_{avg}$ = 1.727 Å (*8*); [MeIn(μ_3-NSiMe$_3$)]$_2$·[Li(Me$_3$Si)N-NH^{t}Bu]$_2$, Si-N$_{avg}$ = 1.743 Å (*9*).

The crystal structure of **1** is depicted in Fig. 3.‡ The compound crystallizes as a lithium dimer, with bridging (trimethylstannyl)(trimethylsilyl)amine moieties and terminal diethyl ether molecules. This species is iso-structural with the bis-trimethylstannyl derivatives possessing coordinating THF (*5*) or ^{t}BuOMe (*10*) molecules. It is noted that the tin and silicon atoms are crystallographically indistinguishable, with an identical interatomic distance of 1.919 Å for all Sn-N and Si-N interactions, which lies between the corresponding interatomic distances observed in the structurally analogous compounds: [(Me$_3$Sn)$_2$NLi·THF]$_2$, Sn-N$_{avg}$ = 2.093 Å (*5*); [(Me$_3$Si)$_2$NLi·Et$_2$O]$_2$, Si-N$_{avg}$ = 1.705 Å (*11*). A summary of all relevant interatomic distances and bond angles for Compounds **1-3** can be found in Table 2. The identity of the heteroleptic species, **1**, that includes both tin and silicon is verified by the elemental analysis and multinuclear NMR.‡

Compounds **2** and **3** are rare examples of Ge- and Pb-nitrogen heterocubanes and represent the first report of either class of compounds that contain the trimethylsilyl moiety. The reaction pathway used to obtain **2** and **3** differs from previous protocol for synthesizing Ge- or Pb-nitrogen cubane assemblies and provides a potential avenue to obtain additional examples of such species.

Cowley Acknowledgement

The authors dedicate this manuscript to Professor Alan Cowley on the wondrous occasion of his birthday, and for his personal impact on both main group chemistry in the U.S., and the career of one of us [William S. Rees, Jr.].

Table 1: Metal-N-Metal and N-Metal-N angles in the Group 14-Nitrogen heterocubane series (*mean angle)

$[M(\mu_3\text{-}NSiMe_3)]_4$	M-N-M*	N-M-N*
M = Ge	96.65(12)°	84.06(12)°
M = Sn	97.26(14)°	82.27(15)°
M = Pb	97.4(3)°	81.8(3)°

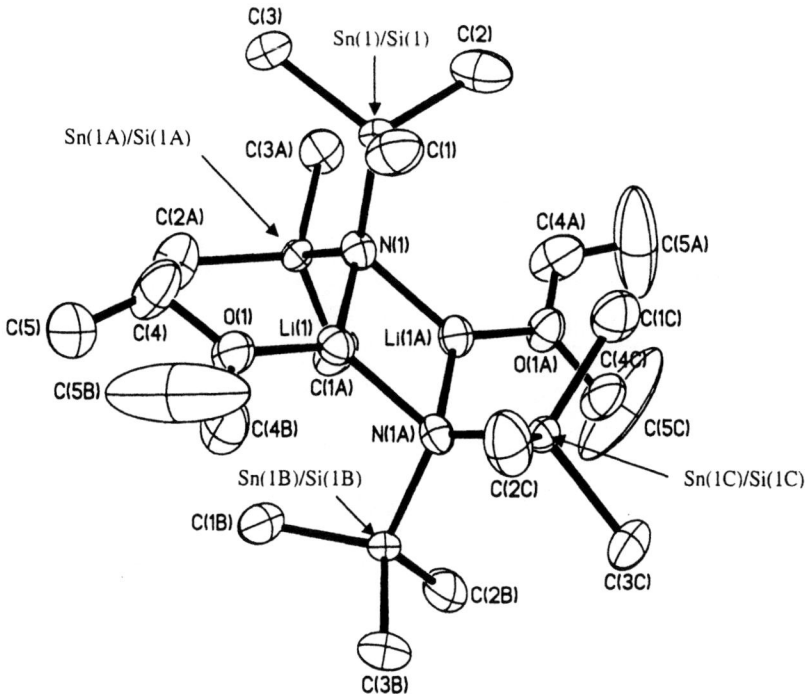

Fig. 3 ORTEP representation (30% probability) of [Li(μ_2-N(SiMe$_3$)(SnMe$_3$)·Et$_2$O]$_2$ (**1**); hydrogen atoms omitted for clarity.

Table 2: Summary of interatomic distances and bond angles for Compounds **1-3**

Atoms	Distance	Atoms	Angle
Ge(1)-N(1)	2.010(3)	Ge(2)-N(1)-Ge(3)	95.52(13)
Ge(1)-N(2)	2.013(3)	Ge(3)-N(2)-Ge(4)	95.72(12)
Ge(1)-N(4)	2.013(3)	Ge(3)-N(2)-Ge(1)	95.94(12)
Ge(2)-N(1)	2.014(3)	Ge(4)-N(2)-Ge(1)	95.76(12)
Ge(2)-N(4)	2.015(3)	Ge(4)-N(3)-Ge(3)	95.71(13)
Ge(2)-N(3)	2.016(3)	Ge(4)-N(3)-Ge(2)	95.83(13)
Ge(3)-N(2)	2.006(3)	Ge(3)-N(3)-Ge(2)	95.52(13)
Ge(3)-N(3)	2.012(3)	Ge(1)-N(4)-Ge(2)	95.35(11)
Ge(3)-N(1)	2.015(3)	Ge(1)-N(4)-Ge(4)	95.61(12)
Ge(4)-N(3)	2.007(3)	Ge(2)-N(4)-Ge(4)	95.55(12)
Ge(4)-N(2)	2.013(3)	Ge(1)-N(1)-Ge(3)	95.78(13)
Ge(4)-N(4)	2.018(3)	N(1)-Si(1)-C(3)	109.7(2)
Si(1)-N(1)	1.740(3)	N(1)-Si(1)-C(2)	109.09(17)
Si(2)-N(2)	1.748(3)	C(3)-Si(1)-C(2)	110.0(2)
Si(3)-N(3)	1.745(3)	N(1)-Si(1)-C(1)	107.91(16)
Si(4)-N(4)	1.735(3)	C(3)-Si(1)-C(1)	109.6(2)
		C(2)-Si(1)-C(1)	110.6(2)
		N(2)-Si(2)-C(4)	108.86(18)
		N(2)-Si(2)-C(5)	108.07(18)
		C(4)-Si(2)-C(5)	109.0(2)
		N(2)-Si(2)-C(6)	108.93(19)
		C(5)-Si(2)-C(6)	111.1(2)
		N(3)-Si(3)-C(7)	108.8(2)
		N(3)-Si(3)-C(8)	109.04(18)
		C(7)-Si(3)-C(8)	110.8(2)
		N(3)-Si(3)-C(9)	107.93(18)
		C(7)-Si(3)-C(9)	110.9(2)
		C(8)-Si(3)-C(9)	109.4(2)
		N(4)-Si(4)-C(10)	108.94(19)
		N(4)-Si(4)-C(11)	108.17(19)
		C(10)-Si(4)-C(11)	110.8(3)
		N(4)-Si(4)-C(12)	108.86(19)
		C(10)-Si(4)-C(12)	109.7(2)
		C(11)-Si(4)-C(12)	110.3(2)
		N(1)-Ge(1)-N(2)	83.84(12)
		N(1)-Ge(1)-N(4)	84.36(11)
		N(2)-Ge(1)-N(4)	84.07(13)
		N(1)-Ge(2)-N(4)	84.22(12)
		N(4)-Ge(2)-N(3)	83.96(13)
		N(2)-Ge(3)-N(3)	83.99(13)
		N(2)-Ge(3)-N(1)	83.90(12)
		N(3)-Ge(3)-N(1)	84.22(11)
		N(1)-Ge(2)-N(3)	84.13(11)
		N(3)-Ge(4)-N(2)	83.95(11)
		N(3)-Ge(4)-N(4)	84.12(11)
		N(2)-Ge(4)-N(4)	83.96(13)
		Ge(1)-N(1)-Ge(2)	95.46(12)

Table 3: Interatomic Distances [Å] and Bond Angles [deg] for Compound 3.

Atoms	Distance	Atoms	Angle
Pb(1)-N(2)	2.298(7)	N(2)-Pb(1)-N(2A)	81.4(4)
Pb(1)-N(2A)	2.305(7)	N(2)-Pb(1)-N(1)	82.3(3)
Pb(1)-N(1)	2.312(6)	N(2A)-Pb(1)-N(1)	82.1(3)
Pb(2)-N(2A)	2.298(7)	N(2A)-Pb(2)-N(2)	81.6(3)
Pb(2)-N(2)	2.298(7)	N(2A)-Pb(2)-N(2B)	81.6(3)
Pb(2)-N(2B)	2.298(7)	N(2)-Pb(2)-N(2B)	81.6(3)
Si(1)-N(1)	1.693(13)	Pb(1)-N(1)-Pb(1B)	97.0(3)
Si(2)-N(2)	1.716(9)	Pb(1)-N(1)-Pb(1A)	97.0(3)
N(1)-Pb(1A)	2.312(6)	Pb(1B)-N(1)-Pb(1A)	97.0(3)
N(1)-Pb(1A)	2.312(6)	Pb(2)-N(2)-Pb(1)	98.0(3)
N(2)-Pb(1B)	2.305(7)	Pb(2)-N(2)-Pb(1B)	97.8(3)
		Pb(1)-N(2)-Pb(1B)	97.5(3)
		N(1)-Si(1)-C(1B)	110.8(4)
		N(1)-Si(1)-C(1)	110.8(4)
		C(1B)-Si(1)-C(1)	108.1(4)
		N(1)-Si(1)-C(1A)	110.8(4)
		C(1B)-Si(1)-C(1A)	108.1(4)
		C(1)-Si(1)-C(1A)	108.1(4)
		N(2)-Si(2)-C(2)	110.0(4)
		N(2)-Si(2)-C(4)	110.6(4)
		C(2)-Si(2)-C(4)	109.7(5)
		N(2)-Si(2)-C(3)	109.7(4)
		C(2)-Si(2)-C(3)	106.8(5)
		C(4)-Si(2)-C(3)	109.9(5)

Table 4: Interatomic Distances [Å] and Angles [deg] for Compound **1**.

Atoms	Distance	Atoms	Angle
Sn(1)-N(1)	1.919(3)	Si(1A)-N(1)-Sn(1A)	0.00(3)
N(1)-Si(1A)	1.919(3)	Si(1A)-N(1)-Sn(1)	120.8(3)
N(1)-Sn(1A)	1.919(3)	Sn(1A)-N(1)-Sn(1)	120.8(3)
N(1)-Li(1)	1.987(9)	Si(1A)-N(1)-Li(1)	119.50(11)
N(1)-Li(1A)	1.987(9)	Sn(1A)-N(1)-Li(1)	119.50(11)
Li(1)-O(1)	1.921(14)	Sn(1)-N(1)-Li(1)	106.91(15)
Li(1)-N(1A)	1.987(9)	Si(1A)-N(1)-Li(1A)	106.91(15)
		Sn(1A)-N(1)-Li(1A)	106.91(15)
		Sn(1)-N(1)-Li(1A)	119.50(11)
		Li(1)-N(1)-Li(1A)	75.0(6)
		O(1)-Li(1)-N(1)	127.5(3)
		O(1)-Li(1)-N(1B)	127.5(3)
		N(1)-Li(1)-N(1B)	105.0(6)

[*] We gratefully acknowledge partial support of this project by the Georgia Institute of Technology Molecular Design Institute, under prime contract N00014-95-1-1116 from the Office of Naval Research.
[**] All correspondence should be directed to this author.

Notes

All manipulations were carried out in a dry atmosphere glovebox or by standard Schlenk techniques. All solvents were dried over Na° metal or P_2O_5 (CH_2Cl_2) and were freshly distilled under an inert atmosphere prior to use. $PbCl_2$ and Me_3SiCl reagents were purchased from Alfa Aesar and used without further purification. $GeCl_2 \cdot$(dioxane) (12) and $(Me_3Sn)_2NLi \cdot THF$ (5), (13) were prepared according to literature procedures. All NMR experiments were performed on a Bruker 400 MHz spectrometer at 300K using C_6D_6 solvent [^{1}H (400 MHz), ^{13}C (100.6 MHz), ^{29}Si (79.5 MHz), ^{119}Sn (149.3 MHz), and ^{7}Li (155.5 MHz)]. ^{1}H, ^{13}C, and ^{29}Si spectra were referenced to TMS. ^{119}Sn and ^{7}Li spectra were externally referenced to Me_4Sn and LiBr respectively. All UV/VIS measurements were performed on an Olis, Cary-14 spectrophotometer in dry hexane solvent.

‡ Selected data for **1**. $C_{20}H_{56}N_2O_2Li_2Si_2Sn_2$: C, 36.17; H, 8.50; N, 4.22%. Found: C, 35.92; H, 8.38; N, 4.28%. ^{1}H NMR: δ 0.28 (s, 18H, CH_3Sn), 0.30 (s, 18H, CH_3Si), 0.98 (t, 12H, CH_3), 3.46 (q, 8H, CH_3); ^{13}C NMR: δ −1.2 (CH_3Sn), 5.9 (CH_3Si), 14.4 (CH_3), 63.8 (CH_2); ^{29}Si NMR: δ −12.6; ^{119}Sn NMR: δ 34.2; ^{7}Li NMR: δ 1.38.

Crystal data for **1**. $C_{20}H_{56}N_2O_2Li_2Si_2Sn_2$: M_r = 664.11 g cm^{-3}, crystal dimensions 0.37 x 0.27 x 0.17 mm, tetragonal, space group P-4n2, a = 9.788(3), c = 17.406(9) Å, β = 90°; V = 1667.7(11) Å^3, Z = 2, ρ_{calcd} = 1.323 g cm^{-3}, Siemens SMART CCD diffractometer, 2.34 ≤ θ ≤ 28.71°, Mo$_{K\alpha}$ radiation (λ = 0.71073 Å), ω scans, T = 193(2) K; of 9,799 measured reflections, 2,074 were independent and 1,739 observed with $I > 2\sigma(I)$, -13 ≤ h ≤ 13, -12 ≤ k ≤ 10, -23 ≤ l ≤ 15; R_1 = 0.0434, wR_2 = 0.0953, GOF = 1.061 for 79 parameters, $\Delta\rho_{max}$ = 0.552 eÅ^{-3}. The structure was solved by direct methods (SHELXS-97) and refined by full-matrix least-squares procedures (SHELXL-97), Lorentz polarization corrections and absorption correction (SADABS) were applied, μ = 1.585 mm^{-1}, min./max. transmission 0.7744/0.5887.

§ Selected data for **2**. $C_{12}H_{36}N_4Ge_4Si_4$: C, 22.55; H, 5.68; N, 8.76%. Found: C, 23.56; H, 5.24; N, 9.10%. ^{1}H NMR: δ 0.31 (s, CH_3); ^{13}C NMR: δ −1.4 (CH_3). UV/VIS: λ_{max} = 280 nm. This crystal data structure was deposited in CCDC, reference number 249797.

Crystal data for **2**. Yield: 0.11 g (29%). $C_{15.5}H_{40}N_4Ge_4Si_4$: M_r = 685.23 g cm^{-3}, crystal dimensions 0.56 x 0.20 x 0.17 mm, monoclinic, space group $C2/c$, a

= 25.748(5), b = 11.878(2), c = 21.478(4) Å, α = 90, β = 110.690°; V = 6145(2) Å^3, Z = 8, ρ_{calcd} = 1.481 g cm^{-3}, Siemens SMART CCD diffractometer, 1.69 ≤ θ ≤ 28.70°, Mo$_{K\alpha}$ radiation (λ = 0.71073 Å), ω scans, T = 193(2) K; of 18,078 measured reflections, 7,251 were independent and 5,531 observed with I > 2σ(I), -29 ≤ h ≤ 33, -14 ≤ k ≤ 15, -28 ≤ l ≤ 12; R_1 = 0.0506, wR_2 = 0.1330, GOF = 1.057 for 269 parameters, $\Delta\rho_{max}$ = 1.412 eÅ^{-3}. The structure was solved by direct methods (SHELXS-97) and refined by full-matrix least-squares procedures (SHELXL-97), Lorentz polarization corrections and absorption correction (SADABS) were applied, μ = 4.041 mm^{-1}, min./max. transmission 0.5466/0.2102. This crystal data structure was deposited in CCDC, reference number 249296.

¥ Selected data for **3**. Yield: 0.25 g (38%). $C_{12}H_{36}N_4Pb_4Si_4$: C, 12.24; H, 3.08; N, 4.76%. Found: C, 12.19; H, 3.09; N, 4.89%. ^{1}H NMR: δ 0.02 (s, CH$_3$); ^{13}C NMR: δ -0.3 (CH$_3$). MS (EI, 274°C): m/z 1178 [M$^+$]. IR (nujol): 1255 (m), 1242 (m), 864 (m), 826 (s), 741 (m), 523 (m) cm.$^{-1}$ UV/VIS: max. absorbance = 320 nm.

Crystal data for **3**. $C_{12}H_{36}N_4Pb_4Si_4$: M_r = 1177.57 g cm^{-3}, crystal dimensions 0.459 x 0.408 x 0.170 mm, trigonal, space group R-3, a = 10.7447(8), c = 47.935(7) Å, α = 90, γ = 120°; V = 4792.6(9) Å^3, Z = 6, ρ_{calcd} = 2.448 g cm^{-3}, Siemens SMART CCD diffractometer, 2.23 ≤ θ ≤ 28.74°, Mo$_{K\alpha}$ radiation (λ = 0.71073 Å), ω scans, T = 193(2) K; of 8,240 measured reflections, 2,473 were independent and 2,127 observed with I > 2σ(I), -14 ≤ h ≤ 11, -12 ≤ k ≤ 14, -54 ≤ l ≤ 63; R_1 = 0.0507, wR_2 = 0.1426, GOF = 1.082 for 82 parameters, $\Delta\rho_{max}$ = 4.103 eÅ^{-3}. The structure was solved by direct methods (SHELXS-97) and refined by full-matrix least-squares procedures (SHELXL-97), Lorentz polarization corrections and absorption correction (SADABS) were applied, μ = 21.167 mm^{-1}, min./max. transmission 0.1234/0.0368. This crystal data structure was deposited in CCDC, reference number 249297.

References

1. Chen, H.; Bartlett, R. A.; Rasika Dias, H. V.; Olmstead, M. M.; Power, P. P. *Inorg. Chem.*, **1991**, *30*, 3390-3394. Grigsby, W .J.; Hascall, T.; Ellison, J. J.; Olmstead, M. M.; Power, P. P. *Inorg. Chem.*, **1996**, *35*, 3254-3261. Veith, M.; Opsolder, M.; Zimmer, M.; Huch, V. *Eur. J. Inorg. Chem.*, **2000**, *6*, 1143-1146. Bashall, A.; Feeder, N.; Harron, E. A.; McPartlin, M.; Mosquera, M. E. G.; Saez, D.; Wright, D. S. *J. Chem. Soc., Dalton Trans.*, **2000**, 4104-4111.
2. Allan, R. E.; Beswick, M. A.; Davies, M. K.; Raithby, P. R.; Steiner, A.; Wright, D. S. *J. Organomet. Chem.*, **1998**, *550*, 71-76.

3. Chen, H.; Bartlett, R. A.; Rasika Dias, H. V.; Olmstead, M. M.; Power, P. P. *Inorg. Chem.*, **1991**, *30*, 3390-3394.
4. Grigsby, W. J.; Hascall, T.; Ellison, J. J.; Olmstead, M. M.; Power, P. P. *Inorg. Chem.*, **1996**, *35*, 3254-3261.
5. J. F. Eichler, O. Just, and W. S. Rees, Jr. *Sulfur, Silicon, Phosphorus, and the Related Elements*, **2004**, *179*, 715-726.
6. Veith, M.; Opsolder, M.; Zimmer, M.; Huch, V. *Eur. J. Inorg. Chem.*, **2000**, *6*, 1143-1146.
7. Uhl, W.; Molter, J.; Koch, R. *Eur. J. Inorg. Chem.*, **1999**, 2021-2027.
8. Kuhner, S.; Kuhnle, R.; Hausen, H. D.; Weidlein, J. *Z. Anorg. Allg. Chem.*, **1997**, *623*, 25-34.
9. Noth H.; Seifert, T. *Eur. J. Inorg. Chem.*, **2002**, 602-612.
10. Neumann, C.; Seifert, T.; Storch, W.; Vosteen, N.; Wrackmeyer, B. *Angew. Chem., Int. Ed.*, **2001**, *40*, 3405-3407.
11. Lappert, M. F.; Slade, M. J.; Singh, A. *J. Am. Chem. Soc.*, **1983**, *105*, 302-304.
12. Reiss, P.; Fenske, D. *Z. Anorg. Allg. Chem.*, **2000**, *626*, 1317-1331.
13. Fjelberg, T.; Haaland, A.; Schilling, B. E. R.; Lappert, M. F.; Thorne, A. J. *J. Chem. Soc., Dalton Trans.*, **1986**, 1551-1560.

Chapter 10

Hydrolysis of Diborane(4) Compounds

R. Angharad Baber, Jonathan P.H. Charmant, M. J. Gerald Lesley, Nicholas C. Norman*, A. Guy Orpen and Jean Rossi

School of Chemistry, The University of Bristol, Bristol, BS8 1TS, United Kingdom

Hydrolysis of diborane(4) compounds including $B_2(NMe_2)_4$ and the amine adduct $[B_2Cl_4(NHMe_2)_2]$ affords either diboronic acid, $B_2(OH)_4$, or the borinane species $B_4O_2(OH)_4$. A crystal structure of the latter species, which co-crystallises with two equivalents of $[NH_2Me_2]Cl$, is described and compared with a previously reported polymorph of the same material. X-ray crystal structures are also described for the amine adducts $[B_2Cl_4(NHMe_2)_2][2\text{-picH}]Cl$ (2-pic = 2-picoline), $[B_2Cl_4(NHMe_2)_2][4\text{-picH}]Cl$ (4-pic = 4-picoline) and $[B_2Cl_4(NHMe_2)_2][NH_2Me_2]Cl$.

Introduction

In a recent publication we described the synthesis and X-ray crystal structure of diboronic acid, $B_2(OH)_4$ (**1**) (*1*) and in a previous paper, we reported the structure of the condensed borinane species $B_4O_2(OH)_4$ (**2**) which was isolated as a co-crystal containing two equivalents of the salt $[NH_2Me_2]Cl$ (*2*). Several synthetic routes to diboronic acid have been studied (*1*) including hydrolysis of B_2Cl_4 (*1,3*), of the amido species $B_2(NMe_2)_4$ (*4,5*), of the alkoxide derivatives $B_2(OEt)_4$ and $B_2(O^iPr)_4$ (*5*) and of $[B_2Br_4(NHMe_2)_2]$ (*1*). The borinane species described in ref. 2 was isolated as a minor product in the preparation of the diborane(4) compound $B_2(1,2\text{-}O_2C_6Cl_4)_2$ from $B_2(NMe_2)_4$. Clearly the factors which determine the nature of the product formed (*ie* **1** or **2**) in hydrolysis reactions of diborane(4) compounds are subtle and as yet poorly understood. Thus, as part of a more detailed study, hydrolysis reactions were performed on a range of diborane(4) precursors. Some preliminary results are described herein.

© 2006 American Chemical Society

A product shown by X-ray crystallography to be diboronic acid (**1**) was initially obtained by hydrolysis of either B_2Cl_4 or $[B_2Br_4(NHMe_2)_2]$ as described in ref. *1*. Subsequent work established that hydrolysis of $B_2(NMe_2)_4$ in aqueous HCl, according to the method of Brotherton (*5*), afforded a colourless crystalline product shown to be **1** by X-ray crystallography and furthermore, that recrystallisation of samples of **2**, as its dimethylammonium chloride co-crystal (see below), from water also afforded crystals of **1**. The original preparation of compound **2** is described in ref. *2*, but samples of the same material, *ie* $B_4O_2(OH)_4 \cdot 2[NH_2Me_2]Cl$, (hereafter **2a**) were obtained in much higher yield from the hydrolysis of $[B_2Cl_4(NHMe_2)_2]$ (*6*) in aqueous acetone. The structure of **2a** was confirmed by X-ray crystallography but was shown to be a different polymorph to the material described in ref. *2*. Fig. 1 shows the molecular units present in the structure of **2a** whilst Fig. 2 presents a comparison of the crystal structures of **2** and **2a** revealing that they differ in the precise arrangement of the intermolecular hydrogen-bonding. Specifically, in **2** the hydrogen bonding between the $[NH_2Me_2]^+$ cations and the borinane species involves an N–H.....OH interaction whereas in **2a**, the hydrogen bond acceptor site is one of the B_4O_2 ring oxygens. In both **2** and **2a** the conformations of the BOH units in the borinane are the same and are presumably determined by the hydrogen-bonding interactions with the chloride ion.

```
    HO    OH          HO       O       OH
      \  /               \    / \    /
       B—B                 B       B
      /  \               |         |
    HO    OH              B       B
                        /    \ /    \
                      HO       O       OH

       1                        2
```

Establishing whether **1** or **2** is present in solution is hampered by the fact that both compounds appear to have the same ^{11}B NMR chemical shift since solutions of both **1** (*1*) and **2** in D_2O exhibit a ^{11}B NMR chemical shift at 30.0 ppm (compound **2** was originally reported to have a ^{11}B NMR shift of 29 ppm in ref. *2*). It is therefore unclear precisely which compound or compounds are present in solution but preliminary hydrolysis studies followed by ^{11}B NMR indicate that **1/2** are formed under a variety of conditions. Thus in addition to the hydrolysis procedures described above and in ref. *1*, several other methods were also examined. Hydrolysis of $B_2(NMe_2)_4$ in aqueous ethanol, acetone or thf resulted in little evidence for the formation of **1/2** after three days at room temperature, but similar reactions using 1,2-$B_2Cl_2(NMe_2)_2$ all revealed a signal at 30.0 ppm in the ^{11}B NMR spectrum. In acetone and thf this was accompanied by a signal at around 19 ppm which is most likely due to boric acid although in ethanol only traces of this product were observed. Boric acid has been reported as a side product arising from oxidation/disproportionation of the B–B bonded precursors in previous studies as described in ref. *1* and refs. therein.

Dissolution of $[B_2Cl_4(NHMe_2)_2]$ in either aqueous ethanol, acetone or thf also resulted in the formation of **1/2** although in these cases with much less boric

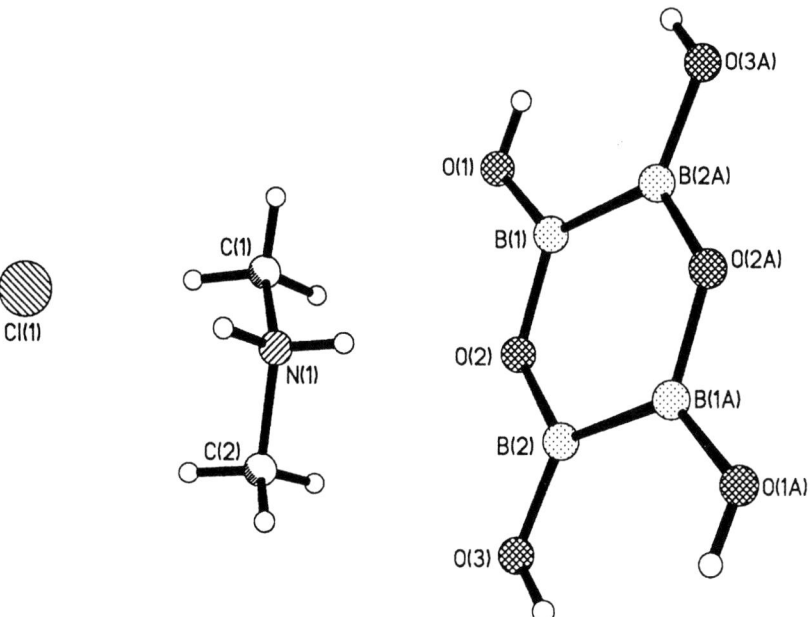

*Figure 1. The molecular structure of the borinane ring and the dimethyl ammonium chloride units in **2a**. Selected bond distances (Å) and angles (deg) include: B(1)–B(2a) 1.728(2), B(1)–O(1) 1.3513(18), B(2)–O(3) 1.3517(19), B(1)–O(2) 1.3886(18), B(2)–O(2) 1.3875(18); O(2)–B(1)–B(2a) 119.92(12), B(1)–B(2a)–O(2a) 119.87(12), B(1)–O(2)–B(2) 120.10(10), O(1)–B(1)–O(2) 113.59(11), O(1)–B(1)–B(2a) 126.47(13), O(3)–B(2)–O(2) 114.22(11), O(3)–B(2)–B(1a) 125.90(13).*

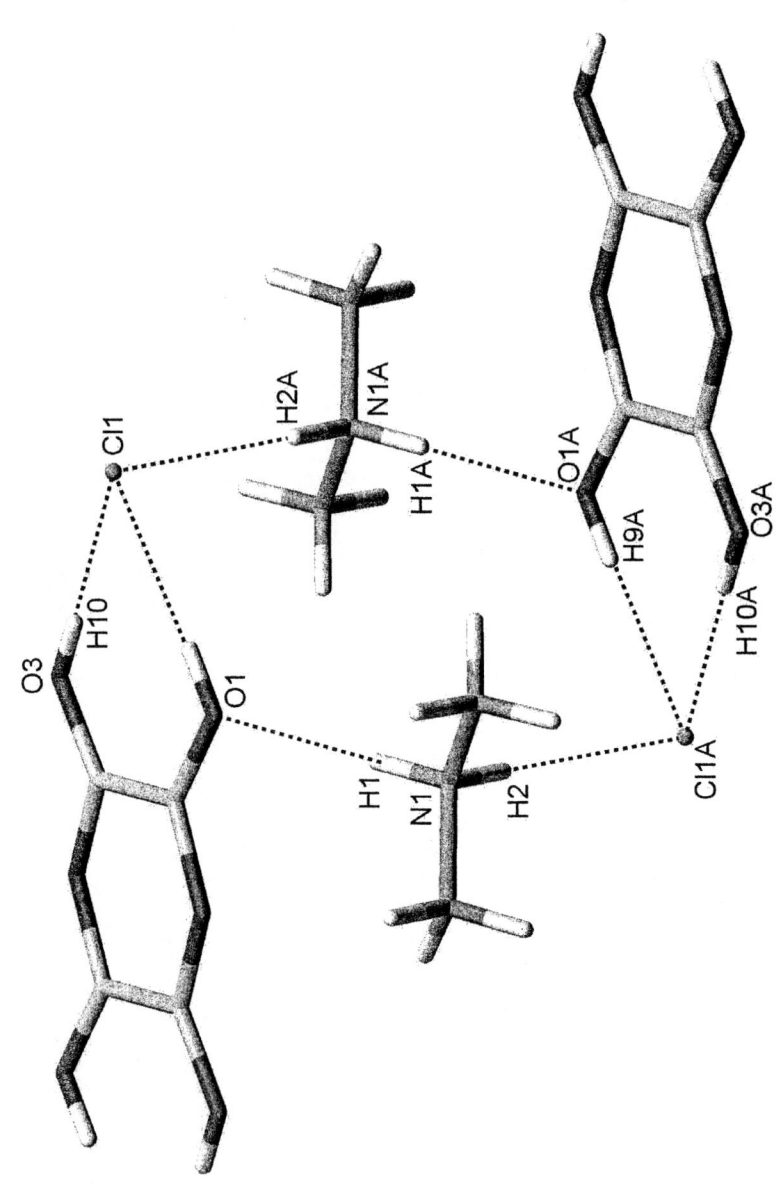

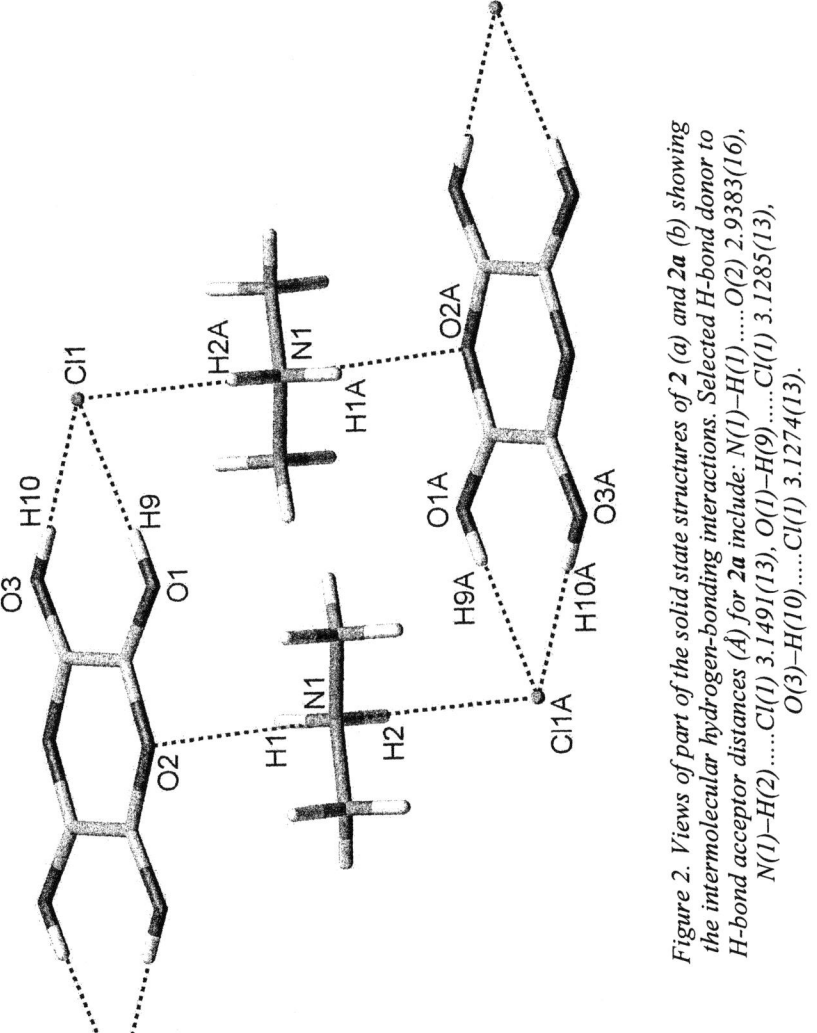

*Figure 2. Views of part of the solid state structures of **2** (a) and **2a** (b) showing the intermolecular hydrogen-bonding interactions. Selected H-bond donor to H-bond acceptor distances (Å) for **2a** include: N(1)–H(1)....O(2) 2.9383(16), N(1)–H(2)....Cl(1) 3.1491(13), O(1)–H(9)....Cl(1) 3.1285(13), O(3)–H(10)....Cl(1) 3.1274(13).*

acid. A preparative scale reaction in acetone resulted in the crystals of **2a** as described above. With smaller amounts of water present, reaction times were longer and more boric acid was observed. Hydrolysis of 1,2-$B_2Br_2(NMe_2)_2$ and [$B_2Br_4(NHMe_2)_2$] in aqueous thf also afforded good yields of **1/2**, the latter affording crystals of **1** (*1*) when carried out on a larger scale (see above). We note in passing that a compound $B_4O_2(NMe_2)_4$, with a structure analogous to the borinane unit found in **2/2a**, is a potential hydrolysis product in these reactions since the related compounds $B_4O_2(NR_2)_4$ (R = ^{i}Pr, ^{i}Bu, ^{s}Bu) have been previously characterised albeit by a route not involving hydrolysis (*7*). No spectroscopic evidence was obtained for such a species, however.

Another aspect of the hydrolysis chemistry described here, and originally in ref. *2*, was that crystals of the borinane $B_4O_2(OH)_4$ were always obtained with co-crystallised [NH_2Me_2]Cl, this salt arising from the presence of amide or amine and chloride in the starting material or reagents used. Attempts to obtain the pure borinane by recrystallisation were unsuccessful. Thus as a possible route to dimethylammonium free products, it was decided to investigate the hydrolysis of precursors containing no NMe_2 units. Reactions between [$B_2Cl_4(NHMe_2)_2$] and either 2- or 4-picoline were therefore performed to explore a potential route to the compounds [$B_2Cl_4(2$-$pic)_2$] (2-pic = 2-picoline) or [$B_2Cl_4(4$-$pic)_2$] (4-pic = 4-picoline). Whilst affording only low yields of the desired products which were therefore of little use in subsequent hydrolysis studies, crystals of the compounds [$B_2Cl_4(NHMe_2)_2$][2-picH]Cl and [$B_2Cl_4(NHMe_2)_2$][4-picH]Cl were isolated which are briefly described here. Views of the molecular structures of both compounds are shown in Figs. 3 and 4 and views of hydrogen bonding interactions present in the solid state structures are presented in Figs. 5 and 6.

A similar crystalline material, [$B_2X_4(NHMe_2)_2$][NH_2Me_2]X, was obtained as a minor product in the synthesis of [$B_2Br_4(NHMe_2)_2$] described in ref. *1* and is also reported here. Although initially formulated as the bromide salt [$B_2Br_4(NHMe_2)_2$][NH_2Me_2]Br, refinement of the X-ray data was more successful in terms of a species containing chloride as indicated by the labels in Fig. 7 which shows the molecular structure; a view of the hydrogen bonding interactions is presented in Fig. 8. Whilst microanalytical data were consistent with the formation of [$B_2Br_4(NHMe_2)_2$] as the major product (*1*), the presence of chloride in a minor product is not unexpected in view of the reaction precursor, ie [$B_2Cl_4(NHMe_2)_2$].

Selected bond lengths and angles for all three [$B_2Cl_4(NHMe_2)_2$][Q]Cl structures, Q = picolinium or ammonium cation, are presented in Table I together with data from the structure of [$B_2Cl_4(NHMe_2)_2$] itself (*8*).

As illustrated in Figs. 5, 6 and 8, the structures of the co-crystals of [$B_2Cl_4(NHMe_2)_2$] with [2-picH]Cl, [4-picH]Cl or [NH_2Me_2]Cl all exhibit significant hydrogen bonding interactions. Specifically, the interactions in all three structures are predominantly between the N–H bonds and the free chloride ions, the latter being better hydrogen bond acceptors than the alternative B–Cl chlorines. Selected H-bond donor to H-bond acceptor distances (Å) include: [$B_2Cl_4(NHMe_2)_2$][2-picH]Cl [N(1)–H(1).....Cl(5) 3.289(14), N(2)–H(2).....Cl(5) 3.297(15), N(3)–H(3).....Cl(5) 2.998(13) Å]; [$B_2Cl_4(NHMe_2)_2$][4-picH]Cl

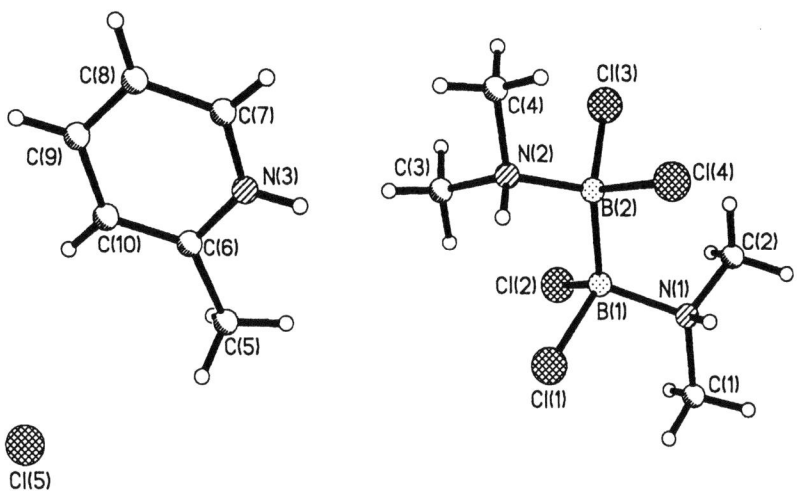

Figure 3. The molecular structure of $[B_2Cl_4(NHMe_2)_2][2\text{-}picH]Cl$

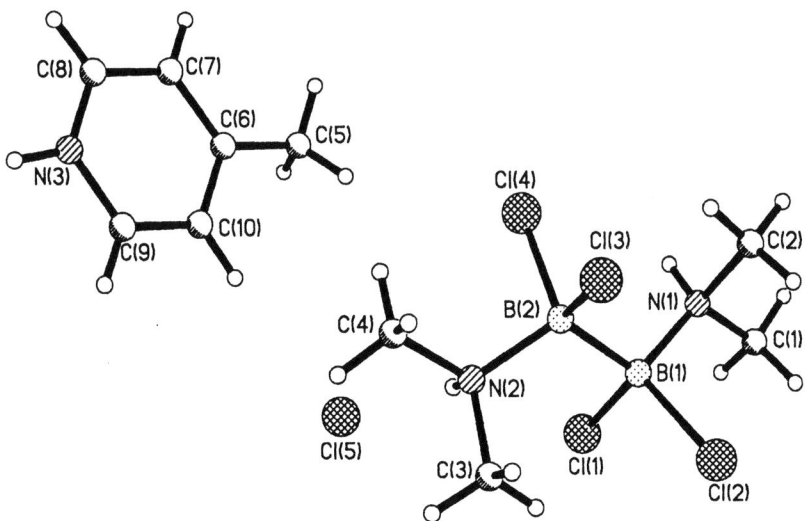

Figure 4. The molecular structure of $[B_2Cl_4(NHMe_2)_2][4\text{-}picH]Cl$

144

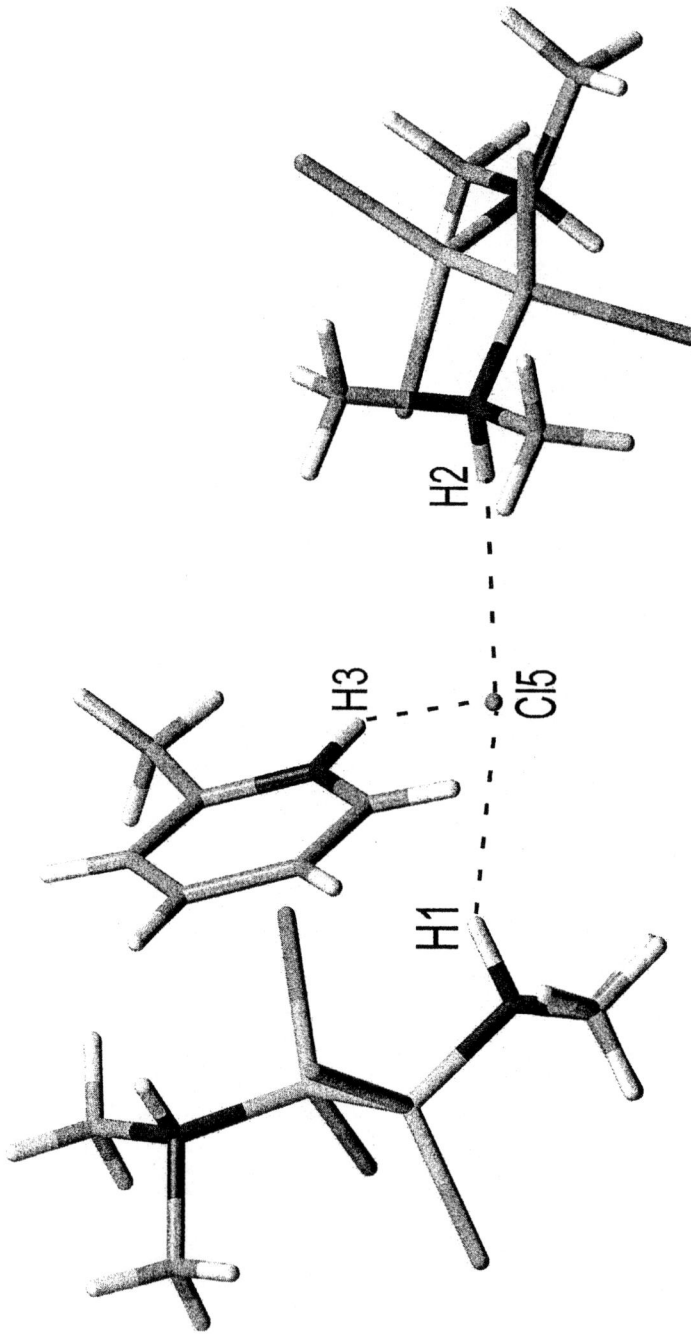

Figure 5. A view of part of the solid state structure of $[B_2Cl_4(NHMe_2)_2]$ [2-picH]Cl showing the intermolecular hydrogen-bonding interactions.

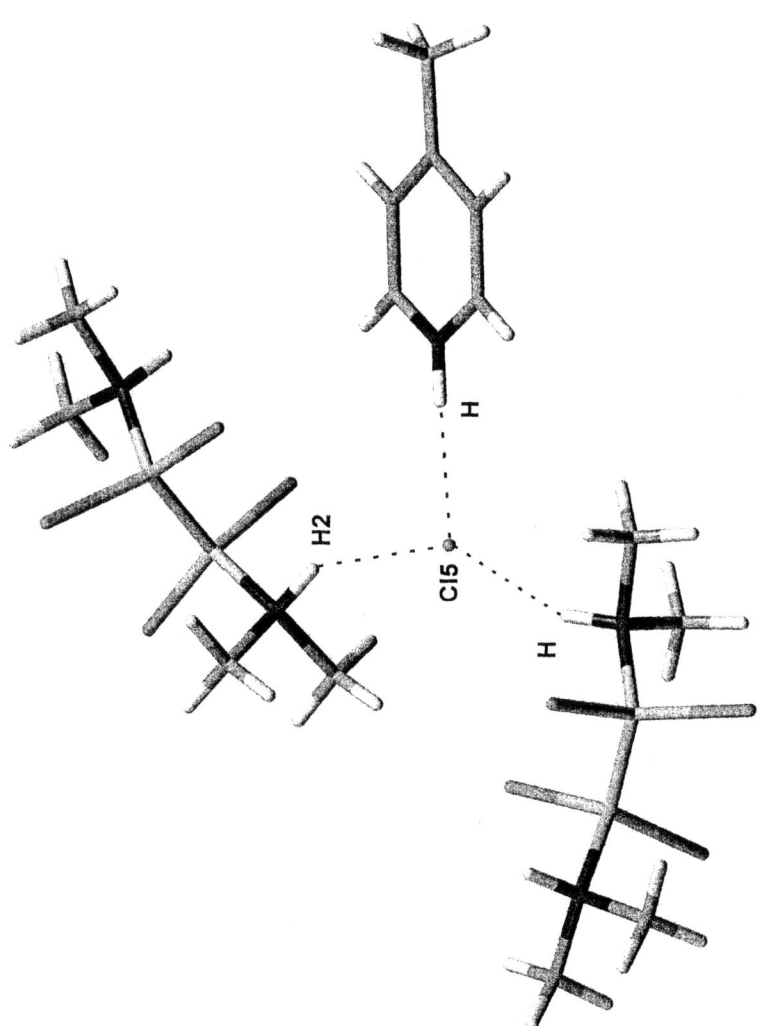

Figure 6. A view of part of the solid state structure of $[B_2Cl_4(NHMe_2)_2]$ [4-picH]Cl showing the intermolecular hydrogen-bonding interactions.

145

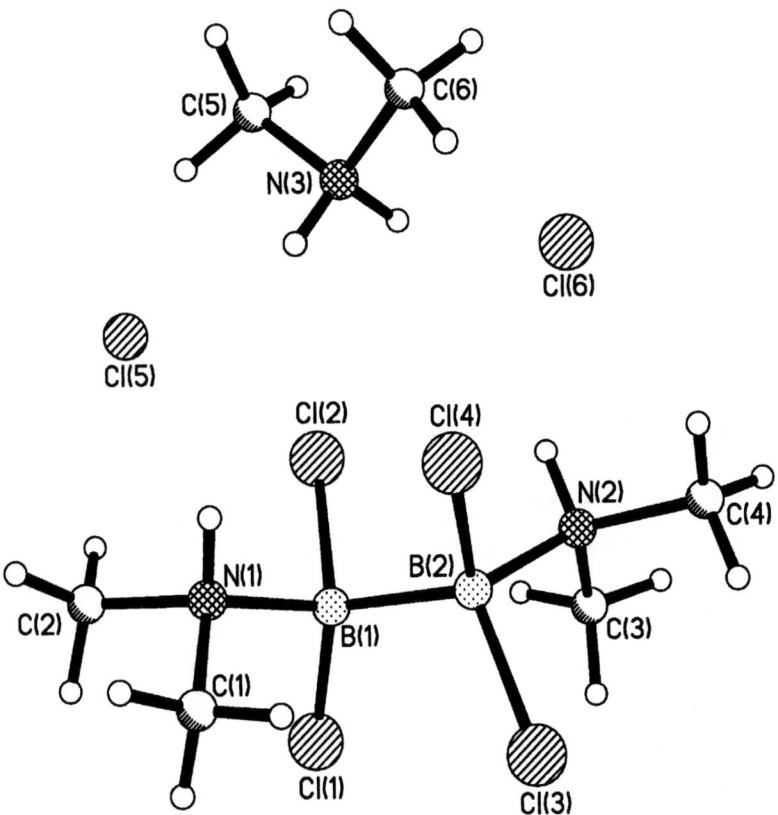

Figure 7. The molecular structure of [$B_2Cl_4(NHMe_2)_2$][NH_2Me_2]Cl

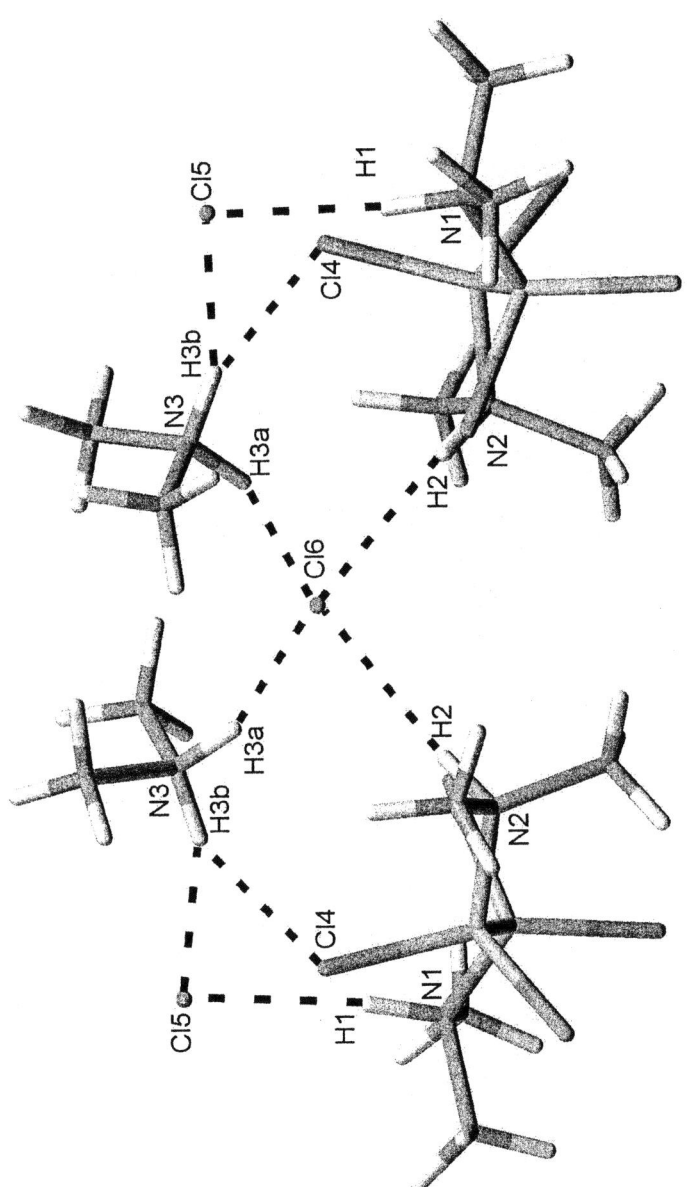

Figure 8. A view of part of the solid state structure of $[B_2Cl_4(NHMe_2)_2][NH_2Me_2]Cl$ showing the intermolecular hydrogen-bonding interactions.

Table I. Selected bond lengths (Å) and angles (°) for crystals containing the diborane(4) compound [$B_2Cl_4(NHMe_2)_2$].

Compound	B–B (Å)	av.B–Cl (Å)	av.B–N (Å)	N–B–BN (°)
[$B_2Cl_4(NHMe_2)_2$] ref. (8)*	1.737(5)	1.885	1.608	-154.2
	1.731(5)			-152.4
	1.740(6)			153.0
[$B_2Cl_4(NHMe_2)_2$][2-picH]Cl	1.78(3)	1.875	1.606	-152.8(13)
[$B_2Cl_4(NHMe_2)_2$][4-picH]Cl	1.735(7)	1.891	1.593	154.6(3)
[$B_2Cl_4(NHMe_2)_2$][NH_2Me_2]Cl	1.78(3)	1.876	1.605	-151.3(13)

* Contains three molecules per asymmetric unit.

[N(1)–H(1).....Cl(5) 3.187(3), N(2)–H(2).....Cl(5) 3.157(3), N(3)–H(3).....Cl(5) 3.019(4) Å]; [$B_2Cl_4(NHMe_2)_2$][NH_2Me_2]Cl [N(3)–H(3).....Cl(5) 3.109(14), N(3)–H(3).....Cl(6) 3.124(14) Å]. Furthermore, the observed hydrogen bonding interactions probably influence the conformation about the B–B bond in the [$B_2Cl_4(NHMe_2)_2$] molecules. Thus the observed conformations allow the chloride ions to approach the N–H bonds in a position which maximises the distance between the chloride ions and the B–Cl chlorine atoms.

Acknowledgement

We thank the EPSRC for financial support.

Experimental

General

All reactions were carried out under a dinitrogen atmosphere, using standard Schlenk line or glove box techniques. Solvents were dried using an Anhydrous Engineering system employing alumina columns and subsequently degassed. ^{11}B-{^{1}H} NMR spectra were obtained using JEOL GX400 and JEOL Eclipse-300 spectrometers.

Synthesis

$B_4O_2(OH)_4.2[NH_2Me_2]Cl$ (2a)

To a sample of $[B_2Cl_4(NHMe_2)_2]$ (0.9 g, 3.54 mmol) dissolved/suspended in acetone (20 cm^3, reagent grade), a degassed solution of H_2O in acetone (25 % v/v, 1.5 cm^3) was added slowly. After stirring for 3 days, the solvent was removed by vacuum. The resulting white solid was stirred in hexane, filtered and dried by vacuum leaving a white powder. Recrystallisation from thf afforded a crop of small crystals one of which was shown to be $B_4O_2(OH)_4.2[H_2NMe_2]Cl$ (2a) by X-ray crystallography, yield 94 %. ^1H NMR (D_2O): δ 2.66 (6H, s, *Me*,), ^{11}B-$\{^1H\}$ NMR: δ 30.4. Microanalysis, Found: C, 16.00; H, 7.45; N, 10.10 %: $C_4H_{20}B_4Cl_2N_2O_6$ requires C, 15.70; H, 6.60; N, 9.15 %).

$[B_2Cl_4(NHMe_2)_2][2\text{-picH}]Cl$

Overnight stirring of a solution of $[B_2Cl_4(NHMe_2)_2]$ (0.73 g, 2.88 mmol) in CH_2Cl_2 (60 cm^3) with an excess of 2-picoline (1.45 cm^3, 14.4 mmol) was followed by removal of all volatile products by vacuum. The resulting solid was washed with hexane (3 x 10 cm^3) but ^1H NMR analysis indicated the presence of mostly 2-picolinium chloride and dimethylammonium chloride as the major products with only small amounts of the desired material $[B_2Cl_4(2\text{-pic})_2]$. Dissolution of the product mixture in thf (2 cm^3) followed by slow diffusion of a layer of hexane (5 cm^3) at $-30°C$ afforded a small crop of colourless crystals one of which was shown to be $[B_2Cl_4(NHMe_2)_2][2\text{-picH}]Cl$ by X-ray crystallography.

$[B_2Cl_4(NHMe_2)_2][4\text{-picH}]Cl$

Crystals of $[B_2Cl_4(NHMe_2)_2][4\text{-picH}]Cl$ were prepared in an analogous manner to those of $[B_2Cl_4(NHMe_2)_2][2\text{-picH}]Cl$ but using 4-picoline as the reagent.

$[B_2Cl_4(NHMe_2)_2][NH_2Me_2]Cl$

Crystals of $[B_2Cl_4(NHMe_2)_2][NH_2Me_2]Cl$ were obtained as a minor product in the preparation of $[B_2Br_4(NHMe_2)_2]$ the synthesis of which, from $[B_2Cl_4(NHMe_2)_2]$ and excess BBr_3, is described in ref. *1*. Attempted recrystallisation of a sample of $[B_2Br_4(NHMe_2)_2]$ from thf afforded a small crop of crystals one of which was shown to be $[B_2Cl_4(NHMe_2)_2][NH_2Me_2]Cl$ by X-ray crystallography. Refinement of the X-ray data was consistent with the presence of chloride as the halide rather than bromide in the crystal analysed.

X-ray crystallography

Crystals were coated in perfluoropolyether oil and mounted on a glass fibres. X-ray measurements were made using a Bruker SMART CCD area-detector diffractometer with Mo-K_α radiation ($\lambda = 0.71073$ Å) (*9*). Intensities were integrated (*10*) from several series of exposures, each exposure covering 0.3° in ω. Absorption corrections were applied, based on multiple and symmetry-equivalent measurements (*11*). The structure was solved by direct methods and refined by least squares on weighted F^2 values for all reflections (12). Complex neutral-atom scattering factors were used (13).

Crystal Data for **2a**: $C_4H_{20}B_4Cl_2N_2O_6$, M = 306.36, triclinic, space group $P-1$, a = 5.9193(6), b = 8.1499(9), c = 8.4416(9) Å, α = 110.755(2), β = 99.073(2), γ = 92.429(2)°, U = 373.91(7) Å^3, Z = 1, D_c = 1.361 Mg m^{-3}, λ = 0.71073 Å, μ(Mo-K_α) = 0.448 mm^{-1}, $F(000)$ = 160, T = 100(2) K, R_1 = 0.0286.

Crystal Data for $[B_2Cl_4(NHMe_2)_2][2\text{-picH}]Cl$: $C_{10}H_{22}B_2Cl_5N_3$, M = 383.18, monoclinic, space group $P2_1$, a = 8.0100(16), b = 12.706(3), c = 9.936(2) Å, β = 113.77(3)°, U = 925.5(3) Å^3, Z = 2, D_c = 1.368 Mg m^{-3}, λ = 0.71073 Å, μ(Mo-K_α) = 0.776 mm^{-1}, $F(000)$ = 392, T = 100(2) K, R_1 = 0.1356.

Crystal Data for $[B_2Cl_4(NHMe_2)_2][4\text{-picH}]Cl$: $C_{10}H_{22}B_2Cl_5N_3$, M = 383.18, monoclinic, space group Cc, a = 16.061(3), b = 10.058(2), c = 12.704(3) Å, β = 114.89(3)°, U = 1861.6(8) Å^3, Z = 4, D_c = 1.367 Mg m^{-3}, λ = 0.71073 Å, μ(Mo-K_α) = 0.772 mm^{-1}, $F(000)$ = 792, T = 100(2) K, R_1 = 0.0504.

Crystal Data for $[B_2Cl_4(NHMe_2)_2][NH_2Me_2]Cl$: $C_{10}H_{20}B_2Cl_5N_3$, M = 333.12, trigonal, space group $P3_121$, a = 9.781(3), c = 30.787(14) Å, U = 2550.7(16) Å^3, Z = 6, D_c = 1.301 Mg m^{-3}, λ = 0.71073 Å, μ(Mo-K_α) = 0.834 mm^{-1}, $F(000)$ = 1032, T = 173(2) K, R_1 = 0.1497.

Crystal data for all four compounds have been deposited at the Cambridge Crystallographic Data Centre (CCDC deposition numbers 242182 - 242185).

References

1. Baber, R.A.; Norman, N.C.; Orpen, A.G.; Rossi, J. *New J. Chem.* **2003**, *27*, 773.
2. Carmalt, C.J.; Clegg, W.; Cowley, A.H.; Lawlor, F.J.; Marder, T.B.; Norman, N.C.; Rice, C.R.; Sandoval, O.J.; Scott, A.J. *Polyhedron* **1997**, *16*, 2325.
3. Wartik, T.; Apple, E.F. *J. Am. Chem. Soc.* **1955**, *77*, 6400; Wartik, T.; Apple, E.F. *J. Am. Chem. Soc.* **1958**, *80*, 6155.
4. Nöth, H.; Meister, W. *Chem. Ber.* **1961**, *94*, 509.
5. Brotherton, R.J.; McCloskey, A.L.; Petterson, L.L.; Steinberg, H. *J. Am. Chem. Soc.* **1960**, *82*, 6242; Brotherton, R.J.; McCloskey, A.L.; Boone, J.L.; Manasevit, H.M. *J. Am. Chem. Soc.* **1960**, *82*, 6245; McCloskey, A.L.; Boone, J.L.; Brotherton, R.J. *J. Am. Chem. Soc.* **1961**, *83*, 1766; McCloskey, A.L.; Brotherton, R.J.; Boone, J.L. *J. Am. Chem. Soc.* **1961**, *83*, 4750; Brotherton, R.J. *Progress in Boron Chemistry*, Eds. Steinberg, H.; McCloskey, A.L. Pergamon Press, Oxford, 1964.
6. Lawlor, F.J.; Norman, N.C.; Pickett, N.L.; Robins, E.G.; Nguyen, P.; Lesley, G.; Marder, T.B.; Ashmore, J.A.; Green, J.C. *Inorg. Chem.* **1998**, *37*, 5282; Nöth, H.; Meister, W. *Z Naturforsch., Teil B.* **1962**, *17*, 714.
7. Meller, A.; Maringgele, W., *Advances in Boron Chemistry*, Ed. Siebert, W., The Royal Society of Chemistry, 1997, p224 and Maringgele, W.; Noltemeyer, M.; Meller, A., *Organometallics*, 1997, *16*, 2276.

8. Clegg, W.; Elsegood, M.R.J.; Lawlor, F.J.; Norman, N.C.; Pickett, N.L.; Robins, E.G.; Scott, A.J.; Nguyen, P.; Taylor, N.J.; Marder, T.B. *Inorg. Chem.* **1998**, *37*, 5289.
9. SMART diffractometer control software, Bruker-AXS Inc., Madison, WI, 2003.
10. SAINT integration software, Bruker-AXS Inc., Madison, WI, 2003.
11. Sheldrick, G.M.; *SADABS: A program for absorption correction with the SMART system;* University of Göttingen: Germany, 2003.
12. *SHELXTL program system version 6.1;* Bruker-AXS Inc., Madison, WI, 2000.
13. *International Tables for Crystallography*, Kluwer, Dordrecht, 1992, vol. C.

Chapter 11

Recent Developments in Boron–Phosphorus Ring and Cage Chemistry

R. T. Paine[1], H. Nöth[2], T. Habereder[1,2], J. F. Janik[1], E. N. Duesler[1], and D. Dreissig[1]

[1]Department of Chemistry, University of New Mexico, Albuquerque, NM 87131
[2]Institut für Anorganische Chemie, Universität München, 8033 München, Germany

The chemical and structural diversity of compounds containing Groups 13/15 element combinations continues to expand at a rapid pace. In this paper we outline some recent results from our groups that illustrate the continued evolution of boron-phosphorus ring and cage constructs.

© 2006 American Chemical Society

The molecular chemistry of boron-phosphorus Lewis acid-base compounds was initially developed in the 1960's; however, the area remained relatively dormant for the next twenty years as other topics in inorganic and organometallic chemistry evolved. Spurred by discoveries in the 1980's of main group element multiple bond conditions (*1*) and rational, low energy syntheses of solid state materials via main group precursors (*2*), attention to Group 13/15 molecular chemistry in general and boron/phosphorus chemistry in particular exploded. In 1995, we summarized the remarkable advances that had occurred in boron-phosphorus chemistry since 1985, and new areas ripe for development in synthesis, structure and reactivity were suggested (*3*). Since that review appeared, the field has continued to prosper in a number of research groups worldwide. No attempt will be made here to comprehensively review the many advances since 1995. Instead, some snapshots of several recent developments from our groups are presented that illustrate the continued excitement in the area.

Phosphine Borane Systems

Boron phosphide (BP), a little studied but potentially important optoelectronic material (*4*) is prepared as a bulk powder by using traditional metallurgical approaches. Several gas mixtures, including B_2H_6/PH_3 (*5*), have been employed to prepare BP films by using CVD methods, but all of the simple reagent mixtures suffer from one or more of the drawbacks common to this methodology. As a result, attention has been given to the development of single source precursor molecules for BP synthesis (*6*). Perhaps the simplest single source precursor to BP would be H_3BPH_3 which ideally could undergo inter- or intra-molecular elimination of H_2 that could be easily and cleanly removed from the preparative system. Although the adduct has been carefully studied and characterized in solid and molten states as well as in the gas phase at -60°C (*7*), it is unstable toward dissociation near 23°C and elimination products containing boron and phosphorus have not been fully characterized due in part to low solubility. Although this adduct may prove to be a suitable low temperature CVD precursor system, it is not particularly useful as a starting material for the synthesis of ring or cage condensation products that could be employed as solution phase precursors for nano-sized materials.

In 1959, Parshall and Lindsey (*8*) reported the formation of $H_3BP(SiMe_3)_3$ as a colorless, crystalline compound from B_2H_6 and $P(SiMe_3)_3$. This reaction system was subsequently studied by Leffler and Teach (*9*), and Nöth and Schrägle (*10*) and found to undergo stepwise Me_3SiH elimination. Products formed at 125°C and at 200°C were proposed to be $H_2BP(SiMe_3)_2$ and $[HBPSiMe_3]_n$, respectively. A reinvestigation of the system in our groups showed that the initial adduct indeed undergoes Me_3SiH elimination in solution even at 23°C and this elimination chemistry is influenced by the presence of either excess phosphine or borane (*11*). Further, the crystalline adduct slowly decomposed in a sealed tube held at 100°C, and the rate of Me$_3$SiH elimination significantly accelerated at 150°C and higher. Two crystalline compounds, **1** and **2**, were isolated from the thermolysis reactions at 150°C and single crystal x-ray diffraction analyses provided unambiguous assignments of the structures which are shown in Figures 1 and 2, respectively. Compound **1** is the trimer of the coordinatively unsaturated monomer H$_2$BP(SiMe$_3$)$_2$ originally proposed by Nöth and Schrägle. It is noteworthy that this six-membered ring containing four coordinate boron and phosphorus atoms is much flatter than the related cyclohexane-like chair shaped structures of (H$_2$BPMe$_2$)$_3$ (*12*) and (H$_2$BPPh$_2$)$_3$ (*13*). The second crystalline compound, **2**, P[μ-H-B$_2$H$_2$]{(H$_2$B)$_2$P(SiMe$_3$)$_2$]$_2$}$_2$ displays a unique spirocyclic structure in which two B$_3$P$_3$ rings share a common P atom. Adjacent BH groups in the two rings are joined via a bridge B-H-B bond. The structure of this molecule suggests that elimination reactions in the B$_2$H$_6$ + P(SiMe$_3$)$_3$ system are not confined to processes that split out Me$_3$SiH, but may also include P(SiMe$_3$) elimination. Subsequent gc-ms studies of the system confirmed that Me$_3$SiH, P(SiMe$_3$)$_3$ and H$_2$ are present in the gas phase over decomposing H$_3$BP(SiMe$_3$)$_3$. Furthermore, several additional non-crystalline intermediate compounds are also formed. As a consequence of the competing elimination processes, the thermal decomposition chemistry of bulk samples is complex. In fact, pyrolysis of gram-sized samples of the adduct at elevated temperatures (> 400°C) gives solid state materials that contain BP and SiC (identified by XRD) and carbon (identified by elemental analyses).

Nöth and Schrägle (*10a*) also reported on the reactions of the haloboranes BX$_3$ (X=Cl, Br, I) with P(SiMe$_3$)$_3$ in hexane solution where they noted formation of crystalline 1:1 adducts. In subsequent reinvestigations of these systems Lube and Wells (*14*) noted that additional products form. By using salt elimination chemistry between BX$_3$ and LiP(SiMe$_3$)$_2$, they isolated and structurally characterized [X$_2$BP(SiMe$_3$)$_2$]$_2$ (X=Cl, Br). The planar four membered (BP)$_2$ rings contain four coordinate B and P atoms. Pyrolyses of the 1:1 adducts and the dimeric elimination products were examined up to 500°C and, under the conditions utilized, none gave boron phosphide as a pure solid state material.

Based upon the prior work by Groshens and Johnson (*6b*) one would not anticipate that organyl phosphines would provide good single source precursors for electronic grade BP since the organyl groups typically provide a source for carbon under high temperature conditions. Nonetheless, reactions of B$_2$H$_6$ and H$_3$B-base adducts with organyl phosphines are of interest for comparison with the chemistry described above. Nöth and Schrägle (*10a*), for example, reported that the reaction of (Me$_3$Si)$_3$P with PhBCl$_2$ gave a stable crystalline adduct Ph(Cl)$_2$BP(SiMe$_3$)$_3$ which upon heating to 150°C produced a golden polymeric

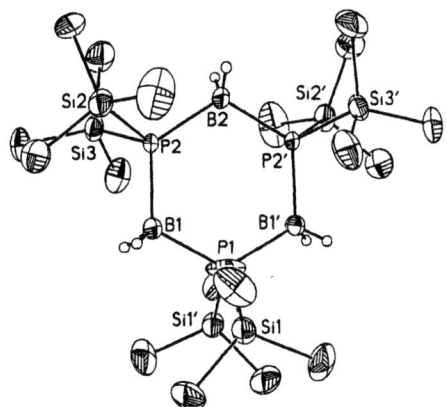

Figure 1. Molecular structures of $[H_2BP(SiMe_3)_2]_3$, 1.

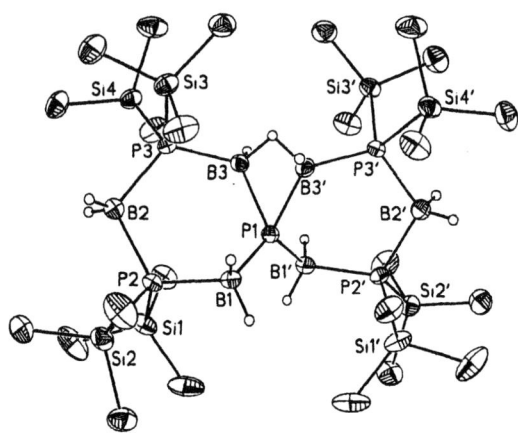

Figure 2. Molecular structure of $P[\mu\text{-}H\text{-}B_2H_2]\{(H_2B)_2P(SiMe_3)_2\}_2$, 2.

solid. We have returned to examine this reaction in greater detail. Indeed, combination of the reagents in a 1:1 ratio in hexane at -78°C leads to immediate formation of a white solid that crystallized from cold toluene. A molecular structure determination confirmed that this initially formed solid 3 is a 1:1 Lewis acid-base adduct with a P-B bond length of 2.059(4) Å as shown in Figure 3. Subsequent heating of a suspension of 3 in refluxing toluene for 3 days gave an orange solution which produced colorless crystals of 4 upon concentration. Single crystal x-ray analysis revealed that 4 is a planar, four-membered ring compound [Ph(Cl)BP(SiMe$_3$)$_2$]$_2$ with a *trans* orientation of the like substituents on the B atoms. The P-B bond lengths of 2.02(1) Å and 2.05(1) Å are similar to those reported by Wells for [Cl$_2$BP(SiMe$_3$)$_2$]$_2$, 2.025(3) Å and 2.024(3) Å, and for [Br$_2$BrP(SiMe$_3$)$_2$]$_2$, 2.03(1) Å. In 4 the single B-Cl bond length, 1.92(1) Å is significantly longer than those in [Cl$_2$BP(SiMe$_3$)$_2$]$_2$, 1.852(3) Å and 1.839(3) Å. Further reflux of the reaction mixture in toluene for several more days led to the formation of an orange colored crystalline compound 5, (PhB)$_5$P$_5$(SiMe$_3$)$_4$. The molecular structure for 5 appears in Figure 5. The compound has a novel cage structure consisting of a tri-fold (PhB)$_3$(PSiMe$_3$)$_2$P crown linked via two B-P bonds and a P-P bond to a (PhB)$_2$(PSiMe$_3$)$_2$ four-membered ring. The tri-fold crown has a B-B-B triangular base, with an average B-B bond length, 1.920 Å, and the average B-P bond lengths to the phosphorus atoms in the crown (P1, P2 and P3) is 1.974 Å. The upper B$_2$P$_2$ ring has an average B-P bond length, 2.005 Å. The two B-P bond lengths connecting the crown and four-membered ring fragments are P1-B4, 1.984(7) Å and P3-B5 1.992(7) Å. The connecting P2-P5 bond length is 2.175(7) Å. The cage is not configured as would be expected for a fragment from a cubic BP unit cell, but it can be considered to be on the way to such a state. Unfortunately, at this time, additional cage intermediates have not been isolated from the reaction mixture, but it can be anticipated that they exist (as evidenced by additional peaks in ^{31}P NMR spectra of the reaction mixture). Further studies will likely provide additional information on the growth of cage molecules from such reactions.

Additional Cage Assembly Chemistry

In prior reports from our group we have described efforts to develop systematic stepwise assembly routes for the formation of B$_x$P$_y$ and B$_x$P$_y$E$_z$ cage molecules (*3, 15-20*). The syntheses have been based upon the use or transient existence of four-membered diphosphadiboretane ring compounds that undergo subsequent substitution and ring closure chemistry. For the known examples, the syntheses have been found to be highly efficient for formation of five and six-membered *closo* cages and several spirocyclic and chained cages. It has become

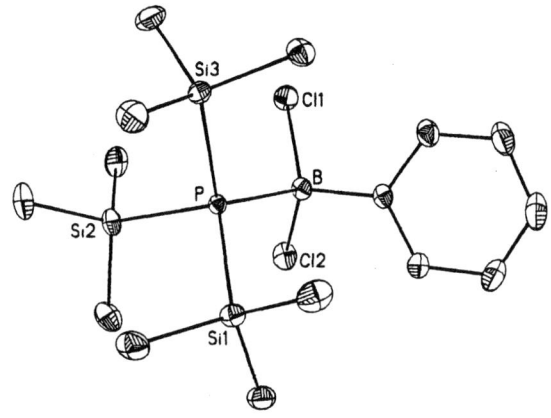

*Figure 3. Molecular structure of PhB(Cl)$_2$P(SiMe$_3$)$_3$, **3**.*

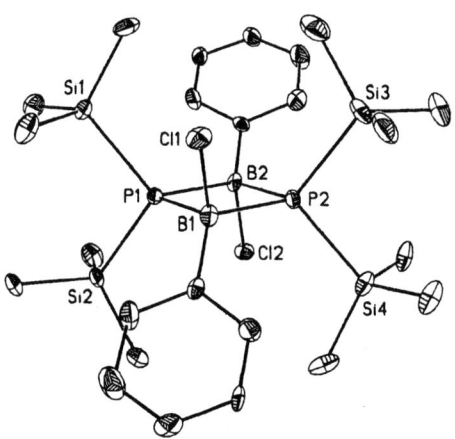

*Figure 4. Molecular structure of [PhB(Cl)P(SiMe$_3$)$_2$]$_2$, **4**.*

clear, however, that these smaller cages represent thermodynamic "sinks" that acyclic systems have a tendency to descend to. In order to obtain larger cages it will likely be necessary to prepare larger building blocks than the four-membered diphosphadiboretanes. Several avenues are understudy and one involves the proposed assembly shown in Scheme 1.

The formation of **6** is known from our prior work, and the existence of **7** also has been substantiated. The subsequent dimerization of **7** to give the eight-membered ring compound **8** has not been confirmed although NMR spectra are consistent with its formation. The subsequent reduction and intramolecular P-P coupling reaction is validated by the isolation of $(THF)(DME)P_2[(i-Pr_2NB)_2PH]_2$. The molecular structure of **9** is shown in Figure 6. The P1-P2 bond length, 2.165(8) Å, in the edge-sharing bicyclic structure is normal for P-P single bonds. The average B-P bond length, 2.00 Å involving P1 and P2, however, is longer than those associated with P3 and P4, 1.84 Å. The P1 atom is coordinated to a lithium which is, in turn, bonded to DME and THF molecules. The reaction of **8** with BuLi was performed in an effort to make the lithiated derivative for subsequent additions of boryl phosphane fragments. Clearly deprotonation occurs, but this is accompanied by P-P bond formation. At the present time, the utilization of **8** and **9** as building units for larger cages remains to be explored, and both molecules provide attractive construction possibilities.

Diborane (4) Building Blocks

Power and coworkers (21) have examined the 2:1 reaction of $LiPPh_2$ with the diborane(4) species $(Me_2N)(Br)BB(Br)(NMe_2)$. They isolated the 1,2-diphosphinodiborane(4), $(Me_2N)(Ph_2P)BB(PPh_2)(NMe_2)$ and reported the formation of chelate complexes with the $Cr(CO)_4$ fragment. The diborane(4) fragment should also serve as a useful construct for B-P ring and cage compounds of interest to our groups. The entry point for these studies is 1,2-dichloro-1,2-bis-(dimethylamino)diborane(4) which was combined, in a 1:2 ratio with $LiP(SiMe_3)_2 \cdot (thf)_x$, $LiP(SnMe_3)_2 \cdot (thf)_x$ and $LiPH_2 \cdot (thf)_x$. In the combinations with $LiP(SiMe_3)_2$ and $LiP(SnMe_3)_2$ crystalline products, $\{[(Me_3Si)_2P](Me_2N)B\}_2$ (**10**) and $\{[(Me_3Sn)_2P](Me_2N)B\}_2$ **11** were isolated, and the molecular structure of **11** was determined by single crystal x-ray diffraction methods. A view of the molecule is shown in Figure 7. Both **10** and **11** are moderately stable in hexane but, dissolved in thf, the compounds slowly evolve $P(SiMe_3)_3$ and $P(SnMe_3)_3$, respectively. In each case at least two condensation products are also formed, but they have not been isolated in a pure form and characterized.

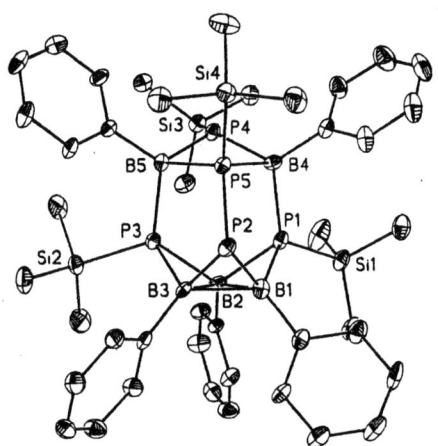

Figure 5. Molecular structure of (PhB)$_5$P$_5$(SiMe$_3$)$_4$, **5**.

Scheme 1.

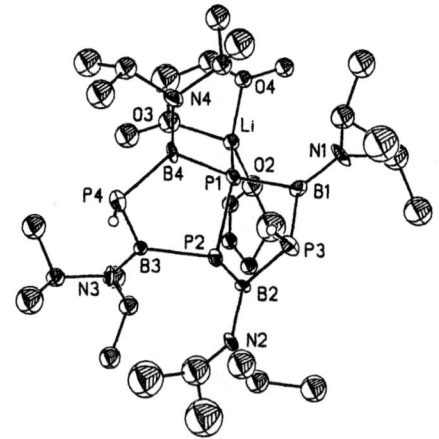

Figure 6. Molecular structure of [(THF)(DME)Li]P$_2$[(i-Pr$_2$NB)$_2$PH]$_2$.

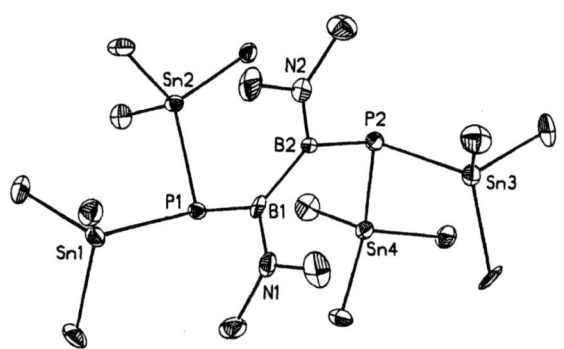

Figure 7. Molecular structure of {[(Me$_3$Sn)$_2$P](Me$_2$N)B}$_2$', 11.

In the case of the product obtained by addition of LiPH$_2$ to [(Me$_2$N)(Cl)B]$_2$, the bis-phosphine was not isolated. Instead a mixture of three products was obtained, one of which, [(Me$_2$N)B-B(NMe$_2$)PH]$_2$, (**12**) has been isolated and structurally characterized by single crystal x-ray diffraction methods. A view of the molecule is shown in Figure 8. It appears that [(H$_2$P)(Me$_2$N)B]$_2$ likely forms in the initial metathesis reaction and this compound must undergo intermolecular PH$_3$ elimination with dimerization giving the six-membered ring. This ring system should be useful for further cage construction reactions although the synthesis must be improved so that **12** is obtained in high yield as a single product without impurities.

In previous work, we have used the 1,2,3,4-diphosphadiboretane (tmp BPH)$_2$ as a building unit to produce five and six vertex *closo* cage compounds. We have attempted to extend that approach as shown in Scheme 2. The yellow, crystalline 1-borylated intermediate (tmpB)$_2$[PB(NMe$_2$)B(Cl)(NMe$_2$)]PH, **13** was isolated and characterized. The ^{31}P NMR spectrum is consistent with the formation of the compound as two equal intensity resonances centered at δ -110 and -104 ppm are observed. The former is a doublet of doublets (J$_{PH}$ = 172 Hz, J$_{PP}$ = 60 Hz) and the latter is a doublet (J$_{PP}$ = 60 Hz). Single crystal x-ray diffraction analysis for **13** confirms the structure of the molecule and a view is shown in Figure 9. The orientation of the B-Cl bond near the P-H group suggests that HCl elimination should provide the *closo* six atom cage molecule, **14**. The intramoleculoar HCl elimination process has been examined with and without base promotion. Compound **13** in hexane appears to be stable toward elimination at 23°C; however, addition of BuLi gives an orange solution and

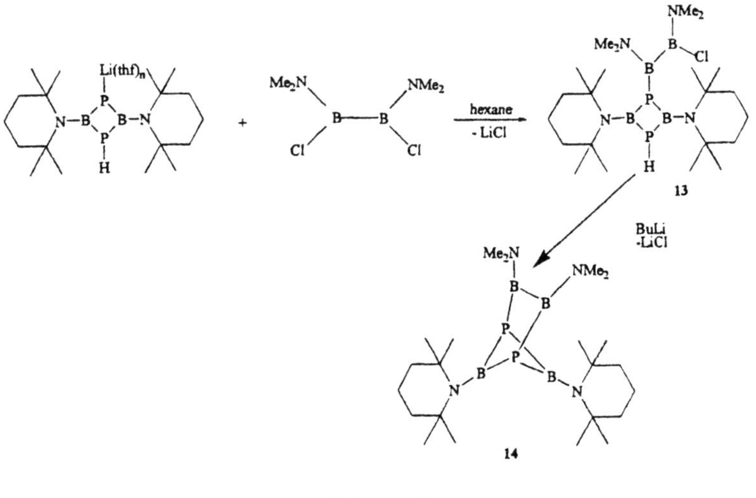

Scheme 2.

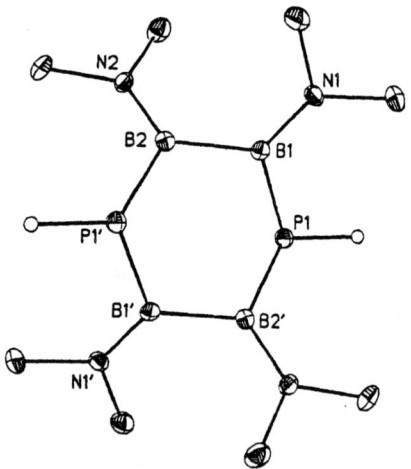

Figure 8. Molecular structure of [(Me$_2$N)B-B(NMe$_2$)PH]$_2$, 12.

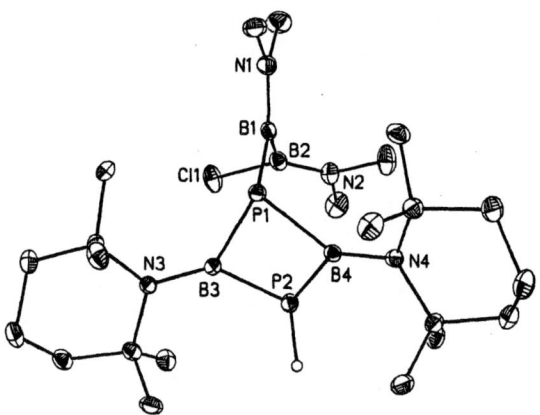

Figure 9. Molecular structure of (tmpB)$_2$[PB(NMe$_2$)B(Cl)(NMe$_2$)]PH, 13.

LiCl. Evaporation of the hexane leaves an orange oil which so far has not yielded single crystals. The ^{31}P NMR spectrum contains a strong, high field resonance at δ 7.7 which falls in the general region observed for $P_2(BNR_2)_3$ five vertex cage compounds. The spectrum also contains resonances at δ 46 and 51 which have not been assigned to a known product.

Conclusion

As pointed out in our 1995 review, molecular boron-phosphorus chemistry still contains many exciting frontiers. The snapshot summaries presented here provide some glimpses at targets under recent study in our groups in Albuquerque and München. Other groups are exploring alternative approaches to B-P ring, cage and oligomer systems including catalyzed dehydrocoupling reactions (22). Others are involved in the generation of novel radical molecules (23) and "heavy atom" ring and cage analogs containing aluminum, gallium and indium (24). We look forward to the continued development of the novel structural and electronic properties of boron-phosphorus ring and cage compounds.

Acknowledgements

At the University of New Mexico, we thank the National Science Foundation (CHE-9983205) for generous support and at the Universität München, we thank Fonds der Chemischen Industrie for generous support. This paper is dedicated to Prof. Alan H. Cowley in celebration of his 70th birthday.

References

1. (a) West, R.; Fink, M. J. *Science* **1981**, *214*, 1343-1344; (b) West, R. *Angew. Chem. Intl. Ed. Engl.* **1987**, *26*, 1201-1211; (c) Cowley, A. H.; Norman, N. C. *Prog. Inorg. Chem.* **1986**, *34*, 1-63.
2. (a) Narula, C. K. *Ceramic Precursor Technology and Applications*; Marcel Dekker:New York, 1995; (b) Cowley, A. H.; Jones, R. A. *Angew. Chem. Intl. Ed. Engl.* **1989**, *28*, 1208-1215; (c) Gleizes, A. N. *Chem. Vapor Dep.* **2000**, *6*, 155-173; (d) Manasevit, H. M.; Hewitt, W. B.; Nelson, A. J.; Mason, A. R. *J. Electrochem. Soc.* **1989**, *136*, 3070-3076.

3. Paine, R. T.; Nöth, H. *Chem. Rev.* **1995**, *95*, 343-379.
4. Kumashiro, Y. *J. Mater. Res.* **1990**, *5*, 2933-2947.
5. Chu, T. L.; Jackson, J. M.; Hyslop, A. E.; Chu, S. C. *J. Appl. Phys.* **1971**, *42*, 420-424.
6. (a) Groshens, T. J.; Higa, K. T.; Nissan, R.; Butcher, R. J.; Freyer, A. J. *Organometallics* **1993**, *12*, 2904-2910; (b) Groshens, T. J.; Johnson, C. E. *J. Organometal. Chem.* **1994**, *480*, 11-14.
7. (a) Rudolph, R. W.; Parry, R. W.; Farron, C. F. *Inorg. Chem.* **1966**, *5*, 723-726; (b) Rudolph, R. W.; Parry, R. W. *J. Am. Chem. Soc.* **1967**, *89*, 1621-1625; (c) Durig, J. R.; Li, Y. S.; Correira, L. A.; Odom, J. D. *J. Am. Chem. Soc.* **1973**, *95*, 2491-2496.
8. Parshall, G. W.; Lindsey, Jr., R. V. *J. Am. Chem. Soc.* **1959**, *81*, 6273-6275.
9. Leffler, A. J.; Teach, E. G. *J. Am. Chem. Soc.* **1960**, *82*, 2710-2712.
10. (a) Nöth, H.; Schrägle, W. *Z. Naturforsch Part B* **1961**, *16B*, 473-474; (b) Nöth, H.; Schrägle, W. *Chem. Ber.* **1965**, *98*, 352-362.
11. Wood, G. L.; Dou, D.; Narula, C. K.; Duesler, E. N.; Paine, R. T.; Nöth, H. *Chem. Ber.* **1990**, *123*, 1455-1459.
12. Hamilton, W. C. *Acta Cryst.* **1955**, *8*, 199-206.
13. Bullen, G. J.; Mallinson, P. R. *J. Chem. Soc. Dalton Trans.* **1973**, 1295-1300.
14. Lube, M. S.; Wells, R. L.; White, P.S. *Inorg. Chem.* **1996**, *35*, 5007-5014.
15. Chen, T.; Duesler, E. N.; Kaufmann, B.; Paine, R. T.; Nöth, H. *Inorg. Chem.* **1997**, *36*, 1070-1075.
16. Chen, T.; Duesler, E. N.; Paine, R. T.; Nöth, H. *Inorg. Chem.* **1997**, *36*, 802-808.
17. Chen, T.; Duesler, E. N.; Paine, R. T.; Nöth, H. *Inorg. Chem.* **1997**, *36*, 1534-1535.
18. Chen, T.; Duesler, E. N.; Paine, R. T.; Nöth, H. *Chem. Ber.* **1997**, *30*, 933-937.
19. Chen, T.; Duesler, E. N.; Paine, R. T.; Nöth, H. *Inorg. Chem.* **1998**, *37*, 490-495.
20. Kaufmann, B.; Nöth, H.; Schmidt, M.; Paine, R. T. *Inorg. Chim Acta* **1998**, *269*, 101-110.
21. Moezzi, A.; Olmstead, M. M.; Pestana, D. C.; Ruhlandt-Senge, K.; Power, P. P. *Main Group Chemistry* **1996**, *1*, 197-206.
22. (a) Jaska, C. A.; Bartole-Scott, A.; Manners, I. *J. Chem. Soc. Dalton Trans.* **2003**, 4015-4021; (b) Horn, D.; Rodezno, J. M.; Brunnhöfer, B.; Rivard, E. Massey, J. A.; Manners, I. *Macromolecules* **2003**, *36*, 291-297.

23. (a) Scheschkewitz, D.; Amii, H.; Gornitzka, H.; Schoeller, W. W.; Bourissou, D.; Bertrand, G. *Science* **2002**, *295*, 1880-1881; (b) Amii, H.; Vranicar, L.; Gornitzka, H.; Bourissou, D.; Bertrand, G. *J. Am. Chem. Soc.* **2004**, *126*, 1344-1345.
24. (a) Driess, M. *Adv. Inorg. Chem.* **2000**, *50*, 236-285; (b) Linti, G.; Schnockel, H. *Coord. Chem. Rev.* **2000**, *206*, 285-319; (c) Kostler, W.; Linti, G. *Eur. J. Inorg. Chem.* **2001**, 1841-1846.

Chapter 12

History of a Paradigm Shift: Multiple Bonds to Silicon

Robert West

Organosilicon Research Center, University of Wisconsin, Madison, WI 53706

In 1981, the first compounds containing Si=C and Si=Si bonds were reported. These discoveries overturned the belief that multiple bonds to the heavier main group elements were unstable (the "double bond rule"), and led to the intense development of chemistry in this field. In this chapter the history leading up to these discoveries is recounted, and some recent developments in the multiple bond chemistry of silicon are described.

Introduction

Science advances in many different ways. Sometimes, as Thomas Kuhn pointed out in his famous book, a paradigm shift can take place, in which a new discovery or insight revolutionizes the pattern of belief in a given area (1). Such a change occurred in main group chemistry in 1981, with the discovery of multiple bonds to silicon.

Early in the 20th century, with multiple bonds between carbon and other first row elements well established, the search for multiple bonds to heavier main group elements attracted the attention of many chemists. Especially for silicon, the immediate congener to carbon, it doubtless seemed only reasonable that multiple bonds should be possible. F. S. Kipping, the pioneer Organosilicon chemist, made numerous attempts to prepare compounds with Si=C, Si=Si, and Si=O bonds between 1900 and 1920. At one time he believed that he had synthesized a disilene, Ph(Et)Si=Si(Et)Ph (*2*). However upon close examination this, and all the compounds obtained in his investigations, proved to be singly bonded oligomers or polymers. Kipping eventually concluded that multiple bonds to silicon were not possible. Similar results were found for other main group elements – thus for instance the structure of the early arsenical chemotherapeutic agent "Salvarsan", active against syphilis, was originally written with an As=As double bond, but later shown to be a mixture of singly-bonded oligomers (*3*). Double bonds were likewise claimed, and later shown not to exist, for other heavier elements, including phosphorus, antimony and germanium.

These uniformly negative results were rationalized in theoretical papers by Mulliken (*4*) and Pitzer (*5*), which showed that π bonds between elements outside the first row of the periodic table should be weak. The experimental and theoretical results led to the consensus embodied in the "double bond rule": "Elements with principal quantum numbers greater than 2 do not form double bonds with themselves or other elements" (*6,7*). Textbooks of inorganic chemistry generally reflected this view until the 1980's.

A paradigm shift took place however in 1981. At the end of the 15th Organosilicon Symposium in Durham, NC, in back-to-back papers, the synthesis of stable compounds containing Si=C (*8*) and Si=Si (*9*) double bonds were announced. Soon after that stable diphosphenes, with P=P double bonds, were reported by Yoshifuji (*10*) and by Cowley (*11*) and their coworkers. These findings led decisively to the overthrow of the double bond rule, and opened the way for a great flowering of multiple-bond chemistry of the heavier main-group elements over the past 23 years.

In this chapter, the events leading up to the discovery of multiply bonded silicon compounds will be outlined, beginning with the Si=C compounds, silenes. Then, some of the recent developments in multiple-bond chemistry of silicon will be described. Finally some yet unknown types of multiple bonds will be mentioned, as challenges and research opportunities for the future.

A number of useful reviews of doubly-bonded silicon compounds have been published (*12, 13*). The early history of this field is covered in a review by Gusel'nikov and Nametkin (*14*).

Silenes, Si=C

The discovery of molecules containing new types of chemical bonding sometimes occurs by the following progression: First, the new molecule is proposed as an unstable intermdiate in chemical reactions. Second, an example of the new species may be isolated in a matrix, at very low temperatures. Finally, the new molecule, stabilized by appropriate substitution, may be obtained as a stable compound at room temperature. The discovery of the silicon-carbon double bond followed exactly this sequence.

The initial experiments, involving the heating of 1,1-dimethyl-1-silacyclobutane (**1**) to 700-1000 K in the gas phase, were carried out by Gusel'nikov and coworkers (Scheme 1) (*15*). The products of this thermolysis are ethene and 1,1,3,3-tetramethyl-1,3-disilacyclobutane. The process mirrors closely the thermolysis of 1,1-dimethylcyclobutane to ethane and isobutene; even the kinetic parameters and activation energies are similar for the two processes. The intermediacy of dimethylsilene (**2**) was further evidenced by trapping reactions, for example with alcohols as shown in Scheme 1.

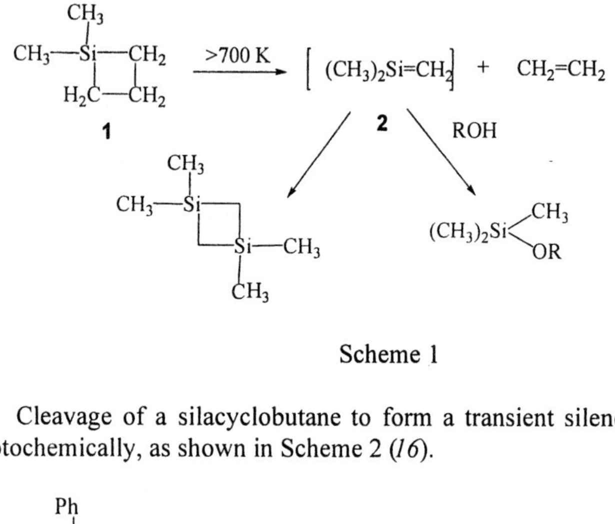

Scheme 1

Cleavage of a silacyclobutane to form a transient silene also takes place photochemically, as shown in Scheme 2 (*16*).

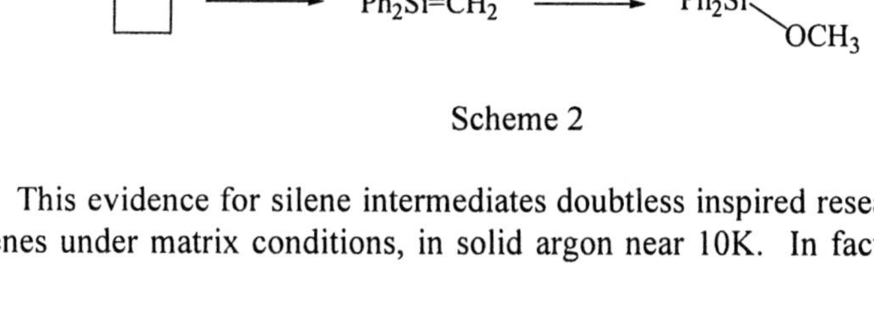

Scheme 2

This evidence for silene intermediates doubtless inspired research to isolate silenes under matrix conditions, in solid argon near 10K. In fact three groups accomplished this at almost the same time, in 1960. Chapman and coworkers (17), and Chedekel et al. (18), generated a silene by photolyzing trimethylsilyldiazomethane, yielding an intermediate carbene which rearranged to the silene (Scheme 3). Mal'tsev and coworkers (19) employed the thermolytic method shown in Scheme 1. Study of the trapped silene by infrared spectroscopy showed a strong band near 1000 cm^{-1}, assigned to the Si=C stretching vibration.

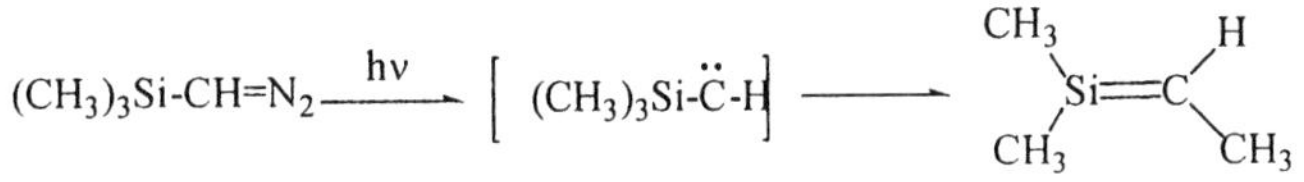

$$(CH_3)_3Si\text{-}CH\text{=}N_2 \xrightarrow{h\nu} \left[(CH_3)_3Si\text{-}\ddot{C}\text{-}H \right] \longrightarrow \begin{array}{c} CH_3 \\ \diagdown \\ CH_3 \end{array} Si\text{=}C \begin{array}{c} H \\ \diagup \\ CH_3 \end{array}$$

Scheme 3

Fifteen years then passed before a stable silene was isolated. The discovery came about from a series of fortunate incidents. In the laboratories of Adrian Brook at the Universty of Toronto, acylsilanes (silyl ketones) were being intensively studied. Photolysis of acylsilanes led to rearrangement, with the silyl group moving from carbon to oxygen to give the a siloxycarbene intermediate, which could be trapped with alcohols as shown in scheme 4. These workers than made the fortunate choice to investigate the photolysis of a disilanyl ketone. In this case, the beta silicon atom rearranged rather than the alpha silicon atom, to give a transient silene.

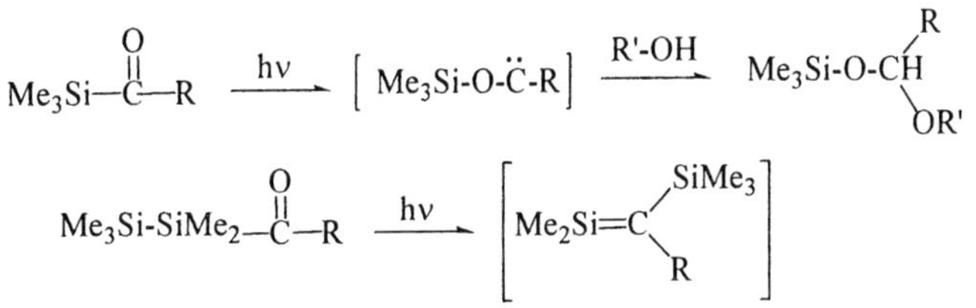

$$Me_3Si\text{-}\overset{O}{\overset{\|}{C}}\text{-}R \xrightarrow{h\nu} \left[Me_3Si\text{-}O\text{-}\ddot{C}\text{-}R \right] \xrightarrow{R'\text{-}OH} Me_3Si\text{-}O\text{-}\overset{R}{\underset{OR'}{\overset{|}{C}H}}$$

$$Me_3Si\text{-}SiMe_2\text{-}\overset{O}{\overset{\|}{C}}\text{-}R \xrightarrow{h\nu} \left[Me_2Si\text{=}C\overset{SiMe_3}{\underset{R}{\diagdown}} \right]$$

Scheme 4

The hunt for a stable silene was now on. Photolysis of a tris(trimethysilyl)silyl ketone gave a silene that self-trapped by an unusual head-to head dimerization. Replacement of the carbon substituent in the starting ketone by a t-butyl group yielded a silene in equilibrium with its head-to-head dimer, in solution. The final step was replacement of the t-butyl group by 1-adamantyl. Photolysis produced the stable silene **3**, an intensely reactive white crystalline substance (Scheme 5) (20).

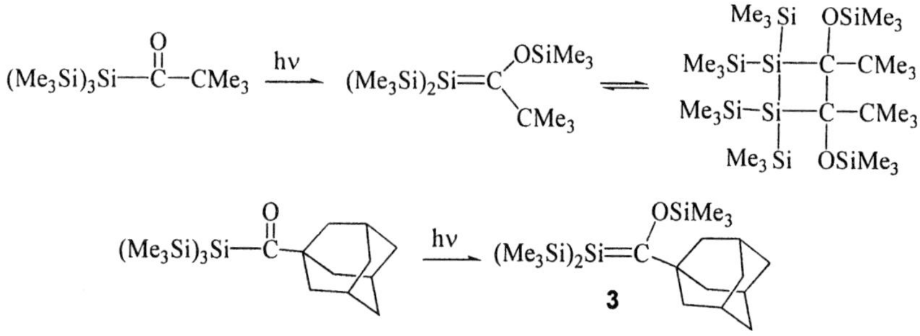

Scheme 5

In compound **3**, the Si=C double bond is strengthened and lengthened by ylidic electron donation by oxygen, as shown in the resonance structure **3a**. X-ray crystallography showed the Si=C distance to be 176.4 pm, somewhat longer than the distance predicted from theoretical calculations, 170 pm (*21*). Later the silene **4**, lacking ylidic character, was synthesized by Wiberg et al., and shown to have a normal Si=C bond length of 170.2 pm (*22*).

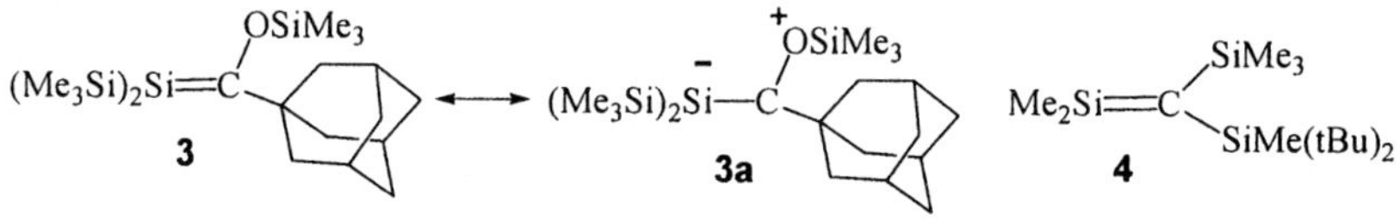

Many reactions of **3** have been investigated by Brook and his group, and were reviewed by him in 1996 (*13d*). Some recent developments on silene chemistry include the synthesis of a cyclic silene (**5**) by rearrangement of a silylene (*23*), and the prepartion of silenes from polysilanyllithium compounds and ketones (Scheme 6) (*24*). This reaction pathway has recently provided the first bis-silene (*25*), as well as a 4-silatriafulvene, **6** (*26*).

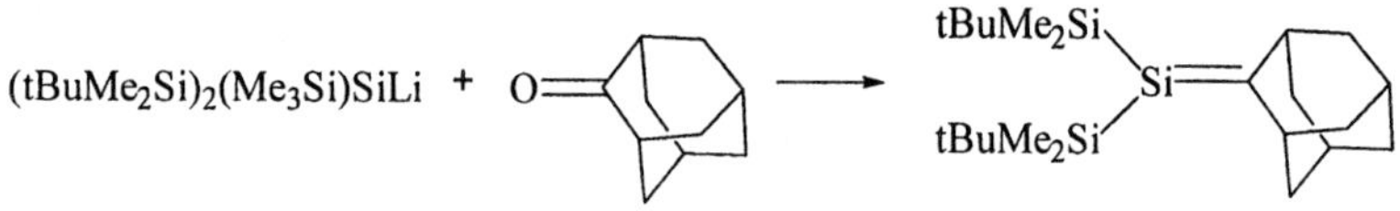

Scheme 6

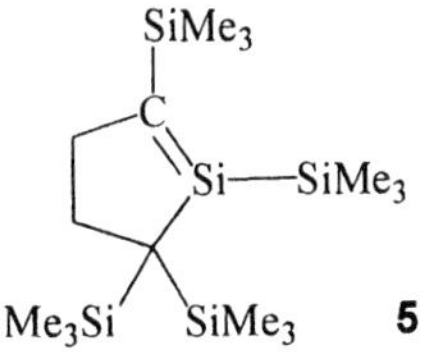

In our recent research at Wisconsin, we have employed silyldilithium compounds to generate novel types of silenes. Reaction of dilithiotetraphenylsilole with 2-adamantanone yields a 5-silapentafulvalene, as shown in scheme 7 (*27*).

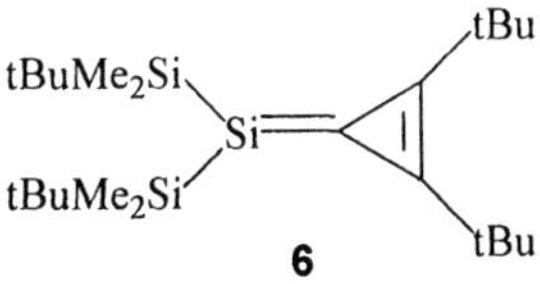

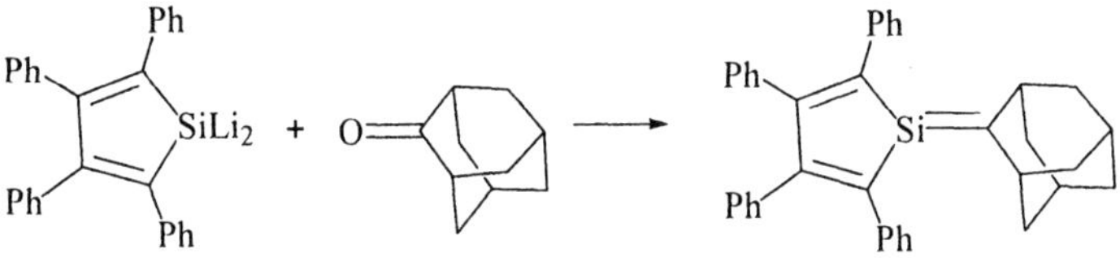

Scheme 7

Pentatriafulvalenes, **7**, are aromatic compounds; charge shift from the three to the five-membered ring gives the doubly aromatic structure **7a**. These compounds are known as "calicenes" from the fancied resemblance of the structure to a wine cup (Greek "calyx", cup). Reaction of the dilithiosilole with di-t-butylcyclopropeneone produced silapentatriafulvalene **8**, the first example of a heterocalicene (scheme 8) (*28*).

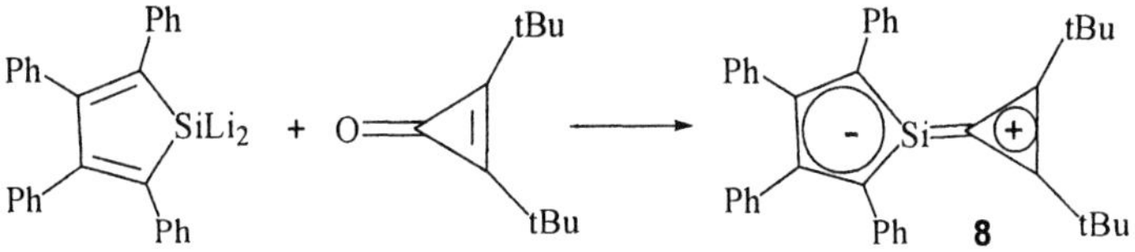

Scheme 8

Aromatic compounds in which a silicon atom occupies a position in the aromatic ring have recently been synthesized by Okazaki and Tokitoh (*29*). Examples are **9** and **10**. These compounds exhibit electronic properties quite similar to those of the all-carbon molecules, indicating that the silicon atoms participate actively in the electron delocalization and aromaticity.

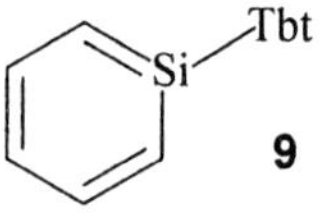

Disilenes, Si=Si

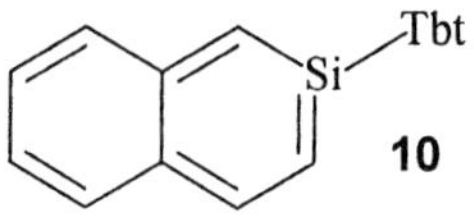

The synthesis of stable disilenes was preceded by evidence for Si=Si compounds as reaction intermediates, but not by matrix isolation of disilenes. The classic early experiments involved thermolysis of a 7,8-disilabicyclo-2,5-diene, leading to a retro-ene reaction forming the disilene (*30*). The latter was trapped with various unsaturated compounds, anthracene, naphthalene or 1,4-diphenyl-1,3-butadiene. The reaction with anthracene is shown in Scheme 9.

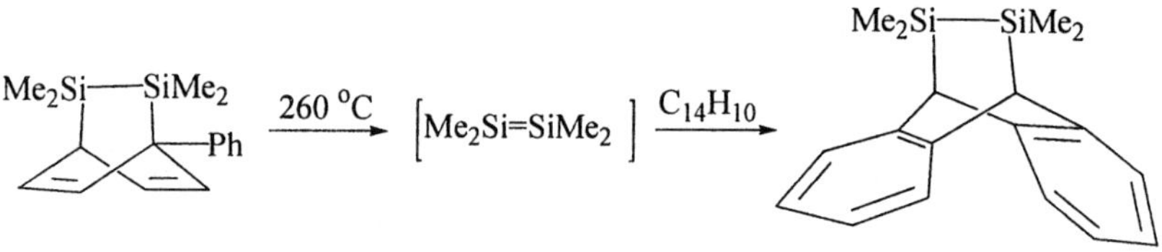

Scheme 9

These experiments provided good evidence for tetramethyldisilene as a transient intermediate, but did not suggest a route to stable Si=Si compounds. The latter were obtained in a somewhat accidental manner, in a collaboration between our group and that of Josef Michl on silylenes, $R_2Si:$. Silylenes may be generated by the photolysis of cyclic or linear polysilanes, as shown by the examples in Scheme 10.

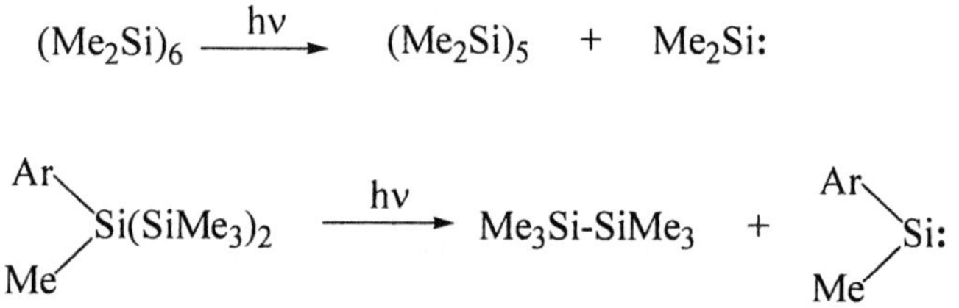

Scheme 10

When we carried out the photolysis of $(Me_2Si)_6$ in solid argon at 10 K, we were able to identify dimethylsilylene, Me_2Si:, trapped in the matrix. This provided the first direct observation of this key intermediate. The electronic and vibrational spectra of this silylene could then be investigated (*31*).

Eventually we found that silylenes could also be isolated in a hydrocarbon glass, 3-methylpentane, at liquid nitrogen temperature. This made their study far more convenient, since no liquid helium cryostat was necessary. We then investigated a series of silylenes with differing substituents, produced by photolysis of cyclosilanes or aryltrisilanes. The silylenes were identified in part by their strong uv-visible absorption, the result of a n-p transition from the lone pair on silicon into the vacant p orbital. The substituents greatly influence the energy of the electronic transition. Aryl substituents, or the presence of bulky groups, lead to a bathochromic shift of the electronic absorption band (*32*). Suitable substitution provided almost a complete rainbow of variously colored silylenes (Table I).

Table I. Absorption Maxima and Colors of Matrix-isolated Silylenes.

Silylene	λ_{max},nm	Color
Me_2Si	453	yellow
Et_2Si	460	yellow
$t\text{-}Bu_2Si$	480	orange
$Ph(Me)Si$	490	red
Ph_2Si	495	red
$Mes(Me)Si$	496	red
$Mes(Ph)Si$	530	purple
Mes_2Si	577	blue

Upon melting of the hydrocarbon matrix, the silylenes generally combine, producing colorless cyclic oligomers or polymers. With dimesitylsilylene however, we observed that upon warming the blue color of the silylene gave way to a yellow solution, which persisted up to room tempereature. In the crucial experiment by Mark Fink, careful evaporation of this solution yielded tetramesityldisilene, **11**, as bright yellow crystals (*33*) (Scheme 11).

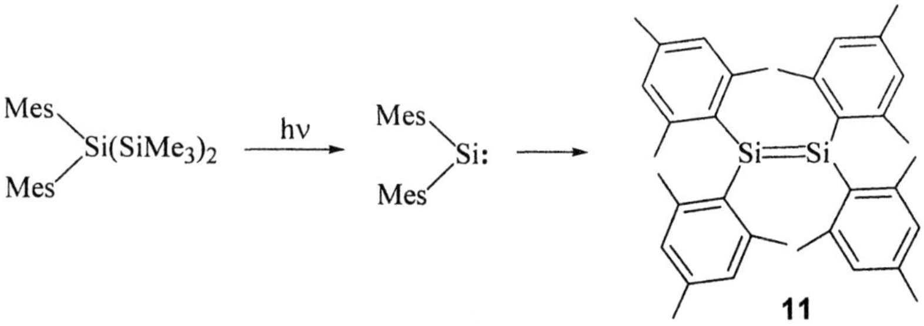

Mes₂Si(SiMe₃)₂ —hv→ Mes₂Si: → **11** (Mes₂Si=SiMes₂)

Scheme 11

An X-ray crystal structure of **11** was soon obtained. The disilene has a Si=Si bond length of 214 pm, about 21 pm shorter than the normal value for a Si-Si single bond. This prototypical disilene has surprisingly complicated solid-state behavior, with at least four different stereoisomeric forms. In all of these the silicon atoms are slightly trans-bent and the Si=Si double bond is slightly twisted (*34*).

Many other disilenes were synthesized following the discovery of **11**. Disilenes with differing substituents, such as tBu(Mes)Si=SiMes(tBu), can exist as Z or E stereoisomers. Studies of the kinetics of E-Z isomerization have been carried out for this and several other disilenes. The rotational barrier, between 25 and 31 kcal/mol., serves as a measure of the strength of the π-bond strength in disilenes (*35*).

The disilenes are brightly colored substances, exhibiting strong electronic absorption bands due to the π-π* transition with an energy of ~3 eV. This is roughly half as large as the corresponding excitation energy in alkenes. Comparing the disilenes with olefins, the π HOMO lies at higher energy and the π* LUMO at lower energy for disilenes. Therefore the disilenes are both more electron-rich and more electron-deficient than alkenes, and consequently they are highly reactive toward both electrophiles and nucleophiles. Reactions of dislienes have given rise to an elaborate chemistry. One result has been the synthesis of a large number of new ring systems, not available by any other means (*12*).

Recently several striking new structures containing silicon-silicon double bonds have been reported by Kira and coworkers, including trisilacyclopropenes **12** (*36*), a pentasilaspiropentadiene **13** (*37*), and a trisilaallene **14** (*38*). In addition the first molecule with two conjugated Si=Si bonds has been synthesized by Weidenbruch et al.(*39*).

Conversations regarding the Si=C and Si=Si discoveries are available online at http://www.sciencestudio.org/www.sciencestudio.org go to chemistry and search for Brook or West.

Other Multiply-bonded Silicon Compounds

The reports of **3** and **11** were followed by the synthesis of many other new types of multiply-bonded compounds of the heavier elements. Known multiply-bonded compounds containing silicon are listed, with the dates they were first prepared and the principal authors or laboratories, in Table II. A striking development was the synthesis of the first compound containing a silicon-silicon triple bond, by Sekiguchi and coworkers in 2004 (*39a*). Also numerous compounds with divalent silicon coordinated to metals have been synthesized, containing at least partial Si=M double bonds.

Table II. Types of Multiply-Bonded Silicon Compounds

Si=C	Brook, 1981	[N=Si=N]⁻	West, 1989
Si=Si	West, Michl, 1981	[P=Si=P]⁻	Niecke, 1989
Si=N	Wiberg, 1985	Si=C=C	West, 1993
Si=P	Bickelhaupt, 1989	Si=C=P	Escudie, 1999
Si=As	Driess, 1992	Si=Se	Tokitoh, 1999
Si=Ge	Baines, Sekiguchi, 1992	Si=Te	Tokitoh, 1999
Si=S	Okazaki, 1994	Si=Si=Si	Kira, 2003
	Si≡Si	Sekiguchi, 2004	

These however represent only a small part of the development of main-group multiple-bond chemistry. Since the paradigm shift of 1981, stable compounds have been obtained for virtually all of the elements of periodic groups 13, 14, 15 and 16.

Possibilites for the Future

Despite the intense activity in the chemistry of multiply-bonded silicon multiply-bonded silicon compounds, many challenges remain.

1. Silanones. Although these is abundant evidence for Si=O doubly-bonded species as reaction intermediates (*40*), and silanones have been isolated in matrix (*41*), no stable silanones have been reported. Theoretical calculations suggest that with appropriate substitution silanones should be isolable (*42*), and it seems likely that they will eventually be synthesized.

2. Triply-bonded silicon compounds. The recent synthesis of the first disilyne, RSi≡SiR, opens the way to a new area of triply-bonded silicon

chemistry. Other triply-bonded silicon species have often been implicated as reaction intermediates, and both HSi≡N and PhSi≡N have been isolated in matrix at cryogenic temperatures (43,44). Silynes HC≡SiF and HC≡SiCl have been identified by neutralization-reionization mass spectrometry (45). We can look forward to the synthesis of new disilynes (46), and of new kinds of triply-bonded silicon compounds, in the future .

3. Silyne and disilyne polymers. The synthesis of a conjugated tetrasilabutadiene (39) by Weidenbruch and coworkers shows that conjugated molecules of this sort are possible and stable. A major challenge now is to extend this conjugation, to make polyacetylene-like polymers, **15** and **16**. These polymers are expected to have electronic properties at least as striking as those of the well-known organic conductor, polyacetylene itself. Conducting and even superconducting behavior is possible for these unknown materials.

4. Reaction mechanisms. The reactivity of many of the new multiply-bonded compounds of the heavier elements has been qualitatively explored. Now, there are great opportunities to study the reaction mechanisms of the new chemical reactions. The reaction pathways for the new multiply-bonded compounds may be quite different from those for carbon compounds. Disilenes, for example, seem to undergo several reactions by unique pathways unknown for alkenes. This is an area of research still in its infancy, which may contain many surprises.

Acknowledgement. Research on multiply-bonded silicon compounds at the University of Wisconsin has been supported by grants from the National Science Foundation.

References

1. Kuhn, T. *The Structure of Scientific Revolutions;* University of Chicago Press, Chicago, IL, 1962.
2. Kipping, F. S., Sands, J. E. *J. Chem. Soc. (Trans.)* **1921,** *119,* 830, 848.
3. Cowley, A. J. *Polyhedron,* **1984,** *3,* 389.
4. Mulliken, R. S. *J. Am. Chem. Soc.,* **1955,** *77,* 884.

5. Pitzer, K. S. *J. Am. Chem. Soc.,* **1948**, *70*, 2140.
6. Jutzi, P. *Angew. Chem., Int. Ed.,***1975**, *14*, 232.
7. Sulfur was considered an exception, since thioketones were known.
8. Brook, A. G., Abdesaken, F., Gutekunst, B., Gutekunst, G., Kallury, K. K. *Abstracts of papers, 15th Organosilicon Symposium, Durham, NC, March 27-28,* **1981**, 45.
9. West, R., Fink, M. J. Michl, J. *Abstracts of papers, 15th Organosilicon Symposium, Durham, NC, March 27-28,* **1981**, 46.
10. Yoshifuji, M., Shima, I., Inamoto, N., Hirotsu, K., Higuchi, T. *J. Am. Chem. Soc.,* **1981**, *103*, 4587.
11. Cowley, A. H.; Kilduff, J. E.; Newman, T. H.; Pakulski, M. *J. Am. Chem. Soc.,* **1982**, *104*, 5820.
12. Okazaki, R.; West, R. *Adv. Organometal. Chem.,***1996**, 39, 232 (disilenes).
13. a) Raabe, G.; Michl, J. *Chem. Revs.* **1985**, *85*, 419. b) Raabe, G.; Michl, J., in Patai, S., Rappoport, Z., Ed's, *The Chemistry of Organic Silicon Compounds*; Wiley, Chichester,UK,1989, Chapter 17, pp. 1015-1142. c) Cowley, A. H. *Acc. Chem. Res.,* **1984**, *17*, 386. d) Brook, A, G.; Brook, M. A. *Adv. Organometal. Chem.,* **1996**, *39*, 71 (silenes). e) Hemme, I.; Klingebiel, U., *Adv. Organometal. Chem.,* **1996**, *39*, 159 (silaimines). f) Driess, M. *Adv. Organometal. Chem.,* **1996**, *39*, 193 (Si=P and Si=As compounds). g) Müller, T.; Ziche, W., Auner, N., in Rappoport, Z., Apeloig, Y., Ed's., *The Chemistry of Organic Silicon, Volume 2,* Wiley, Chichester, UK, 1998, Chapter 16 (Si=C and Si=N compounds). h) Tokitoh, N.; Okazaki, R., in Rappoport, Z., Apeloig,Y., Ed's., *The Chemistry of Organic Silicon, Volume 2,* Wiley, Chichester, UK, 1998, Chapter 17, pp. 1063-1103 (Si=chalcogen compounds).
14. Gusel'nikov, L. E.; Nametkin, N. S. *Chem. Revs.,* **1979**, *79*, 529.
15. a) Gusel'nikov, L.E.; Flowers, M. C. (1967) *Chem. Commun.,* **1967**, 864. b) Flowers , M. C.; Gusel'nikov, L. E.*J. Chem Soc. B,* **1968**, *419*, 1396.
16. Boudjouk , P.; Sommer, L. E., *Chem. Commun.,* **1973**, 54.
17. Chapman, O. L.; Chang, C.-C.; Kole, J.; Jung, M. E.; Lowe, J. A.; Barton, T. J.; Tumey, M. L. *J. Am. Chem. Soc.,* **1976**, *98*, 7844.
18. Chedekel, M. R.; Skoglund, M.; Dreeger, R. L.; Schechter, H. *J. Am. Chem. Soc.,* **1976**, *98*, 7846.
19. Mal'tsev, A. K.; Khabashesku, V. N.; Nefedov, O. M. *Izv. Akad. Nauk SSSR, Ser. Khim.,* **1976** 1193.
20. Brook, A. G.; Nyburg, S. C.; Abdesaken, F.; Gutekunst, B.; Gutekunst, G.; Kallury, R. K.; Poon, Y. C.; Chang, Y. M.; Wong-Ng, W. *J. Am. Chem. Soc.,* **1982**, *104*, 5667.
21. Grev, R. *Adv. Organometal. Chem.,* **1991**, *33*, 125.
22. Wiberg, N.; Wagner, G.; Müller, G. *Angew. Chem., Int. Ed.,* **1985**, *24*, 229.
23. Kira, M.; Ishida, S.; Iwamoto, T.; Kabuto C. *J. Am.Chem. Soc.,* **1999**, *121*, 9722.

24. Apeloig, Y.; Bendikov, M.; Yuzefovich, M.; Nakash, M.; Bravo-Zhivotovskii, D.; Blaser, D.; Boese, R. *J. Am. Chem. Soc.*, **1996**, *118*, 12228.
25. Bravo-Zhivotovskii, D.; Korogodsky, G.; Apeloig, Y. *J. Organomet. Chem.*, **2003**, *686*, 58.
26. Sakamoto, K.; Ogasawara, J.; Kon, Y.; Sunagawa, T.; Kabuto, C.; Kira, M. *Angew. Chem., Int. Ed.*, **2002**, *41*, 232.
27. Toulokhonova, I.; Guzei, I. A., West, R. *J. Am. Chem. Soc.*, **2004**, *126*, 5336.
28. Toulokhonova, I.; West, R. unpublished research.
29. a) Wakita, K.; Tokitoh, N.; Okazaki, R.; Nagase, S.; von Schleyer, P.; Jiao, H. *J. Am. Chem. Soc.*, **1999**, *121*, 11336. b) Wakita, K.; Tokitoh, N.; Okazaki, R.; Nagase, S. *Angew. Chem., Int. Ed.*, **2000**, *39*, 634.
30. Roark, D. N.; Peddle, G. J. D. *J. Am. Chem. Soc.*, **1972**, *94*, 5837.
31. a) Drahnak, T. J.; Michl, J.; West, R. *J. Am. Chem. Soc.*, **1979**, *101*, 5427. b) Arrington, C. A.; Klingensmith, K. A.; West, R., Michl., J. *J. Am. Chem. Soc.*, **1984**, *106*, 525.
32. Gaspar, P. P.; West, R., in Rappoport, Z.; Apeloig, Y., Ed's., *The Chemistry of Organic Silicon Compounds, Vol. 2, Part 3*, Wiley, Chichester, UK, 1998, Chapter 43, pp.2463-2568.
33. West, R.; Fink, M. J.; Michl, J. *Science (Washington DC)* **1981**, *214*, 2343.
34. Leites, L. A.; Bukalov, S. S.; Mangette, J. E.; Schmedake, T. A.; West, R. *Spectrochim. Acta*, **2003**, *59A*, 1975.
35. Michalczyk, M. J.; West, R.; Michl, J. *Organometallics*, **1985**, *4*, 826.
36. Iwamoto, T.; Tamura, M.; Kabuto, C.; Kira, M. *Organometallics*, **2003**, *22*, 2342.
37. Iwamoto, T.; Tamura, M.; Kabuto, C.; Kira, M. *Science (Washington DC)* **2000**, *290*, 504.
38. Ishida, S.; Iwamoto, T.; Kabuto, C.; Kira, M. *Nature (London, UK)* **2003**, *421*, 725.
39. Weidenbruch, M.; Willms, S.; Saak, W.; Henkel, G. *Angew. Chem, Int. Ed.* **1997**, *36*, 1433.
39a. Sekiguchi, A.; Kinjo, R.; Ichinohe, M. *Science* **2004**, *305*, 1755.
40. Fattakhova, D. S.; Jouikov, V. V.; Voronkov, M. G. *J. Organomet. Chem.* **2000**, *613*, 170.
41. Kabashesku, V. N.; Kudin, K. N.; Margrave, J. L.; Freidin, L. *J. Organomet. Chem.*, **2000**, *595*, 248.
42. Kimura, M.; Nagase, S. *Chem. Lett.*, **2001**, 1098.
43. Maier,G.; Glathaar, J. *Angew. Chem., Int. Ed.*, **1994**, *33*, 473.
44. Kuhn, W.; Sander, W. *Organometallics*, **1998**, *17*, 4776.
45. Karni, M.; Apeloig,Y.; Schröder, D.; Zummack, W.; Rabezzana, R.; Schwartz, H. *Angew. Chem., Int. Ed.*, **1999**, *38*, 332.
46. Kobayashi, K.; Takagi, N.; Nagase, S. *Organometallics*, **2001**, *20*, 234.

Chapter 13

Stable Derivatives of New Isomeric Forms of Heavier Group 14 Element Alkene Analogues

Philip P. Powers

Department of Chemistry, University of California, One Shields Avenue, Davis, CA 95616

The stabilization of new isomeric forms of heavier group 14 element (Si, Ge, Sn or Pb) analogues of ethylene are described. The simplest such compounds, involving hydrogen substituents only, were calculated by Trinquier to have structures that were quite different from the familiar planar one found in ethylene. The most stable arrangement for the disilene Si_2H_4 and the digermene Ge_2H_4 were similar to ethylene except that silicon and germanium had pyramidal instead of planar coordination. For Sn_2H_4 and Pb_2H_4, however, the trans-doubly bridged structures $HE(\mu-H)_2EH$ (E = Sn or Pb) were calculated to be the most stable. Moreover, the calculations also showed that several other minima could exist, for example, $HEEH_3$ or $HE(\mu-H)EH_2$, which had stabilities close to those of the trans-pyramidal or bridged structures. Nonetheless, up to the year 2000, the only stable examples of these structural forms that had been isolated were those that corresponded to the trans-pyramidal isomer. In this chapter, the synthesis and characterization of the first stable derivatives of the $HE(\mu-H)_2EH$, $HEEH_3$ and $HE(\mu-H)EH_2$ isomers are described.

Introduction

In the early 1990s calculations by Trinquier (*1,2*) showed that the heavier group 14 element analogs of ethylene could adopt at least five different isomeric forms as illustrated by the line drawings II-V in Figure 1.

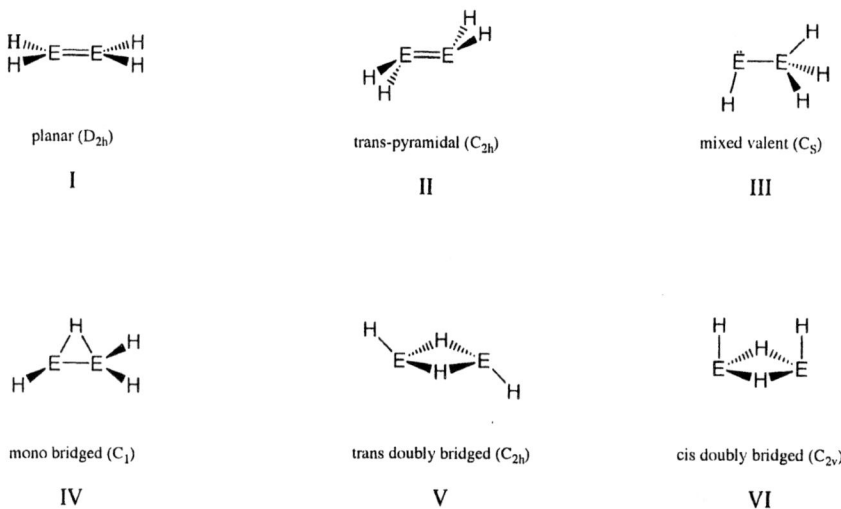

Figure 1. *Isomeric forms of E_2H_4 (E = C—Pb) identified as minima on the potential surfaces by Trinquier (1,2).*

The planar form I is observed as a minimum on the potential surface for carbon only. In contrast, several or all of the other five isomers are minima for each of the heavier elements E = Si, Ge, Sn and Pb. The planar form I is now known never to be a minimum for these elements, although the trans-pyramidal form II lies close to it in energy in the case of silicon and germanium. In sharp contrast, the trans-doubly-bridged isomer V was calculated to be the most stable for tin and lead. Uniquely, all five isomers II-VI were found to be minima on the potential surface for tin (Figure 2), including the singly bridged isomer IV, which was also calculated to be a minimum for lead (2). Recent calculations for Si$_2$H$_4$ with more sophisticated basis sets have shown that singly

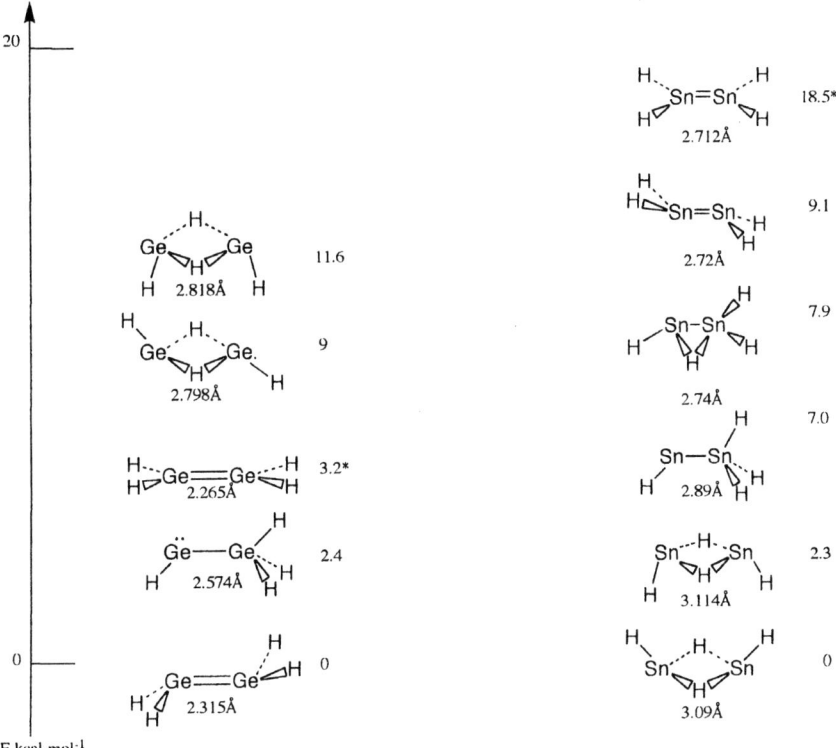

Figure 2. Relative energies (kcal mol^{-1}) and E-E bond lengths (E = Ge or Sn) calculated by Trinquier (1,2). *signifies that the planar isomer is not a minimum on potential surface.

bridged IV is also a minimum for silicon (3). It is only 7.16 kcal mol^{-1} less stable than the trans-pyramidal global minimum structure II. Apart from the structures the most striking feature of the computational data concerns the relatively small energy separations between the isomers as shown in Figure 2 for germanium and tin. For these elements just 11.6 and 9.0 kcal mol^{-1} separate the highest and lowest calculated energy minima (1,2). For the corresponding silicon derivatives, the hydrogen substituted isomers can be detected spectroscopically (3). Despite this a range of stable structural models for isomers of the heavier elements was unknown. As recently as the year 2000, the only isolated stable derivatives of the isomeric forms II-VI were those of the trans-pyramidal II. Four bulky alkyl, aryl, or silyl groups were generally used to achieve stability in these 'dimetallenes' and, since the beginning of the work in the 1970s (*4*), the structures of over 40 R_2EER_2 compounds (E = Si-Pb, R = bulky group) have been determined (*5*). The data confirmed an increasingly trans-bent geometry and weaker E-E bonds as atomic number increases Si $\longrightarrow$ Pb. Given the availability of theoretical data concerning hydrogen substituted models of isomeric forms III-VI since 1990, it is odd that no stable derivatives were reported. It seems probable that many workers believed that the isolation of stable bridged structures of type IV-VI was unlikely because no stable hydrogen bridged compounds of any kind were known for group 14 elements. Yet, parallel work on germanium, tin and lead alkyne analogues, hydrides and low valent species, which seemed unrelated to structures III-VI at the time, led to vital breakthroughs (see below) that allowed substituted analogues of all the remaining isomers except VI, to be isolated. In this review, aspects of the isolation of the first stable examples of substituted derivatives of III, IV and V are now described.

Derivatives of the Mixed Valence Isomer III

The first stable derivative of the mixed valent isomer III were discovered serendipitously. We were investigating the reactions of Ar*SnCl (Ar* = C_6H_3-2,6(C_6H_2-2,4,6-Pr^i_3)) with various organolithium reagents (LiR; R = Me, Bu^t, Ph, etc) in order to produce a series of derivatives of formula Ar'SnR for spectroscopic (^{119}Sn NMR, UV-Vis, etc.,) examination. It was found that the reaction with LiMe afforded, not the expected symmetric product Ar*(Me)SnSn(Me)Ar*, but the unsymmetric Ar*SnSn(Me)$_2$Ar* as shown in eq(1) (*6*).

$$2Ar^*SnCl + 2LiMe \longrightarrow Ar^*SnSn(Me)_2Ar^* + 2LiCl \quad (1)$$

Further reaction with LiMe yielded LiAr*(Me)SnSn(Me)$_2$Ar* eq(2)

$$Ar^*SnSn(Me)_2Ar^* + LiMe \longrightarrow LiAr^*(Me)SnSn(Me))_2 \quad (2)$$

Both compounds were characterized by X-Ray crystallography and by ^{119}Sn NMR spectroscopy. The latter readily distinguished the two tin centers in each compound and the X-Ray crystal data confirmed the unsymmetric structure of Ar*SnSn(Me)$_2$Ar* (Figure 3) which is, in effect, a substituted derivative of III. Steric arguments and bonding considerations supply a ready justification for why it is favored over the symmetric isomer. The steric argument is illustrated in eq(3).

$$RR'SnSnR'R \longrightarrow RSnSn(R')_2R \tag{3}$$

In the unsymmetric stannylstannylene form on the right of eq(3) is sterically disfavored if both R and R' are sterically demanding. This is because excessive crowding is induced on the triorganosubstituted tin center by the three large substituents. This explains why only the symmetric distannene structure on the left was observed in earlier studies (5) where homoleptic bulky ligand set was usually employed to stabilize the compounds. Even in the few heterdeptic

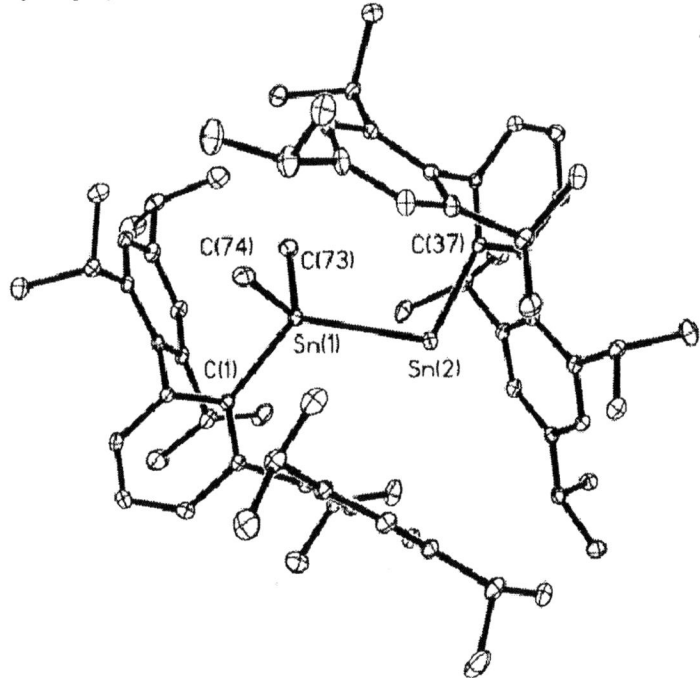

Figure 3. X-ray crystal structure of the unsymmetric Ar*SnSn(Me)$_2$Ar. The Sn-Sn distance is 2.8909(2)Å (6).

examples that are currently known, a combination of bulky ligands was always employed to achieve stability. In contrast, it can be seen that, if one of the ligands (R') used is relatively small (eg., Me), the stannylstannylene structure becomes sterically feasible since there would be no undue crowding introduced at the trionganosubstitued tin by an extra methyl group. However, this requires that the other ligand R be particularly bulky to stabilize the structure, and the terphenyl group Ar* fulfills this requirement. The lone pair effect, ie, the tendency of s-valence electrons to remain non-bonding in heavier p-block elements, explains the preference for structure III instead of II in the tin species (7).

The corresponding reaction of LiMe with Ar*GeCl yielded the symmetric trans-bent structure for Ar*(Me)GeGe(Me)Ar* analogous to II (8). This finding is in agreement with the calculations of Trinquier where structure II was predicted to be the most stable for germanium. The reaction of LiMe with Ar*PbBr produced only a monomer Ar*PbMe (9). Trinquier's calculations showed that the unsymmetric model species $HPbPbH_3$, III, is 17.2 kcal mol^{-1} less stable than the doubly bridged form V (which is the global minimum). It seems likely that III is disfavored for this reason. Furthermore, it seems probable that the doubly bridged structure V is not observed because of the lower bridging tendency of Me in comparison to H. In the case of tin, however, the unsymmetric structure III is only 7.0 kcal mol^{-1} less stable than the doubly bridged form V and is sufficiently stable to be observed.

Derivatives of the Trans-Doubly Bridged Isomer V

The calculations which predicted that $HE(\mu-H)_2EH$ would be the global minimum for tin and lead raised the possibility of isolating these compounds. However, no stable divalent hydride derivatives of any of group 14 elements were then known. In parallel work, we had shown that the reduction of Ar*PbBr(Ar* = C_6H_3-2,6-(C_6H_2-2,4,6-Pr^i_3)$_2$) with LiAlH$_4$ afforded Ar*PbPbAr* — the first stable heavier group 14 element alkyne analogue (10). Presumably, the reaction proceeds through a Ar*PbH intermediate which readily decomposes to give Ar*PbPbAr*. Subsequent experiments designed to synthesize the corresponding tin compound Ar*SnSnAr*, via treatment of Ar*SnCl with LiAlH$_4$ or HalBu$^i_2$, led instead to the isolation of Ar*Sn(μ-H)$_2$SnAr* – the first stable divalent hydride of a group 14 element (11). The bridged structure is illustrated in Figure 4 and it is consistent with Trinquier's prediction that V is the most stable hydride isomer for tin. The synthesis of the corresponding germanium hydride was then attempted to test Trinquiers calculations, which predicted that, unlike tin, it should have a trans-pyramidal (type II) structure. It was found that Al-H reducing agents were unsuitable for the reduction of Ar*GeCl to the hydride as they produced elemental germanium. Further investigation showed that the use of LiHBBU$^s_3$ gave the desired Ar'(H)GeGe(H)Ar' (Ar' = C_6H_3-2,6(C_6H_3-2,6-Pr^i_2)$_2$ species in moderate yield

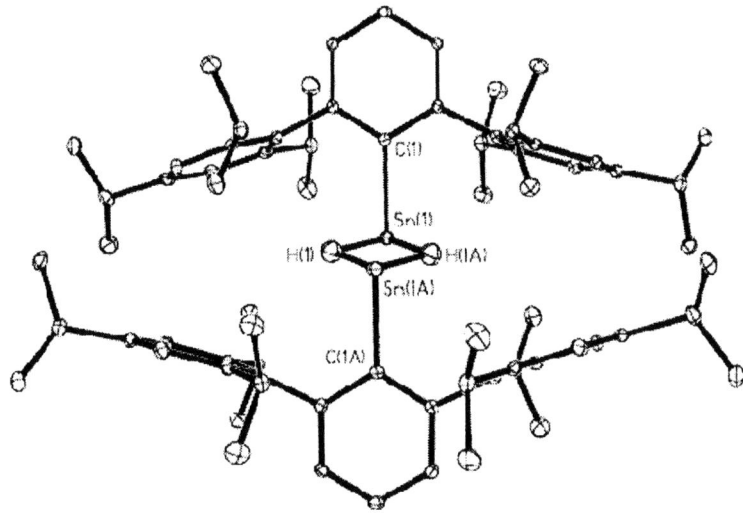

Figure 4. Illustration of the structure of Ar*Sn(μ-H)$_2$SnAr* (11) was the first stable group 14 divalent hydride to be isolated.

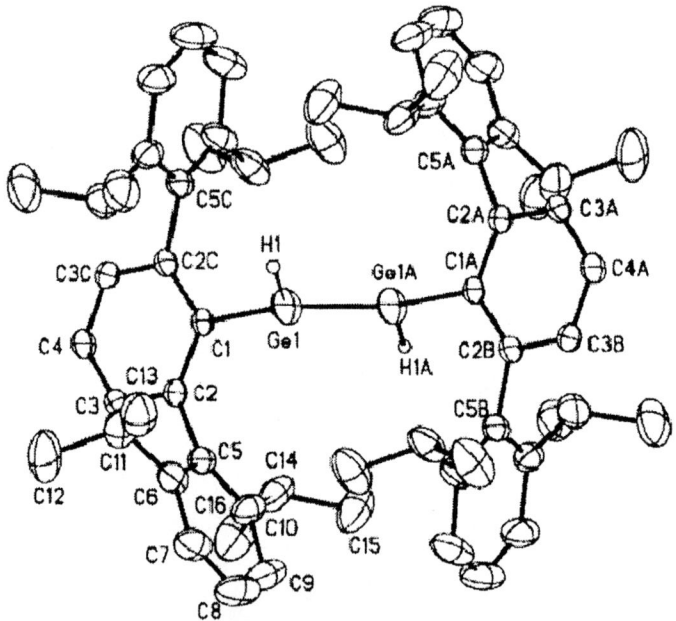

Figure 5. Illustration of the structure of Ar'(H)GeGe(H)Ar' (Ge – Ge = 2.372(1)Å) which, unlike the related tin species (Figure 4), has a trans-pyramidal configuration (12).

(12). X-ray crystallographic data revealed that this compound had, indeed, a trans-pyramidal structure (Figure 5) consistent with Trinquier's prediction. Further experiments, involving the addition of excess PMe₃ to the reaction mixture to remove BBu$^s_3$ byproduct by forming Me₃PBBu$^s_3$, afforded the unique adduct (Me₃P)Ar*GeGe(H)₂Ar* illustrated in Figure 6. In this compound, a hydrogen has been displaced from one germanium to the other to give an unsymmetric species of type III which is stabilized by PMe₃ donor. The ease with which this reaction occurs underlines the closeness of the stabilities of the two isomeric forms II and III for germanium (see Figure 2).

Derivatives of the Singly Bridged Isomer IV

Theoretical exploration of the potential energy surfaces of Sn_2H_4 and Pb_2H_4 led to the prediction that structure IV, in which there is a single E-H-E bridge as

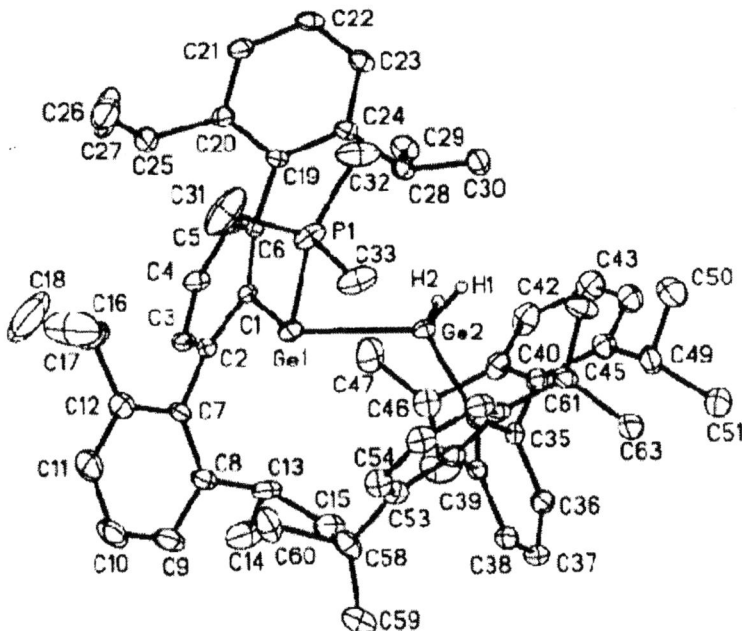

Figure 6. The addition of PMe₃ to Ar'(H)GeGe(H)Ar' (Figure 5) displaces a hydrogen from Ge(1) to Ge(2) to afford a phosphine stabilized unsymmetric type III structure (12).

well as a short E-E distance, lay within 8 (Sn) or 15(Pb) kcal mol^{-1} of the doubly hydrogen bridged global minima. More recent calculations by Schaefer (*3*) and co-workers for Si$_2$H$_4$ showed that the monobridged form IV had comparable stability to the unsymmetric III. Furthermore, the bridged species could be detected by microwave spectroscopy (*3*). In the earlier studies, Trinquier had pointed out that the monobridged structure represents an intermediate step between II and III. There is little doubt that II and III can be interconverted as the above mentioned interaction of PMe$_3$ with Ar'(H)GeGe(H)Ar' demonstrates. Other experimental results also support the intermediacy of a monobridged species. For example, treatment of Ar*SnCl with LiPh at or below –30°C afforded the unsymmetric Ar*SnSn(Ph)$_2$Ar*, Warming the green solution of this species to ca. 25°C resulted in a color change to red due to dissocation to two

Ar*SnPh monomers (*13*), which shows that the Ph group must be transferred between the tins. This monomer can be crystallized above ca. 10°C, and the equilibrium between these two species can be followed by variable temperature ^{119}Sn NMR spectroscopy. It is probable that the rearrangement from two Ar*SnPh to Ar*SnSn(Ph)$_2$Ar* monomers occurs via a weak Sn-Sn to give a symmetric structure Ar*(Ph)SnSn(Ph)Ar* (most R$_2$SnSnR$_2$ species exist in solution in equilibrium with the monomer SnR$_2$) which rapidly rearranges via a bridged intermediate Ar*Sn(μ-Ph)Sn(Ph)Ar* to Ar*SnSnPh$_2$Ar*. Nonetheless, this intermediate could not be trapped or even detected by NMR spectroscopy. However, related experiments involving the generation of benzyl substituted distannenes in accordance with

2Ar'SnCl + BrMgCH$_2$C$_6$H$_4$-4-Pr^i ⟶

Ar'Sn(μ-Br)Sn(Ar')CH$_2$C$_6$H$_4$-4-Pr^i + MgClBr   (4)

The product had the structure featured in Figure 7 (*14*). It displays a unique monobridged arrangement in which bromine bridges two tins that are separated by a relatively long Sn-Sn distance of 2.9407(4)Å. The bridging Sn-Br distances, 2.7044(5)Å (Sn(1)) and 2.7961(5)Å (Sn(2)), differ slightly. ^{119}Sn

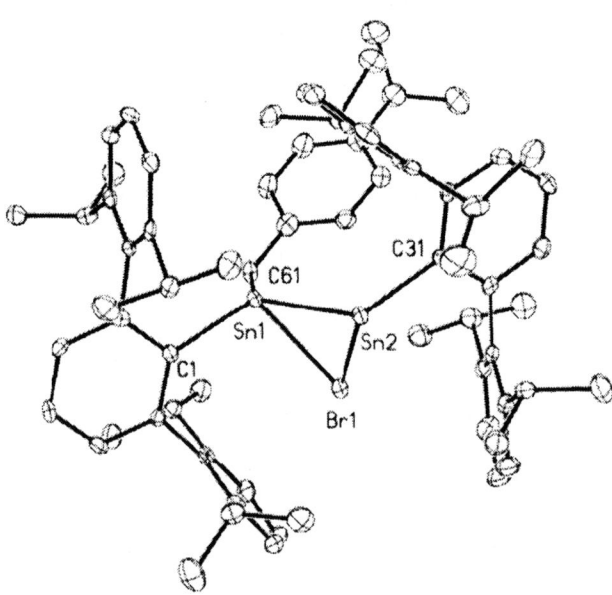

Figure 7. Illustration of the bridged structure of Ar'Sn(μ-Br(Sn(Ar')CH$_2$C$_6$H$_4$-4-Pr^i. Selected structural data are given in the text (14).

NMR spectroscopy affords two signals at δ 1399.8ppm (Sn(1)) and 2274.3ppm (Sn(2)). A separate experiment involving the reaction of a 1:1 ratio of Ar'SnCl and $BrMgCH_2C_6H_4$-4-Pr^i to give $[Sn(Ar')CH_2C_6H_4$-4-$Pr^i]_2$ which has one ^{119}Sn NMR signal with a chemical shift of 1120 ppm. This shift indicated that a symmetric Sn-Sn bonded type II structure for this species existed in solution. Unfortunately, this compound did not afford crystals that were suitable for X-ray crystallography. However, the closely related species $[Sn(Ar')CH_2C_6H_4$-4-$Bu^t]_2$, in which there is a Bu^t instead of an Pr^i group at the para position of the benzyl ring, confirmed the dimeric structure. It had an Sn-Sn distance of 2.7705(8)Å and it had an almost identical ^{119}Sn NMR chemical shift (1205.7 ppm) to that of the Pr^i substituted species in solution. It may be concluded, therefore, that the bromine bridged species also remains intact in solution, as its dissociation is expected to give a mixture of Ar'SnBr and $(Sn(Ar')CH_2C_6H_4$-4-$Pr^i)_2$, and no signals that corresponded to these species could be observed. The compound Ar'Sn(μ-Br)Sn(Ar')$CH_2C_6H_4$-4-Pr^i is thus a stable model species for the elusive singly-bridged type IV. At present, no other compounds of this type are known, but it is probably that other examples, including a hydrogen bridged species can be synthesized by suitable manipulation of the substituents.

Conclusions

The results described above have shown that it is possible to isolate the first examples of heavier group 14 element species whose structures correspond to the model hydrides III, IV and V. Only the cis-doubly bridged structural type VI remains unmodelled. In addition, this account has emphasized the role that serendipity played in their discovery. The importance of Trinquier's calculations on the simple model species, which accurately predicted structures almost ten years prior to the isolation of stable derivatives, cannot be underestimated

Acknowledgements

The financial support of the National Science Foundation is gratefully acknowledged. The author thanks his coworkers, named in the references, whose skill and dedication made these results possible. In addition, the author thanks Professors M. Lattman and R. Kemp for their invitation to present results both here and at the Main Group Symposium honoring Professor A.H. Cowley at the Spring 2004 American Chemical Society meeting at Anaheim.

References

1. Trinquier, G., *J. Am. Chem. Soc.,* **1990**, *112*, 2130.
2. Trinquier, G.; Barthelat, J-C., *J. Am. Chem. Soc.,* **1990**, *112*, 9121.
3. Sari, L.; McCarthy, M.C.; Schaefer, H.F.; Thaddeus, P. *J. Am. Chem. Soc.,* **2003**, *115*, 11409.
4. Goldberg, D.E.; Harris, D.H.; Lappert, M.F.; Thomas, K.M., *Chem. Commun.,* **1976**, 261.
5. Weidenbruch, M., *Organometallics,* **2003**, *22*, 4348.
6. Eichler, B.E.; Power, P.P., *Inorg. Chem.,* **2000**, *39*, 5444.
7. Hall, M.B., *Inorg. Chem.,* **1978**, *17*, 2261.
8. Stender, M.; Pu, L.; Power, P.P., *Organometallics,* **2001**, *20*, 1820.
9. Pu, L.; Twamley, B.; Power, P.P., *Organometallics,* **2000**, *19*, 2874.
10. Pu, L.; Twamley, B.; Power. P.P., *J. Am. Chem. Soc.,* **2000**, *122*, 3524.
11. Eichler, B.E.; Power, P.P., *J. Am. Chem. Soc.,* **2000**, *122*, 8785.
12. Richards, A.F.; Phillips, A.D.; Olmstead, M.M.; Power, P.P., *J. Am. Chem. Soc.,* **2003**, *125*, 3204.
13. Phillips, A.D.; Hino, S.; Power, P.P., *J. Am. Chem. Soc.,* **2003**, *125*, 7520.
14. Stanciu, C.; Richards, A.F.; Power, P.P., *J. Am. Chem. Soc.,* **2004**, *126*, 4106

Coordination Chemistry

Chapter 14

β-Diketiminates of Some Main Group Elements: New Structural Motifs

Laurence Bourget-Merle[1], Yanxiang Cheng[1], David J. Doyle[1], Peter B. Hitchcock[1], Alexei V. Khvostov[1], Michael F. Lappert[1,*], Andrey V. Protchenko[1], and Xue-hong Wei[1,2]

[1]The Chemistry Laboratory, University of Sussex, Brighton, UK BN1 9QJ
United Kingdom
[2]The Institute of Modern Chemistry, Shanxi University, Taiyuan 030006, Peoples Republic of China

A number of crystalline metal (Li, K, Al, Sn) β-diketiminates have been prepared and X-ray-characterized. Each shows unusual metal-to-ligand bonding modes. In the case of the reduction of Li[{N($C_6H_3Pr^i_2$-2,6)C(H)}$_2$CPh] by lithium in the presence of TMEDA, C–N homolysis with ligand fragmentation was observed.

Introduction

β-Diketiminates are useful spectator ligands. They bind strongly to metals, have readily tuneable steric and electronic demands on the metal, and show a variety of bonding modes. The β-diketiminato ligands are able to stabilize compounds not only in an unusually low state of molecular aggregation but also in a low metal oxidation state, as cations, and complexes containing metal multiply bonded co-ligands. Many β-diketiminatometal complexes have a useful role as catalysts or biomolecular mimics. The topic has been reviewed.[1] Our studies in this field date back to 1994,[2] and our most recent publication is found in ref. 3.

β-Diketiminato ligands which are widely used are close relatives of acac⁻, with the two oxygen atoms of acac⁻ replaced by NR moieties, as in **1**. We have been much involved with ligands of formula **2**; another featured in this paper has formula **3**. Each of **1 – 3** is shown in its monoanionic π–delocalized form.

Ar = $C_6H_3Pr^i_2$-2,6

2a R^1 = Ph = R^2
2b R^1 = C_6H_4Me-4 = R^2
2c R^1 = $C_6H_4Bu^t$-4 = R^2
2d R^1 = Ph, R^2 = Bu^t

Ar = $C_6H_3Pr^i_2$-2,6

The β-diketiminato ligand binds to metals either in a terminal or a bridging mode; examples are shown in **a – n** of Figure 1. The former is usually N,N'-chelating and is attached to the metal either (i) in a four-electron σ-bond fashion, the metallacycle being planar as in **a**,[4] or has the metal out of the NCCCN plane as in **b**,[5] or (less usually) (ii) in a six-electron 2σ + 4π η⁵-fashion with the MNCCCN ring adopting a boat conformation as in **c**.[2] Rarer are the monodentate N- (as in **d**[6]) or C- (as in **e**, X = Cl[7] or **f**, X = Ph[8]) centered metal β-diketiminates. Similarly, the bridging β-diketiminates may be terminal as in **g**[9] or, more frequently, chelating as in **h**,[2] **i**,[10] **j**,[11] or **k**.[12]

Of the more than 500 known β-diketiminatometal complexes, in every case but three the ligand has been monoanionic. The exceptions were the ytterbium complexes **l**,[13] **m**,[13] and **n**.[14] In **l** and **m** the ligand **2a** (abbreviated as $L^{Ph, Ph}$) was regarded as dianionic. As shown in **n** of Figure 1, two of the ligands were assigned on the basis of detailed structural data as trianionic, one bound to Yb(II) and the other to Yb(III), while the third was attached to Yb(II). The structures of the crystalline complexes **l** to **n** were consistent with the ligand charge distribution shown in valence bond terms for $[L^{Ph, Ph}]^{2-}$ and $[L^{Ph, Ph}]^{3-}$ in **4** and **5**, respectively.[13, 14] These structures also illustrate thus far unique bonding modes: both N, N'-bridging and chelating in **l** and **m**, and multicentered chelating and bridging for the trianionic ligands in **n**.

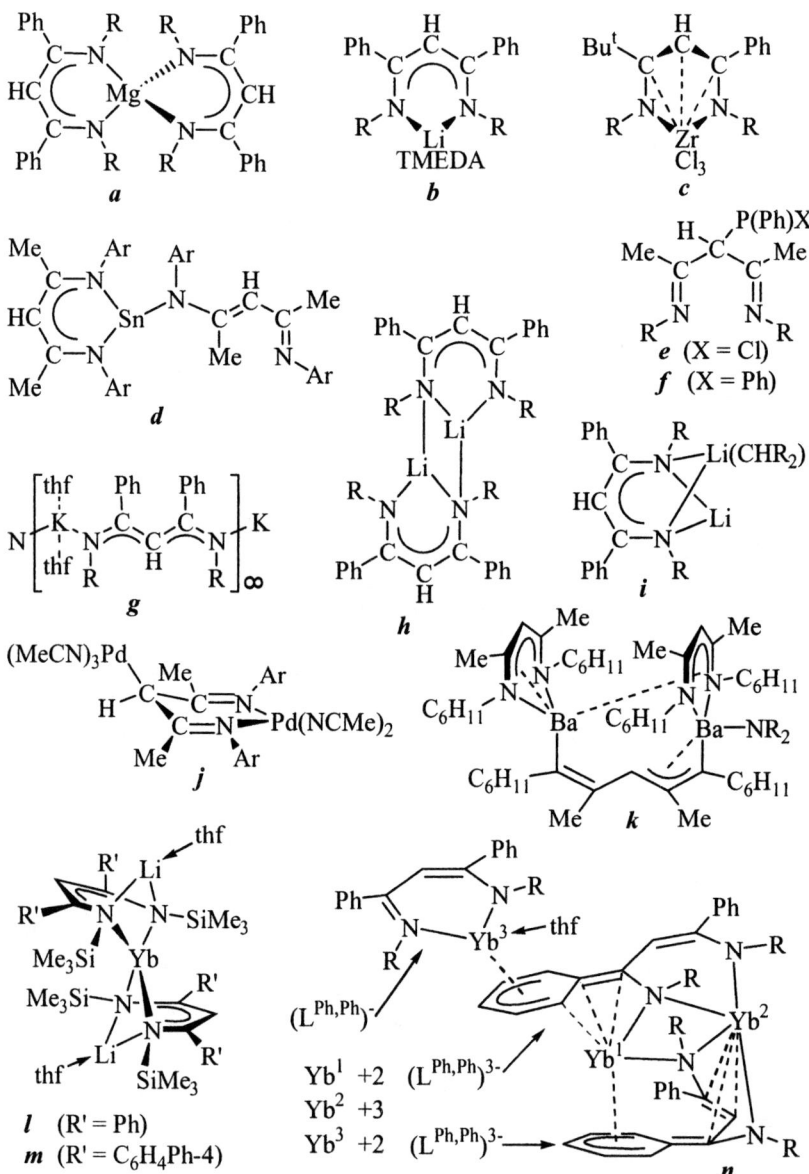

Figure 1. Examples (*a* – *n*) of metal β-diketiminates, each showing different ligand-to-metal bonding modes (R = SiMe$_3$, Ar = C$_6$H$_3$Pr$^i_2$-2,6).

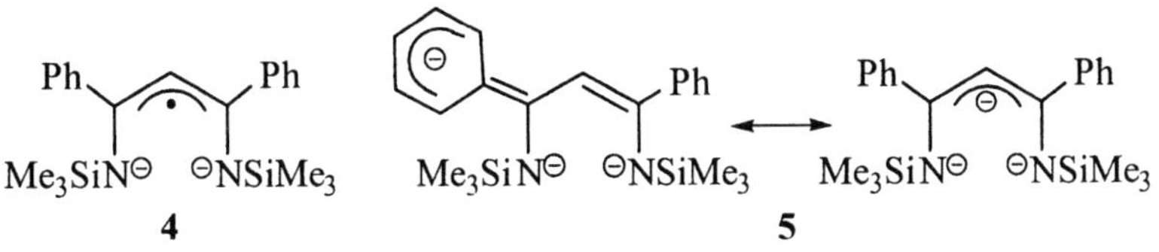

Results and Discussion

The reaction of an alkali metal (M) trimethylsilylmethyl with an α–H-free nitrile is exemplified in Scheme 1 for the Li(CHR$_2$)/PhCN system (R = SiMe$_3$).[5] Noteworthy features are the following. (i) In the presence of a strong neutral donor ligand (TMEDA in Scheme 1), the 1:1 reaction yielded the lithium enamide, whereas in its absence, the 1,2-adduct the lithium β-diketiminate was obtained. (ii) The latter was also prepared from the enamide and PhCN, via the thermodynamically less stable 1,3-diazaallyl isomer as the intermediate. (iii) The reactions of Scheme 1 demonstrate not only C-C coupling but also 1,3-shifts of SiMe$_3$ from carbon to nitrogen; furthermore, the latter were preferred to migrations of hydrogen from C→N.

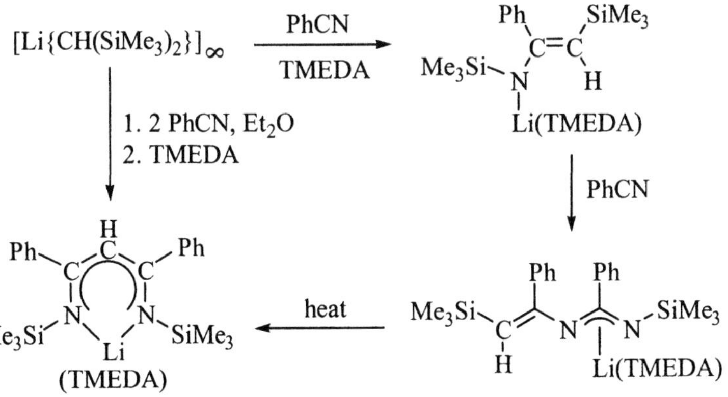

Scheme 1. Reactions of Li[CH(SiMe$_3$)$_2$] with PhCN.

Lithium β-Diketiminates from Li[CH(SiMe$_3$){SiMe$_{3-n}$(OMe)$_n$}] (n = 1 or 2)[15]

The reactions of [Li{CH(SiMe$_3$)(SiMe$_2$OMe)}]$_8$[16] (the monomer is abbreviated as LiL^1) with various nitriles are summarized in Scheme 2. These have yielded the alkali metal β-diketiminates **6 - 9**, as well as the β-diketimine

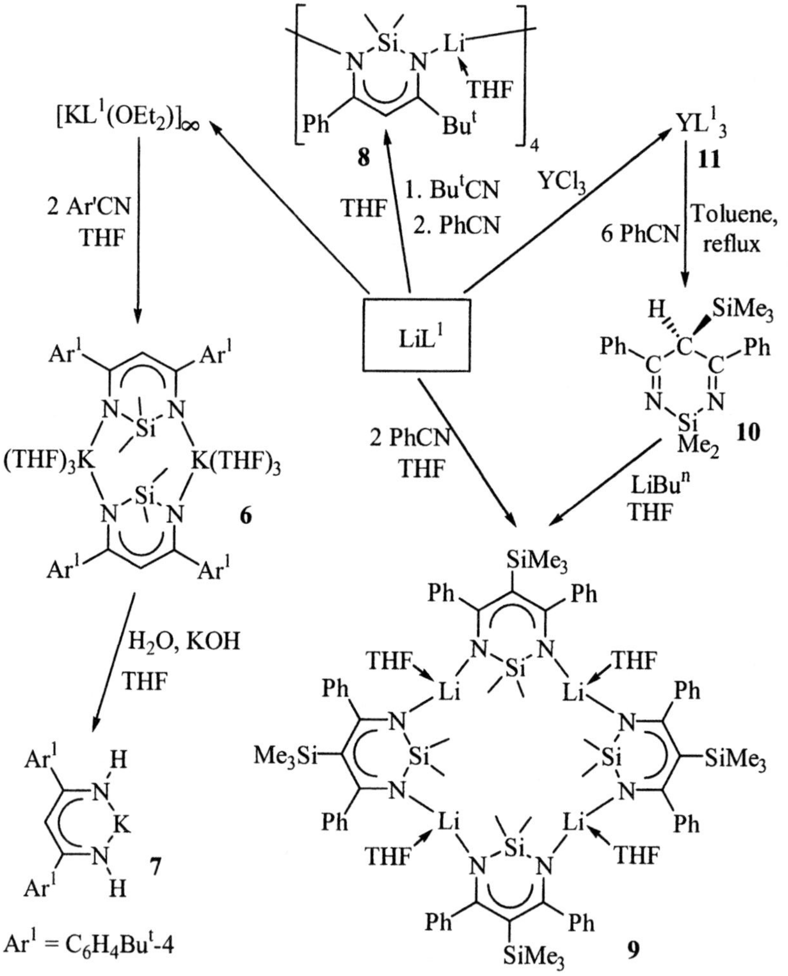

Scheme 2. Li and K β-diketiminates from M[CH(SiMe$_3$)(SiMe$_2$OMe)] (ML^1).

10, via the yttrium β-diketiminate **11**. An analogue of **6**, having the ligand **12a** had previously been prepared by a different procedure.[9] The following features are significant. (i) New β-diketiminato ligands **12a – 12d** have been generated in which the nitrogen atoms are bridged by $SiMe_2$ groups. (ii) The 1,3-silyl C→N migration is regiospecific, with $SiMe_2OMe$ preferred over $SiMe_3$. (iii) Compounds **8** and **9** show that under different reaction conditions 1,3-carbon-to-nitrogen migrations as between H or $SiMe_3$ are not regioselective; this is in contrast to the $SiMe_3$ preference shown in Scheme 1. (iv) The Si–N bonds are sensitive to hydrolysis (*cf.*, **7** → **8**). Details of some of these reactions have recently been published.[17]

12a $R^1 = Ph = R^2, R^3 = H$
12b $R^1 = Ph = R^2, R^3 = SiMe_3$
12c $R^1 = C_6H_4Bu^t\text{-}4 = R^2, R^3 = H$
12d $R^1 = Ph, R^2 = Bu^t, R^3 = H$

The proposed pathway to the β-diketimine **10** is shown in Scheme 3, and those[17] to **9** and the isomer **13** (an analogue of **8**) in Scheme 4. Each of the crystalline lithium β-diketiminates has been structurally characterized. From the geometric data for **8**, **9**, and **13**, it is evident that whereas **8** and **13** have π-delocalized structures, the central carbon atom (C2) in **9** has significant carbanionic character, attributed to its stabilization by the exocyclic $SiMe_3$ group at C2.

The reactions of the lithium alkyl $[Li\{CH(SiMe_3)(SiMe(OMe)_2)\}]_\infty$[16] with benzonitrile under different conditions furnished the binuclear lithium bis(β-diketiminate)s **14** and **15**, as illustrated in Scheme 5. In both compounds there is a central Si–O–Si bridge. It is evident that, as for **8** and **9**, **14** and **15** differ in that the 1,3-migration from C→N is indiscriminate as between H and $SiMe_3$, although the reactions are regiospecific in that $SiMe_{3-n}(OMe)_n > SiMe_3$ in migratory aptitude ($n = 1$ or 2).

Proposed pathways to compounds **14** and **15** are shown in Scheme 6. Structural data for the two crystalline compounds again (as for **9**) show that **15**,

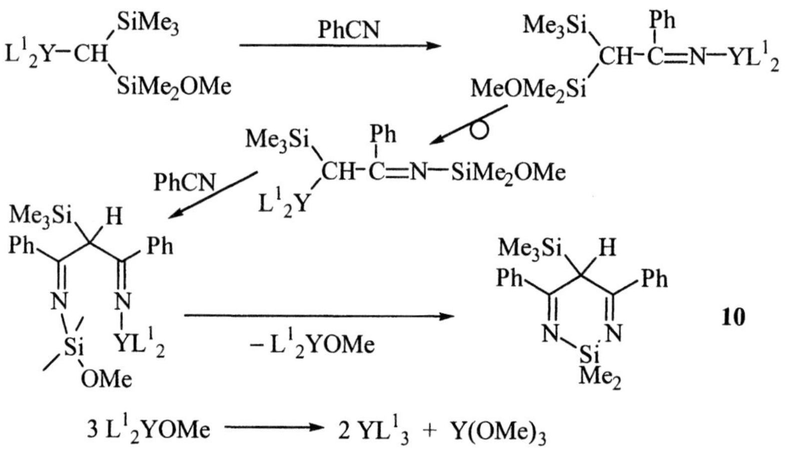

Scheme 3. Proposed pathway to the β-diketimine **10** [L^1 = $CH(SiMe_3)(SiMe_2OMe)$].

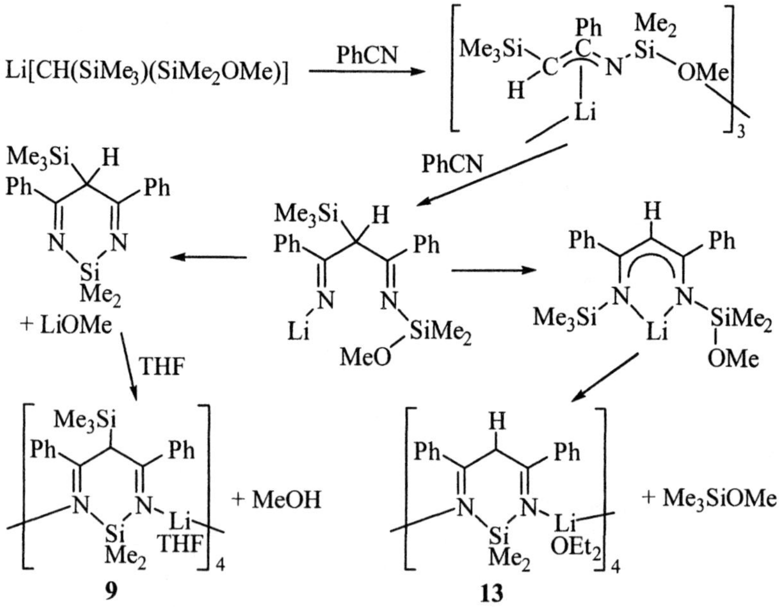

Scheme 4. Proposed pathway to $[Li\{N(SiMe_2)C(Ph)C(R^3)C(Ph)N\}(D)]_4$ [R^3 = H, D = OEt_2 (**13**) or R = $SiMe_3$, D = THF (**9**)].

the compound with an exocyclic SiMe₃ groups at C2, shows significant carbanionic character at C2.

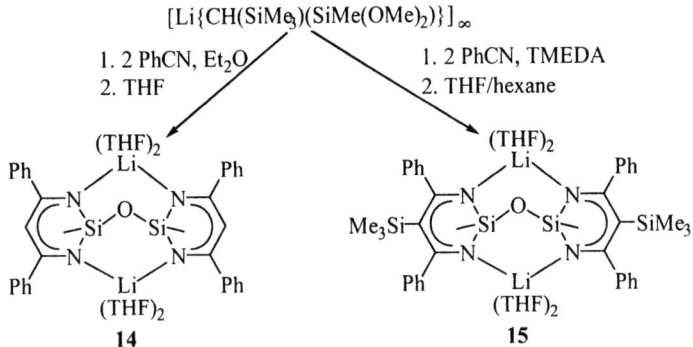

Scheme 5. *Lithium β-diketiminates from [Li{CH(SiMe₃)(SiMe(OMe)₂)}]∞ and PhCN.*

Lithium and Aluminum CH$_2$-Bridged Bis(β-diketiminate)s Having Diverse Structures[18]

The lithium β-diketiminate derived from the ligand **2a** was shown to react with dibromomethane to yield the CH$_2$-bridged bis(β-diketimine) **16a**.[2] The *p*-tolyl analogue **16b** has been prepared in similar fashion from the lithium salt of the ligand **2b**; the crystal structure of **16a** has been determined.[2, 19] Reaction of **16a** or **16b** with a two molar portion of n-butyllithium in hexane afforded the crystalline bicyclic dilithium compound **17a** or **17b**; **16b** with 2AlMe₃ furnished the crystalline binuclear CH$_2$-bridged bis[(β-diketiminato)dimethylalane] **18**. The latter was also obtained from **17b** and 2Al(Cl)Me₂. These experiments are summarized in Scheme 7.

Features of interest regarding the metal-ligand bonding modes are the following. (i) Compounds **16a** and **16b** are rare examples of the β-diketiminate behaving as a *C*-centered moiety. (ii) Compounds **17a** and **17b** are 4,6-dilithio-3,5,7,9-tetraazabicyclo[5.5.1]tridecane derivatives, in which each of the two β-diketiminato ligands acts as an *N,N'*-centered bridge between the two lithium atoms and the central carbon atoms of each ligand are joined through a methylene bridge. (iii) The crystalline binuclear aluminum compound **18** is C_2-

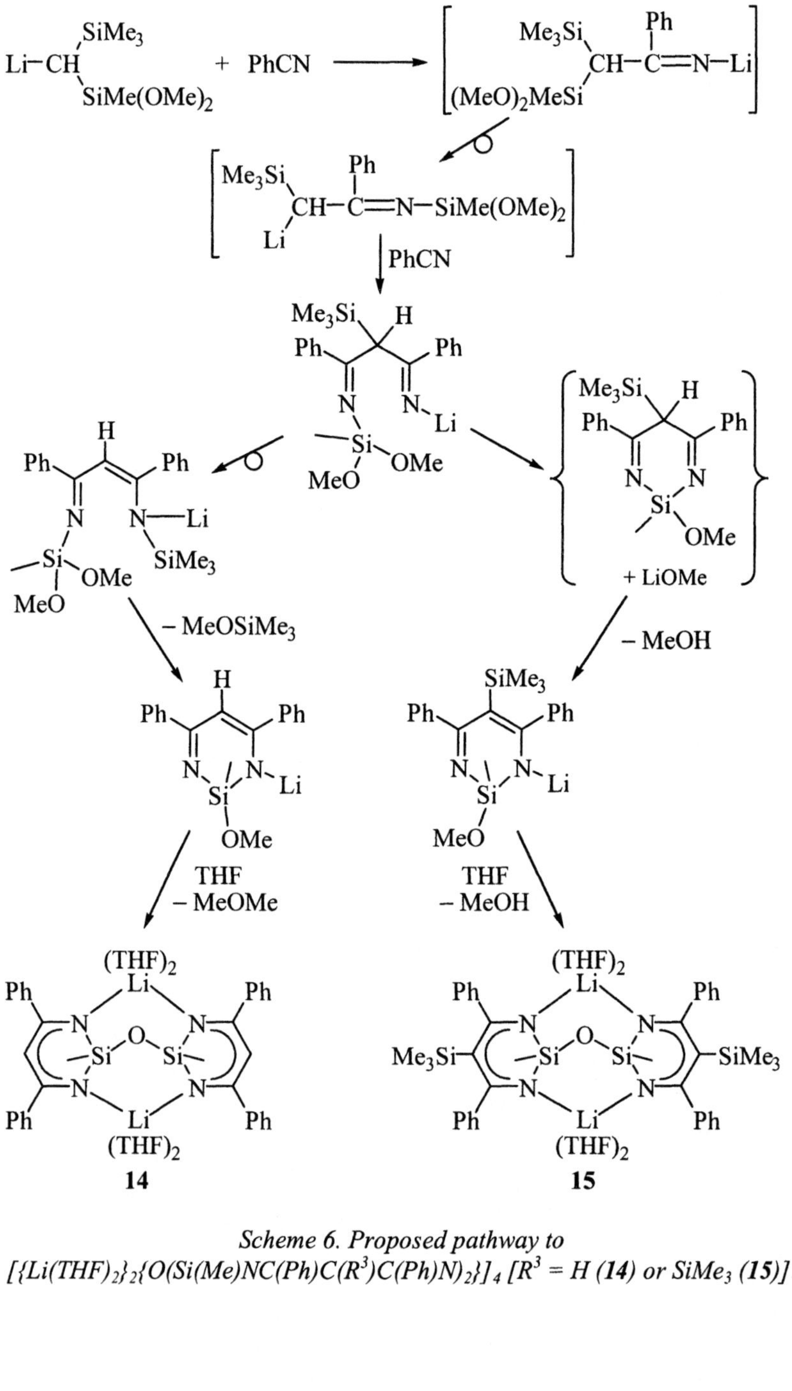

Scheme 6. Proposed pathway to
[{Li(THF)$_2$}$_2${O(Si(Me)NC(Ph)C(R^3)C(Ph)N)$_2$}]$_4$ [R^3 = H (14) or SiMe$_3$ (15)].

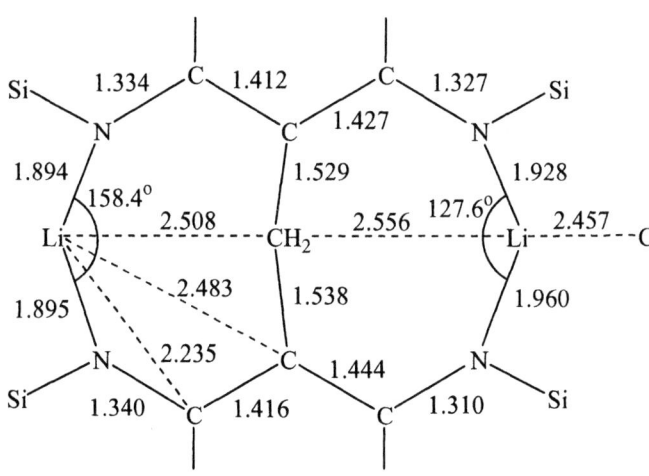

Scheme 7. Syntheses of compounds **16a**, **16b**, **17a**, **17b**, and **18** (R = SiMe$_3$).

Figure 2. Selected geometrical parameters for **17b**.

symmetric and is the first example of an *ansa*-CH$_2$-bridged bis(β-diketiminatometal) derivative. As for (i), recent examples of *C*-centered β-diketiminato-element compounds are shown in **e**[7] and **f**[8] of Figure 1.

The X-ray structures of compounds **17a** and **17b** show that the environment of the two lithium atoms is very different, as illustrated for **17b** schematically in Figure 2; thus one of the lithium atoms has a reasonably close Li...C intermolecular contact.

The reaction pathway for the formation of the dilithio compound **17a** or **17b** from the CH$_2$-bridged bis(β-diketimine) **16a** or **16b** may follow the route shown in Scheme 8. The preparation of **18** from **16b** and AlMe$_3$, on the other hand, is unexceptional since no rearrangement is implicated. However, that **18** is also obtained from Al(Cl)Me$_2$ and **17b** indicates that the isomer **18'** is thermodynamically less favored than **18** and if **18'** is first formed then its rearrangement into **18** requires a process similar in reverse of the final two steps of Scheme 8 with the lithium atoms replaced by AlMe$_2$ moieties in **18'**.

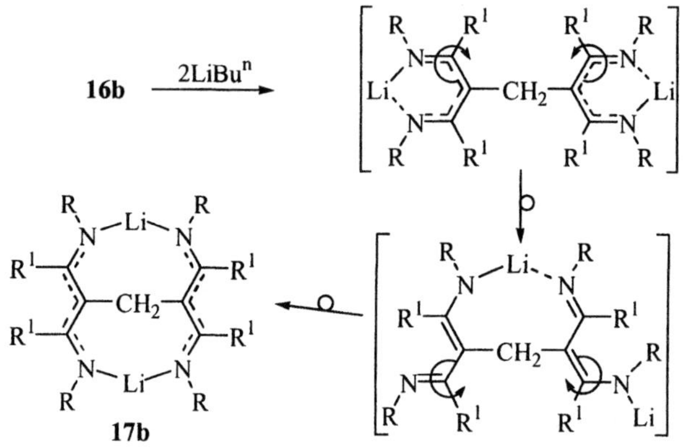

Scheme 8. Proposed pathway for the conversion of **16b**→**17b** ($R^1 = C_6H_4Me$-4).

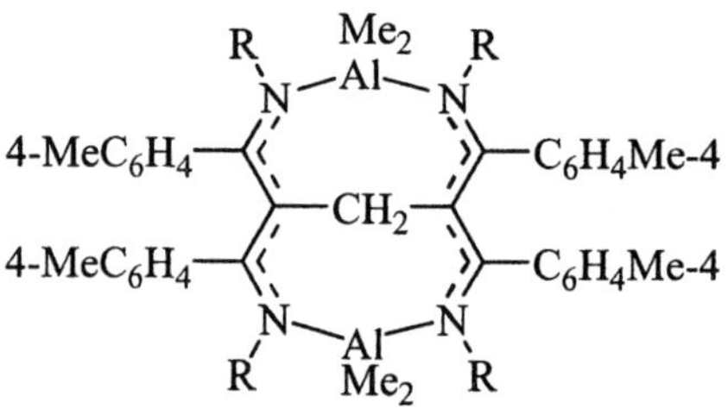

18'

The data presented in this section are presented in greater detail in ref. 19.

Di- and Tri-anionic β-Diketiminates[20]

We recently showed that monoanionic ligands of the type 2 could be reduced to the dianionic or the trianionic state, giving rise to the heterotrimetallic (Yb/Li$_2$)[13] or the homotrimetallic (Yb$_3$)[14] β-diketiminates, as illustrated in **l** and **m** or **n**, respectively of Figure 1.

The paramagnetic dilithium β-diketiminates **19a** and **19b** and the diamagnetic trilithium derivative **20** were prepared by reduction of the appropriate mono-lithium precursor, as shown in Scheme 9.

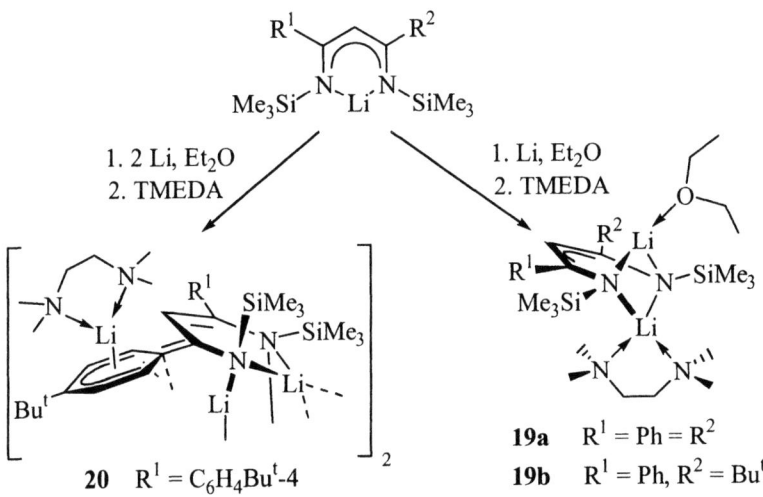

Scheme 9. Synthesis of compounds 19a, 19b, and 20.

The structures of the crystalline complexes **19a** and **20**, shown schematically in the Scheme 9, were established by X-ray diffraction. The β-diketiminato ligand in **19a** both chelates each lithium atom and bridges the two, one of which is above and the other below the NCCCN plane. The bonding mode of the ligand in the dimeric complex **20** is more complicated; a schematic representation of the skeletal atoms is illustrated in Figure 3 and shows the *cis*-disposition of the two TMEDA-bound lithium atoms. This feature was consistent with the ^{6}Li- and ^{7}Li-NMR spectra of **20** in toluene-d$_8$ at 228 K, exhibiting four inequivalent lithium nuclei. At progressively higher temperatures fluxional processes were set in train until at 338 K all ^{7}Li signals merged. The ^{1}H-NMR spectra revealed a dynamic process, corresponding to $\Delta G^{\#}_{264\,K} = 53$ kJ mol^{-1}, eq (1).

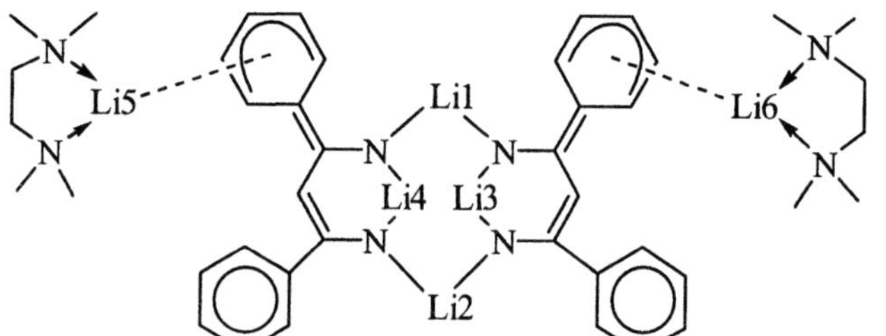

Figure 3. Schematic representation of the skeletal atoms of 20.

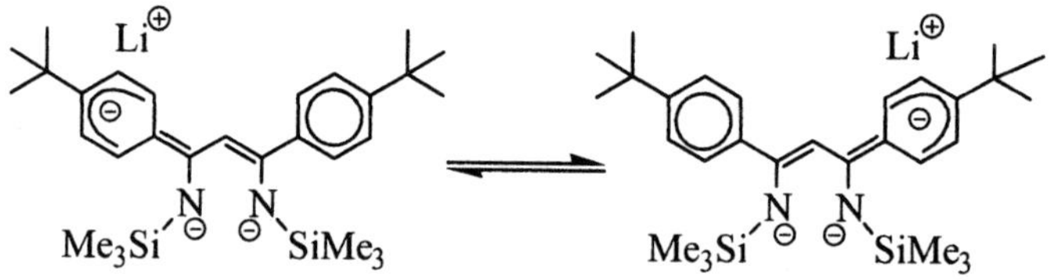

(1)

More details of the data presented in this section will be published.[21]

Experiments with Alkali Metal N,N'-Bis(diisopropylphenyl)-β-Diketiminates

Unusual bonding modes have been observed in a number of studies involving the β-diketiminato ligands **1** and **3**, and a selection of these is described in this section.

The crystalline N-bonded potassium β-diketiminate **21** was obtained as shown in eq (2).[22] The molecular structure shows that the skeletal atoms of the ligand are essentially π-delocalized. The two aryl groups are arranged in a transoid fashion. The K–N distance is 2.861(3) Å, and the potassium coordination environment is completed by contacts of 2.794 to 2.943 Å to the six oxygen atoms of the crown ether.

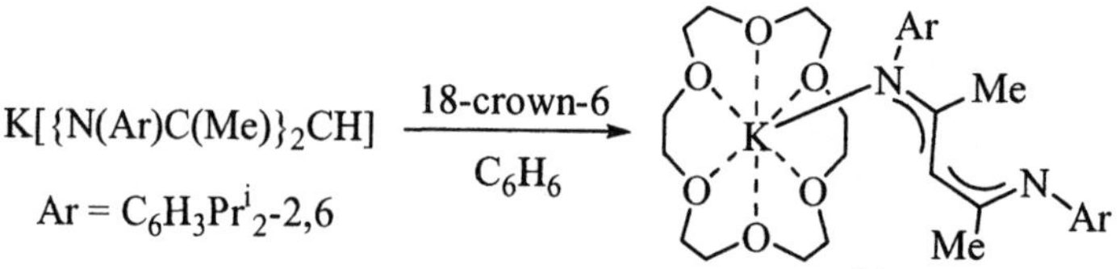

K[{N(Ar)C(Me)}$_2$CH] $\xrightarrow{\text{18-crown-6}}$ **21**   (2)

Ar = C$_6$H$_3$Pr$^i_2$-2,6   C$_6$H$_6$

The crystalline tin(IV) β-diketiminate **22** was prepared as shown in eq (3).[23] It is evident that the ligand **1** behaved as its tautomer **1′** in order to account for the conversion of the *C*-methyl substituent of the substrate into the *C*-triphenylstannylmethyl moiety in the product **22**. The X-ray structure of **22** indicates that the π-bonding in the NCCCN unit is largely localized. A previous example of a Me-activated metal β-diketiminate is **23**, obtained by dehydrochlorination of [Sc(Cl)(NR$_2$){(N(CH$_2$CH$_2$NEt$_2$)C(Me))$_2$CH}].[24]

Attempts to reduce the lithium β-diketiminate of the ligand **3**[25] with metallic lithium under the conditions of eq (4) afforded in low yield the crystalline 1,6-dilithio-1,6-diazahexa-2,4,6-triene derivative **24** and 1,2-bis(2′,6′-diisopropylphenyl)hydrazine **25**.[26] These products are reasonably accounted for by assuming that the paramagnetic dilithio-β-diketiminate was a transient intermediate, which underwent homolysis of a C–N bond, the two radicals then coupling to furnish the observed products **24** and **25**.

Acknowledgements

We are grateful for the award of fellowships to the European Community (Marie Curie Fellowship for L. B.-M.), The Royal Society (Y. C. and X.-H. W.), E.P.S.R.C. (A. V. K.) and the Leverhulme Trust (A. V. P.), and for support for D. J. D. to Dr. Yu. K. Gun'ko and Trinity College, Dublin.

References

1. Bourget-Merle, L.; Lappert, M. F.; Severn, J. R. *Chem. Rev.* **2002**, *102*, 3031.
2. Hitchcock, P. B.; Lappert, M. F.; Liu, D.-S. *J. Chem. Soc. Chem. Commun.* **1994**, 1699, 2637.
3. Avent, A. G.; Caro, C. F.; Hitchcock, P. B.; Lappert, M. F.; Li, Z.; Wei, X.-H. *Dalton Trans.* **2004**, 1567.
4. Caro, C. F.; Hitchcock, P. B.; Lappert, M. F. *Chem. Commun.* **1999**, 1433.
5. Hitchcock, P. B.; Lappert, M. F.; Layh, M.; Liu, D.-S.; Sablong, R.; Shun, T. *J. Chem. Soc. Dalton Trans.* **2000**, 2301.
6. Ding, Y.; Roesky, H. W.; Noltemeyer, M.; Schmidt, H.- G.; Power, P. P. *Organometallics* **2001**, *20*, 1190.
7. Hitchcock, P. B.; Lappert, M. F.; Nycz, J. E. *Chem. Commun.* **2003**, 1142.
8. Ragogna, P. J.; Burford, N.; D'eon, M.; McDonald, R. *Chem. Commun.* **2003**, 1052.
9. Hitchcock, P. B.; Lappert, M. F.; Liu, D.-S.; Sablong, R. *Chem. Commun.* **2002**, 1920.
10. Hitchcock, P. B.; Lappert, M. F.; Layh, M. *Chem. Commun.* **1998**, 201.
11. Feldman, J.; McLain, S. J.; Parthasarathy, A.; Marshall, W. J.; Calabrese, J. C.; Arthur, S. D. *Organometallics* **1997**, *16*, 1514.
12. Clegg, W.; Coles, S. J.; Cope, E. K.; Mair, F. S. *Angew. Chem. Int. Ed. Engl.* **1998**, *37*, 796.
13. Avent, A. G.; Khvostov, A. V.; Hitchcock, P. B.; Lappert, M. F. *Chem. Commun.* **2002**, 1410.
14. Eisenstein, O.; Hitchcock, P. B.; Khvostov, A. V.; Lappert, M. F.; Maron, L.; Perrin, L.; Protchenko, A. V. *J. Am. Chem. Soc.* **2003**, *125*, 10790.
15. Experiments of Dr. X.-H. Wei.
16. Antolini, F.; Hitchcock, P. B.; Lappert, M. F.; Wei, X.-H. *Organometallics* **2003**, *22*, 2505.

17. Hitchcock, P. B.; Lappert, M. F.; Wei, X.-H. *J. Organomet. Chem.* **2004**, *689*, 1342.
18. Experiments of Dr. L. Bourget-Merle.
19. Bourget-Merle, L.; Hitchcock, P. B.; Lappert, M. F. *J. Organomet. Chem.* **2004**, in press.
20. Experiments of Drs. A. V. Khvostov and A. V. Protchenko.
21. Avent, A. G.; Hitchcock, P. B.; Khvostov, A. V.; Lappert, M. F.; Protchenko, A. V. *Dalton Trans.* **2004**, 2272.
22. Experiments of Dr. A. V. Protchenko.
23. Experiments of D. J. Doyle.
24. Neculai, A. M.; Roesky, H. W.; Neculai D.; Magull, J. *Organometallics*, **2001**, *20*, 5501.
25. Spencer, D. J. E.; Aboelella, N. W.; Reynolds, A. M.; Holland, P. L.; Tolman, W. B. *J. Am. Chem. Soc.* **2002**, *124*, 2108.
26. Experiments of Dr. Y. Cheng.

Chapter 15

Fluoride Ion Complexation by Chelating 1,8-diborylnaphthalene Lewis Acids and Their Isoelectronic Dicarbocationic Analogs

Huadong Wang, Stéphane Solé, and François P. Gabbaï

Department of Chemistry, Texas A&M University,
College Station, Texas 77843-3255

This article deals with the chemistry of 1,8-diborylnaphthalene and 1,8-bis(methylium)naphthalene derivatives as bidentate Lewis acids. The synthesis, structure and reactivity of these two isoelectronic classes of derivatives toward fluoride anion will be presented.

Introduction and background

1,8-Diborylnaphthalenes are prototypical examples of bidentate Lewis acids which have been studied for the complexation of both neutral and electron rich substrates (*1-5*). 1,8-Bis(dimethylboryl)naphthalene (**1**) is one of the simplest representatives of this class of compounds. It reacts with potassium hydride in THF to form a kinetically and thermodynamically stable 1:1 hydride complex featuring a B-H-B 3c-2e bond (*1,2*). Further exploration of the chemistry of this derivative showed that fluoride is also readily chelated (Scheme 1). Although the crystal structure of the resulting fluoride chelate complex has not been determined, NMR data strongly support the formation of a symmetrical B-F-B bridge. This complex is very stable and fluoride anion decomplexation can only be effected via the use of a strong Lewis acid such as BF_3. This situation contrasts with that of monofunctional boranes which form more labile fluoride complexes; for example, tris(9-anthryl)borane forms an isolable fluoride complex which readily dissociates in the presence of water (*6*). Based on the above, it can be concluded that 1,8-diborylnaphthalenes could be used as molecular recognition units for the design of fluoride sensors.

Scheme 1

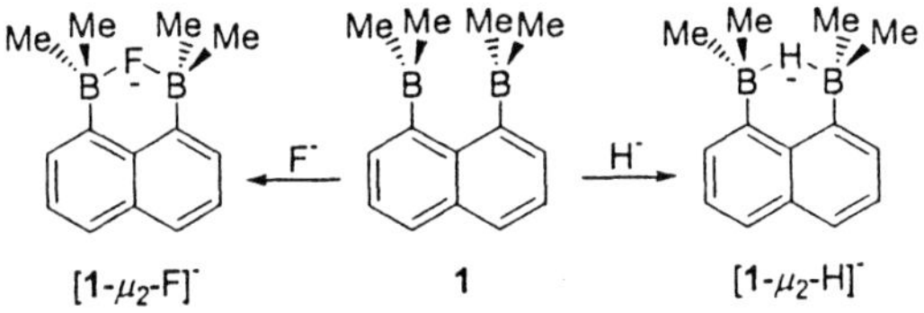

Stable methylium cations, which also constitute powerful Lewis acids (7) are isoelectronic analogs of boranes. On the basis of this analogy, a new class of bidentate Lewis acids with methylium moieties as eletrophilic sites may be envisaged. For example, it is conceivable that a derivative featuring two methylium moieties connected by a *peri*-substituted naphthalene backbone would display chemistry similar to that of the aforementioned 1,8-di(boryl)naphthalenes (*1-5*). If this is indeed the case, the complexation of fluoride by a 1,8-bis(methylium)naphthalenediyl bidentate Lewis acid of type A could afford a C-F-C bridged species (Scheme 2). As part of our continuing efforts in the chemistry of naphthalene based Lewis acids (*5,8-10*), we have recently engaged in an exploration of this paradigm.

Scheme 2

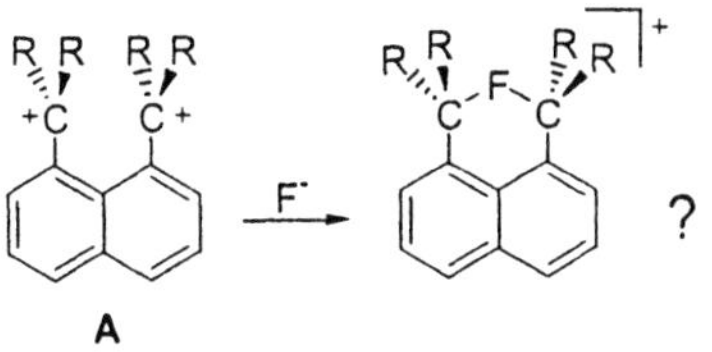

In this article, we present some of our recent results concerning the chemistry of 1,8-diborylnaphthalene and 1,8-bis(methylium)naphthalenediyl derivatives as bidentate Lewis acids. In particular, we will focus on the synthesis and structure of these two isoelectronic classes of derivatives as well as their reactivity toward the fluoride anion. A large portion of the material presented herein is extracted from two of our recent papers (*11,12*).

1,8-bis(diphenylboryl)naphthalene and its isoelectronic dicarbocation

The reaction of 1,8-dilithionaphthalene with two equivalents of diphenylboronbromide affords 1,8-bis(diphenylboryl)-naphthalene (2) (*10*).

The isoelectronic dicarbocationic analog of this diborane has also been prepared by reaction of the known 1,8-bis(diphenylhydroxymethyl)naphthalene with a mixture of [HBF$_4$]$_{aq}$ and (CF$_3$CO)$_2$O which affords 1,8-bis(diphenylmethylium)naphthalenediyl (**3^{2+}**) as the [BF$_4$]$^-$ salt (Scheme 3) (*11,13*). Both of these compounds have been fully characterized. As shown by single crystal X-ray analysis, **2** and **3^{2+}** display very similar structure (Figure 1). The boron centers of **1** as well as the methylium centers of **3^{2+}** adopt a trigonal planar arrangement in agreement with a sp^2 hybridization. Although carbon has a smaller covalent radius than boron, the separation between vicinal methylium centers in **3^{2+}** (3.112(4) Å) is greater than that observed between the boron centers of **2** (3.002(2) Å) which possibly results from electrostatic repulsions present in the former. As shown by DFT calculations, the empty p$_z$ orbitals of the boron centers in **2** and the empty p$_z$ orbitals of the methylium carbon atoms in **3^{2+}**, largely contribute to the LUMO of these species and are oriented toward one another in a transannular fashion. Hence, the electronic and structural features of **2** and **3^{2+}** present some important parallel and nicely substantiate the isoelectronic relationship that exists between these derivatives.

Scheme 3

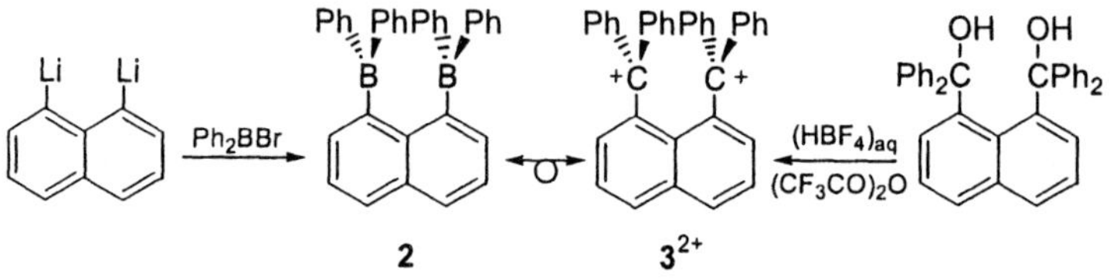

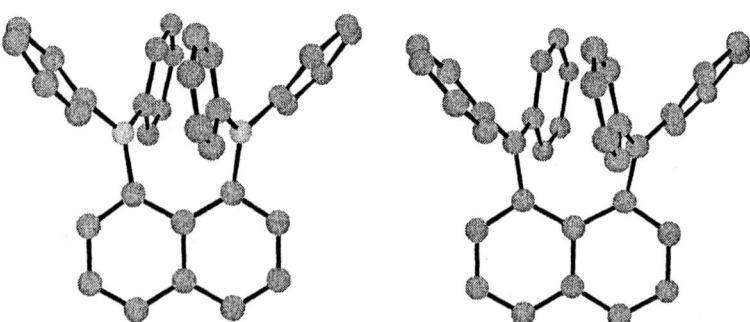

*Figure 1: Crystal structure of **2** (left) and **3^{2+}** (right) shown side by side. color code: C, grey; B, light grey. H atoms omitted for clarity.*

Bidentate boranes as colorimetric fluoride ion sensor

Taking into account the importance of the fluoride anion in the treatment of osteoporosis (14) and in dental care (15), a great deal of effort has been devoted to the design of selective fluoride sensors. In addition to receptors capable of hydrogen bonding with the anionic guest (16), Lewis acidic receptors that can covalently bind fluoride ion have also been reported (17). In this domain, the study of colorimetric fluoride sensors based on boron-containing π-electron systems is especially noteworthy (6,18). In such derivatives, fluoride ion complexation to the boron center leads to a perturbation of the frontier orbitals thereby altering the photophysical properties of the receptor. Since only monotopic complexation of fluoride occurs in these boron based sensors, we decided to investigate the synthesis and properties of a colorimetric fluoride sensor based on a bidentate Lewis acidic borane.

Scheme 4

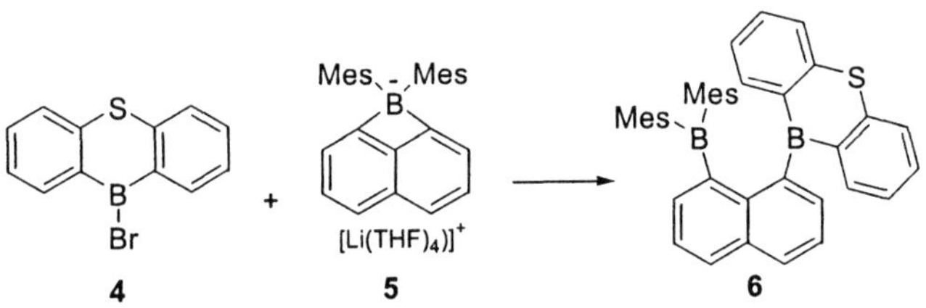

The successful design of such a sensor necessitated the incorporation of a chromophoric boron moiety in a bidendate Lewis acid. To this end, 10-bromo-9-thia-10-boranthracene (**4**) was prepared by reaction of bis-(2-trimethylsilylphenyl)sulfide with boron tribromide and was allowed to react with dimesityl-1,8-naphthalenediylborate (**5**) to afford diborane **6** (Scheme 4) (12). This bright yellow diborane is soluble in chloroform, THF and pyridine. The ^{11}B NMR spectrum of **6** shows two resonances at 56 and 74 ppm confirming the presence of two different boron centres. The structure of **6** was computationally optimized using DFT methods. The optimized geometry is close to that observed for other diboranes bearing a dimesitylboryl group (9). As in **2**, examination of the DFT orbitals reveals that the boron p_z orbitals contribute to the LUMO and are oriented toward one another in a transannular fashion (Figure 2) as observed for 1,8-bis(diphenylboryl)naphthalene. Also, the UV-Vis spectrum of this derivative features a broad band centered at 363 nm, ε = 17400 mol^{-1} cm^{-1}. As indicated by a time-dependent DFT calculations, electronic excitations from the HOMO, HOMO-1 and HOMO-2 to the LUMO are the major contributors to this broad band. Hence, any events leading to the disruption of the LUMO should greatly affect the absorption spectrum of compound **6**.

*Figure 2: DFT orbital picture of **6** showing the LUMO (surface isovalue = 0.04). H atoms omitted for clarity. Adapted with permission from reference 12. Copyright 2004 Royal Society of Chemistry.*

In the presence of $[Me_3SiF_2]^-[S(NMe_2)_3]^+$ in THF, compound **6** readily complexes fluoride anions (Scheme 5). This reaction is accompanied by a rapid loss of the yellow color and affords the anionic chelate complex $[3 \cdot \mu_2\text{-F}]^-$ which has been fully characterized. The ^{19}F NMR resonance of the bridging fluoride appears at -188 ppm. As confirmed by single crystal X-ray analysis (Figure 3), the fluorine atom is bound to both boron centers and forms B-F bonds of comparable lengths (F-B(1) 1.633(5) Å, F-B(2) 1.585(5) Å) (Fig. 2). The B(2)–F–B(1) bond angle is equal to 126.0(3)°. Owing to the bridging location of the fluoride anion, these bonds are slightly longer than terminal B-F bonds observed in other borate anions (1.47 Å).[18c] The sum of the coordination angles ($\Sigma_{(C-B1-C)} = 347.8°$, $\Sigma_{(C-B1-C)} = 341.2°$) indicates that both boron centers are substantially pyramidalized.

Scheme 5

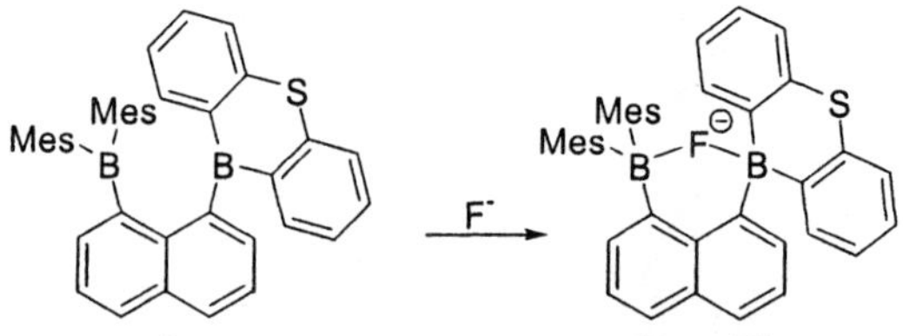

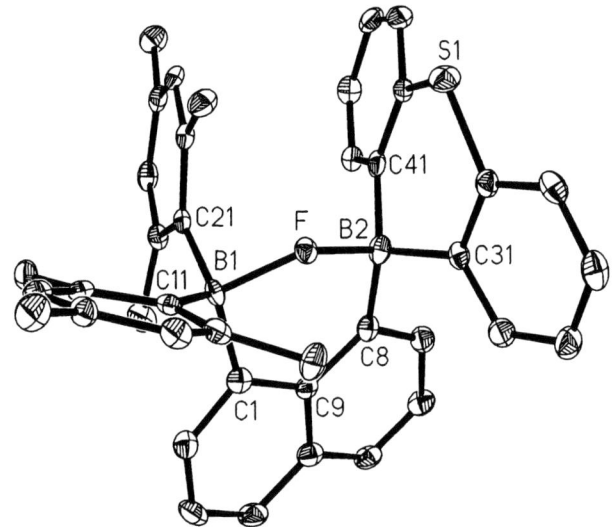

Figure3: Ortep view of the borate anion [6•µ$_2$-F]⁻ (50% ellipsoids). H atoms omitted for clarity. Adapted with permission from reference 12. Copyright 2004 Royal Society of Chemistry.

Fluoride complexation leads to population of the LUMO of **6** and is logically accompanied by an instantaneous loss of the yellow color. Remarkably, no changes in the color of the solution or in the NMR of diborane **6** are observed in the presence of chloride, bromide and iodide anions indicating that the larger halides are not complexed. As previously proposed, the size of the binding pocket provided by this bidentate borane can be held responsible for this phenomenon. As determined by a UV-Vis titration experiment, **6** complexes fluoride anions with a binding constant of at least 5×10^9 M^{-1} which exceeds that observed for monofunctional borane receptors by 3 to 4 orders of magnitude (*6,18*). Addition of water does not lead to decomplexation of the fluoride anion as typically observed for fluoride adducts of monofunctional boranes (*6*). This difference substantiates the chelating ability of **6** which leads to the formation of a thermodynamically more stable fluoride complex. Altogether, the charge neutrality of this sensor as well as the short space available between the boron centers makes it highly selective for fluoride. By virtue of its bidentate nature, the fluoride association is remarkably high and by far exceeds that measured for monofunctional borane receptors.

Formation of a cation containing a C-F→C bridge by reaction of the 1,8-bis(diphenylmethylium)naphthalenediyl dication with fluoride

Formally, a species featuring a C-F-C bridge corresponds to a fluoronium cation whose existence, in the condensed state, remains to be confirmed. Indeed, we note that while halonium ions of the type [R_3C-X-CR_3]$^+$ (X = halogen, R = alkyl or aryl) are known for chlorine, bromine, and iodine (*19,20*), the corresponding fluorine species are remarkably elusive and have not been isolated nor structurally characterized (*21*). In comparison to its heavier group 17 congeners, fluorine possesses the highest one-electron ionization energies for the outer-shell s and p orbitals (*22*). These intrinsic characteristics are responsible for the high electronegatvity of fluorine and also provide a rationale for the reluctance of this element to form fluoronium ions. Although the latter can be observed as gas phase species (*23,24*), the mere intermediacy of fluoronium ions in the condensed state has long been questioned (*25*) and often refuted (*19,26*). Recent studies, however, provide strong support for the intermediacy of such cations in intramolecular fluorine transfer reactions (*27*). As mentioned in the introductory section, it occurred to us that the complexation of fluoride by a 1,8-bis(methylium)naphthalenediyl bidentate Lewis acid such as 3^{2+} could afford a C-F-C bridged species.

Scheme 6

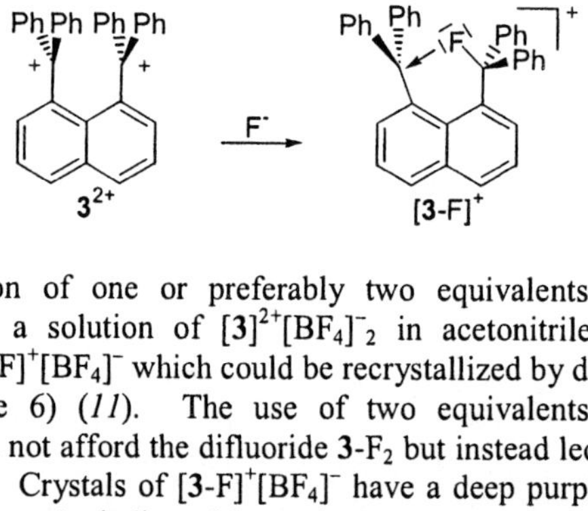

Addition of one or preferably two equivalents of [Me$_3$SiF$_2$]$^-$[S(NMe$_2$)$_3$]$^+$ to a solution of [3]$^{2+}$[BF$_4$]$^-_2$ in acetonitrile results in the formation of [3-F]$^+$[BF$_4$]$^-$ which could be recrystallized by diffusion of ether vapors (Scheme 6) (*11*). The use of two equivalents of [Me$_3$SiF$_2$]$^-$[S(NMe$_2$)$_3$]$^+$ did not afford the difluoride 3-F$_2$ but instead led to higher yield of [3-F]$^+$[BF$_4$]$^-$. Crystals of [3-F]$^+$[BF$_4$]$^-$ have a deep purple color and are very air-sensitive. As indicated by a single crystal X-ray analysis (Figure 4), the [3-F]$^+$ and [BF$_4$]$^-$ ions are well separated. The cation [3-F]$^+$ is C$_1$ symmetrical, with the bridging fluorine atom preferentially bound to one of the two former methylium centers. Indeed, inspection of the structure shows that the C(01) carbon center is tetrahedral and forms a regular C-F bond of 1.424(2) Å with the bridging fluorine atom. By contrast, the C(02) carbon atom retains a formal sp^2 hybridization ($\Sigma_{(C-C(02)-C)}$ = 359.5°) and forms a contact of 2.444(2) Å with the bridging fluoride atom. While the C(02)-F linkage is markedly longer than the C(01)-F single bond, it remains less than the sum of the van der Waals radii of the two elements (r_{vdw}(F) = 1.30 – 1.38 Å, r_{vdw} (C) = 1.7 Å) (28). Moreover, the C(01)-F-C(02) angle is 111.11(9)°, a value expected for a formally sp³ hybridized fluorine atom with one of its lone pair pointing directly toward the C(02) center (Figure 4). In agreement with this view, the C(31)-C(02)-F (90.4(1) °), C(41)-C(02)-F (99.6(1)°) and C(8)-C(02)-F (86.7(1)°) angles are close to 90°, which indicates that the fluorine atom is positioned along the direction of the p-orbital of the C(02) carbon center.

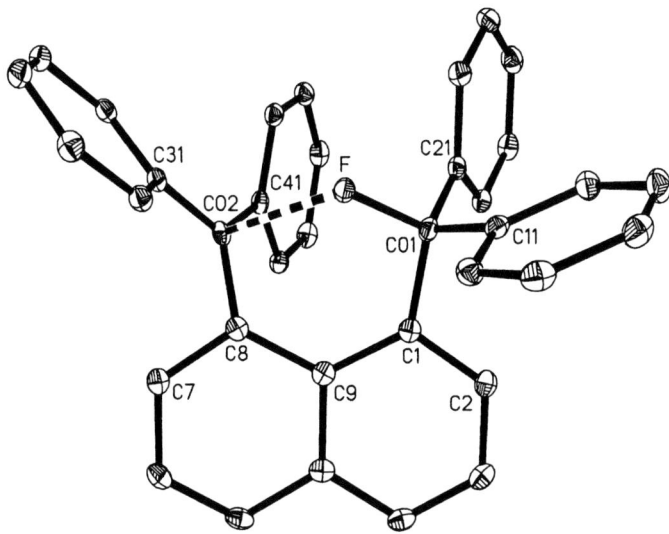

Figure 4: : Ortep view of [3-F] ⁺ (50% ellipsoids). H atoms omitted for clarity.

The structure of [3-F]⁺ was optimized using B3LYP-DFT. The lowest energy structure of the ion has C_1 symmetry and corresponds very closely to the experimental geometry determined by single crystal X-ray diffraction. In order to elucidate the nature of the bonding interactions involving the unsymmetrically coordinated fluorine atom, an Atoms In Molecules (AIM) analysis (29) was performed on the density of the optimized geometry of the C_1 minimum. The AIM analysis indicates bond paths and bond critical points (CP) between the fluorine atom F and both central carbon atoms C(01) and C(02) thus confirming the bonding nature of both the C(01)-F and C(02)-F interactions. This analysis also suggested that the C(02)-F interaction is dative in character. Further evidence of the delocalized electron sharing between F and C(02) was shown by the Boys localized orbital (Figure 5) (30) which corresponds to this interaction. The lone pair of the F atom is clearly extended in the direction of C(02) and mixes with the p-orbital of this carbon atom to form a dative bond.

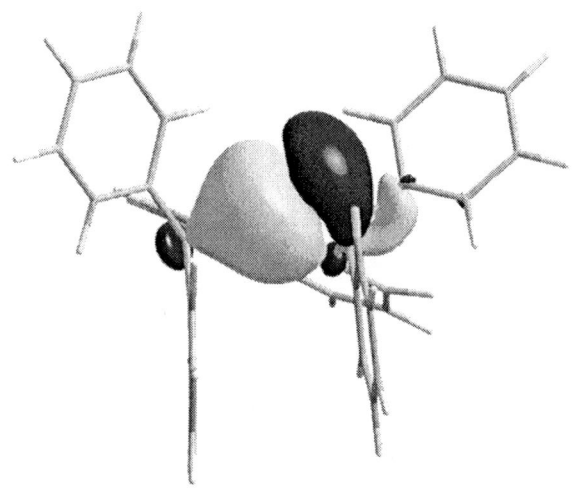

Figure 5. Boys localized orbital (at 0.01 isodensity value) in [3-F]$^+$ showing he delocalization of the F 2p "lone pair" on the methylium C(02) atom (viewed perpendicular to the naphthalene ring).

In agreement with its bridging location, the ^{19}F NMR resonance of the bridging fluorine atom in [3-F]$^+$ (–117.0 ppm at 25°C in CD$_3$CN) appears downfield from that of Ph$_3$CF (–126.7 ppm at 25°C in CD$_2$Cl$_2$) *(31)* The low temperature carbon spectrum of [3-F]$^+$ at –60°C in CHCl$_3$ indicates that the fluorine bridge remains unsymmetrical in solution at this temperature. Indeed two distinct ^{13}C NMR signals are detected for the former methylium centers of 3^{2+}. The resonance of the C(01) (105.1 ppm, J$_{CF}$ = 178.5 Hz) and C(02) (209.3 ppm, J$_{CF}$ = 11.8 Hz) centers both appear as a doublet. While the chemical shift of the C(02) resonance indicates that this atom retains considerable methylium character, the magnitude of the coupling constant is too large for a five-bond spin-spin coupling interaction that would proceed through the naphthalene backbone. In turn, this coupling possibly reflects the presence of a direct C(02)-F interaction. In accordance with the above observations, the ^{1}H spectrum of [3-F]$^+$ at –60 °C in CDCl$_3$ displays six distinct resonances for the naphthalene H-atoms. Upon elevation of the temperature, however, the peaks become broader and eventually coalesce (Figure 6). These results can be interpreted on the basis of a fluxional system in which the fluoride anion oscillates between the C(01) and C(02) center (Scheme 7) with apparent activation parameters of $\Delta H^\ddagger$ = 52 (±3) kJ·mol^{-1} and $\Delta S^\ddagger$ = –18 (±9) J·K^{-1}·mol^{-1} as derived from line shape analysis. As confirmed by calculations, the transition state corresponds to a fluoronium ion with a symmetrical C-F-C bridge (Scheme 7). The ^{19}F NMR resonance of the [BF$_4$]$^-$ anion appears at -150 ppm. This chemical shift is identical to that typically observed for free [BF$_4$]$^-$ anion.

Moreover, this resonance does not change as a function of temperature, thus indicating that the [BF$_4$]$^-$ anion is not involved in the observed dynamic process.

Scheme 7

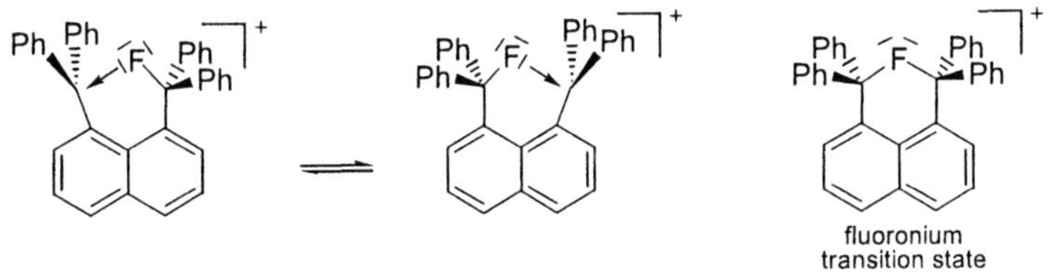

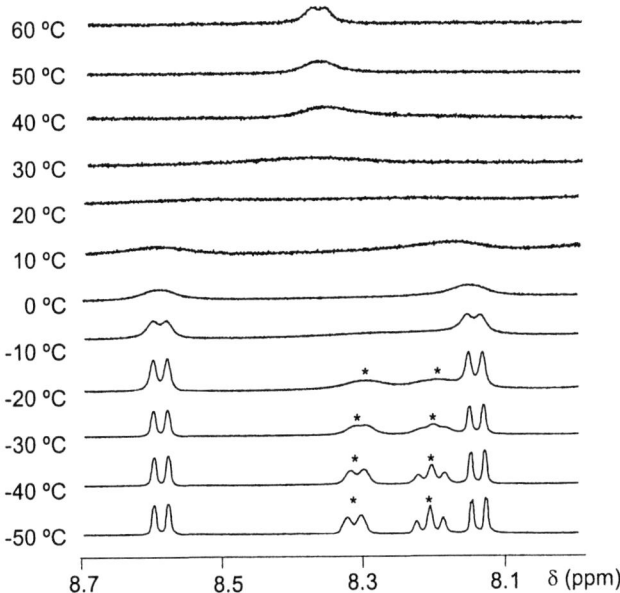

Figure 6. Variable ^1H NMR of [3-F]$^+$ in CDCl$_3$. Portion of the spectrum showing the 2- and 7-CH$_{naphthalene}$ resonances and their coalescence. The resonances marked by * correspond to CH$_{phenyl}$ resonances.

The structure of [3-F]$^+$ can be largely accounted for by considering the methylium description **B** (Scheme 8) which is supported by the approximate trigonal planar arrangement of the C(02) carbon atom as well as its ^{13}C NMR chemical of 209.3 ppm. However, by virtue of the bonding interaction between C(02) and F, which is supported by both experiment and theory, the unsymmetrical fluoronium description **C** (Scheme 8) must also be contributing. The variable temperature solution studies of this ion allow us to witness the oscillation of a fluorine atom between two carbon centers on the NMR time scale. Taking into account the strength of the carbon-fluorine bond and its chemical inertness, the occurrence of this transfer process, which involves the concomitant lengthening and shortening of a C-F bond, is rather unusual and suggests that cationic pathways are available for the activation of primary C-F bonds. This assertion corroborates the finding of Lectka and coworkers who have proposed the formation of a fluoronium intermediate in the intramolecular reaction of a carbocation with a C-F bond (27). The fluxionality of the structure of [3-F]$^+$ also indicates that symmetrical fluoronium ion species constitute low energy transition states for such processes. These results are of possible relevance to C-F bond activation chemistry, an area that has witnessed a spur of interest caused by the environmental accumulation and nuisance of fluorocarbons.

Scheme 8

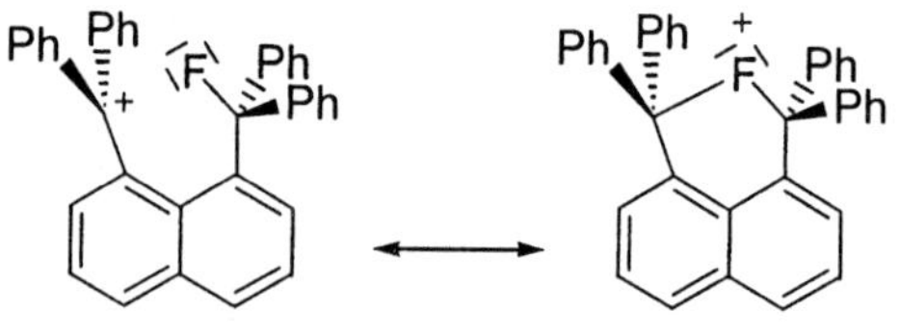

Conclusions

This article outlines some of the parallels and differences encountered in the chemistry of 1,8-diborylnaphthalenes and their isoelectronic 1,8-bis(methylium)naphthalenediyl dicarbocationic analogs. As demonstrated by single crystal X-ray analysis and theoretical calculations carried out on 1,8-bis(diphenylboryl)-naphthalene (**2**) and 1,8-bis(diphenylmethylium)naphthalenediyl (**3^{2+}**), these two classes of compounds present very similar electronic and structural features. In particular, the Lewis acidic boron centers in 1,8-diborylnaphthalenes and the methylium centers in 1,8-bis(methylium)naphthalenediyl dicarbocations are separated by 3-3.3 Å (8,9). This relatively short distance as well as the rigidity of the backbone favors cooperative effects between the Lewis acidic centers. This cooperativity becomes evident in the binding of fluoride which is readily complexed by both 1,8-diborylnaphthalenes and 1,8-bis(methylium)naphthalenediyl dicarbocations. While this had long been known in the case of 1,8-diborylnaphthalenes, the results presented in this article provide structural confirmation for the existence of a symmetrical B-F-B bridge in 1,8-diborylnaphthalene-μ_2-fluoride adducts such as [**6**•μ_2-F]$^-$. By contrast, the adduct [3-F]$^+$ obtained by addition of fluoride to 1,8-bis(diphenylmethylium)naphthalenediyl (**3^{2+}**), features a highly unsymmetrical C-F→C bridge in which both C-F bonds differ by approximately 1 Å. This difference can be rationalized qualitatively by first considering the electronegativity of boron, carbon, and fluorine. In the case of a B-F-B bridged species of the type $[R_3B\text{-}F\text{-}BR_3]^-$ (R = aryl), the large difference in the electronegativity of boron and fluorine ($\Delta\chi_{F,B} = \chi_F\text{-}\chi_B = 2$) suggests that the B-F bonds will remain polar with a substantial amount of negative charge on the fluorine atom. As a result, the fluorine atom in a symmetrical B-F-B bridge should not acquire a prohibitive amount of positive character. In the case of a $[R_3C\text{-}F\text{-}CR_3]^+$ species (R = aryl), the difference in the electronegativity of carbon and fluorine ($\Delta\chi_{F,C} = \chi_F\text{-}\chi_C = 1.4$) is smaller than $\Delta\chi_{F,B}$, thus suggesting that the fluorine will acquire a greater positive character. Keeping in mind that fluorine is the most electronegative element, the positive charge development at fluorine in $[R_3C\text{-}F\text{-}CR_3]^+$ may very well exceed the tolerable limit. This situation will be further intensified by the cationic nature of a $[R_3C\text{-}F\text{-}CR_3]^+$ species. These arguments constitute a possible rationalization for the existence of an unsymmetrical C-F→C bridge in $[3\text{-}F]^+$.

Acknowledgment: This work would not have been possible without the financial support of the National Science Foundation and the Welch Foundation that we gratefully acknowledge. We would also like to thank Michael B. Hall, Charles Edwin Webster, Lisa M. Pérez for their help and collaborative efforts on some of the computations.

References

1. Katz, H. E. *J. Am. Chem. Soc.* **1985**, *107*, 1420-1421.
2. Katz, H. E. *J. Org. Chem.* **1985**, *50*, 5027-5032.
3. Katz, H. E. *Organometallics* **1987**, *6*, 1134-1136.
4. Gabbaï, F. P. *Angew. Chem. Int. Ed. Engl.* **2003**, *42*, 2218-2221.
5. Hoefelmeyer, J. D.; Schulte, M.; Tschinkl, M.; Gabbaï, F. P. *Coord. Chem. Rev.* **2002**, *235*, 93-103.
6. Yamaguchi, S.; Akiyama, S.; Tamao, K. *J. Am. Chem. Soc.* **2001**, *123*, 11372-11375.
7. Freedman, H. H. in *Carbonium Ions*, Vol. IV, Olah, G. A.; Schleyer, P. v. R. eds., Wiley-Interscience, New York, 1973, Chapter 28.
8. Hoefelmeyer, J. D.; Solé, S.; Gabbaï, F. P. *Dalton Trans.* **2004**, 1254-1258.
9. Hoefelmeyer, J. D.; Gabbaï, F. P. *Organometallics* **2002**, *21*, 982-985.
10. Hoefelmeyer, J. D.; Gabbaï, F. P. *J. Am. Chem. Soc.* **2000**, *122*, 9054-9055.
11. Wang, H.; Webster, C. E.; Pérez, L. M.; Hall, M. B.; Gabbaï F. P. *J. Am. Chem. Soc.* **2004**, *126*, 8189-8196.
12. Solé, S.; Gabbaï F. P. Chem Commun. **2004**, 1284-1285.
13. Wang, H.; Gabbaï F. P. *Angew. Chem. Int. Ed. Engl.* **2004**, *43*, 184-187.
14. Briancon, D. *Rev. Rheum.* **1997**, *64*, 78-81.

15. Matuso, S.; Kiyomiya, K.; Kurebe, M. *Arch. Toxicol.* **1998**, *72*, 798-806.
16. see for example (a) Cho, E. J.; Moon, J. W.; Ko, S. W.; Lee, J. Y.; Kim, S. K.; Yoon, J.; Nam, K. C. *J. Am. Chem. Soc.* **2003**, *125*, 12376-12377. (b) Mizuno, T.; Wei, W.-H.; Eller, L. R.; Sessler, J. L. *J. Am. Chem. Soc.* **2002**, *124*, 1134-1135.
17. Martínez-Máñez, R.; Sancenón, F. *Chem. Rev.* **2003**, *103*, 4419-4476.
18. (a) Yamaguchi, S.; Akiyama, S.; Tamao, K. *J. Am. Chem. Soc.* **2000**, *122*, 6335-6336. (b) Kubo, Y.; Yamamoto, M.; Ikeda, M.; Takeuchi, M.; Shinkai, S.; Yamaguchi, S.; Tamao, K. *Angew. Chem., Int. Ed. Engl.* **2003**, *42*, 2036-2040.
19. Olah, G. A. *Halonium Ions*; Wiley: New York, 1975. Olah. G. A.; Prakash, G. K. S.; Sommer, J. *Superacids*, John Willey & Sons: New York, 1985. Olah, G. A.; DeMember, J. R.; Mo, Y. K.; Svoboda, J. J.; Schilling, P.; Olah, G. A. *J. Am. Chem. Soc.* **1974**, *96*, 884-892.
20. Rathore, R.; Lindeman, S. V.; Zhu, C.-J.; Mori, T.; Schleyer, P. v. R.; Kochi, J. K. *J. Org. Chem.* **2002**, *67*, 5106-5116.
21. Olah, G. A.; Rasul, G.; Hachoumy, M.; Burrichter, A.; Prakash, G. K. S., *J. Am. Chem. Soc.* **2000**, *122*, 2737-2741.
22. Allen, L. C. *J. Am. Chem. Soc.* **1989**, *111*, 9003-9014.
23. Beauchamp, J. L.; Holtz, D.; Woodgate, S. D.; Patt, S. L. *J. Am. Chem. Soc.* **1972**, *94*, 2798-2807.
24. Viet, N.; Cheng, X.; Morton, T. H. *J. Am. Chem. Soc.* **1992**, *114*, 7127-7132. Nguyen, V.; Mayer, P. S.; Morton, T. H. *J. Org. Chem.* **2000**, *65*, 8032-8040. Leblanc, D.; Kong, J.; Mayer, P. S.; Morton, T. H. *International Journal of Mass Spectrometry* **2003**, *222*, 451-463.
25. Peterson, P. E.; Bopp, R. J. *J. Am. Chem. Soc.* **1967**, *89*, 1283-1284. Clark, D. T. *Special Lectures of the XXIII International Congress of Pure and Applied Chemistry*, Boston, 1971, Vol. I, Butterworth, London, pp. 31.
26. Olah, G. A.; Prakash, G. K. S.; Krishnamurthy, V. V. *J. Org. Chem.* **1983**, *48*, 5116-5117.
27. Ferraris, D.; Cox, C.; Anand, R.; Lectka, T. *J. Am. Chem. Soc.* **1997**, *119*, 4319-4320.
28. Nyburg, S. C.; Faerman, C. H. *Acta Crystallogr. Section B.*, **1985**, *41*, 274-279. Caillet, J.; Claverie, P. *Acta Cryst.* **1975**, *A31*, 448-461.
29. Bader, R. F. W. *Atoms In Molecules: A Quantum Theory*, Oxford, 1990.
30. Boys, S. F. *Rev. Mod. Phys.* **1960**, *32*, 296-299; Boys, S. F. *In Quantum Theory of Atoms, Molecules, and the Solid State*, Lowdin, P. O. Ed.; Academic Press: New York, 1966; pp 253-262.
31. Weigert, F. J. *J. Org. Chem.* **1980**, *54*, 3476-3483.

Chapter 16

New Complexes of Lanthanides with Unusual Main Group Ligands

Richard A. Jones[1,*], Xiaoping Yang[1], Abdul Waheed[1], Michael Wiester[1], and Lilu Zhang[2]

[1]The University of Texas at Austin, Department of Chemistry and Biochemistry, 1 University Station A5300, Austin, TX 78712
[2]University of Pennsylvania, Department of Chemistry, Philadelphia, PA 19104

We describe the synthesis and structures of new multinuclear lanthanide complexes which are formed from conventional "salen" style Schiff base ligands, derivatives of these ligands, or from vanallin based ligands.

As part of a study aimed at generating new large supramolecular complexes of both main group and transition metals, we recently began the investigation of modified Schiff-base ligands such as **1** shown in Figure 1. The reaction of **1** with $Zn(OAc)_2 \cdot 2H_2O$ in THF followed by $Co_2(CO)_8$ gave the unusual Zn_3Co_8 complex **2** (Figure 2) (1).

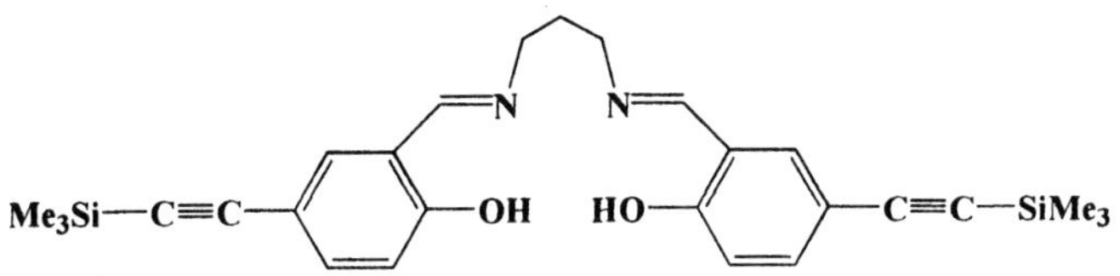

*Figure 1. Coventional "salen" style Schiff-base ligand **1** modified with $-C{\equiv}CSiMe_3$ units.*

© 2006 American Chemical Society

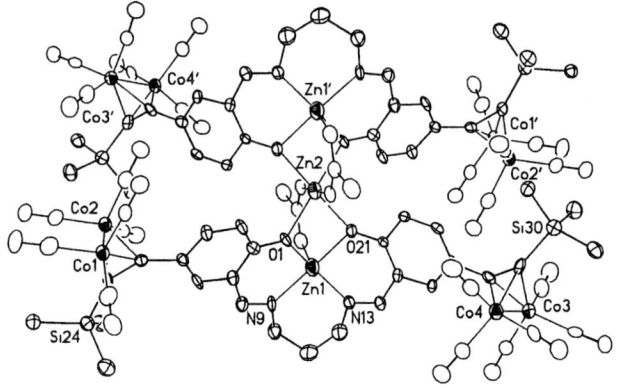

*Figure 2. The unusual Zn_3Co_8 complex **2**.*
(*Chemical Communications*, **2004**, 2986.)
Reproduced by permission of the Royal Society of Chemistry.

In addition to the unusual linear Zn_3 core, the four outer —C≡C—SiMe$_3$ groups of **2** each added $Co_2(CO)_6$ units. One may view the central core of **2** as consisting of $Zn(OAc)_2$ units sandwiched between two "ZnL" groups (L = Schiff base **1**). We also noted that Nabeshima and coworkers recently reported the conversion of a Zn_3 complex into a mixed metal "(3d-4f)" compound with a Zn_2Eu core (2). Since 3d-4f complexes (3) are of interest for their magnetic (4) and luminescent properties (5) we explored the idea of using a Schiff-base complex of Zn(II) as an unusual "main group" ligand system for the coordination of lanthanide ions.

Rather than relying on the fortuitous presence of acetate ligands in order to facilitate the coordination of the central metal as in **2**, we initially focused on modified Schiff-base ligands such as **3** shown in Figure 3.

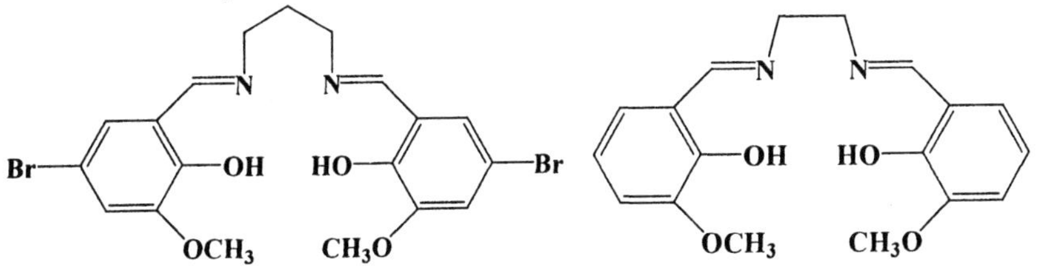

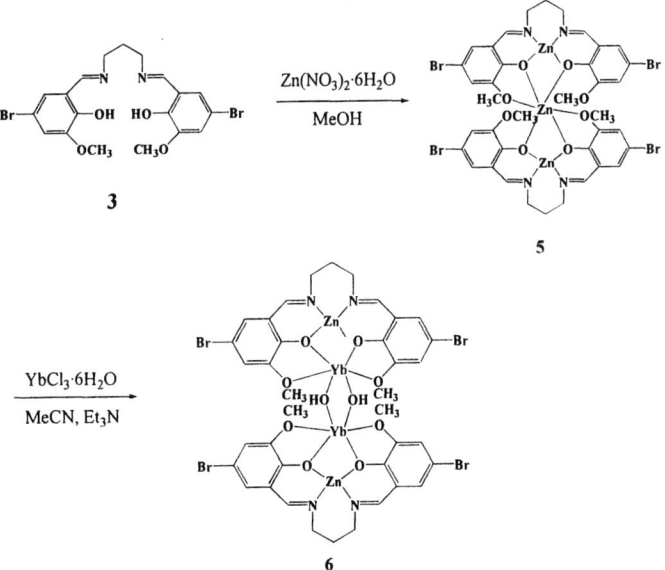

Figure 3. *Schiff-base ligands for the stabilization of 3d-4f complexes.*

Pioneering work by Costes and coworkers has shown that coordination of lanthanide ions is achieved via the extra OMe groups of **4** (Figure 3) (6). In our hands the reaction of **3** with $Zn(NO_3)_2 \cdot 6H_2O$ in MeOH in a 2:3 molar ratio afforded the trinuclear Zn_3 complex **5** as shown in Figure 4 (7). This compound reacts with $YbCl_3 \cdot 6H_2O$ (1:2 molar ratio) in acetonitrile in the presence of Et_3N to give the tetrameric Zn_2Yb_2 complex **6** (Figure 4).

Figure 4. *Synthesis of the Zn_3 complex **5** and conversion to the Zn_2Yb_2 dimer **6**. (Chloride and nitrate ligands omitted for clarity).*

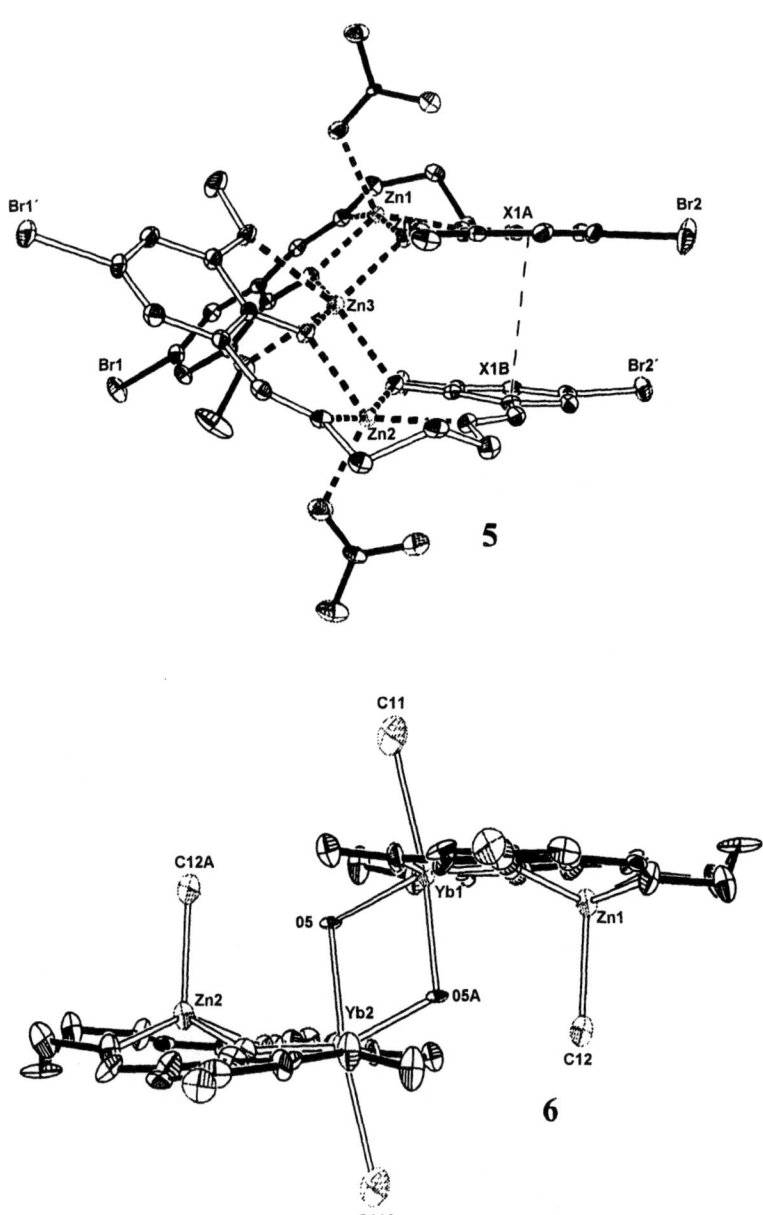

Figure 5. Views of the X-ray crystal structures of **5** *and* **6**.

The X-ray structures of **5** and **6** are shown in Figure 5. In **5** the molecule displays an interesting intramolecular π-π stacking interaction between two aryl rings (dihedral angle 4.8°) with the distance between the centers of each ring of 3.577 Å. The formation of **6** is accomplished by the replacement of the central zinc atom of **5** by two Yb^{3+} ions, which are bridged by two hydroxides. These OH units give the molecule an overall slipped sandwich configuration.

The photophysical properties of the lanthanides are of considerable interest for their applications in biology, medicine and materials science (8). Recently, because of the potential applications in bioassays and laser systems, attention has focused on the near-infrared (NIR) emissive properties of complexes of Yb(III), Nd(III) and Er(III) (9). In the case of **6** we were able to observe the NIR luminescence at 977 nm assigned to the $^2F_{5/2} \rightarrow {}^2F_{7/2}$ transition upon excitation of the ligand centered absorption band either at 275 or 350 nm. The relevant absorption and NIR luminescence spectra are shown in Figure 6.

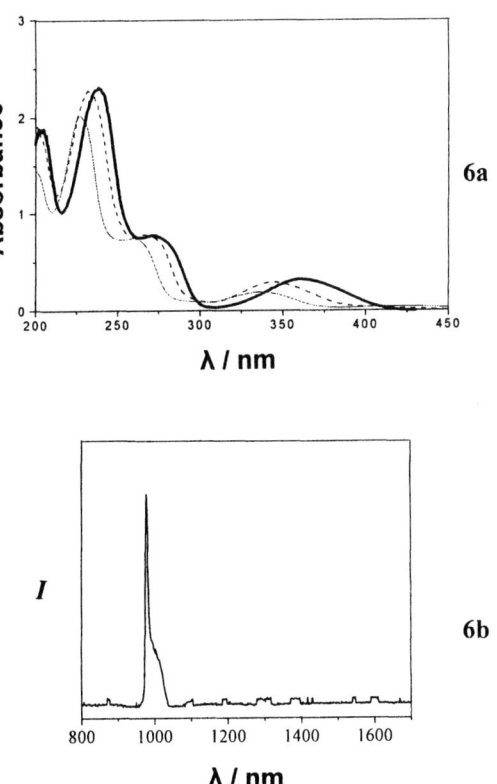

Figure 6a. Absorption spectra of 3 (thin), 5 (dotted) and 6 (solid).
Figure 6b. The NIR luminescence of 6 in MeCN(10^{-5} M) at room temperature.

Here we are no doubt using the main group ligand system as the chromophore to act as the antenna or sensitizer to facilitate the fluorescence of Yb(III) ions. The composition of the final product in these 3d-4f complexes seems to depend to some extent on the nature of the counter anions of both metals. Thus with $Zn(OAc)_2 \cdot H_2O$ as the 3d starting material, with $Yb(NO_3)_3 \cdot 6H_2O$, complexes which display a simple 1:1 (3d:4f) metal ratio are obtained. Figures 7a and 7b show the X-ray structures of two representative examples **7** and **8** formed from Schiff base ligands with C_4 and C_3 backbones respectively.

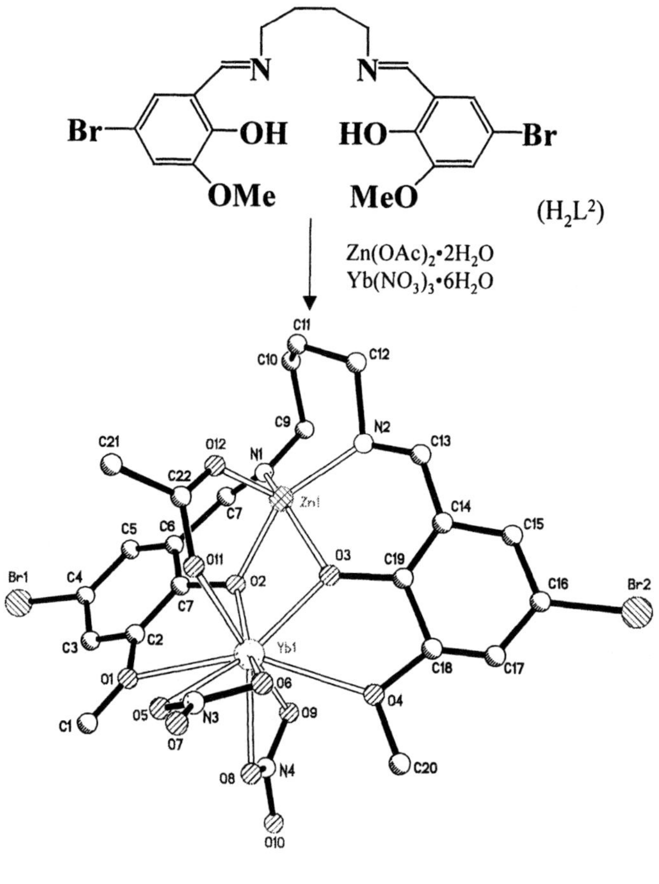

7 $[ZnYbL^2(NO_3)_2(OAc)] \cdot C_2H_5OH$

*Figure 7a. 3d-4f Complex **7** with 1:1 metal ratio.*

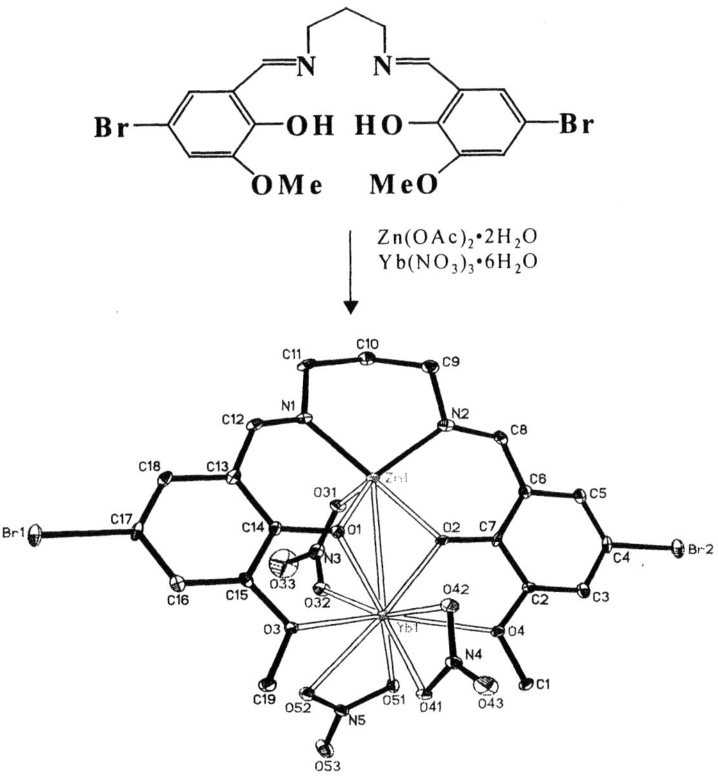

8 [ZnLnL³(NO₃)₃] (Ln = Nd, Eu, Tb and Yb)

Figure 7b. Complex **8** with 1:1 metal ratio.

In order to explore the results obtained by further variations of the backbones of these Schiff base ligands, we investigated the reaction of the vanallin derivative 5-bromo-3-methoxysalicylaldehyde with 1,2-diaminobenzene. At room temperature in ethanol the conventional Schiff base ligand is formed, however at 78 °C a rearrangement occurs to give the benzimidazole derivative as shown in Figure 8. This kind of rearrangement is well known (10). Reaction of the salen type Schiff-base with TbCl₃·6H₂O (4:3 ratio) in MeCN/MeOH leads to the unusual "tetra-decker" complex [Tb₃L₄(H₂O)₂]Cl (**9**) shown in Figure 9 along with its excitation and emission spectra. Interestingly the benzimidazole derivative (Figure 8) is also capable of forming a variety of different complexes with metal ions via the imizadole N, OH and OMe groups (Type I, Figure 8) or the two-OH and two OMe groups (Type II, Figure 8). Examples of such complexes (**10, 11**) involving the coordination of Zn₂ and ZnEu₂ groups are shown in Figure 10. Mixed metal (3d-

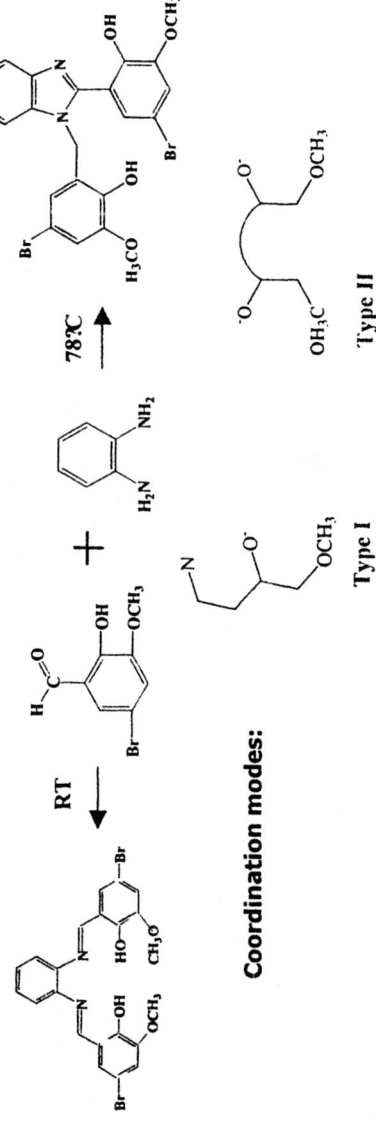

Figure 8. Ligands derived from 1,2-diaminobenzene

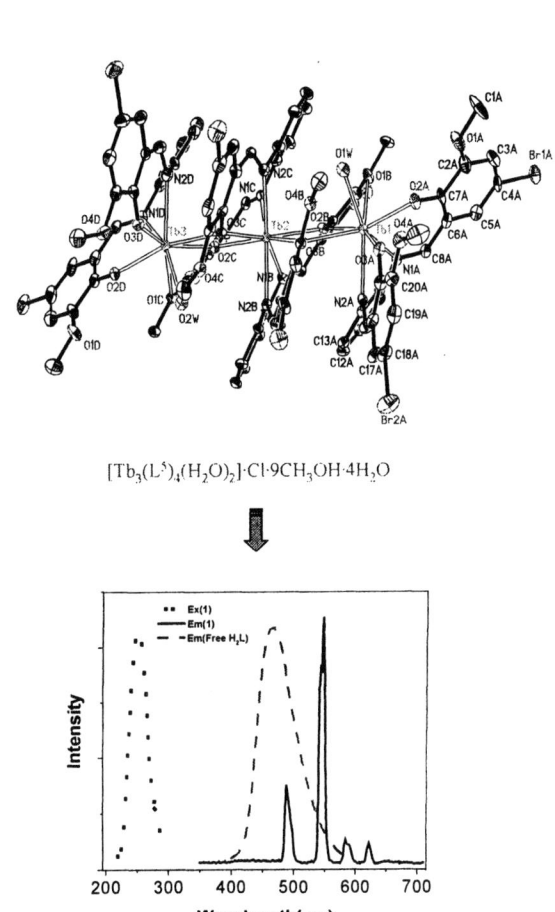

Figure 9. The tetra decker [Tb3L4]⁺ complex **9** and its fluorescence spectrum.

230

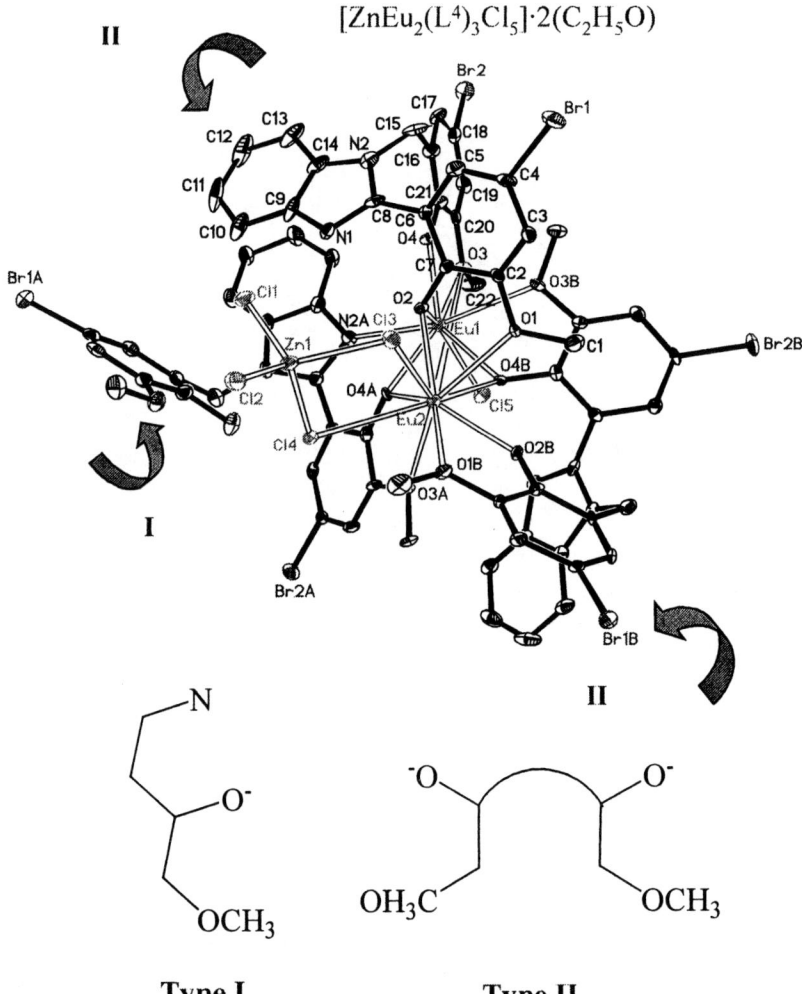

Type I

Type II

Figure 10a. Complex **10**: *Example of coordination mode for the benzimidazole based ligand.*

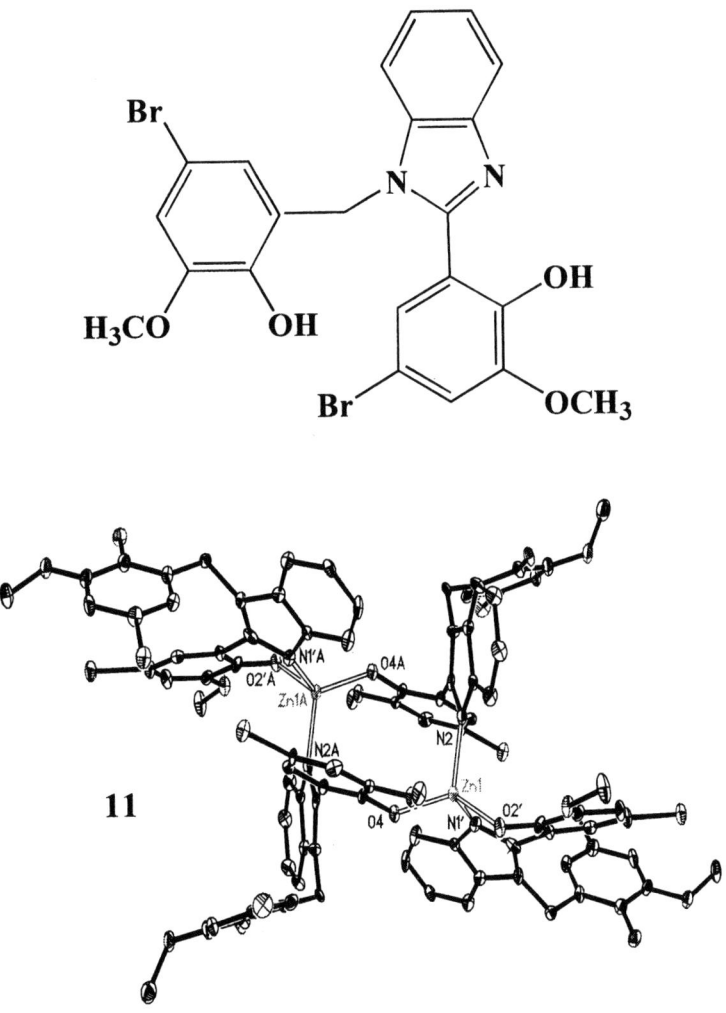

$Zn_2(L^4)_4]\cdot 3(C_2H_5OH)$

*Figure 10b. Complex **11**: Example of coordination mode for the benzimidazole based ligand.*

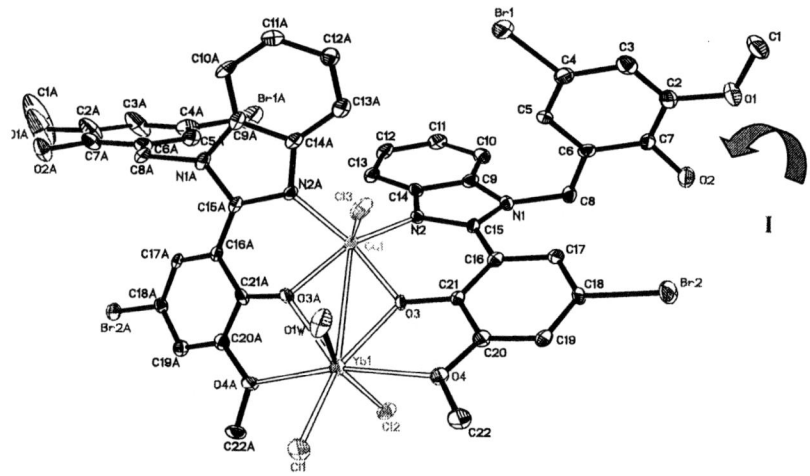

12 [CuYb(L^4)$_2$Cl$_3$]·3(C$_2$H$_5$OH)

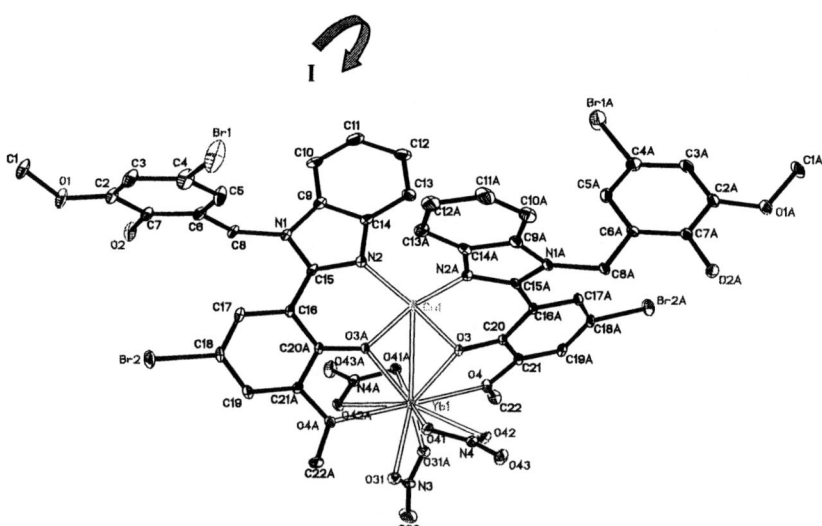

13   [CuLn(L^4)$_2$(NO$_3$)$_3$] (Ln = Yb, Eu, Tb and Er)

Figure 11. Mixed metal 3d-4f complexes found with the benzimidazole based ligand.

4f complexes can also be prepared with this ligand. The X-ray structures of two dinuclear examples (Cu(II)-Yb) (**12**) and Cu(II)-(Yb, Eu, Tb and Er) (**13**) are shown in Figure 11.

We were also interested in the possibility of forming mixed metal (3d-4f) complexes with the vanillin based starting materials themselves. However, the reaction of 5-bromo-3-methoxysalicylaldehyde with $TbCl_3 \cdot 6H_2O$ and $Zn(OAc) \cdot 2H_2O$ in methanol gave the unusual Tb_{10} complex (**14**) shown in Figure 12, and which contained in no Zinc (!!). Complex **14** has several interesting features. The overall Tb_{10} core is stabilized by eight μ_3-OH groups and has a

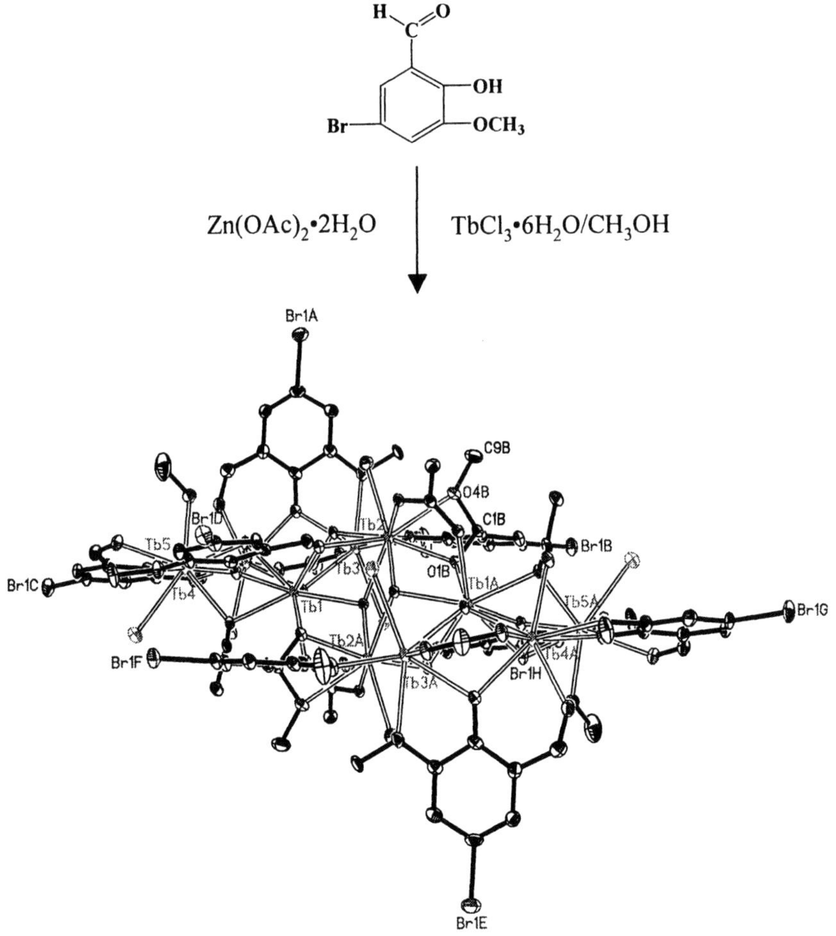

14 $[Tb_{10}(L^8)_6(L^9)_2(OH)_8(OAc)_6Cl_4(CH_3OH)_2(H_2O)_2] \cdot 2Cl \cdot 2CH_3OH$

Figure 12. Structure of the Tb_{10} complex
(*Dalton Trans.*, **2004**, 1787.)
Reproduced by permission of the Royal Society of Chemistry.

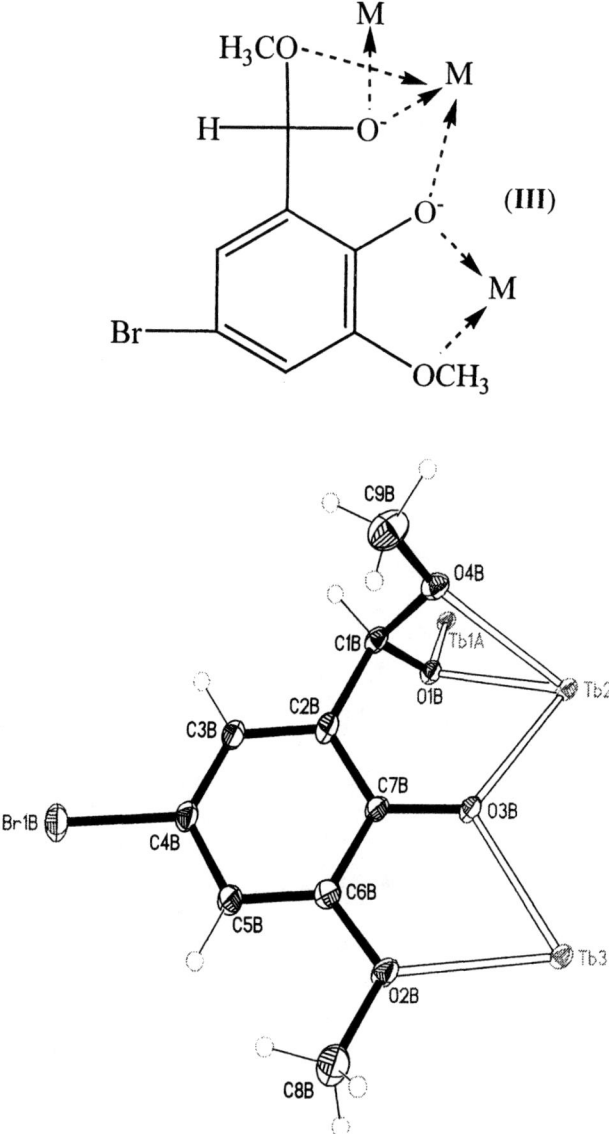

*Figure 13. Detail of the coordinated methyl hemiacetal unit in **14**
(Dalton Trans., **2004**, 1787.)*
Reproduced by permission of the Royal Society of Chemistry.

slipped sandwich configuration formed by the partial overlap of two nearly planar Tb_5 units. The compound also features an unusual example of a coordinated methyl hemiacetal group as illustrated in Figure 13.

Although the mechanism of formation of **14** is not known at present, the use of $Zn(OAc)_2 \cdot 2H_2O$ is important in its formation. Under similar conditions in the absence of zinc acetate similar reactions with $LnCl_3 \cdot 6H_2O$ (Ln = Tb, Er) gave the isostructural trinuclear complexes **15** and **16**. A view of the X-ray structure of **15** and **16** is shown in Figure 14. The central cores of the complexes are similar to that of the trinuclear Gd_3 complex $[L^lGd_3(OH)_3(NO_3)_2(H_2O)_4]^{2+}$ (L^l = deprotonated 3-methoxysalicylaldehyde) recently reported by Costes and coworkers (12).

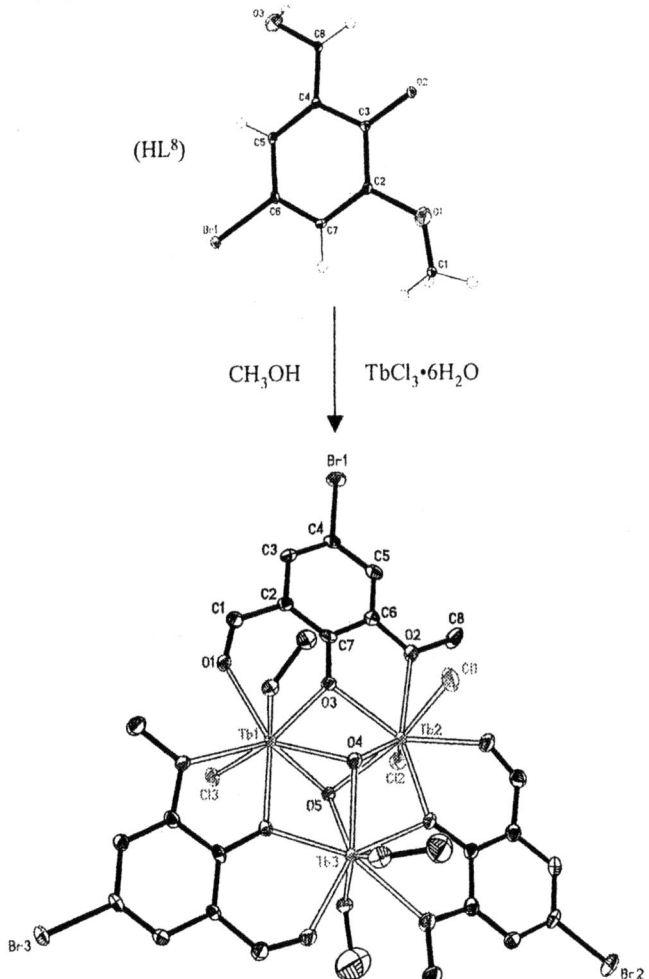

$[Ln_3(L^8)_3(OH)_2Cl_3(CH_3OH)_2H_2O]\cdot Cl\cdot 3CH_3OH$ (Ln = Tb and Er)

Figure 14. Trinuclear complexes of Tb (15) and Er (16).
*(Dalton Trans., **2004**, 1787.)*
Reproduced by permission of the Royal Society of Chemistry.

Acknowledgements

We are grateful to the Robert A. Welch Foundation for financial support (Grant F-816).

References

1. Zhang, L; Jones, R. A.; Lynch, V. M.; *Chem. Comm.*, **2002**, 2986.
2. Akine, S,; Taniguchi, T.; Nabeshima, T., *Angew. Chem. Int. Ed*, **2002**, *41*, 4671
3. (a) Sakamoto, M; Manseki, K.; Okawa, H, *Coord. Chem. Rev.*, **2001**, *219*, 379. (b) Costes, J.-P.; Dahan, F; Dupuis, A., *Inorg. Chem.*, **2000**, *39*, 165.
4. Winpermy, R. E., *Chem. Soc. Rev.*, **1998**, *27*, 447.
5. Wong, W.-K.; Liang, H.; Wong, W.-Y.; Cai, Z.; Li, K.-F.; Cheah, K.-W., *New J. Chem.*, **2002**, *26*, 275. See also Yang, X.-P., Su, C.-Y.; Kang B.-S.; Fong, X.-L.; Xiao, W.-L.; Liu, H.-Q., *J. Chem. Soc., Dalton Trans.*, **2000**, 3253. Yang, X.-P.; Kang, B.-S., Wong, W.-K., Su, C.-Y.; Liu, H.-Q., *Inorg. Chem.*, **2003**, *42*, 169.
6. Costes J.-P.; Laussac, J.-P.; Nicodème, F., *J. Chem. Soc., Dalton Trans.*, **2002**, 2731.
7. Yang, X.-P.; Jones, R. A.; Lynch, V. M.; Oye, M. M.; Holmes, A. L., *Dalton Trans.*, 2005, XXXX.
8. (a) McGehee, M. D.; Bergstedt, T.; Zhang, C.; Saab, A. P., O'Regan, M. B.; Bazan, G. C.; Strdanov, V. I.; Heeger, A., *J. Adv. Mater.*, **1999**, *11*, 1349; (b) Piguet, C.; Edder, C.; Rigault, S.; Bernardinelli, G.; Bünzli, J.-C. G.; Hopfgartner, G., *J. Chem Soc., Dalton Trans.*, **2000**, 3999, (c) Sabbatini, N.; Guardigli, M.; Lehn, J. M., *Coord. Chem. Rev.*, **1993**, *123*, 201.
9. (a) Werts, M. H. V., Hofstraat, J. W.; Geurts, F. A. J.; Verhoeven, J. W., *Chem. Phys. Lett.*, **1997**, *276*, 196; (b) Hasegawa, Y., Ohktbo, T.; Sogabe, K.; Kawamura, Y.; Wada, Y.; Nakashima, H.; Yanagida, S., *Angew. Chem.*, **2000**, *112*, 365 and *Angew. Chem. Int. Ed.*, **2000**, *39*, 357.
10. Kitazume, T.; Ishikawa, N., *Bull. Chem. Soc. Jap.*, **1974**, *47*, 785.
11. Yang, X.-P.; Jones, R. A.; Wiester, M. J., *Dalton Trans*, **2004**, 1787.
12. Costes, J.-P.; Dahan, F.; Nicodème, F., *Inorg. Chem.*, **2001**, *40*, 5285.

Chapter 17

Main Group Element Calixarenes: Molecular Constraint and Flexibility

Michael Lattman

Chemistry Department, Southern Methodist University, Dallas, TX 75275

The class of compounds commonly referred to as calixarenes possess a combination of constraint and flexibility. This combination of features allows the calixarene backbone to conform to the geometrical needs of a central atom or atoms while at the same time stabilizing the geometry. The smaller calix[4]arene is ideally suited for inserting a single main group element and adapting to geometry changes of that central atom. The larger calix[5]arene can accommodate two atoms within the cavity and provide sufficient constraint, in certain cases, to control the interaction between the two atoms.

Calixarenes (*1*) are a unique class of macrocyclic compounds possessing a combination of constraint and flexibility. The constraint is due to the size of the calixarene cavity, while the flexibility arises from the conformational mobility of the phenolic rings. This is particularly true for the smaller members of the series, calix[4]- and calix[5]arenes. The calix[4]arenes are ideally suited to supporting a single main group element in its central cavity and adapting to various coordination changes of that central atom. The larger calix[5]arene can accommodate two atoms within the cavity and, in certain cases, allows control of the interaction between the two atoms. This review describes our work, as well as related work by others, with main group element calix[4]- and [5]arenes.

Calix[4]arenes

The ability of the calix[4]arene framework to support a variety of coordination geometries around a single main group element is best illustrated with phosphorus (*2, 3, 4, 5*). A single phosphorus atom can be inserted into $[4]^{tBu}$ (See Figure 1 for abbreviations, etc.) via treatment with hexamethylphosphorustriamide (Figure 2). The reaction proceeds via elimination of two moles of dimethylamine yielding the zwitterionic six-coordinate phosphorus derivative **1**. Removal of the last mole of dimethylamine can be accomplished via treatment of **1** with acid or by heating. This results in cleavage of one of the P–O bonds yielding the three-coordinate phosphite, **2**, containing a "dangling" phenolic group. The phosphorus in **2** can be alkylated to give the four-coordinate phosphonium salt **3**. Finally, deprotonation produces the five-coordinate phosphorane **4**.

The x-ray crystal structures of **1** (R = H), **3**, and **4** (Figure 3) are illustrative of the conformational changes that the calixarene undergoes to support these different geometries. We have not obtained the x-ray structure of **2**; however, the geometrical requirements of this compound should be similar to **3**. The structure of **1** shows the geometry around the phosphorus atom to be a slightly distorted octahedron with the phosphorus lying just above the plane of the four oxygens and the hydrogen directed "inside" the basket. The calixarene backbone is in the cone conformation (see Figure 1 for definitions). The structure of the four-coordinate phosphonium cation shows that the phosphorus is a distorted tetrahedron. The free phenolic ring [O(4)] is flipped "up" leading to an approximate partial cone conformation. In addition, one of the P–O bonds ([P(1)–O(1)]) shows an "upward twist" relative to the other two bonds. This twist places the actual conformation of the calix[4]arene backbone between the partial cone and 1,2-alternate conformations. Finally, the structure of the pentacoordinate phosphorus is very close an ideal trigonal bipyramid: the O(1)–P–O(3) angle is 177°, all of the axial/equatorial bonds are within 2° of 90°, and the sum of the three equatorial angles is 360°. The calix[4]arene backbone is in an approximate partial cone conformation.

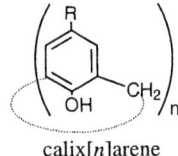

calix[n]arene

The symbols $[n]^R$ will be used throughout with n and **R** defined in the calix[n]arene structure at left. *In most cases R = t-Bu, unless otherwise indicated.*

Idealized conformations of calix[4]arenes

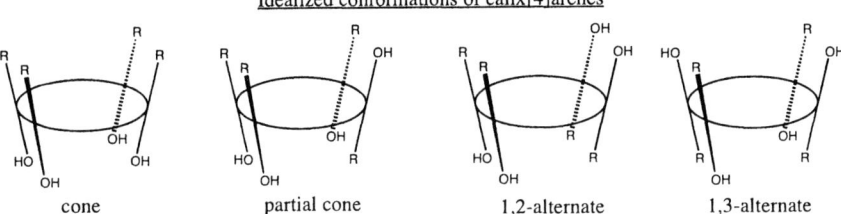

cone   partial cone   1,2-alternate   1,3-alternate

Figure 1. Abbreviations and idealized conformations of calix[4]arenes. Note that the four idealized conformations of calix[5]arenes are the same as for the calix[4]arenes.

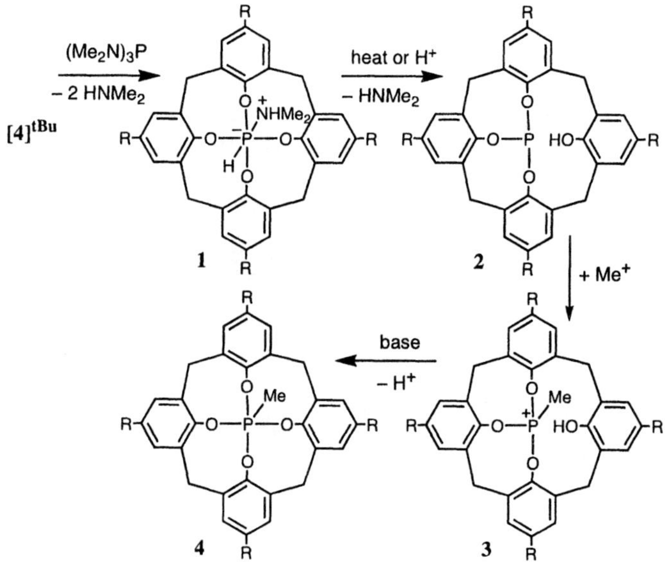

Figure 2. The calix[4]arene framework can support a six-, five-, four-, and three-coordinate phosphorus atom. Similar reactivity is observed for R = H.

It is interesting to postulate how **4** is formed from **3**. For example, after deprotonation, P-O(1) might twist down and the phenoxy group simply rotates to form the fourth P-O bond. Alternatively, the phenoxy group might flip and then form the fourth P-O bond. The difference between these two pathways is that O(1) and O(3) in **3** become axial groups in **4** in the former route, whereas they end up at equatorial positions in the latter.

The asymmetry created by the upward twist of the P-O bonds in **3** and **4** is not observed in the solution ^{1}H NMR spectra of these compounds: the phosphonium cation exhibits C_s symmetry while the phosphorane shows C_{4v} symmetry.

Similar, but more limited, coordination changes are observed for silicon-containing calix[4]arenes (Figure 4) (*5, 6*).

When the calixarene oxygens are not all bound to the central atom, for example in **3** and **5**, there exists the possibility of conformational isomerism. In fact, we do find that both of these derivatives exist as partial cone and cone isomers. These can be isolated and separately characterized. We measured the interconversion barrier for the silicon dervative and found an activation energy of about 120 kJ/mol with a negligible enthalpy difference between the conformers (*5*). This is about twice the value of the inversion barrier of the free calix[4]arene, **[4]**^{R}. For the R = H derivative, this isomerism even occurs in the solid state: the partial cone conformer of **5** (R = H) undergoes an irreversible phase change to the cone conformer at 230 °C, well below the melting point. No such conversion is observed for the R = t-Bu derivative.

5-cone ⇌ **5-partial cone**
(k_1, k_{-1})

The observation and isolation of these isomers also illustrates the power of ^{1}H NMR spectroscopy to help determine conformation. Both conformers of **5** have C_s symmetry, so their spectra have the same number and multiplicities of resonances. However, the orientation of the free phenolic group has a marked effect on the position of the (silyl) methyl resonance. In the cone conformation (R = H), this resonance appears at 0.72 ppm, while in the partial cone, this same resonance is at − 0.29 ppm, a full ppm upfield. The large upfield shift is due to the shielding of the methyl group in the partial cone conformation, since this group lies "above" the aromatic ring in this orientation.

Insertion of the larger arsenic into calix[4]arene proceeds smoothly (*7*). However, in this case, all three moles of dimethylamine are lost, and the product obtained is the arsenite, **7**. No hexacoordinate compound analogous to **1** is observed. While it is tempting to assume that the heavier congener is too large

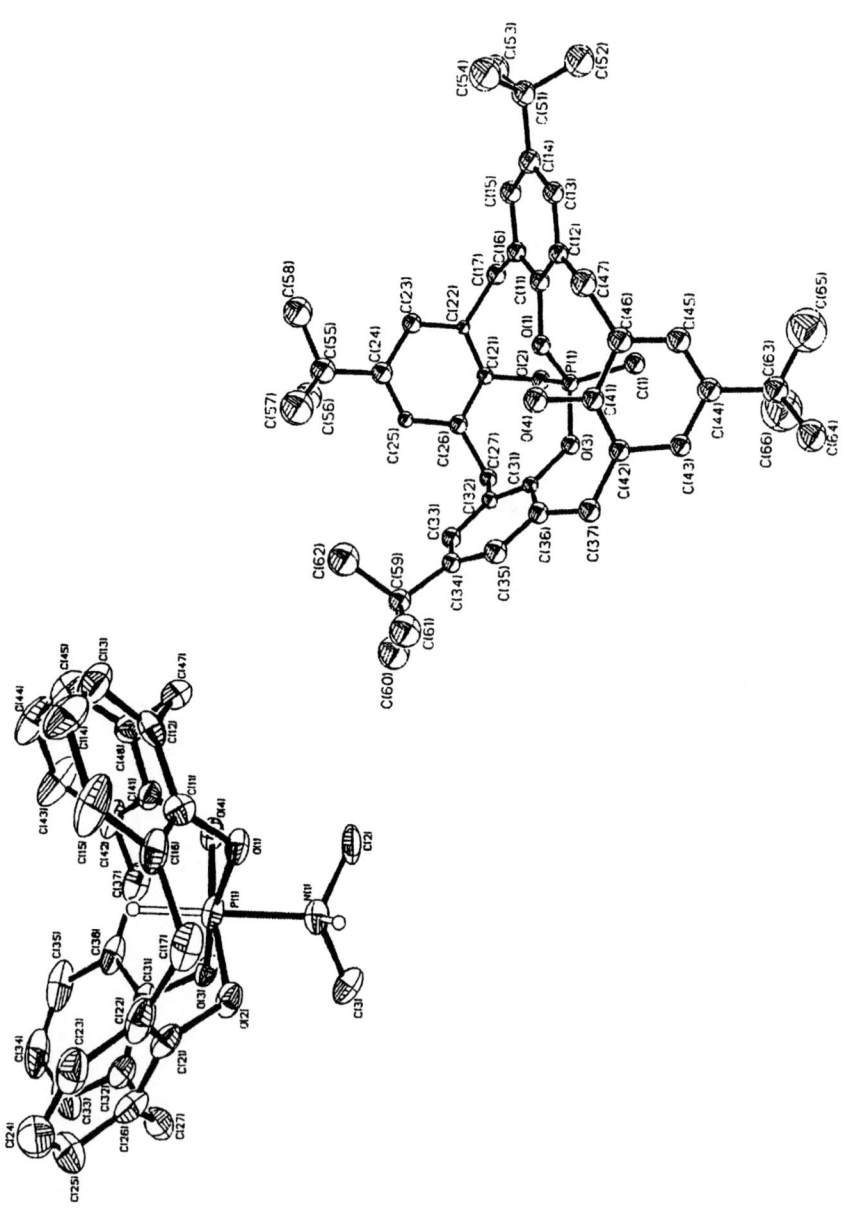

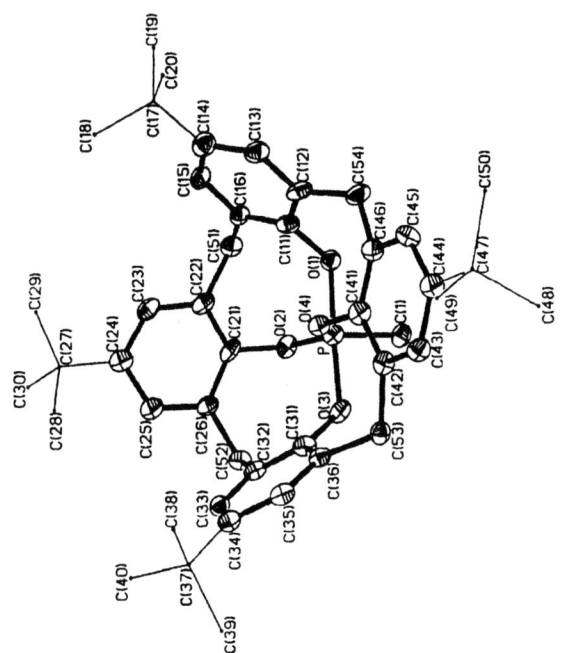

Figure 3. X-ray crystal structures of **1** (R = H), **3**, and **4** illustrating the conformational changes that the calix[4]arene framework undergoes due to the coordination at the central main group atom. (Adapted from references 4 and 5. Copyright 1994, 2000 American Chemical Society.)

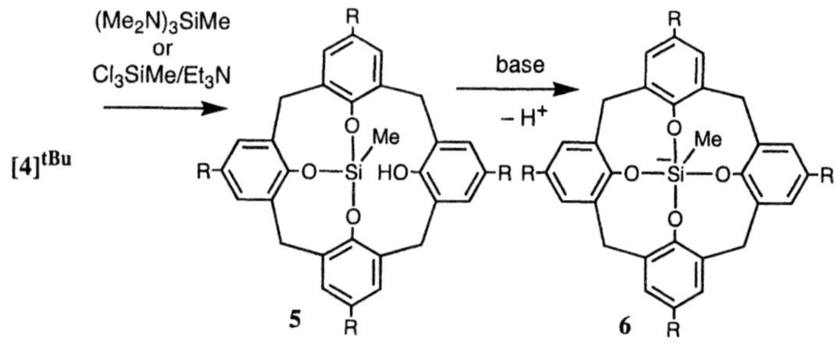

Figure 4. Synthesis of four- and five-coordinate silicon calix[4]arenes.

to fit into the base of the basket when bound to all four oxygens, the fact that transition metals can be inserted and bind to all four oxygens argues against this. Perhaps the more significant factor in the inability to observe the hexacoordinate arsenic is the As–H bond energy which is substantially lower than that of the P–H, by about 75 kJ/mol.

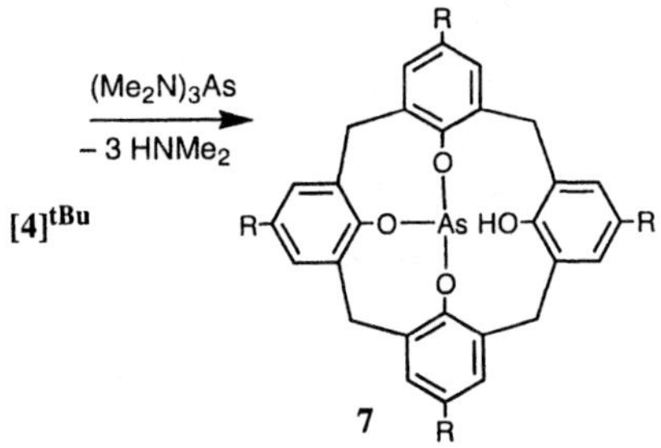

Attempts to use some of our silicon-containing calix[4]arenes to direct substitution at specific oxygen groups has met with limited success. We have succeeded in using the silicon in **5** as a protecting group for the generation of monosubstituted calix[4]arenes according to Figure 5 (*6*). Unfortunately, attempts to synthesize "1,2-disubstituted" calixarenes using the dimethylsilyl protecting group led to an apparent disproportionation (Figure 6) (*8*). However, Miyano and coworkers, reasoning that such a disproportionation might arise due to "...the lability of the dimethylsilyl moiety per se as well as by the ring strain..." did succeed in synthesizing 1,2-disbustituted calixarenes by use of "...a silyl bridge of proper chain length...", specifically tetraisopropyldisiloxane (TIPDS) according to Figure 7 (*9*).

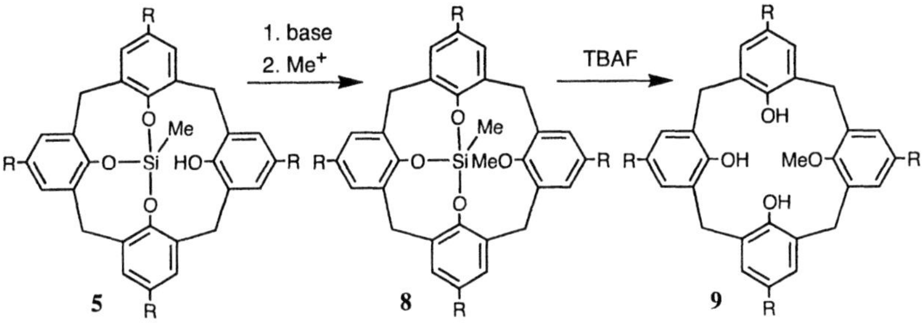

Figure 5. Synthesis of monosubstituted calix[4]arenes via 5.

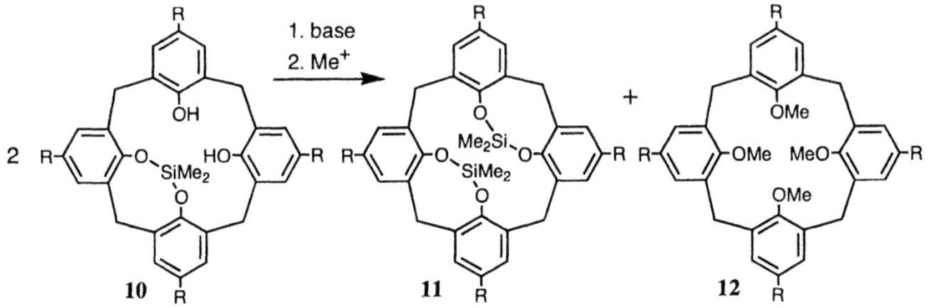

Figure 6. Apparent disproportionation attempting to synthesize 1,2-disubstituted calix[4]arenes.

Y = i-Pr$_2$SiOSii-Pr$_2$

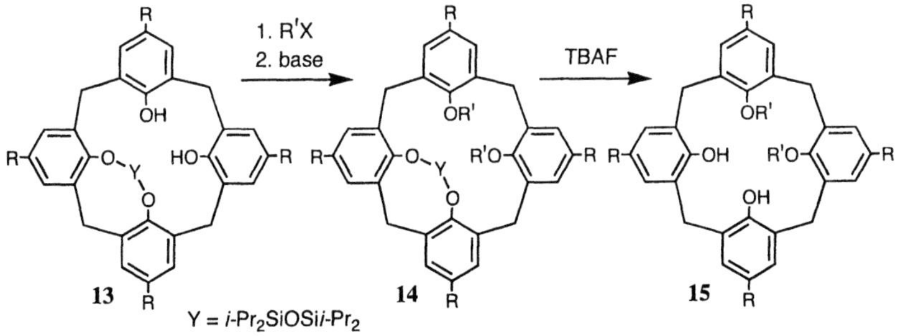

Figure 7. Successful use of the i-Pr$_2$SiOSii-Pr$_2$ protecting group to synthesize 1,2-disubstituted calix[4]arenes.

Several groups have used some of our silyl and phosphorus calix[4]arene compounds (as well as derivatives synthesized from them) in catalytic applications. For example, Ladipo and coworkers have used **16**, obtained by the treatment of **10** with TiCl$_4$, for the highly regioselective

16

cyclotrimerization of alkynes (*10*). We have used a series of silyl-substituted calix[4]arenes as external electron donors in Ziegler-Natta polypropylene catalysts (*11*). Pringle and coworkers have synthesized a variety of metal complexes of **2** (R = *t*-Bu, H). In addition, they reported a catalytic study of rhodium complexes of these ligands in the hydroformylation of 1-hexene (*12, 13*). Concurrently, van Leeuwen and coworkers reported the use of derivatives of **2** in the rhodium-catalyzed hydroformylation of 1-octene (*14*).

The complexes of **2** mentioned in the previous paragraph utilize this phosphorus calixarene as a simple monodentate phosphorus ligand. However, the free phenolic group might allow for binding of both the phosphorus and oxygen to a single metal center. Moreover, the constraint provided by the calixarene backbone might allow steric control of the ligand-metal interaction. In an effort to synthesize such a derivative, we treated **2** with butyllithium followed by addition of Cl$_3$TiCp. This does lead to **17**, a species containing a Ti–O bond (*15*). However, the x-ray crystal structure of this complex (Figure 8) reveals that the calix[4]arene is in an approximate partial cone conformation placing the phosphorus and metal on "opposite" sides of the calix[4]arene, thus prohibiting any phosphorus/titanium interaction. If the reason for this orientation is the steric crowding that would occur if both the phosphorus and metal were on the same side of the small calix[4]arene cavity, then moving to a slightly larger calixarene might lead to complexes where both the phosphorus and oxygen might bind to a single metal.

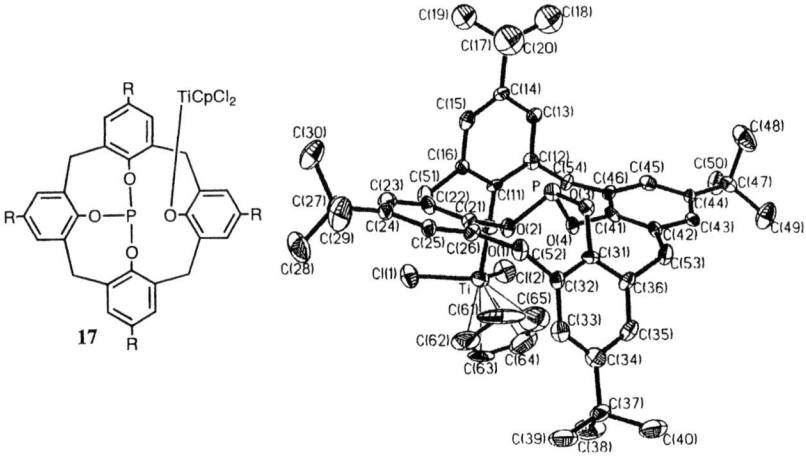

*Figure 8. Complex **17** and its x-ray crystal structure. (Adapted with permission from reference 15. Copyright 1999 Taylor and Francis.)*

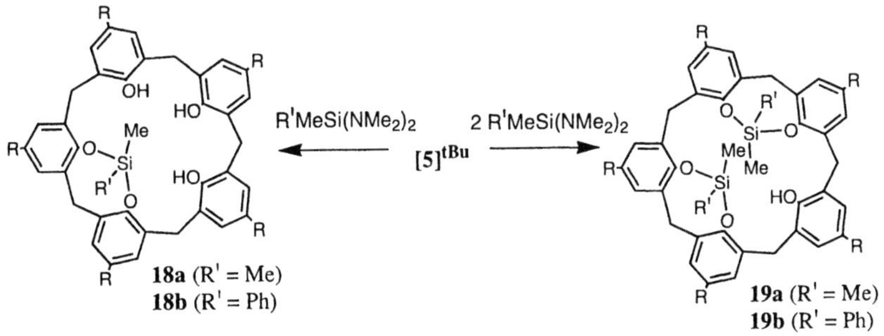

Figure 9. Insertion of one and two bridging silyl groups into calix[5]arene.

Calix[5]arenes

Examining models of the larger calixarenes, we felt that the calix[6]arene would be too large and floppy to provide any significant constraint. The calix[5]arene, on the other hand, appeared to have the right combination of constraint and flexibility to allow insertion of both a (phosphorus) ligand and metal inside and to be able to study and control their interaction.

In a study to determine substitution patterns in the calix[5]arene (*16*), we were able to insert one and two bridging silyl groups (Figure 9). The insertion of the single silyl group is best accomplished by using a large excess of the calixarene (0.35:1 silane:calixarene). Otherwise, significant amounts of the disubstituted **19** are formed which leads to mixtures that are difficult to separate.

Tricoordinate phosphorus can also be inserted into calix[5]arene (Figure 10). However, control of substitution is more difficult. We have only succeeded in isolating a monophosphorus deriviative with the Me$_2$NP group, not with PhP (*17*).

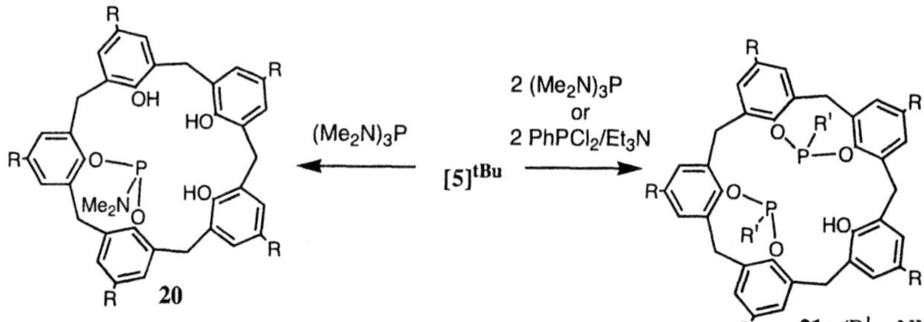

Figure 10. Insertion of one and two bridging phosphorus groups into calix[5]arene.

Calixarene **20** seemed to be the ideal candidate to test our hypothesis that the calix[5]arene framework is ideal to control the interaction between a ligand and metal. Tungsten can be inserted via the mixed imido/amido W(VI) reagent according to Figure 11 (*18*). The P---W distance in **22** is 3.15 Å, outside the range of a P–W bond. The x-ray crystal structure (*18*) shows the phosphorus lone pair to be pointing directly at the vacant coordination site of the tungsten. Several reasons could account for the long distance: the calix[5]arene framework is holding the two atoms apart; the soft phosphorus lone pair and the hard tungsten (VI) center are a poor match; the amido and imido ligands place a lot of electron density back onto the tungsten, so no extra density is needed. To test the importance of the latter reason, we treated **22** with triflic acid to replace the excellent π-backbonding amido ligand with a much weaker binder, triflate. The resulting product, **23**, has a P---W distance of 2.74 Å, on the order of a normal P–W bond, demonstrating that the calix[5]arene backbone is indeed the right framework to study and control the interaction between a ligand and metal.

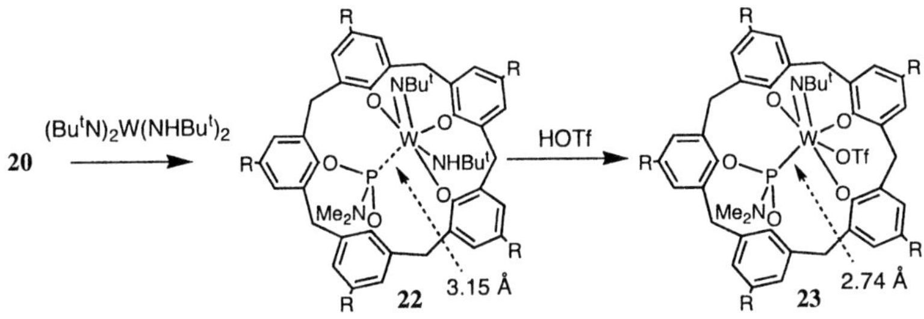

Figure 11. Insertion of tungsten and control of the phosphorus/metal interaction.

We have expanded the range of phosphorus-containing calix[5]arene ligands, two examples of which are illustrated in Figure 12 (*17*).

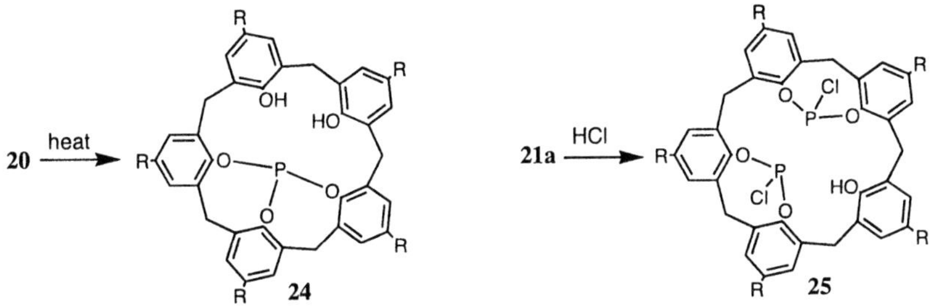

Figure 12. Further examples of phosphorus-containing calix[5]arene ligands.

The derivatives containing two trivalent phosphorus atoms, such as **21a**, **21b**, and **25** should be ideal as bidentate ligands toward transition metals. However, the specifics of the binding may be significantly different. This is illustrated in Figure 13 which shows the x-ray crystal structures of **21a** and **25**. In **21a**, the calix[5]arene is in an approximate cone conformation, while **25** is in an approximate 1,2-alternate conformation.

These compounds are expected to be excellent ligands toward transition metals, particularly in light of work on metal binding of larger calixarenes (*19*). Both phosphorus lone pairs are oriented to bind to a single metal in **21a** and **25** with, perhaps, the metal "outside" the calixarene cavity with **21a** and "inside" the cavity with **25**. In addition, the free phenolic oxygen may also interact. The monophosphorus compound **24** may serve a similar function as **20**, where both the phosphorus and oxygens bind to the metal.

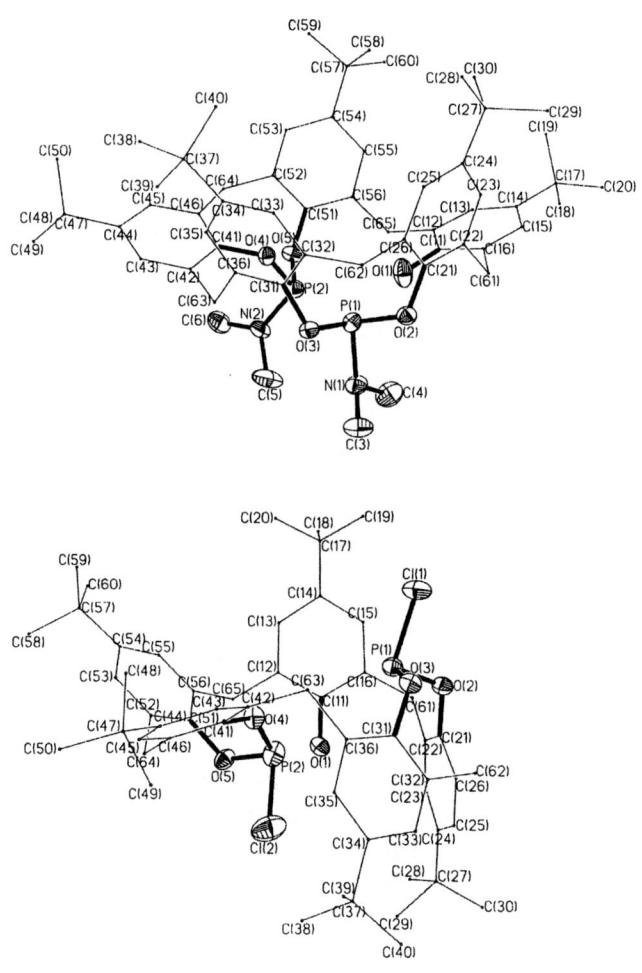

*Figure 13. X-ray crystal structures of **21a** (top) and **25** (bottom). (Reproduced from reference 17. Copyright 2004 American Chemical Society.)*

Acknowledgment

This work was supported in part by funding from the the National Science Foundation (CHE-9522606) and the Robert A. Welch Foundation. This paper is dedicated to my postdoctoral advisor, Alan H. Cowley, F. R. S., on the occasion of his 70th birthday, for his continuing inspiration and support throughout my career.

References

1. See, for example, *Calixarenes 2001*; Asfari, Z.; Böhmer, V.; Harrowfield, J.; Vicens, J., Eds.; Kluwer: Dordrecht, 2001., and references cited therein.
2. Khasnis, D. V.; Lattman, M.; Gutsche, D. *J. Am. Chem. Soc.* **1990**, *112*, 9422.
3. Khasnis, D. V.; Burton, J. M.; Lattman, M.; Zhang, H. *J. Chem. Soc., Chem. Commun.* **1991**, 562.
4. Khasnis, D. V.; Burton, J. M.; McNeil, J. D.; Santini, C. J.; Zhang, H.; Lattman, M. *Inorg. Chem.*, **1994**, *33*, 2657.
5. Fan, M.; Shevchenko, I. V.; Voorhies, R. H.; Eckert, S. F.; Zhang, H.; Lattman, M. *Inorg. Chem.*, **2000**, *39*, 4704-4712.
6. Shang, S.; Khasnis, D. V.; Burton, J. M.; Santini, C. J.; Fan, M.; Small, A. C.; Lattman, M. *Organometallics*, **1994**, *13*, 5157.
7. Shang, S.; Khasnis, D. V.; Zhang, H.; Small, A. C.; Fan, M.; Lattman, M. *Inorg. Chem.*, **1995**, *34* , 3610.
8. Fan, M., Zhang, H.; Lattman, M. *Organometallics*, **1996**, *15*, 5216.
9. Narumi, F.; Morohashi, N.; Matsumura, N.; Iki, N.; Kameyama, H.; Miyano, S. *Tetrahedron Letters* **2002**, *43*, 621.
10. Ozerov, O. V.; Ladipo, F. T.; Patrick, B. O. *J. Am. Chem. Soc.* **1999**, *121*, 7941.
11. Kemp, R. A.; Brown, D. S.; Lattman, M.; Li, J. *J. Mol. Catal. A: Chem.*, **1999**, *149*, 125.
12. Cobley, C. J.; Ellis, D. D.; Orpen, A. G.; Pringle, P. G. *J. Chem. Soc. Dalton Trans.* **2000**, 1101.
13. Cobley, C. J.; Ellis, D. D.; Orpen, A. G.; Pringle, P. G. *J. Chem. Soc. Dalton Trans.* **2000**, 1109.
14. Parlevliet, F. J.; Kiener, C.; Fraanje, J.; Lutz, M.; Spek, A. L.; Kamer, P. C.; van Leuwen, P. W. N. M. *J. Chem. Soc. Dalton Trans.* **2000**, 1113.
15. Fan, M., Zhang, H.; Lattman, M. *Phosphorus Sulfur and Silicon*, **1999**, *144-6*, 257.
16. Sood, P.; Zhang, H.; Lattman, M. *Organometallics*, **2002**, *21*, 4442.
17. Sood, P.; Koutha, M.; Fan, M.; Klichko, Y.; Zhang, H.; Lattman, M. *Inorg. Chem.* **2004**, *43*, 2975.
18. Fan, M., Zhang, H.; Lattman, M. *J. Chem. Soc., Chem. Commun.*, **1998**, 99.
19. Redshaw, C. *Coord. Chem. Rev.* **2003**, *244*, 45.

Chapter 18

N-Heterocyclic Carbene Adducts of High-Oxidation-State Metal Halides: An Unexpected Instance of Lewis Acidic $C_{carbene}$ Character

Mark D. Spicer[1,*], Christopher A. Dodds[1], John P. Culver[2], and Colin D. Abernethy[2,*]

[1]Department of Pure and Applied Chemistry, University of Strathclyde, Glasgow G1 1XL, United Kingdom
[2]Department of Chemistry, Western Kentucky University, Bowling Green, KY 42101

N-heterocyclic carbenes (NHCs) are well-known as strong Lewis bases. When complexed to high-oxidation-state metal centers, however, the NHC carbenic carbon atoms display an unexpected Lewis acidic character. In such complexes, close interligand contacts occur which result from the donation of neighboring ligand lone-pair electron density into the formally vacant $C_{carbene}$ p-orbital. These contacts have been observed in mono-NHC complexes of both high-oxidation-state early transition metals (Ti(IV), V(V), and Re(V)) and main group elements (Si(IV) and Sn(IV)).

N-Heterocyclic Carbenes

Since the first isolation and complete characterization of a stable, crystalline N-heterocyclic carbene (NHC) by Arduengo *et al.* in 1991 (*1*), interest in these remarkable species has continued to grow (*2*). The powerful Lewis basicity of NHCs allows them to coordinate strongly to even relatively weakly Lewis acidic centers, forming in the process very strong and largely inert bonds. The importance of NHCs in organometallic chemistry arises from the dramatic increases in both stability and reactivity observed when they are used to replace tertiary phosphine ligands in many of the transition metal based catalysts that are used for effecting carbon-carbon bond formation (*3*). The significance of NHCs is not, however, restricted to their use as ligands in complexes featuring transition metal centers. The powerful electron-donating properties of NHCs have also been employed to stabilize novel main group moieties; for instance, NHCs have been used to prepare thermally stable adducts of InH_3, which could not be fully stabilized by weaker Lewis bases such as phosphines (*4*).

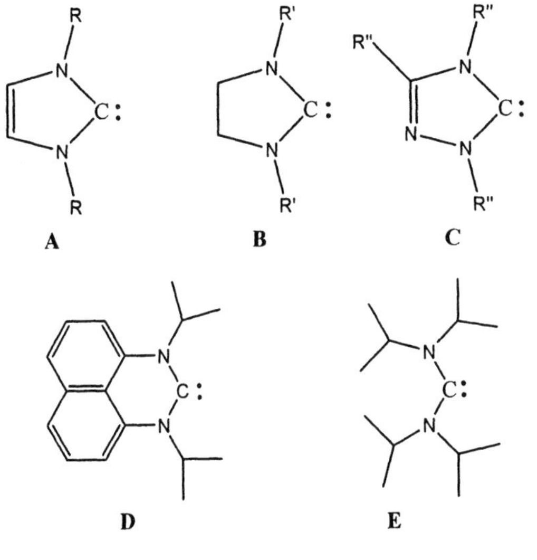

Figure 1. Typical structural motifs for stable carbenes. **A**: *NHC based on an unsaturated five-membered diamino-heterocyclic ring (R = alkyl or aryl).* **B**: *NHC based on a saturated five-membered diamino-heterocyclic ring (R' = 2,4,6-trimethylphenyl).* **C**: *NHC based on an unsaturated five-membered triamino-heterocyclic ring (R'' = phenyl).* **D**: *NHC based on an unsaturated six-membered heterocyclic ring.* **E**: *acyclic diaminocarbene.*

The first stable carbenes to be reported were the NHCs based on a five-membered unsaturated diamino-heterocyclic ring (Figure 1, **A**), subsequently, NHCs based on saturated five-membered diamino-heterocyclic rings (Figure 1, **B**), five-membered unsaturated triamino-heterocyclic rings (Figure 1, **C**), or unsaturated six-membered heterocyclic rings (Figure 1, **D**), as well as acyclic diaminocarbenes (Figure 1, **E**), have also been prepared (2). The stability and general chemical behavior of carbenes is determined by the electronic state of their carbenic carbon center. All of the stable diaminocarbenes prepared to date exist in a singlet electronic state in which the carbenic carbon possesses a lone-pair of electrons contained in an in-plane σ-orbital together with a formally vacant p_π-orbital which is orthogonal to the N-$C_{carbene}$-N plane (Figure 2).

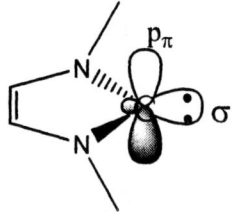

Figure 2. Electronic configuration of NHCs.

The remarkable stability of these molecules arises from the interaction of the $C_{carbene}$ center with the adjacent π-donating and σ-withdrawing amino moeties. Significant N→C←N π-donation serves to stabilize the singlet state as this raises the energy of the formally vacant $C_{carbene}$ p_π-orbital, which is antibonding with respect to the N-C-N π-interaction (Figure 3) (5).

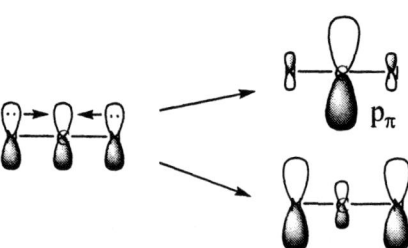

Figure 3. N→C←N π-interactions which serve to raise the energy of the p_π-orbital of the carbenic carbon atom.

The infinite stability of NHCs in the absence of air or moisture is testimony to the high energy and relative inaccessibility of the p_π-orbital on the $C_{carbene}$ in these systems. Any Lewis acidic character of the carbenic carbon in the free carbenes would result in dimerization, forming an electron rich olefin. The reaction would follow a pathway involving mutual attack of the p_π-orbitals by the σ-lone-pairs of two $C_{carbene}$ atoms (Figure 4) (*6*).

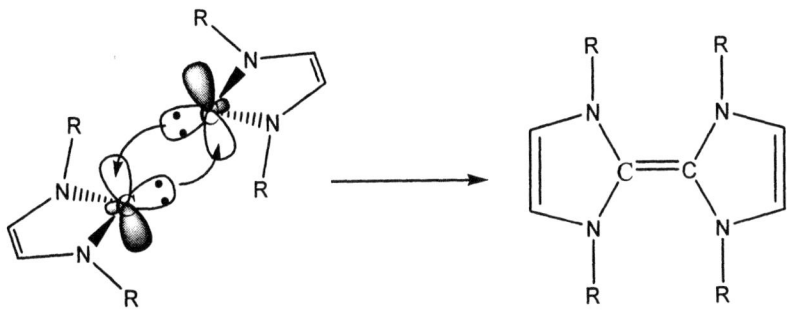

Figure 4. Proposed pathway for the dimerization of NHCs to form electron-rich olefins.

The absence of such a reaction confirms the inaccessibility of the p_π-orbital on the $C_{carbene}$ in free NHCs, and led to the early acceptance that these species act as σ-donor only ligands with little or no tendency to form π-backbonds. Such a view was reinforced by the successful synthesis of many stable main group metal NHC complexes as well as initial density functional theory (DFT) calculations and photoelectron spectroscopic studies of palladium and platinum d^{10} bis(carbene) complexes (*7*). Although recent (2004) DFT calculations have shown that π-backbonding interactions may contribute up to 30% of an NHC's total orbital interaction energy with an electron-rich low-valent late transition metal such as silver(I), copper(I) or gold(I) (*8*), our initial assumption that NHCs were strong σ-only ligands led us to investigate their reactivity towards high-oxidation-state early transition metal Lewis acids.

N-Heterocyclic Carbene Complexes of Transition Metal Halides in High Oxidation States

Given the strong σ-donor properties of NHCs together with the strength of NHC-metal bonds and their resistance towards oxidation, it is surprising that the importance of complexes featuring NHC ligands coordinated to transition metal ions in high oxidation states has only recently been recognized (2003). The first study of such systems, conducted by Herrmann *et al.* in 1994, involved the preparation of bis-(1,3-dimethylimidazol-2-ylidene) adducts of Group 4 metal tetrachlorides (M = Ti, Zr, and Hf) together with the complex (1,3-dimethylimidazol-2-ylidene)-methyl-trioxo-rhenium (*9*). However, single-crystal X-ray diffraction studies were not performed on any of the reported compounds. A second early paper by Kuhn *et al.* (1995) reported the synthesis of mono adducts of $TiCl_4$ with the NHCs, 1,3-dialkyl-4,5-dimethylimidazol-2-ylidene (alkyl = methyl, ethyl, or isopropyl) (*10*). Unfortunately, single crystals suitable for X-ray structure analysis could not be obtained, so structural data for these complexes are also lacking. However, controlled hydrolysis of (1,3-diisopropyl-4,5-dimethylimidazol-2-ylidene)titanium tetrachloride yielded the oxo-bridged di-nuclear NHC-titanium(IV) chloro species, bis-{(1,3-diisopropyl-4,5-dimethylimidazol-2-ylidene)$TiCl_3$}(μ_2-O) (**1**), which was characterized by X-ray diffraction (*vide infra*) (*10*).

In order to further our understanding of the coordination chemistry of NHCs towards high-oxidation-state early transition metal Lewis acids, we decided to prepare a mono NHC adduct of vanadium(V) oxo trichloride; as the standard potential of the V(V)/V(IV) couple is +1.00 V, coordination to vanadium(V) would test the stability of the NHC when bonded to a strongly oxidizing metal center. Reaction of equimolar amounts of 1,3-bis-(2,4,6-trimethylphenyl)imidazol-2-ylidene and $V(O)Cl_3$ in toluene resulted in the formation of a 1:1 adduct (**2**) (*11*). In contrast to the species reported by Herrmann *et al.* and Kuhn *et al.*, which were highly moisture sensitive, **2** proved remarkably stable (solid samples have been stored in air for over 2 years with no observable decomposition).

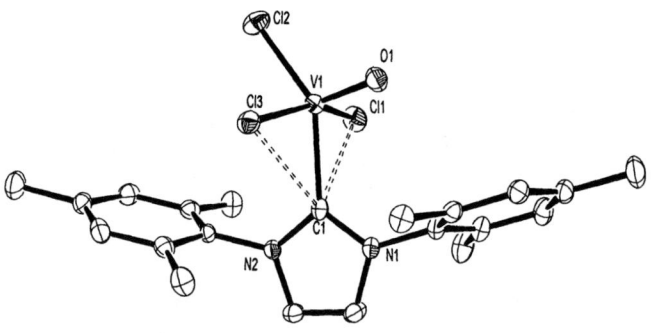

Figure 5. The X-ray crystal structure of **2**.

An X-ray structure determination of **2** revealed a remarkable feature of this molecule (Figure 5). Two chloride ligands lie almost perpendicular to the plane of the carbene's heterocyclic ring and are oriented towards the $C_{carbene}$ atom with acute Cl-V-C angles (Cl(1)-V-$C_{carbene}$ 81.0° and Cl(3)-V-$C_{carbene}$ 82.2°). The distances of these chloride ligands from the $C_{carbene}$ (Cl(1)-$C_{carbene}$ 2.85 Å and Cl(3)-$C_{carbene}$ 2.89 Å) are well within the sum of the van der Waals radii for carbon and chlorine (3.45 Å); such close proximity indicates that strong attractive interactions exist between the $C_{carbene}$ and adjacent chloride ligands. The nature of these interactions were investigated by performing DFT calculations (6-31 G + + (d, p) basis set, B3LYP method) on the model complex (1,3-dimethylimidazol-2-ylidene)V(O)Cl$_3$ (**3**). The optimized minimum-energy structure yielded bond distances and angles in good agreement with those obtained from the X-ray diffraction study of **2** (Cl(1)-V-$C_{carbene}$ 78.9°, Cl(3)-V-$C_{carbene}$ 79.0°, $C_{carbene}$ (Cl(1)-$C_{carbene}$ 2.81 Å, and Cl(3)-$C_{carbene}$ 2.81 Å). The acute Cl-V-$C_{carbene}$ angles and close Cl-$C_{carbene}$ distances obtained by the DFT calculations demonstrate that the observed interactions are independent of the steric bulk of the coordinated NHC and are unrelated to any crystal-packing forces. Analysis of the molecular orbitals generated by the DFT calculations indicates a bonding overlap between lone-pair containing orbitals on the adjacent chloride ligands and the p_π-orbital on the $C_{carbene}$. The HOMO-8 of **3** is shown in Figure 6 as an example. It is also worthy to note the significant vanadium d-orbital contribution to this molecular orbital.

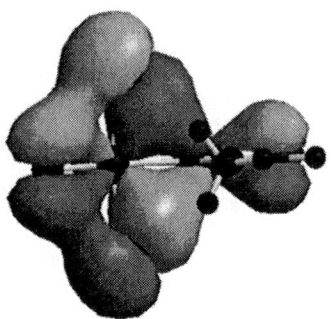

Figure 6. The HOMO-8 of 3 showing overlap of chloride p-orbitals, vanadium d-orbital and $C_{carbene}$ p_π-orbital.

The occurrence of Cl···$C_{carbene}$ interactions in **2** led us to re-examine the X-ray structural data for **1** (Figure 7) (*10*). The coordination around both of the titanium(IV) centers in **1** are similar, each displaying short contacts between the carbenic carbon atoms and neighboring chloride ligands situated above and below the planes of each of the NHC heterocyclic rings. As in the case of **2**, the

Cl-M-C$_{carbene}$ angles are acute (Cl(2)-Ti(1)-C(1) 81.9°, Cl(3)-Ti(1)-C(1) 81.2°, Cl(5)-Ti(2)-C(21) 82.7°, and Cl(6)-Ti(2)-C(21) 81.2°) and the Cl-C$_{carbene}$ distances (Cl(2)-C(1) 2.94 Å, Cl(3)-C(1) 2.91 Å, Cl(5)-C(21) 2.96 Å, and Cl(6)-C(21) 2.93 Å) are within the sum of the van der Waals radii for these atoms.

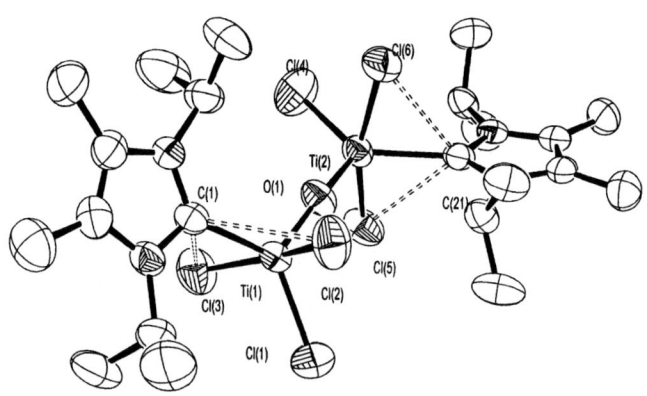

Figure 7. X-ray Crystal Structure of **1**.

In order to determine whether the Cl···C$_{carbene}$ interactions in **1** were due to chloride-C$_{carbene}$ p$_\pi$ orbital overlaps similar to those observed in **2**, DFT calaculations were performed (*12*). The dimeric nature of **1** made calculations on this compound using a high level of theory impractical, so the hypothetical (1,3-dimethylimidazol-2-ylidene)TiCl$_4$ (**4**) was chosen as a model system. The minimum energy structure calculated for **4**, using the B3LYP method with the 6-31 G + + (d, p) basis set, reveals the expected close C$_{carbene}$ contacts with two chloride ligands situated symmetrically above and below the plane of the NHC heterocyclic ring (Cl(1 & 2)-Ti-C$_{carbene}$ 79.2°, Cl(1 & 2)-C$_{carbene}$ 2.87 Å). The molecular orbitals generated for **4** clearly indicate significant bonding overlap between lone-pair containing orbitals on the adjacent chloride ligands and the p$_\pi$-orbital on the C$_{carbene}$ facilitated by the participation of the vacant titanium d-orbitals. The HOMO-8 of **4** is shown in Figure 8 as an example of these orbital interactions.

Figure 8. The HOMO-8 of 4 showing overlap of chloride p-orbitals, titanium d-orbital and $C_{carbene}$ p_π-orbital.

In 2003 Braband and Ulrich published details of a series of complexes featuring the triazacyclic NHC, 1,3,4-triphenyl-1,2,4-triazol-5-ylidene, coordinated to Re(V) (*13*). Although not commented upon by the authors, the X-ray structure of one of the compounds described, (1,3,4-triphenyl-1,2,4-triazol-5-ylidene)Re(≡N)Cl$_2$(PMe$_2$Ph)$_2$ (**5**, Figure 9), features a chloride ligand situated approximately perpendicular to the plane of the heterocyclic ring of the carbene. Although the size of this complex precluded a detailed DFT study, the acute Cl-Re-C$_{carbene}$ angle and short Cl-C$_{carbene}$ distance (Cl(1)-Re-C$_{carbene}$ 81.61°, Cl(1)-C$_{carbene}$ 3.02 Å) indicate that a similar attractive Cl···C$_{carbene}$ interaction also occurs in this Re(V), d^2, system.

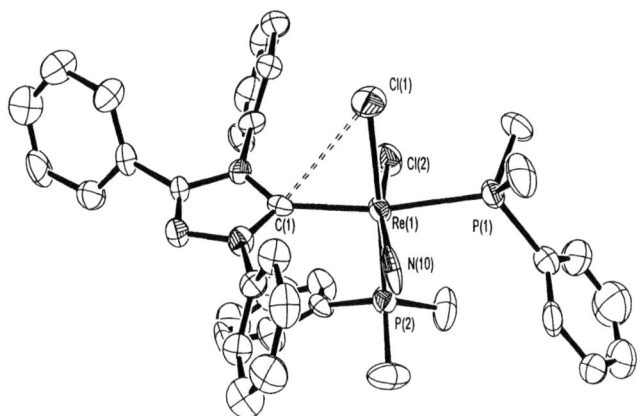

Figure 9. The X-ray Crystal Structure of 5.

In 2004 Cowley et al. presented details of a monomeric NHC complex of titanium(IV) containing two chloride ligands, (1,3-bis-(2,4,6-trimethylphenyl)imidazol-2-ylidene)TiCl$_2$(NMe$_2$)$_2$ (**6**, Figure 10) (*14*). The molecular structure of **6**, determined by single-crystal X-ray diffraction, shows that the two chloride ligands are positioned *cis* to the NHC orthogonal to the plane of the heterocycle. Once again, these are oriented towards the NHC ligand with the acute Cl-M-C$_{carbene}$ angles and short Cl$\cdots$C$_{carbene}$ characteristic of these systems (Cl(1)-Ti-C$_{carbene}$ 83.6°, Cl(2)-Ti-C$_{carbene}$ 83.1°, Cl(1)-C$_{carbene}$ 3.12 Å, and Cl(2)-C$_{carbene}$ 3.10 Å).

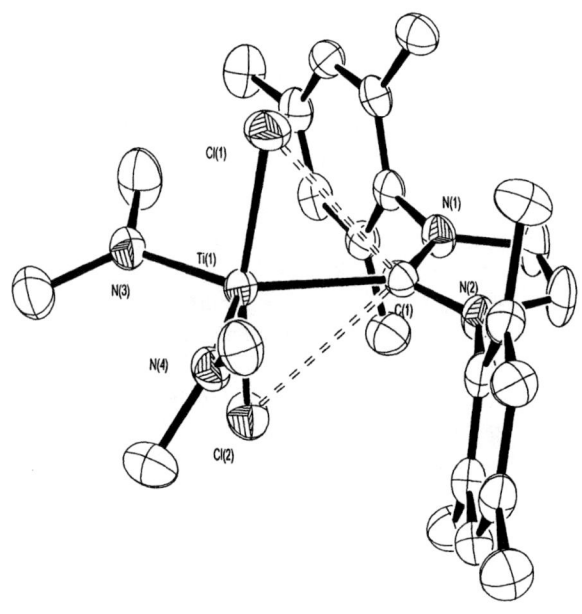

*Figure 10. X-ray crystal structure of **6**.*

Cowley et al. performed DFT calculations (B3LYP method, 6-31 G + + (d, p) basis set) on the model compound, (1,3-dimethylimidazol-2-ylidene)TiCl$_2$(NMe$_2$)$_2$ (**7**). The minimum energy structure calculated for **7** is in good agreement with the experimental structure observed for **6** with acute Cl-Ti-C$_{carbene}$ angles and short Cl-C$_{carbene}$ contacts (Cl(1)-Ti-C$_{carbene}$ 83.6°, Cl(2)-Ti-C$_{carbene}$ 83.1°, Cl(1)-C$_{carbene}$ 3.12 Å, and 3.10 Å). The molecular orbitals calculated for **7** (Figure 11) show the same Cl-M-C$_{carbene}$ orbital overlaps seen in those of **3** and **4**.

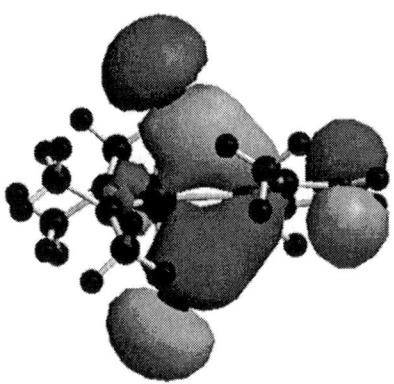

Figure 11. The HOMO-7 of 7 showing overlap of chloride p-orbitals, titanium d-orbital and $C_{carbene}\ p_\pi$-orbital.

The experimental structural data for **1**, **2**, **5**, and **6** indicate that short intramolecular $Cl\cdots C_{carbene}$ contacts are a pervasive feature of mono NHC complexes containing high-oxidation-state early transition metal ions and chloride ligands. Analysis of the molecular orbitals generated by DFT calculations on **3**, **4**, and **6** reveals that these contacts result from the donation of chloride lone-pair electron density into the formally vacant $C_{carbene}\ p_\pi$-orbital. The participation of the d-orbitals of the transition metal center is also evident in the molecular orbitals that involve the $Cl(p)\cdots C_{carbene}(p_\pi)$ overlap. This led us to ask whether the participation of metal d-orbitals was necessary to facilitate such interactions. If metal d-orbitals were necessary for such overlap to occur then close $Cl\cdots C_{carbene}$ contacts would not be expected in NHC complexes of main group element halides. We therefore conducted a search of the literature to find NHC complexes of main group element halides with similar geometries to those of the transition metal complexes discussed above.

N-Heterocyclic Carbene Complexes of High-Oxidation-State Main Group Halides

The search for main group analogs of the early transition metal NHC complexes described above led us to the work of Kuhn *et al.*, who prepared a series of NHC complexes of $SiCl_4$, $SnCl_4$ and $SnCl_2Ph_2$ (*15*). X-ray diffraction structural data were available for (1,3-diethyl-4,5-dimethylimidazol-2-ylidene)$SiCl_4$ (**8**, Figure 12) and (1,3-diisopropyl-4,5-dimethylimidazol-2-

ylidene)SnCl$_2$Ph$_2$ (**9**, Figure 13). These structures reveal that short Cl$\cdots$C$_{carbene}$ interactions do indeed occur in NHC complexes of main group elements. In both **8** and **9** the two chloride ligands *cis* to the NHC are oriented towards the C$_{carbene}$ atom. Compound **8** crystallizes with two crystallographically distinct molecules in the unit cell. The significant atomic distances and angles for the first molecule are: Cl(1)-Si(1)-C(1) 89.1°, Cl(4)-Si(1)-C(1) 85.4°, Cl(1)-C(1) 2.86 Å, and Cl(4)-C(1) 2.84 Å; and those of the second: : Cl(7)-Si(2)-C(11) 87.7°, Cl(8)-Si(2)-C(11) 87.5°, Cl(7)-C(11) 2.88 Å, and Cl(8)-C(11) 2.86 Å. In **9** the relevant metrical parameters are: Cl(1)-Sn-C$_{carbene}$ 82.5°, Cl(2)-Sn-C$_{carbene}$ 83.3°, Cl(1)-C$_{carbene}$ 3.12 Å, and Cl(2)-C$_{carbene}$ 3.17 Å.

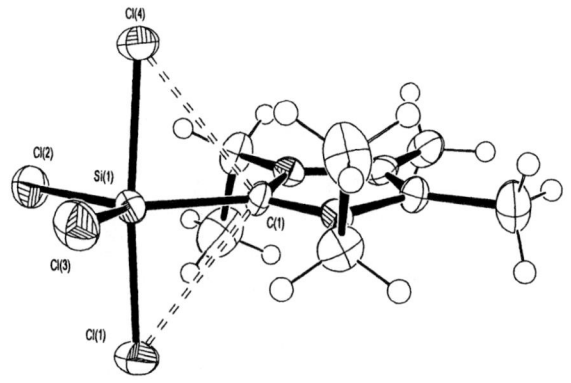

Figure 12. X-ray crystal structure of 8.

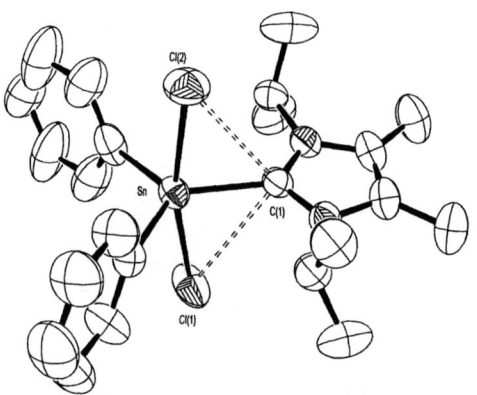

Figure 13. X-ray crystal structure of 9.

In order to understand the nature of the Cl···$C_{carbene}$ interactions in these complexes with main group central atoms, we performed DFT calculations, using the B3LYP method, on the hypothetical model compounds (1,3-dimethylimidazol-2-ylidene)SiCl$_4$ (**10**, with a 6-31 G + + (d, p) basis set), and (1,3-dimethylimidazol-2-ylidene)SnCl$_2$Ph$_2$ (**11**, with a LANL2DZ basis set) (*12*). The calculated minimum-energy structures for **10** and **11** were in good agreement with the structures experimentally determined for **8** and **9**, respectively (For **10**: Cl(1)-Si-$C_{carbene}$ 85.0°, Cl(2)-Si-$C_{carbene}$ 85.0°, Cl(1)-$C_{carbene}$ 2.85 Å, and Cl(2)-$C_{carbene}$ 2.85 Å. For **11**: Cl(1)-Sn-$C_{carbene}$ 81.4°, Cl(2)-Sn-$C_{carbene}$ 81.4°, Cl(1)-$C_{carbene}$ 3.15 Å, and Cl(2)-$C_{carbene}$ 3.15 Å). Importantly, analysis of the molecular orbitals generated for **10** and **11** show that in both of these compounds there is significant bonding overlap between the occupied chloride p-orbitals and the formally vacant $C_{carbene}$ p_π-orbital. Furthermore, the Cl(p)···$C_{carbene}(p_\pi)$ bonding overlap does not include the participation of any orbitals of the central atom of the complex. The HOMO-7 of **10** (Figure 14) and the HOMO-9 of **11** (Figure 15) are shown as examples of molecular orbitals containing Cl(p)···$C_{carbene}(p_\pi)$ bonding.

*Figure 14. The HOMO-7 of **10** illustrating the Cl(p)···$C_{carbene}(p_\pi)$ orbital interaction.*

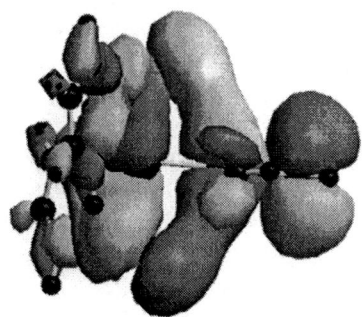

Figure 15. The HOMO-9 of 11 showing the overlap of the chloride p-orbitals with the $C_{carbene}$ p_π orbital.

Conclusion

Short Cl···$C_{carbene}$ contacts occur in NHC complexes of both early transition metal (Ti(IV), V(V), and Re(V)) and main group (Si(IV), Sn(IV)) elements in high oxidation states. These intramolecular interactions arise through the donation of chloride lone-pair electron density into the formally vacant $C_{carbene}$ p_π-orbital. This electron-donation was unexpected as this orbital is high in energy and remains vacant in free NHCs. The donation should result in a weakening of the N-C-N bonds within the NHC (though this effect is not expressed as an observable increase in the lengths of these bonds, it must be remembered that in the case of carbonyl complexes it is not generally possible to correlate bond distance with stretching frequency). In the transition metal containing examples, metal d-orbitals participate in the bonding overlap between the chloride p–orbitals and the $C_{carbene}$ p_π-orbital. However, such metal d-orbital participation is not required to facilitate Cl(p)···$C_{carbene}(p_\pi)$ orbital overlap; the Cl···$C_{carbene}$ interactions observed in the NHC complexes of silicon and tin(IV) occur *via* direct Cl(p)···$C_{carbene}(p_\pi)$ orbital overlap without the participation of any of the orbitals of the central main group metal. It appears, therefore, that it is complexation to a strongly Lewis acidic center that serves to lower the energy of the $C_{carbene}$ p_π-orbital sufficiently to render the $C_{carbene}$ atom of NHCs sufficiently Lewis acidic to engage in bonding interactions with the Lewis basic neighboring chloride ligands. Lewis acidic $C_{carbene}$ centers may well be a pervasive feature of the NHC adducts of all strong Lewis acids. Our current research is directed towards discovering the extent of this phenomenon. In any event, the observed Cl···$C_{carbene}$ interactions in the compounds described above reveals a new aspect of the truly fascinating chemistry of *N*-heterocyclic carbenes.

Acknowledgement

C. D. A. gratefully acknowledges The Western Kentucky University Faculty Scholarship Council for financially supporting this work through a Junior Faculty Scholarship Grant.

References

1. Arduengo, A. J.; Harlow, R. L.; Kline, M. *J. Am. Chem. Soc.* **1991**, *113*, 361-363.
2. Bourissou, D.; Guerret, O.; Gabbai, F. P.; Bertrand, G. *Chem. Rev.* **2000**, *100*, 39-91.
3. Trnka, T. M.; Grubbs, R. H. *Acc. Chem. Res.* **2001**, *34*, 18-29.
4. Abernethy, C. D.; Cole, M. L.; Jones, C. *Organometallics* **2000**, *19*, 4852-4857.
5. Denk, M. K.; Thadani, A.; Hatano, K.; Lough, A. J. *Angew. Chem., Int. Ed. Engl.* **1997**, *36*, 2607-2609.
6. Jacobsen, H.; Ziegler, T. *J. Am. Chem. Soc.* **1994**, *116*, 3667-3679.
7. Green, J. C.; Scurr, R. G.; Arnold, P. L.; Cloke, F. G. N. *J. Chem. Soc., Chem. Commun.* **1997**, 1963-1964.
8. Hu, X.; Castro-Rodriguez, I.; Olsen, K.; Meyer, K. *Organometallics* **2004**, *23*, 755-764.
9. Herrmann, W. A.; Öfele, K.; Elison, M.; Kühn, F. E.; Roesky, P. W. *J. Organomet. Chem.* **1994**, *480*, C7-C9.
10. Kuhn, N.; Kratz, T.; Bläser, D.; Boese, R. *Inorg. Chim. Acta* **1995**, *238*, 179-181.
11. Abernethy, C. D.; Codd, G. M.; Spicer, M. D.; Taylor, M. K. *J. Am. Chem. Soc.* **2003**, *125*, 1128-1129.
12. Spicer, M. D.; Cowley, A. H.; Culver, J. P.; Abernethy, C. D. manuscript in preparation.
13. Braband, H.; Abram, U. *Chem. Commun.* **2003**, 2436-2437.
14. Shukla, P.; Johnson, J. A.; Vidovic, D.; Cowley, A. H.; Abernethy, C. D. *Chem. Commun.* **2004**, 360-361.
15. Kuhn, N.; Kratz, T.; Bläser, R.; Boese, R. *Chem. Ber.* **1995**, *128*, 245-250.

Chapter 19

Imidazol-2-ylidenes and Their Reactions with Small Reagents

Taramatee Ramnial and Jason A. C. Clyburne*

Department of Chemistry, Simon Fraser University, 8888 University Drive, Burnaby, British Columbia, V5A 1S6 Canada

The reactivity of *N*-heterocyclic carbenes (**NHCs**) with small reagents has been explored and the focus of this review is mainly on those reactions involving some type of redox process. Results from our studies include the formation of neutral radicals derived from hydrogen atom addition to the carbeneic center (*1*), electrochemical and chemical reduction of imidazolium ions to produce **NHCs** (*2*), formation of radical cations upon treatment of **NHCs** with one-electron oxidizing agents (*3*) and identification of unconventional [C-H$^{\delta+}$•••H$^{\delta-}$-B] dihydrogen bonds (*4*).

Carbenes are two-coordinate carbon compounds that have two non-bonding electrons and no *formal charge* on the carbon. The formal divalent state exhibited by the carbon site is accountable for the high reactivity of these molecules, and in turn, is partly responsible for their elusive nature. In 1988, Bertrand reported the first stable singlet carbene, [*bis*(diisopropylamino)-phosphino]-trimethylsilylcarbene] (*5*), which was soon followed by the synthesis and characterization of the first imidazol-2-ylidenes (*aka* *N*-heterocyclic carbenes, **NHCs**) by Arduengo in 1991 (*6*). Recognizing the impact that the isolation of these highly reactive molecules has had on synthetic chemistry, reviews have covered the quest for isolable carbenes (*7*), main group element carbene complexes (*8*), carbene analogues (*9*), transition metal carbene complexes and catalysts (*10,11*). Our research group has focused on the reactivity of **NHCs** with small reagents, and our results are summarized in Scheme 1.

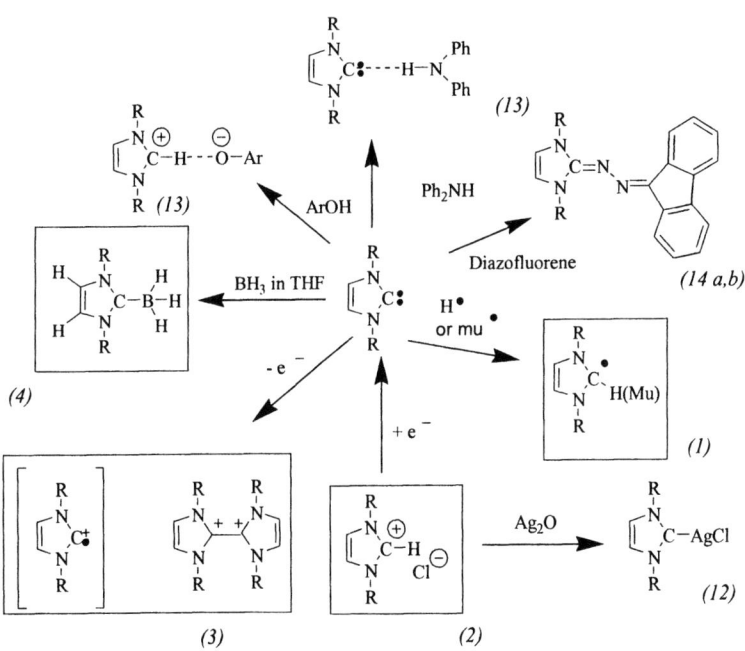

Scheme 1: Reactivity of NHCs with small reagents. Molecules of interest to this review are highlighted in boxes. References are given below the structures.

Summary of Reactivity

1A. Hydrogen Atom Addition to Imidazol-2-ylidenes

Recently, in collaboration with the Davidson research group from the University of Bath, we showed that reaction of **NHCs** with simple protic reagents such as phenols and secondary amines provided a facile route to hydrogen-bonded imidazolium salts with unique [C-H•••O] and [C•••H-N] interactions *(13)*. Treatment of an **NHC** with a proton produced a carbocation derivative. It was obvious to us that if a *neutral* hydrogen atom could be added to an **NHC**, the tricoordinate radical produced would be analogous to the methyl radical. This reaction was also attractive since, to date, reactions of **NHCs** with radical species have not yet been reported.

Addition of a hydrogen atom to 1,3-*bis*(2,4,6-trimethylphenyl)imidazol-2-ylidene, **1**, and 1,3-di(isopropyl)-4,5-dimethylimidazol-2-ylidene, **2**, was studied by means of density functional calculations *(1)*. NHCs **1** and **2** were chosen because of their high symmetry and the availability of isotopically ^{13}C labeled samples (see section 1C).

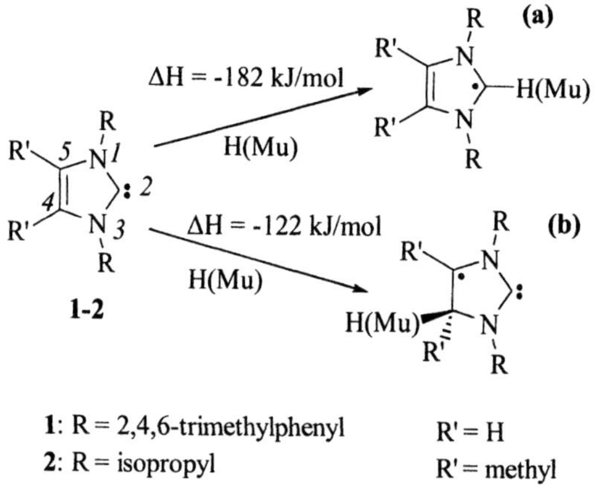

1: R = 2,4,6-trimethylphenyl   R' = H
2: R = isopropyl   R' = methyl

Scheme 2: Possible reactions of imidazol-2-ylidenes with hydrogen atoms or muonium.

The addition of a hydrogen atom to **NHCs 1** and **2** could occur at two probable sites, either at the carbeneic carbon to give **1a** and **2a** or at the alkeneic carbon to produce **1b** and **2b** as shown in Scheme 2. Calculations showed the formation of **1a** to be more exothermic than the formation of **1b**, and therefore this mode of reactivity is thermodynamically favored. **1a** was also kinetically favored due to the absence of an activation barrier for the addition at the carbeneic center.

The structure of **2a**, obtained from theoretical calculations, was found to be different from those calculated for the parent carbene as shown in Table 1. The C_2-N_1 or C_2-N_3 bonds and the N_1-C_2-N_3 angle were observed to be larger for the radical than the carbene. The imidazole ring was no longer planar but was puckered at N_1 and N_3. The radical center was found to be pyramidal, and to our knowledge, this was the first reported addition reaction to a stable carbene that did not produce a planar tricoordinate carbon center.

Table 1. Selected bond lengths (Å) and angles (°) calculated for carbene 2, and its radical 2a, derived from the addition of H.

	2	*2a*
(N_3-C_2), (N_1-C_2) (Å)	1.377	1.430
(N_1-C_5), (N_3-C_4) (Å)	1.415	1.418
(C_4-C_5) (Å)	1.367	1.363
(N_1-C_2-N_3) (°)	102.4	104.6
(C_5-N_1-C_2), (C_2-N_3-C_4) (°)	112.8	108.6
(N_1-C_5-C_4), (N_3-C_4-C_5) (°)	106.0	108.5

1B. Experimental Confirmation Using Muonium

Due to the inherent difficulty in the generation of hydrogen atoms, a hydrogen atom surrogate in non-aqueous media, muonium (Mu) (*15*) was used to confirm our computational studies. Muonium is a one-electron "atom" whose nucleus is the positive muon; it is chemically similar to hydrogen, but has one-ninth the mass and a lifetime of 2.2 μs. Muon spin rotation and muon level–crossing spectroscopy (*15*) were used to determine muon, ^{13}C and ^{14}N hyperfine coupling constants (hfcs) for **2a** (*16,17*).

Exposure of **NHCs** to the muonium beam was performed at TRIUMF, Canada's meson facility. Only one type of radical was observed, evident from the characteristic pair of frequencies above and below the muon Larmor frequency as shown in Figure 1. The muon hfcs were determined by transverse field muon spin rotation (TF-μSR) at 298 K and it was observed to be 286.69 MHz for **1a** and 246.43 MHz for **2a**. Hyperfine coupling constants for the other nuclei in the radical, namely ^{13}C and ^{14}N, were observed in the μLCR spectrum with the ^{13}C resonance observable only for a 40% ^{13}C enriched at C_2 sample for **2**.

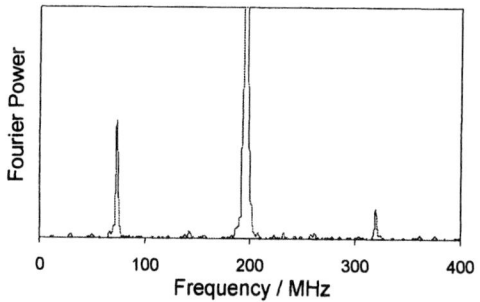

*Figure 1. Transverse field μSR spectrum at 14.4 kG from **2** in THF at 298 K. The pair of peaks at ca. 73 and 320 MHz are due to a single muoniated radical. (Reproduced with permission from reference 1. Copyright 2003 American Chemical Society.)*

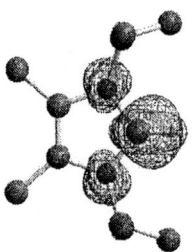

*Figure 2. Unpaired electron spin density in **2a**. For clarity, the hydrogen atoms on the alkyl substituents are not shown. (Reproduced with permission from reference 1. Copyright 2003 American Chemical Society.)*

The calculated hyperfine coupling constants of **1a** and **2a** were compared to the experimentally determined hyperfine coupling constants and these were in good agreement. It is interesting to note that most of the unpaired spin density was localized on the carbeneic carbon, C_2, with very little on the adjacent nitrogens as shown in Figure 2.

Finally, given the agreement between measured and predicted hfcs, as well as relative energies of the possible radicals, we conclude that hydrogen atoms and muonium add exclusively at the carbeneic center for **NHC**s.

1C. Synthesis of a ^{13}C Labeled Imidazol-2-ylidene

One of the major advances that we have made has been the synthesis of ^{13}C labeled **NHC 1** and **2**. Labeled **[1-H][Cl]** was used to prepare a monomeric silver (I) carbene complex, **[1-AgCl]**, which was fully characterized by solution and solid state ^{109}Ag and ^{13}C CP/MAS NMR spectroscopy (*12*). ^{13}C labeled **NHC 2** was used to observe ^{13}C resonances for muon addition and muon-level-crossing spectroscopy (*1*).

Labeled **NHC 2** was synthesized as shown in Scheme 2 (*1*) and we report here for the first time details of its preparation. Two equivalents of isopropyl amine were reacted with ^{13}C labeled carbon disulfide in benzene. The product obtained from this reaction was isolated and subsequently heated to reflux in hexanol for twenty-four hours in the presence of 3-hydroxy-2-butanone. The solid from this reaction was recrystallized in anhydrous ethanol to give a thione. Reduction of which with potassium metal gave the labeled **NHC 2** in 80% yield. ^{1}H and ^{13}C NMR data were consistent with the reported values of the unlabeled compound (*18*).

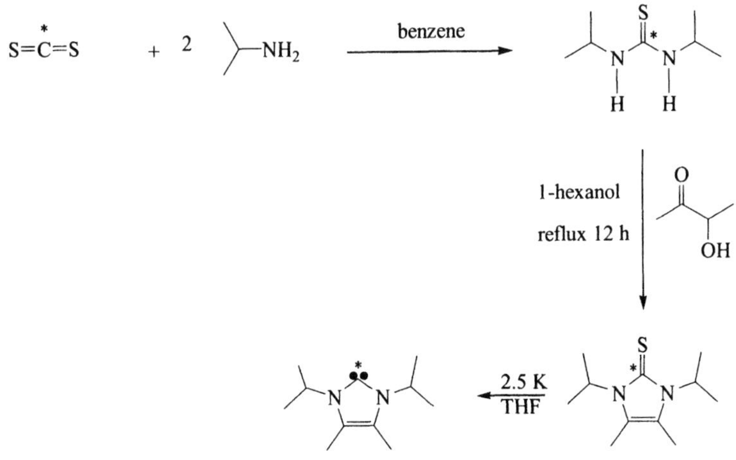

*Scheme 2: Synthesis of ^{13}C labeled **NHC 2**.*

1D. Electrochemical and Chemical Reduction of Imidazolium Ions

Ionic liquids are a new class of solvent that have come to the fore as a component of "Green Chemistry" (*19*). Their applications have extended into niche areas and lately into bulk synthetic chemistry (*20*). The most extensively studied ionic liquids are based on imidazolium ions. The relationship between imidazolium ions and imidazol-2-ylidenes, being the same as that of water and the hydroxide ion, led us to investigate the chemistry and electrochemistry of imidazolium ions with strong reducing agents.

Electrochemical reduction of **[1-H][Cl]** exhibited an irreversible reduction at $Ep_{(red)}$ = -2.28 V *vs* SCE. The CV also showed the presence of a significant peak on the return scan at –0.19 V corresponding to the oxidation of a new chemical species that is generated from the one-electron reduction of **[1-H][Cl]** as shown in Equation 1. This peak lies at the same potential as measured for the *oxidation* of **1**(see section E), suggesting that it is the carbene that is produced by reducing the imidazolium cation.

Equation 1. Reduction of imidazolium salt [1-H][Cl].

We propose that **[1-H]**• is initially formed in the electrochemical reduction process and this radical is related to the radical described in section 1B. There are two possible fates of this radical and these include dimerization of the **[1-H]**• radical or decomposition of this radical to give a hydrogen molecule and carbene **1**. However, dimerization is unlikely since the H-H bond (436 kJ/mol) is stronger than a C-C bond (348 kJ/mol).

To probe the chemical reduction of the imidazolium salt we performed a bulk reduction of **[1-H][Cl]** by heating a THF solution of **[1-H][Cl]** with potassium to reflux for three hours under nitrogen. This gave a brick red solution with a brown precipitate. The solution was filtered, dried under vacuum and washed with cold hexane to give white solid **1** in 89% yield (*6*). This carbene is typically generated by deprotonation of **[1-H][Cl]** (*21*), but the reaction involves several steps and often produces impure materials that are not useful for further reactions. Our new method provides a fast, one-step, aprotic procedure to prepare large quantities of **NHC 1** without resorting to the use of, for example, liquid ammonia.

This study clearly shows the role of carbenes in the chemistry of ionic liquids. Although evidence for the existence of carbenes in ionic liquids has been suggested by oxidative addition reactions of imidazolium ions to formally metal (0) sites (*22*), facile conversion of an imidazolium to a imidazol-2-ylidene through electrochemical methods has not been reported.

1E. One Electron Oxidation of Carbenes

Electrochemical studies on the reduction of carbenes have been previously reported (*23*), however a simple one electron oxidation of carbene has not been identified. **NHCs** are good carbon based electron donors and we decided to investigate their reactivity with carbon based electron acceptors, such as **TCNE**.

1: R = Mes, R_1 = H
2: R = *i*Pr, R_1 = CH_3

$[1\text{-}1]^{2+}$: R = Mes, R_1 = H
$[2\text{-}2]^{2+}$: R = *i*Pr, R_1 = CH_3

Equation 2: Reaction of NHCs 1 and 2 with TCNE. (Reproduced with permission from reference 3. Copyright 2003 Royal Society of Chemistry.)

Treatment of **NHCs 1** or **2** with **TCNE** in THF solution resulted in the formation of insoluble red materials. Spectroscopic data and elemental analyses were consistent with a material containing 1:1 ratio of **1** or **2** with **TCNE** as shown in Equation 2. The IR data showed strong ν_{CN} absorptions at 2161 and 2147 cm^{-1} consistent with the formation of the dianion π-stacked $[TCNE]_2^{2-}$ (ν_{CN} at 2163, 2147 cm^{-1}) and these are different from those observed for neutral

TCNE (v_{CN} at 2257 (s) and 2219 (s) cm^{-1}) (24). The formation of [TCNE]$_2^{2-}$ was confirmed in the ^{13}C NMR spectra, where peaks corresponding to the ethyleneic carbon moved upfield from 110 ppm to 68 ppm (25). Once the anion had been identified, we focused on the structure of the cation. The products were diamagnetic and the ^{1}H NMR spectrum showed downfield shifts for the 2,4,6-trimethylphenyl and isopropyl protons from the corresponding resonances in **1** and **2**. Distinct peaks at 304 and 180 amu were observed in the positive mode of electrospray mass spectrometry data, and pure samples of **1** or **2**, under the same conditions, exhibit peaks for the [M+H]$^+$ ions. The data taken in total suggested that the cations were the symmetrical dication [1-1]$^{2+}$ and [2-2]$^{2+}$. Dication [2-2]$^{2+}$ was previously reported by Kunh et al. (26) and our spectroscopic data were consistent with those reported.

Formation of these dications suggests the intermediacy of a radical cation. Although not detected, computational studies were used to shed light on the structure of such an intermediate carbene cation. An interesting feature of the radical cation is the high electron spin density located on C$_2$. The shape of the SOMO (Figure 3) also suggests the possibility of dimerization through *C-C* bond formation.

At least two routes leading to the formation of **[1-1][TCNE]$_2$** and **[2-2][TCNE]$_2$** are possible. *Route 1* involves the formation of the radical cations **[1]$^{+\bullet}$** and **[2]$^{+\bullet}$** followed by dimerization. *Route 2* is a stepwise reaction involving formation of a radical cation followed by reaction with **1** or **2** to produce **[1-1]$^{+\bullet}$** or **[2-2]$^{+\bullet}$** which were then oxidized to give dications.

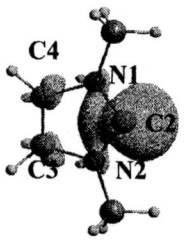

Figure 3: Total electron spin density on [2]$^{+\bullet}$. (Reproduced with permission from reference 3. Copyright 2003 Royal Society of Chemistry.)

Reaction of **1** or **2** with ferrocenium salts such as [Cp$_2$Fe][PF$_6$] or [Cp$_2$Fe][BF$_4$] did not produce dications but rather imidazolium salts such as [1-H][PF$_6$], [1-H][BF$_4$], [2-H][PF$_6$] and [2-H][BF$_4$] as shown in Scheme 3. Ferrocene was identified in the washings.

The different reactivity of the **NHCs 1** and **2** with ferrocenium salts compared to that observed for **TCNE** can be explained by noting that the reaction with ferrocenium is *heterogeneous* and *slow*. Under these conditions there is only a very small amount of the radical cation formed at any given time. Hence, instead of undergoing dimerization reactions it abstracts a hydrogen atom from other sources, such as solvent.

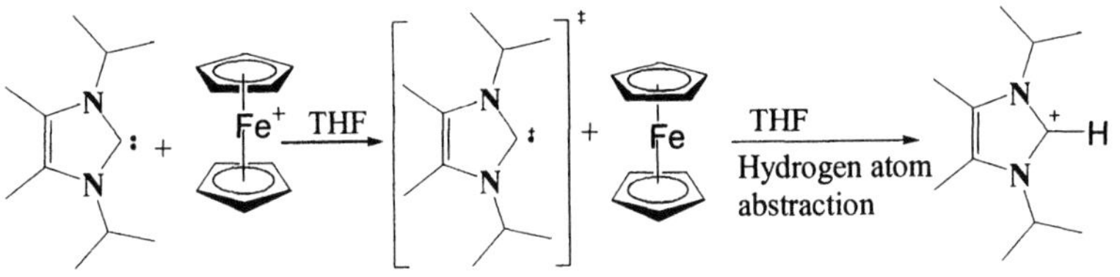

Scheme 3: Reaction of **2** with *[Cp₂Fe]⁺*. (Reproduced with permission from reference 3. Copyright 2003 Royal Society of Chemistry.)

1F. Imidazol-2-ylidene Borane Complex Exhibiting Intermolecular Dihydrogen Bonding

Hydrogen bonding plays a central role in biological molecules (*27*) and has been extensively used in "crystal engineering" since it can influence the structure of molecules in the solid, liquid and gaseous states (*28*). Recently, we reported the formation of extremely short [C-H•••O] hydrogen bonds, which resulted from the protonation reaction of the **NHC 1** with phenols (*13*). The shortness of this unconventional hydrogen bond can be explained by "charge assistance" between the cationic imidazolium ion and the anionic phenoxide.

Imidazolium salts have three potentially acidic *C-H* sites and the shortest hydrogen bonds are typically observed from C_2. Expecting the formation of **[1-H][BH₄]** we attempted to prepare an imidazolium-based, charge-assisted *dihydrogen* bond by ion exchange of the **[1-H][Cl]** with NaBH₄. This molecule could possibly exhibit an exotic [H•••H] interaction between the cation and anion pair. We found however, that the putative [H•••H] bond in this system was thermodynamically unstable with respect to hydrogen gas evolution, resulting in coupling of the resultant carbene with **BH₃** as shown in Scheme 4. Analytical and spectral data were consistent with the formation of **1·BH₃** (*4*).

We were somewhat discouraged by this result but we noted that the melting point of crystalline **1·BH₃** was *ca.* 300 °C, significantly higher than that of **1** (150 °C) or the related alane adduct **1·AlH₃** (246-247 °C) (*29*). This observation

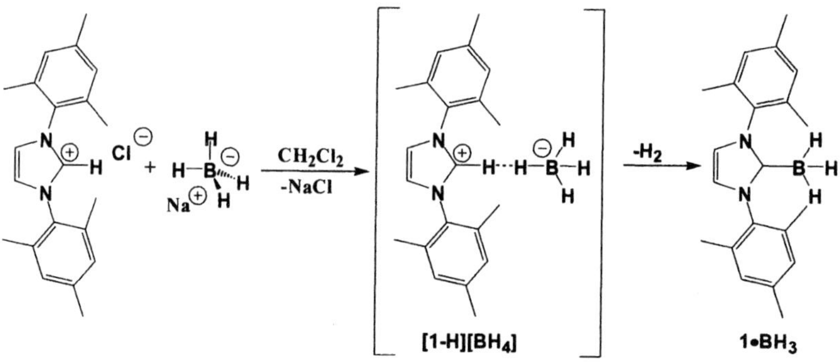

Scheme 4: Synthesis of [1·BH₃]. (Reproduced with permission from reference 4. Copyright 2003 Royal Society of Chemistry.)

suggested that there were strong intermolecular interactions in **1·BH₃**. The solid-state structure of **1·BH₃** was determined by X-ray crystallography and the results are shown in Figure 4. The structural parameters obtained for **1·BH₃** were consistent with previously reported carbene-borane adducts (*30,8*). The closest intermolecular contact occurred through a novel [C-H···H-B] bond whose [H···H] distance (2.24 Å) was well within their sum of van der Waals radii (2.65 Å) (*31*). These dihydrogen bonds are perhaps the weakest hydrogen bonds with the best-known example being observed in solid [NH₃-BH₃].

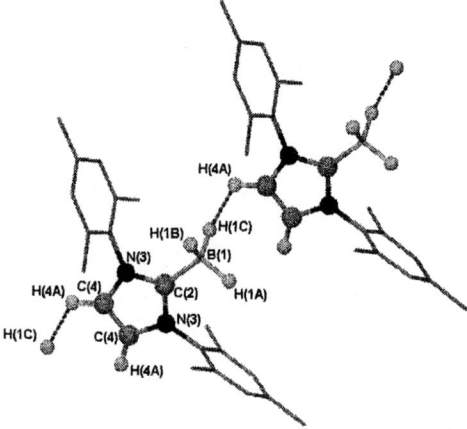

*Figure 4: X-ray structure of carbene–borane adduct **1·BH₃**. (Reproduced with permission from reference 4. Copyright 2003 Royal Society of Chemistry.)*

Calculations on **1·BH₃** suggested a partial negative charge on the hydrogen atom attached to boron, consistent with its hydridic nature, and a slight partial positive charge on H (4A) as shown in Figure 5. While individually such interactions are weak, collectively they can significantly influence the macroscopic properties, as observed by the regular solid-state head-to-tail alignment of **1·BH₃**.

It is possible that [C-H•••H-B] dipolar interactions are also present in solution, for example between amine•borane adducts and solvent with slightly acidic hydrogen atoms (*32*). Dihydrogen bonds between acidic solvent molecules and hydridic hydrogen sites of amine•borane adducts may account for slow hydroboration reaction rates observed in dichloromethane compared to other solvents, such as diglyme, ether and THF. We note that the estimated partial charge on the hydrogen atoms in CH_2Cl_2 is +0.2 (*4*), close to the value calculated for the acidic hydrogen in **1·BH₃**. We suggest that in solutions containing amine•borane adducts significant dihydrogen bonding will occur, thus accounting for sluggish hydroboration reactions.

While the [C-H•••H-B] dihydrogen bond may seem exotic, it is related to the **[NH₃-BH₃]** adduct where the estimated partial charges on the hydrogen atoms attached to boron and nitrogen are -0.07 and +0.31 respectively (*33*). The larger charges for **[NH₃-BH₃]** suggest stronger dihydrogen bonds in amine•borane adducts than in **1·BH₃**.

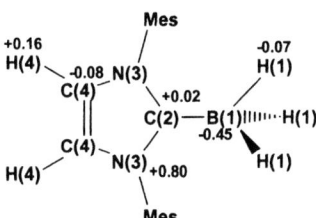

*Figure 5: Estimated partial charges on **1·BH₃** adduct. (Reproduced with permission from reference 4. Copyright 2003 Royal Society of Chemistry.)*

Summary

In this review we have presented the reactivities of **NHCs** with the simplest of reagents, including hydrogen atoms, and other organic reagents such as tetracyanoethylene, ferrocenium salts and sodium borohydride. This study provides us with an understanding of the fundamental chemistry of carbenes and opens up new appreciation of the potential role of **NHC**s in ionic liquids.

Acknowledgments

We appreciate the help of our collaborators namely Dr. Iain McKenzie, Dr. Paul Percival and Dr. Jean-Claude Brodovitch for the muonium related work and DFT calculations; Brian Gorodetsky for the electrochemical studies; Dr. Charles MacDonald and Dr. Michael Jennings for the X-ray crystallographic studies; Dr. Neil Branda for his useful discussions, and for providing the instruments necessary for the electrochemistry. Gratitude is also extended to Howard Jong and Emily Tsang for their contribution to the synthesis of the carbene borane and carbene tetracyanoethylene complexes.

1. McKenzie, I. D.; Brodovitch, J.-C.; Percival, P. W.; Ramnial, T.; Clyburne, J. A. C. *J. Am. Chem. Soc.* **2003**, *125*, 11565.
2. Gorodetsky, B.; Ramnial, T.; Branda, R. N.; Clyburne, J. A. C. *Chem. Commun*, **2004**, in press.
3. Ramnial, T.; McKenzie, I. D.; Gorodetsky, B.; Tsang, M. W. E.; Clyburne, J.A. C. *Chem. Commun.* **2004**, 1054.
4. Ramnial, T.; Jong, H.; McKenzie, I. D.; Jennings, M.; Clyburne, J. A. C. *Chem.Commun..* **2003**, 1722.
5. Igau, A.; Grutzmacher, H.; Baceiredo, A.; Bertrand, G. *J. Am. Chem. Soc.* **1988**, *110*, 6463.
6. Arduengo, A. J., III; Harlow, R. L.; Kline, M. *J. Am. Chem. Soc.* **1991**, *113*, 361.
7. Arduengo, A. J., III. *Acc. Chem. Res.* **1999**, *32*, 913.
8. Carmalt, C. J.; Cowley, A. H. *Adv. Inorg. Chem.* **2000**, *50*, 1.
9. Bourissou, D.; Guerret, O.; Gabbaï, F. P.; Bertrand, G. *Chem. Rev.* **2000**, *100*, 39.
10. Herrmann, W. A.; Köcher, C. *Angew. Chem. Int. Ed.* **1997**, *36*, 2162.
11. Herrmann, W. A. *Angew. Chem. Int. Ed.* **2002**, 1290.
12. Ramnial, T.; Abernethy, C. D.; Spicer, S. D.; McKenzie, I. D.; Gay, I. D.; Clyburne, J. A. C. *Inorg. Chem.* **2003**, *42*, 1391.
13. Cowan, J. A.; Clyburne, J. A. C.; Davidson, M. G.; Harris, R. L. W.; Howard, J. A. K.; Küpper, P.; Leech, M. A.; Richards, S. P. *Angew. Chem. Int. Ed.* **2002**, *41*, 1432.
14. (a) Choytun, D. D.; Langlois, L. D.; Johansson, T. P.; MacDonald, C. L. B.; Leach, G. W.; Weinberg, N.; Clyburne, J. A. C. *Chem. Commun.* **2004,**

1842. (b) Hopkins, I. M.; Bowdridge, M.; Clyburne, J. A. C.; Robertson, K. N.; Cameron, T. S.; Jenkins, H. A. *J. Org. Chem.* **2001**, *66*, 5713.
15. Roduner, E. *The Positive Muon as a Probe in Free Radical Chemistry*; Lecture Notes in Chemsitry 49; Springer-Verlag: Berlin, 1988.
16. Cox, S. F. J. *Solid State Phys.* **1987**, *20*, 3187.
17. Brewer, J. H. In *Encyclopedia of Applied Physics*; VCH: New York, 1994; Vol. 11, 23.
18. Kuhn, N.; Kratz, T. *Synthesis* **1993**, 561.
19. *Ionic Liquids: Industrial Applications to Green Chemistry*; Rogers, R. D.; Seddon, K. R., Eds.; American Chemical Society: Washington, 2002.
20. Freemantle, M. *Chem. Eng. News* **2001**, 21.
21. Arduengo, A. J., III; Dias, H. V. R.; Harlow, R. L.; Kline, M. *J. Am. Chem. Soc.* **1992**, *114*, 5530.
22. McGuinness, D. S.; Cavell, K. J.; Skelton, B. W.; White, A. H. *Organometallics* **1999**, *18*, 1596.
23. Enders, D.; Breuer, K.; Raabe, G.; Simonet, J.; Ghanimi, A.; Stegmann, H. B.; Teles, H. J. *Tet. Lett.* **1997**, *38*, 2833.
24. Johnson, M. T.; Campana, C. F.; Foxman, B. M.; Desmarais, W.; Vela, M. J.; Miller, J. S. *Chem. Eur. J.* **2000**, *6*, 1805.
25. Mohajer, D.; Dehghani, H. *Bull. Chem. Soc. Japan* **2000**, *73*, 1477.
26. Kuhn, N.; Weyers, G.; Blaser, D.; Boese, R. *Z. Naturforsch* **2001**, *56b*, 1235.
27. Jeffrey, G. A.; Saenger, W. *Hydrogen Bonding in Biological Structures*; Springer: Berlin, 1994.
28. Whitesides, G. M.; Simanek, E. E.; Mathias, J. P.; Seto, C. T.; Chin, D.; Mammen, M.; Gordon, D. M. *Acc. Chem. Res.* **1995**, *28*, 37.
29. Arduengo, A. J., III; Dias, H. V. R.; Calabrese, J. C.; Davidson, F. *J. Am. Chem. Soc.* **1992**, *114*, 9724.
30. Kuhn, N.; Henkel, G.; Kratz, T.; Kreutzberg, J.; Boese, R.; Maulitz, H. A. *Chem. Ber.* **1993**, *126*, 2041.
31. Padilla-Martinez, I.; Jesus Rosalez-Hoz, M.; Tlahuext, H.; Camacho-Camacho, C.; Ariza- Castolo, A.; Contreras, R. *Chem. Ber.* **1996**, *129*, 441.
32. Kanth, J. V. B.; Brown, H. C. *Tet. Lett.* **2000**, *41*, 9361.
33. Klooster, W. T.; Koetzle, T. F.; Siegbahn, P. E. M.; Richardson, T. B.; Crabtree, R. H. *J. Am. Chem. Soc.* **1999**, *121*, 6337.

Chapter 20

Coordination Chemistry of Phosphorus(III) as a Lewis Acceptor

Neil Burford* and Paul J. Ragogna

Department of Chemistry, Dalhousie University,
Halifax, Nova Scotia, B3H 4J3, Canada
*Corresponding author: telephone: (902) 494-3190; fax: (905) 494-1310;
Neil.Burford@dal.ca

The traditional view of phosphines as Lewis bases or ligands in transition metal complexes is challenged by the series of adducts involving phosphorus(III) Lewis acids. A variety of ligands form complexes with the phosphadiazonium cation (Mes*NP$^+$) or phosphenium cation (Ph$_2$P$^+$). Structural features are compared with related systems to highlight new directions in structure, bonding and reactivity. Complexes involving homoatomic P-P coordinate bonds are susceptible to ligand exchange reactions, a versatile new approach for element-P bond formation. Phosphenium cations can be associated using tethered diphosphine ligands.

Introduction

Compounds containing electron rich (lone pair bearing) phosphorus(III) (phosphines) are classic Lewis donors (ligands) in transition metal coordination chemistry. Nevertheless, a series of complexes involving Lewis acceptor phosphorus(III) centers[1] are known with carbene,[2-4] arene,[5] amine,[6;7] imine,[7-10] chalcogenourea,[11] phosphine[12-24] and gallane[25] ligands. The complex anion PBr$_4^-$ represents a prototypical example of a phosphine (PBr$_3$) behaving as an acceptor for bromide or cyanide,[26] and the disphenoidal structure is consistent with retention of a lone pair on phosphorus. Reactions of halophosphines with AlCl$_3$

were first shown by ^{31}P NMR spectroscopy to give compounds with P-P bonds that have been designated as coordination complexes of type **1**.[12;13;16] In addition, complex cations with two phosphine ligands **2**[15] or a DBN (1,5-diazabicyclo[4.3.0]non-5-ene) ligand **3**[8] have been structurally characterized, as well as a series of complexes involving intramolecular interaction of a pendant amine with phosphorus, e.g. **4**.[6]

In this context, coordinatively unsaturated phosphorus(III) centers are suited for coordination chemistry as Lewis acceptors and cationic species are likely to exhibit the greatest Lewis acidity by virtue of molecular charge. We have exploited these features in complexes of the phosphadiazonium cation **5**,[27] and phosphenium cations **6**,[28] which are listed in **Tables 1** and **2**, respectively.

Coordination complexes of the phosphadiazonium cation

Iminophosphines with the general formula RNPX (R = Mes* = 2,4,6-tri-t-butylphenyl; X = OTf,[29] AlCl$_4$,[27] GaCl$_4$,[5] Ga$_2$Cl$_7$[5]) can be considered as salts of phosphadiazonium cations, [RNP][X], **5X**, representing analogues of diazonium salts. Consistently, the solid state structure of [Mes*NP][OTf] reveals a very short NP distance (147pm) and an almost linear C-N-P (176°).[29] In addition, vibrational spectra for derivatives of [Mes*NP][X] exhibt a very strong NP stretch[30] that is interpreted in terms of an NP triple bond.[27]

Table 1. NMR [δ ^{31}P{^{1}H}] and structural parameters for complexes of [Mes*NP]$^+$ 5 (R = Mes*) and related compounds

Acid[a]	L[b]	δ ^{31}P (ppm)	NP (pm)	P(E) (pm)	PO (pm)	CNP (°)	Ref.
[RNP][OTf]	-	50	147	-	192	176	29
[RNP][GaCl$_4$]	benz	76	148	[c] 300	-	176	5
[RNP][GaCl$_4$]	tol	76	153	[c] 304	-	177	5
[RNP][Ga$_2$Cl$_7$]	benz	93	146	[c] 282	-	179	5
[RNP][Ga$_2$Cl$_7$]	tol	95	146	[c] 277	-	179	5
[RNP][Ga$_2$Cl$_7$]	mes	91	147	[c] 269	-	176	5
[R'$_2$CP][AlCl$_4$][e]	Ph$_3$P	317	-	(P) 227	-	-	17
[RPP][BPh$_4$]	Ph$_3$P	334	-	(P) 221	-	-	18
[RNP][OTf][d]	Ph$_3$P	[f]	149	(P) 263	230	169	19
[RNP][OTf]	Quin	144	152	(N) 193	270	144	7
[RNP][OTf]	Pyr	71	147	(N) 196	271	162	7
[RNP][OTf]	OIm	77	149	(O) 177	277	160	11
[RNP][OTf]	OU	62	149	(O) 179	294	166	11
[RNP][OTf]	Im	50	157	(C) 185	295	116	4
[RNP][OTf]	SIm	156	150	(S) 227	>340	174	11
[RNP][OTf]	SeIm	182	150	(Se) 241	>340	176	11
[RNP][OTf]	Bipy	54	150	(N) 207	>340	169	10

[a]R = Mes*, R' = Me$_3$Si; [b]L = ligand, benz = benzene, tol = toluene, mes = mesitylene, Im =1,3-diisopropyl-4,5-dimethyl-imidazol-2-ylidene, Quin = quinuclidine, Pyr = pyridine, Bipy = 2,2'-bipyridine, OU = 1,3-dimethyldiphenylurea, OIm = 1,3-diisopropyl-4,5-dimethyl-1,3-dihydro-2-imidazol-2-one, SIm = 1,3-diisopropyl-4,5-dimethylimidazol-2-thione, SeIm = 1,3-diisopropyl-4,5-dimethylimidazol-2-selenone; [c]P-arene(centroid); [d]OTf = trifluoromethanesulfonate or triflate; [e] R' = Me$_3$Si; [f] dissociated

Tetrachloroaluminate,[27] tetrachlorogallate[5] and heptachlorodigallate[5] salts of [Mes*NP]$^+$ isolated from solutions in benzene, toluene or mesitylene involve an arene molecule η^6 π-coordinated to the phosphorus centre 7.[5] The P–C(arene centroid) distances in the digallate salts correlate with the π-donor strength of the arene, decreasing in the order, benzene [282.0(4) pm] > toluene [276.7(7) pm] > mesitylene [268.7(7) pm]. The complexes are resilient in solution, but under dynamic vacuum, crystals of the tetrachloroaluminate and tetrachlorogallate salts release the arene and lose their integrity.[31] The cations 7 represent models of the π-coordination complex intermediates that are postulated to form during electrophilic aromatic substitution reactions.[5]

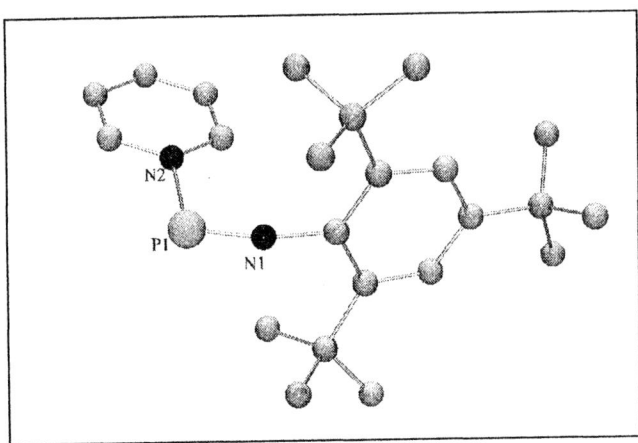

          7   8

Classical two electron liagnds (L) such as amines and phosphines react with [Mes*NP][GaCl$_4$] to give Mes*NPCl and the L-GaCl$_3$ adduct. In contrast, the triflate anion is innocent in [Mes*NP][OTf]. Consequently, ligands interact with the phosphorus center, which is the most Lewis acidic site, as illustrated for the pyridine complex in **Figure 1**. The triflate salts listed in **Table 1** are presented in order of increasing P-O(triflate) distance revealing a correlation between the Lewis basicity of the ligand and the nucleophilic displacement of the triflate anion.

Figure 1: *Solid state structure of the cation in [Mes*NP(Pyr)][OTf]*

The P-E dstances are longer than typical single bonds in all cases (E = element at donor site) and coordination of the ligands effect minimal adjustment of the short (Mes*)N-P distance with respect to the free Lewis acid [Mes*NP][OTf] (first entry in **Table 1**). For example, a very short (Mes*)NP bond is retained in the chelate complex of 2,2'-bipyridine **8**[10] despite two nitrogen donors, which adopt identical N_{ligand}-P distances [206.6(4) pm, 206.5(4) pm] (**Figure 2**) that are slightly longer than those in the pyridine [195.8(8) pm] and quinuclidine [N-P 193.3(2) pm] complexes.

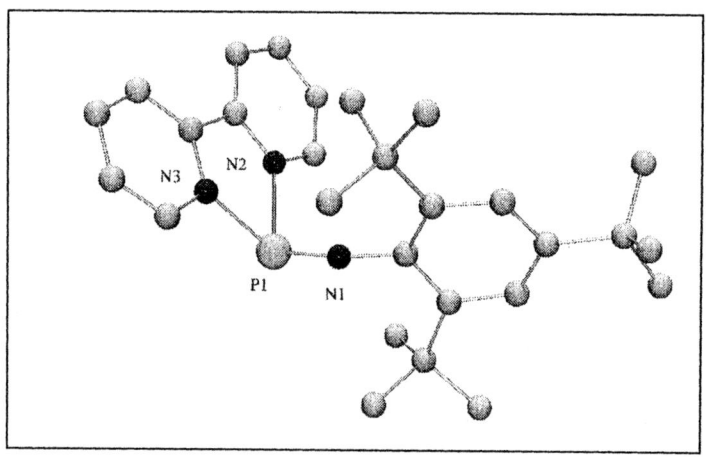

Figure 2: *Solid state structure of the cation in [Mes*NP(Bipy)][OTf]*

The P-P coordinate interaction in [Mes*NP(PPh$_3$)] **9** [OTf] presents an interesting comparison (**Table 1**) with the unique structures observed for [Mes*PP(PPh$_3$)] **10** [BPh$_4$][18] and [(Me$_3$Si)$_2$CP(PPh$_3$)] **11** [AlCl$_4$],[17] which include P-P bond distances that are typical of a single bond [cf. Ph$_2$P-PPh$_2$ 221.7(1) pm].[32] The small (Mes*)C$_{ipso}$PP bond angle in **10**[18] [98.8(2)°] implicates a diphosphene-phosphonium bonding arrangement, rather than a coordination complex.

Coordination complexes of the diphenylphosphenium cation

Amino-[13;28;33-36] or thia-phosphenium[37;38] cations are readily prepared by reaction of the corresponding halophosphine with a Lewis acid such as $AlCl_3$, and are stabilized by effective π–interaction between phosphorus and nitrogen or sulfur, respectively. In contrast, analogous reactions of dialkyl- or diarylhalophosphines with $AlCl_3$[16] or $GaCl_3$[20] typically form phosphine-alane or phosphine-gallane adducts involving a P-Al or P-Ga coordinate bond, as shown for aluminum in **12**. Nevertheless, in the presence of excess halophosphine, non-symmetric P-P bonded compounds are observed by ^{31}P NMR ($^1J_{PP}$ >200 Hz) spectroscopy (**Table 2**),[12;13;16] and [**13**][$GaCl_4$] has been structurally characterized for R = Ph (**Figure 3**).[23]

```
                              [AlCl4]⊖
    Cl       ·Cl      R2PCl      Cl                      Cl
    |        |                    |         ⊕           ⊕|
  R—P——————►Al—Cl    ——►       R—P—————————►P—R      R—P————P—R
    |        |                    |          |           |    |
    R        Cl                   R          R           R    R

     12                                13a                13b
```

These compounds have been recognized for more than 30 years[39-42] and can be considered as phosphine-phosphenium complexes **13a**, representing rare examples of homoatomic coordinate bonding. The more traditional phosphinophosphonium bonding model **13b** rationalizes the structural features, but the phosphine-phosphenium designation **13a** identifies a synthetic source of diphenylphosphenium. Facile ligand (chlorophosphine) exchange occurs on addition of more basic trialkyl- or triaryl-phosphines to deirivatives of **13** to give a variety of phosphine-phosphenium cations **1** with either tetrachlorogallate or triflate anions (**Table 2**).

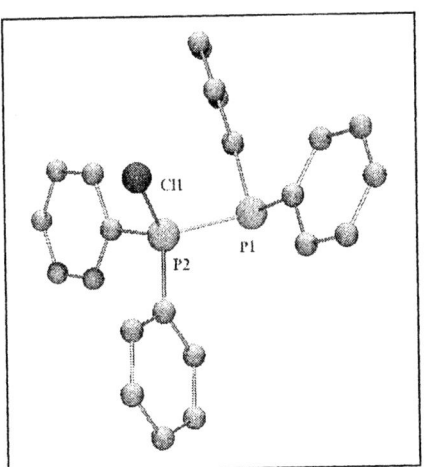

Figure 3: *Solid state structure of the cation in [$Ph_2(Cl)P$-PPh_2][$GaCl_4$]*

The phosphorus centers of the ligands exhibit a narrow ^{31}P NMR chemical shift range, with the bis(phosphine) complex 2 exhibiting the only unusual data. The P-P distances are all typical of a single P-P bond, except in the weak complex of [Mes*NPP(PPh$_3$)] 9 [OTf], which dissociates completely in solution.

All derivatives of 1 adopt a staggered conformation about the P-P bond in the solid state (**Figure 3**), with a distorted tetrahedral geometry for the tetracoordinate phosphorus centers and a distinctly pyramidal geometry with standard phosphine-like bond angles (90-105°) for the tricoordinate phosphorus centers. In this context, any electronic features of the coordinate P-P bond that may be distinct from P-P covalent bonds in compounds such as Ph$_2$PPPh$_2$,[32] do not express themselves in the metrical parameters. Solid state ^{31}P CP-MAS NMR data are consistent with the data obtained for solutions.

The homoatomic P→P coordinate bonding model for derivatives of 1 and 13 highlights the opportunity for ligand exchange as a versatile[1] synthetic methodology.[7] This has been exploited to obtain polyphosphorus dications composed of bis-(1,2-diphenylphosphino)ethane (dppe), bis-(1,2-dimethylphosphino)ethane (dmpe) or bis-(1,2-diphenylphosphino)hexane (dpph), which link [Ph$_2$P]$^+$ Lewis acids (**Table 2**).[24] Quantitative reactions occur at room temperature for 1:2 stoichiometric combinations of the diphosphine with 1 or 13 to give complexes in which the ligand tethers two phosphenium cations, as illustrated for [Ph$_2$P-dpph-PPh$_2$][OTf] in **Figure 4**.

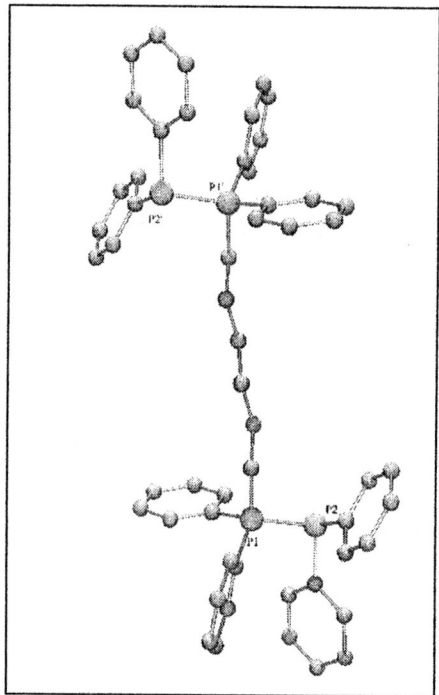

Figure 4: *Solid state structure of the dication in [Ph$_2$P-dpph-PPh$_2$][OTf]*

Bifunctional amines have been used to tether neutral phosphines,[7] and the phosphadiazonium cation 5 forms chelate complexes 8.[10] Therefore, the isolation and characterisation of otherwise synthetically inaccessible polyphosphorus dications highlights the versatility of ligand exchange at the $[Ph_2P]^+$ Lewis acceptor.

Table 2. NMR data [$\delta\ ^{31}P\{^1H\}$ and J_{PP}] and P-(E) distances for complexes of $[Ph_2P]^+$ and related compounds

Acid	L^a	$\delta\ ^{31}P$ (ppm)	J_{PP} (Hz)	P(E) (pm)	Ref.
$[Ph_2P][GaCl_4]$	Ph_2PCl	78, 3	^{b}	(P) 220	23
$[Ph_2P][OTf]$	Ph_3P	15, -10	350	(P) 223	23
$[Ph_2P][GaCl_4]$	Ph_3P	13, -12	340	(P) 222	23;43
$[Ph_2P][OTf]$	Cy_3P	25, -21	361	(P) 222	23
$[Ph_2P][OTf]$	Me_3P	15, -23	289	(P) 219	23
$[BAAP]^c[OTf]$	Me_3P	139, -6	504	(P) 231	22
$[P][AlCl_4]$	Ph_3P	30, -174	502	(P) 214	15
$[TMS_2CP][AlCl_4]$	Ph_3P	317, 20	455	(P) 227	17
$[Mes^*PP][BPh_4]$	Ph_3P	334	384	(P) 221	18
$[Mes^*NP][OTf]^d$	Ph_3P	^{e}	406^f	(P) 263	19
$[Ph_2P][OTf]$	Dppe	21, -33	322	(P) 222	24
$[Ph_2P][OTf]$	Dmpe	21, -32	300	(P) 222, 221	24
$[Ph_2P][OTf]$	Dpph	20, -24	318	(P) 221	24
$[Ph_2P][AlCl_4]$	Im	-27	-	(C) 181	2
$[Ph_2P][GaCl_4]$	Im	-27	-	(C) 184	3
$[Ph_2P][OTf]$	DMAP	88	-	(N) 179	7
$[Ph_2P][OTf]$	Ga(I)	-58	-	(Ga) 231	25

^{a}L = ligand, DMAP = 4-dimethylaminopyridine, Im = 1,3-diisopropyl-4,5-dimethylimidazole, Ga(I) = 1-galla-2,4-diaza-3-5-dimethylcyclohexadiene; ^{b}Not observed; ^{c}BAAP = bis(arylamino)phosphenium; ^{d}OTf = trifluoromethanesulfonate or triflate; ^{e}dissociated; ^{f}solid state.

The bifunctional diphosphine ligands, 1,2-bis(diphenylphosphino)benzene (dppb) and 1,2-bis(di-*tert*-butylphosphino)benzene (dbpb) also react quantitatively with derivatives of **1**. However, the products represent structural alternatives of the expected diphosphine-phosphenium complexes **14**. A 'segregated' diphosphine-phosphonium cation **16** is formed from dppb, which is composed of a tetraarylphosphonium centre and a tetraaryldiphosphine 'segregated' by a bridging benzo- unit. We envision the rearrangement of **14** to involve intramolecular phenyl transfer from the tetracoordinate phosphonium to the second phosphine *via* a bridging Wheland intermediate **15**. Such facile isomerism demonstrates the relative thermodynamic instability of the P-P bonded phosphine-phosphenium **1** given a route to a more favourable bonding alternative. Although the isolated P-P bonded phosphine-phosphenium derivatives (**1**) are perhaps only kinetically stable, thermodynamically favoured alternative structures are not apparent.

14 ⇒ **15** ⇒ **16**

The bicyclic di(phosphino)phosphonium cation **17** (**Figure 5**), formed from dbpb, is reminiscent of the proposed cyclic intermediate **15** in the formation of **16**, resulting from an elaborate dehydrogenation of the [Ph$_2$P]$^+$(dbpb) adduct. The ^{31}P NMR spectra of reaction mixtures show coincidental formation of Ph$_3$P (δ = -5ppm), Ph$_2$PH (δ = -40ppm) and [Ph$_3$PH]$^+$ (δ = 2ppm) resulting from hydride abstraction by a phosphenium cation and deprotonation by Ph$_3$P.

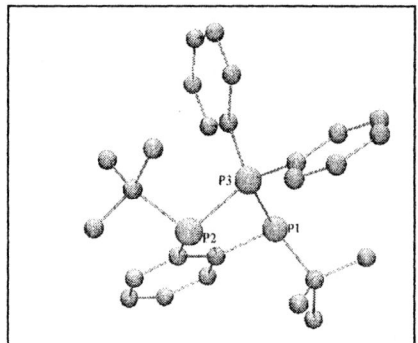

Figure 5: *Solid state structure of cation 17*

2Ph$_2$P-PPh$_3$$^+$ ⇒ − 2Ph$_3$PH$^+$ − Ph$_3$P − 2Ph$_2$PH

17

A more general application of ligand exchange on the diphenylphosphenium cation provides a new synthetic approach to element-P bond formation. Reactions of derivatives of **1** or **13** with Im, DMAP or Ga(I) proceed quantitatively as evidenced by the distinctive chemical shift for the free phosphine in the ^{31}P NMR spectrum as the only by product. The resulting cations **18** and **19** represent complexes involving carbon and nitrogen donors, respectively, on phosphorus(III) (**Table 2**). The aluminate salt of **18** is also formed by displacement of chloride from Ph$_2$PCl in halide abstraction conditions (with AlCl$_3$) to give [Ph$_2$P(carbene)][AlCl$_4$].[2]

18   **19**   **20**

Formation of the (gallane)Ga→P(phosphine) complex **20** represents the most profound application of the ligand exchange reaction.[25] The solid state structure (**Figure 6**) confirms the Ga-P connectivity and can be considered an example of a 'coordination chemistry umpolung' (an inverse of the traditional coordinate bond) in that the metal centre (gallium) behaves as the Lewis donor (ligand) and the electron-rich non-metal centre (phosphorus) behaves as the Lewis acceptor.

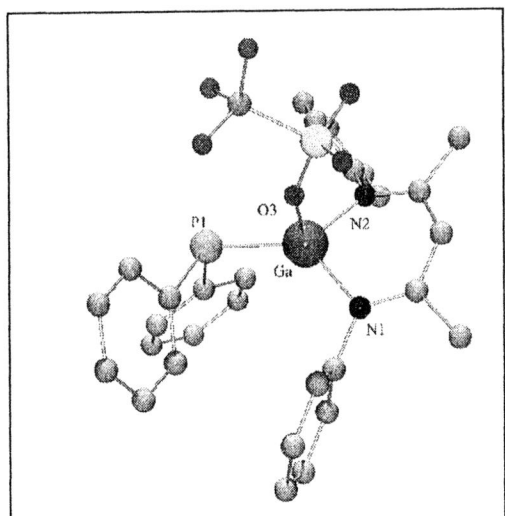

Figure 6: Solid state structure of 20. Hydrogen atoms and ^{i}Pr groups have been removed.

Conclusions

Phosphine centers can behave as Lewis acids in spite of the presence of a lone pair of electrons at phosphorus. Comprehensive characterisation data is available for complexes of phosphines with carbene, arene, amine, imine, phosphine and gallane ligands. Cationic phosphines are naturally the most effective acceptors, although the coordination chemistry of neutral phosphines is developing. Phosphines bearing good leaving groups are prone to anion displacement on interaction with a ligand. Complexes involving all electron-rich donors (elements of groups 15, 16 and 17) will be accessible and the diversification of the coordination chemistry of phosphines offers opportunities for novel catalysis and the discovery of new structure and bonding.

References

(1) Burford, N.; Ragogna, P. J. *Dalton Trans.* **2002**, 4307-4315.
(2) Kuhn, N.; Fahl, J.; Bläser, D.; Boese, R. *Z.Anorg.Allg.Chem.* **1999**, *625*, 729-734.
(3) Burford, N.; Cameron, T. S.; Ragogna, P. J.; Ocando-Mavarez, E.; Gee, M.; McDonald, R.; Wasylishen, R. E. *J.Am.Chem.Soc.* **2001**, *123*, 7947-7948.
(4) Burford, N.; Cameron, T. S.; LeBlanc, D. J.; Phillips, A. D.; Concolino, T. E.; Lam, K. C.; Rheingold, A. L. *J.Am.Chem.Soc.* **2000**, *122*, 5413-5414.
(5) Burford, N.; Clyburne, J. A. C.; Bakshi, P. K.; Cameron, T. S. *Organometallics* **1995**, *14*, 1578-1585.
(6) Kaukorat, T.; Neda, I.; Schmutzler, R. *Coord.Chem.Rev.* **1994**, *137*, 53-107.
(7) Burford, N.; Losier, P.; Phillips, A. D.; Ragogna, P. J.; Cameron, T. S. *Inorg.Chem.* **2003**, *42*, 1087-1091.
(8) Bouhadir, G.; Reed, R. W.; Réau, R.; Bertrand, G. *Heteroat.Chem.* **1995**, *6*, 371-375.
(9) Jones, V. A.; Sriprang, S.; Thornton-Pett, M.; Kee, T. P. *J.Organomet.Chem.* **1998**, *567*, 199-218.
(10) Burford, N.; Cameron, T. S.; Robertson, K. N.; Phillips, A. D.; Jenkins, H. A. *Chem.Commun.* **2000**, 2087-2088.
(11) Burford, N.; Phillips, A. D.; Spinney, H. A.; Robertson, K. N.; Cameron, T. S. *Inorg.Chem.* **2003**, *42*, 4949-4954.
(12) Schultz, C. W.; Parry, R. W. *Inorg.Chem.* **1976**, *15*, 3046-3050.

(13) Thomas, M. G.; Schultz, C. W.; Parry, R. W. *Inorg.Chem.* **1977**, *16*, 994-1001.
(14) Schmidpeter, A.; Lochschmidt, S.; Sheldrick, W. S. *Angew.Chem.Int.Ed.* **1982**, *21*, 63-64.
(15) Schmidpeter, A.; Lochschmidt, S.; Sheldrick, W. S. *Angew.Chem.Int.Ed.* **1985**, *24*, 226-227.
(16) Shagvaleev, F. Sh.; Zykova, T. V.; Tarasova, R. I.; Sitdikova, T. Sh.; Moskva, V. V. *Zh.Obshch.Khim.* **1990**, *60*, 1775-1779.
(17) David, G.; Niecke, E.; Nieger, M.; Radseck, J.; Schoeller, W. W. *J.Am.Chem.Soc.* **1994**, *116*, 2191-2192.
(18) Romanenko, V. D.; Rudzevich, V. L.; Rusanov, E. B.; Chernega, A. N.; Senio, A.; Sotiropoulos, J. M.; Pfister-Guillouzo, G.; Sanchez, M. *Chem.Commun.* **1995**, 1383-1385.
(19) Burford, N.; Cameron, T. S.; Clyburne, J. A. C.; Eichele, K.; Robertson, K. N.; Sereda, S.; Wasylishen, R. E.; Whitla, W. A. *Inorg.Chem.* **1996**, *35*, 5460-5467.
(20) Burford, N.; Cameron, T. S.; LeBlanc, D. J.; Losier, P.; Sereda, S.; Wu, G. *Organometallics* **1997**, *16*, 4712-4717.
(21) Burford, N.; LeBlanc, D. J. *Inorg.Chem.* **1999**, *38*, 2248-2249.
(22) Abrams, M. B.; Scott, B. L.; Baker, R. T. *Organometallics* **2000**, *19*, 4944-4956.
(23) Burford, N.; Ragogna, P. J.; McDonald, R.; Ferguson, M. *J.Am.Chem.Soc.* **2003**, *125*, 14404-14410.
(24) Burford, N.; Ragogna, P. J.; McDonald, R.; Ferguson, M. J. *Chem.Commun.* **2003**, 2066-2067.
(25) Burford, N.; Ragogna, P. J.; Robertson, K. N.; Cameron, T. S.; Hardman, N. J.; Power, P. P. *J.Am.Chem.Soc.* **2002**, *124*, 382-383.
(26) Sheldrick, W. S.; Schmidpeter, A.; Zwaschka, F.; Dillon, K. B.; Platt, A. W. G.; Waddington, T. C. *Dalton Trans.* **1981**, 413-418.
(27) Niecke, E.; Nieger, M.; Reichert, F. *Angew.Chem.Int.Ed.* **1988**, *27*, 1715-1716.
(28) Cowley, A. H.; Kemp, R. A. *Chem.Rev.* **1985**, *85*, 367-382.
(29) Niecke, E.; Detsch, R.; Nieger, M.; Reichert, F.; Schoeller, W. W. *Bull.Soc.Chim.Fr.* **1993**, *130*, 25-31.
(30) Burford, N.; Clyburne, J. A. C.; Silvert, D.; Warner, S.; Whitla, W. A.; Darvesh, K. V. *Inorg.Chem.* **1997**, *36*, 482-484.
(31) Curtis, R. D.; Schriver, M. J.; Wasylishen, R. E. *J.Am.Chem.Soc.* **1991**, *113*, 1493-1498.
(32) Dashti-Mommertz, A.; Neumüller, B. *Z.Anorg.Allg.Chem.* **1999**, *625*, 954-960.
(33) Kopp, R. W.; Bond, A. C.; Parry, R. W. *Inorg.Chem.* **1976**, *15*, 3042-3046.

(34) Sanchez, M.; Mazières, M.-R.; Lamandé, L.; Wolf, R. Phosphorus compounds with coordination number 2; In *Multiple bonds and low coordination in phosphorus chemistry*; Regitz, M., Scherer, O. J., eds. 1990.
(35) Gudat, D. *Eur.J.Inorg.Chem.* **1998**, 1087-1094.
(36) Gudat, D. *Coord.Chem.Rev.* **1997**, *163*, 71-106.
(37) Burford, N.; Royan, B. W.; Linden, A.; Cameron, T. S. *Inorg.Chem.* **1989**, *28*, 144-150.
(38) Burford, N.; Dipchand, A. I.; Royan, B. W.; White, P. S. *Inorg.Chem.* **1990**, *29*, 4938-4944.
(39) Summers, J. C.; Sisler, H. H. *Inorg.Chem.* **1970**, *9*, 862-869.
(40) Frazier, S. E.; Nielsen, R. P.; Sisler, H. H. *Inorg.Chem.* **1964**, *3*, 292-294.
(41) Spangenberg, S. F.; Sisler, H. H. *Inorg.Chem.* **1969**, *8*, 1006-1010.
(42) Ramirez, F.; Tsolis, E. A. *J.Am.Chem.Soc.* **1970**, *92*, 7553-7558.
(43) Gee, M.; Wasylishen, R. E.; Ragogna, P. J.; Burford, N.; McDonald, R. *Can.J.Chem.* **2002**, *80*, 1488-1500.

Chapter 21

Metallacarboranes of Main Group, Transition, and Lanthanide Elements: Syntheses, Structures and Reactivities

Jianhui Wang[1], John A. Maguire[2], and Narayan S. Hosmane[1,*]

[1]Department of Chemistry and Biochemistry, Northern Illinois University, DeKalb, Illinois 60115
[2]Department of Chemistry, Southern Methodist University, Dallas, TX 75275
*Corresponding author: (email: nhosmane@niu.edu)

The carboranes in the $[nido\text{-}R_2C_2B_4H_4]^{2-}$ and $[nido\text{-}R_2C_2B_9H_9]^{2-}$ cage systems have been used extensively as ligands for various metals.[1] Interest in these systems stems from the fact that the primary metal binding orbitals of the ligands are three π-type molecular orbitals directed above the C_2B_3 open pentagonal face of the carboranes, that are quite similar to those found in the cyclopentadienide $[C_5R_5]^-$ ligands. Our research in this area covers synthetic, structural, reactivity and theoretical studies on the full- and half-sandwich metallacarboranes derived from the interactions of $[nido\text{-}2\text{-}(SiMe_3)\text{-}n\text{-}(R)\text{-}2,n\text{-}C_2B_4H_4]^{2-}$ [n = 3, 4; R = SiMe$_3$, n-Bu, t-Bu, Me, H] with main group,[2] d-group,[3] and f-group metals.[4] The isomer where n = 3, the so-called carbons adjacent isomer, is made directly from the reaction of pentaborane(9), B_5H_9, with substituted acetylenes (Me$_3$SiCCR),[5] while the isomer where n = 4 (carbons apart isomer) is obtained by a oxidative cage closure of the carbon adjacent cage, followed by a reductive cage opening sequence.[3b] At one time the B_5H_9 could be obtained from US-government surplus sources at no cost. At present, that source of B_5H_9 is no longer available, nor is there a commercial source to take its place. Fortunately, a convenient laboratory method for the synthesis has recently been described.[6] With a steady source of starting materials available, we have extended our studies of the carbons apart metallacarboranes.[7-9] We have also investigated the synthesis of the $C_{(cage)}$-appended alkyl-, silylamido- and alkyloxo-derivatives of the larger C_2B_9-cage systems and have studied their

Dedicated to Professor Alan H. Crowley on the occasion of his 70th birthday.

© 2006 American Chemical Society

reactivities toward group 4 and group 14 metals. This has led to the preparation of metallacarboranes with new geometries that could function as precursors to catalysts or possibly exhibit catalytic activity themselves.[8-10] Some of theses studies are detailed in the following pages.

A New Class of Constrained-Geometry Metallocenes: Synthesis and Crystal Structure of a Carboranyl-Thiol-Appended Half-Sandwich Titanocene and Its Conversion to Halotitanocene

A carboranyl-thiol-appended cyclopentadiene ligand, 1-SH-2-[HCpCH(Ph)]-*closo*-1,2-$C_2B_{10}H_{10}$ was prepared in 85% yield from the reaction of the dilithium salt of 1-SH-*closo*-1,2-$C_2B_{10}H_{11}$[11] with 6-phenylfulvene, followed by hydrolysis (see **Scheme 1**).[12] The corresponding titanium complex,

Scheme 1

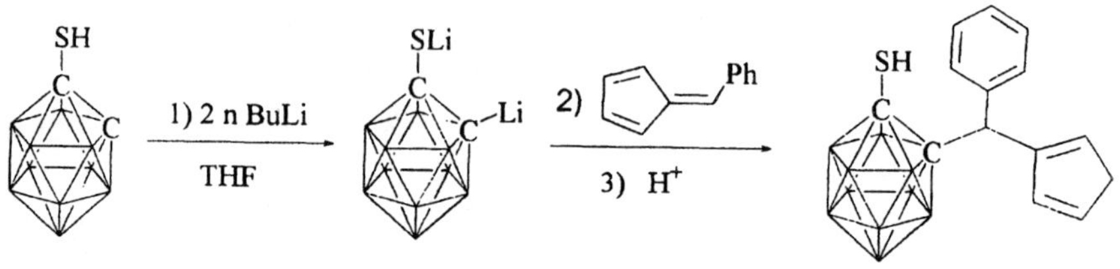

Scheme 2

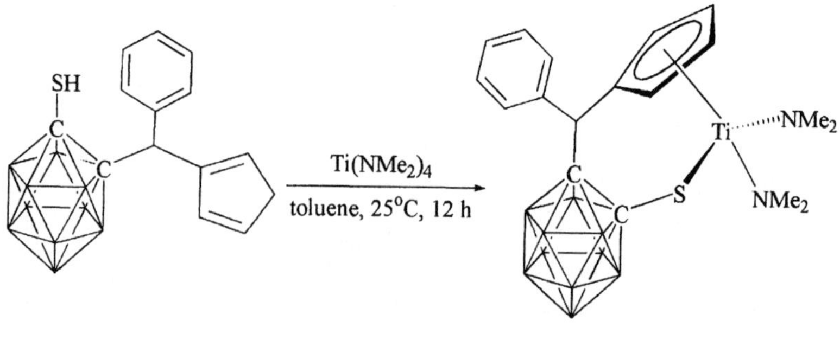

[1-(σ-S)-2-(η^5-C_5H_4CH(Ph))-1,2-$C_2B_{10}H_{10}$]Ti(NMe$_2$)$_2$ was obtained in good yield (83%) *via* amine elimination reaction between Ti(NMe$_2$)$_4$ and 1-SH-2 -

[HCpCH(Ph)]-*closo*-1,2-$C_2B_{10}H_{10}$ in toluene (**Scheme 2**).[12] The reaction of [1-(σ-S)-2-(η^5-C_5H_4CH(Ph))-1,2-$C_2B_{10}H_{10}$)]Ti(NMe$_2$)$_2$ with an excess of Me$_3$SiCl led to [1-(σ-S)-2-(η^5-C_5H_4CH(Ph))-1,2-$C_2B_{10}H_{10}$)]TiCl(NMe$_2$) in 71 % yield. On the other hand, even with a carborane to Me$_3$NHCl molar ratio of 1:2 the chlorocomplex could only be made in modest yield (41%) (**Scheme 3**).[12] [1-(σ-S)-2-(η^5-C_5H_4CH(Ph))-1,2-$C_2B_{10}H_{10}$)]Ti(NMe$_2$)$_2$ was structurally characterized by single-crystal X-ray diffraction. **Figure 1**, shows the compound to be a monomeric complex in which the Cp, S and the two NMe$_2$ groups surround the Ti in a nearly tetrahedral fashion; the Cp(centroid)-Ti-S angle is 113.4(1)° and the N_2-Ti-N_1 angle is 107.27(13)°. The Cp(centroid)-Ti distance of 2.054 Å found in the complex is quite close to the value of 2.00(1)Å reported for [C_5H_4PhCH(PhO)]TiCl$_2$,[13] but is considerably shorter than the analogous distances found in (Me$_4$CpPhO)Ti(CH$_2$Ph)$_2$ (2.36(1)Å)[14] and Me$_2$Si(Me$_4$C$_5$)(^{t}BuN)TiCl$_2$ (2.36(1)Å),[15] indicating a tighter ligand-metal binding in [1-(σ-S)-2-(η^5-C_5H_4CH(Ph))-1,2-$C_2B_{10}H_{10}$)]Ti(NMe$_2$)$_2$. In addition, the Cp(centroid)-Ti-S angle of 113.4(1) is larger than the Cp-Ti-O angles found in [C_5H_4PhCH(PhO)]TiCl$_2$ (110.7°)[13] and (Me$_4$CpPhO)Ti(CH$_2$Ph)$_2$ (107.7°)[14], as well as the Cp-Ti-N angles in Me$_2$Si(Me$_4$C$_5$)(^{t}BuN)TiCl$_2$ (107.6°)[15a], Me$_2$Si(C$_5$H$_5$)(^{t}BuN)TiCl$_2$ (107.0°)[15b] and Me$_2$Si(C$_5$H$_5$)(^{t}BuN)Ti(NMe$_2$)$_2$ (105.5°)[15b]. These structural parameters reflect the influence of a carboranylthiol group on the metal binding to a Cp ligand.

Scheme 3

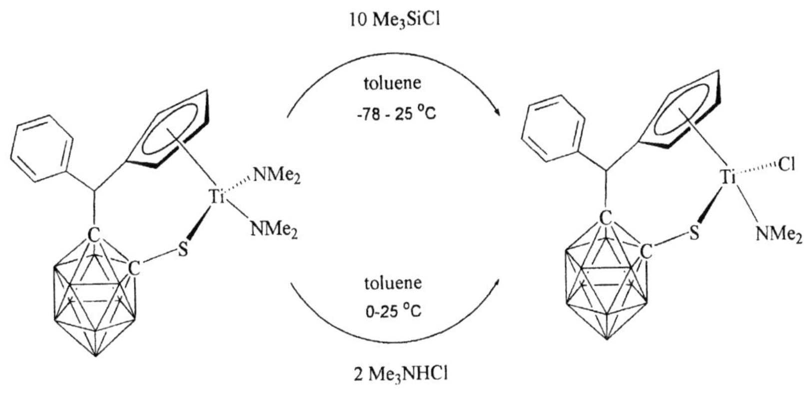

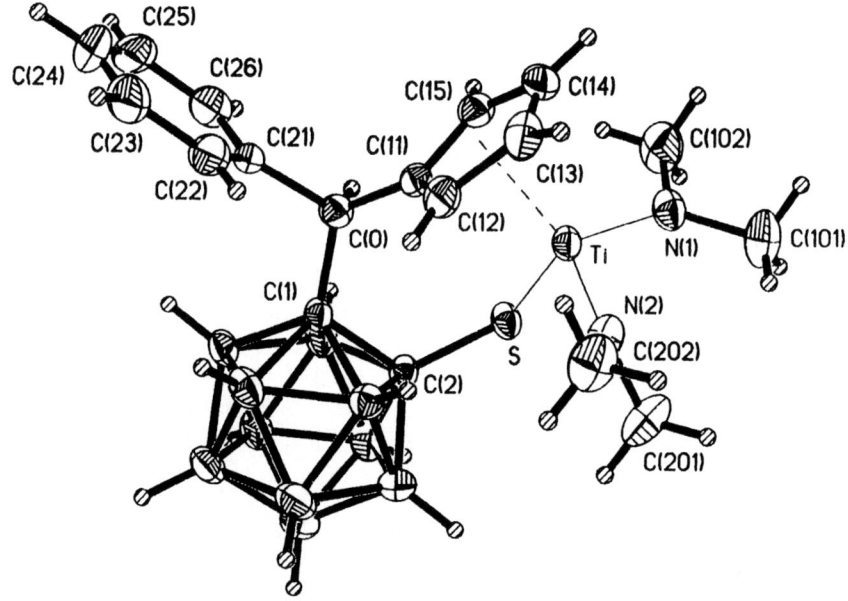

Figure 1. Carboranyl-Thiol-Appended Half-Sandwich Titanocene

Synthetic, Structural and Reactivity Studies on Lanthanacarboranes with Two and Three "Carbons Apart"-Carborane Cages Bonding to Ln(III) Metal [Ln(III) = Nd, Gd, Dy, Ho, Er, Tb and Lu]

The reactions of closo-exo-5,6-Na(THF)$_2$-1-Na(THF)$_2$-2,4-(SiMe$_3$)$_2$-2,4-C$_2$B$_4$H$_4$ with anhydrous LnCl$_3$ (Ln = Nd, Gd, Dy, Ho, Er, Tb and Lu), in molar ratios of 2:1 in dry benzene at elevated temperature (60 °C) produced the full-sandwiched lanthanacarborane complexes, 2,2',4,4'-(SiMe$_3$)$_4$-3,5',6'-[(m-H)$_3$Na[(X)$_n$(Y)$_m$]-1,1'-commo-Ln(η^5-2,4-C$_2$B$_4$H$_4$)$_2$ {Ln = Nd, X = THF, n = 2, Y = none; Gd, X = THF, n = 3, Y = none; Dy, X = THF, n = 1; Y = TMEDA, m = 1; Ho, X = DME, n = 1, Y = none; Er, X = THF, n = 1, Y = none; Tb, X, Y = none; and Lu, X = THF, n = 2, Y = none; in 88, 86, 70, 88, 76, 93 and 88% yields, respectively, as outlined in **Scheme 4**.[7i,k,16] These results differ markedly from those obtained in the carbons adjacent system where a similar procedure gave exclusively the trinuclear clusters of the half-sandwiched lanthanacarboranes and lithiacarboranes.[17-19] These clusters were obtained from the room temperature reactions of LnCl$_3$ with closo-exo-5,6-Li(THF)$_2$-1-Li(THF)$_2$-2,3-(SiMe$_3$)$_2$-2,3-C$_2$B$_4$H$_4$.[20] It is difficult to see how either the lithium cation or lower reaction temperature would favor the half-sandwich clusters.

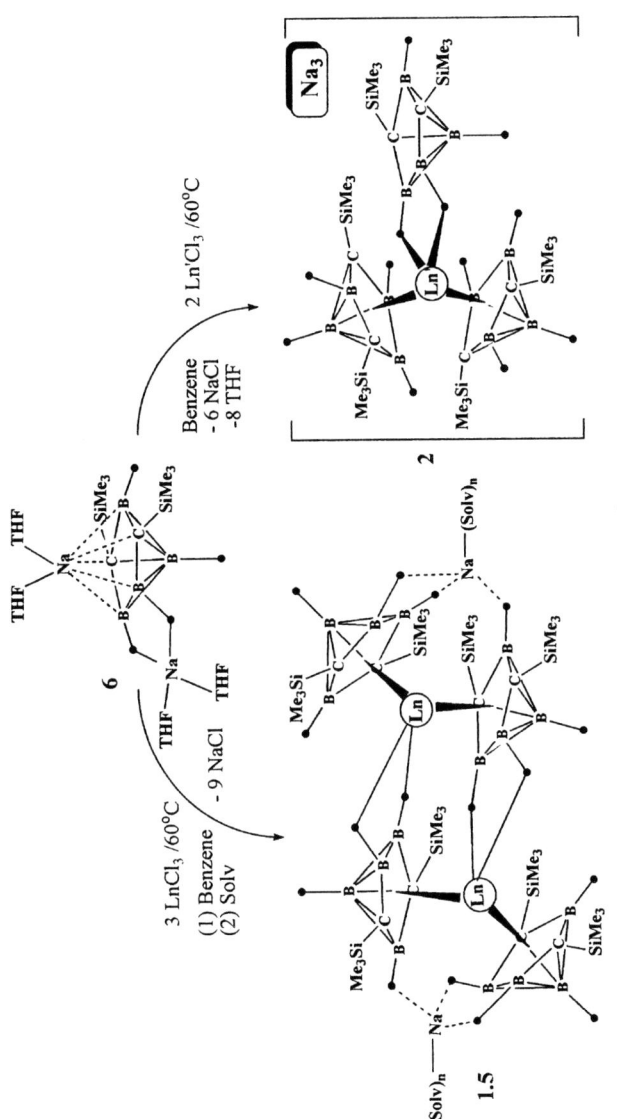

Scheme 4. Syntheses of "Carbons-Apart" Lanthanacarborane Complexes.

Ln = Nd, Gd, Tb, Dy, Ho, Er, Lu; Ln' = Dy, Er; Solv = THF and/or TMEDA; n = 1-3

However, the carbons adjacent carborane requires the use of *t*-BuLi as a deprotonating agent. Since both *t*-BuLi and lanthanide compounds are known to degrade THF and other oxygen-containing compounds,[21,22] it is likely that degradation reactions produced the methoxide and oxide products that effectively prevented the formation of the expected full-sandwich compounds.[20] The results described in **Scheme 4**, in which the full sandwich complexes were formed under similar conditions but in the absence of *t*-BuLi, further support this contention. These findings are also consistent with the fact that when the TMEDA-solvated lithiacarborane, *closo-exo*-5,6-[(μ-H)$_2$Li(TMEDA)-1-Li(TMEDA)-2,3-(SiMe$_3$)$_2$-2,3-C$_2$B$_4$H$_4$, was reacted with LnCl$_3$ only the full-sandwich complexes were formed.[4c] It is of interest that the reaction of the larger, *nido*-C$_2$B$_9$-carborane anion with LnCl$_3$ in a 2:1 molar ratio produced only the expected full-sandwiched lanthanacarboranes.[23]

The effect of the reaction stoichiometry on the nature of the products was probed by studying the reactions of *closo-exo*-5,6-Na(THF)$_2$-1-Na(THF)$_2$-2,4-(SiMe$_3$)$_2$-2,4-C$_2$B$_4$H$_4$ with anhydrous LnCl$_3$ (Ln = Dy, Er), in molar ratios of 3:1, under the same experimental conditions as that used in the above preparation (see **Scheme 4**). The products were the mixed metallacarborane complexes, [Na$_3$][1,1'-{5,6-(μ-H)$_2$-*nido*-2,4-(SiMe$_3$)$_2$-2,4-C$_2$B$_4$H$_4$}-2,2',4,4'-(SiMe$_3$)$_4$-1,1'-*commo*-Ln-(2,4-C$_2$B$_4$H$_4$)$_2$] [Ln = Dy, Er], obtained as yellow crystalline solids in 78 and 82% yields, respectively (see **Scheme 4**).[7j,16] In these compounds three carbons apart carborane ligands were found to be associated with each lanthanide metal center, two through η^5-bonding modes and one through a set of two Ln-H-B bonding interactions (see **Figure 2**). Although these are stoichiometric analogues of tris(cyclopentadienyl) lanthanide complexes, their structures and bonding modes are quite different.[24] In the lanthanocenes, all three Cp$^-$ ligands are η^5-bonded to the lanthanide, giving a neutral, trigonal planar (Cp)$_3$Ln geometry,[24] while in the lanthanacarboranes only two of the three carborane ligands are η^5-bonded. It may be that, even in the presence of the three Na$^+$ counter-ions, the high total negative charge due to the three carborane ligands (–6) prevents a (tris) η^5-bonding interaction. Steric considerations may also be important in such cases. However, a number of sterically crowded (C$_5$Me$_4$R)$_3$La (R = Me, Et, *i*-Pr and SiMe$_3$) complexes have been synthesized and structurally characterized,[25] so it is difficult to assess the relative importance of the size of the carborane ligands in destabilizing tris-complexes. The bonding motif shown in **Figure 2** is a commonly encountered one, with each lanthanide metal atom associated with three carboranes, two of which are η^5-bonded and the other is η^2-bonded. When excess carborane ligand is present, it is incorporated as the η^2-bonding group, otherwise dimeric compounds are formed.[7j,16]

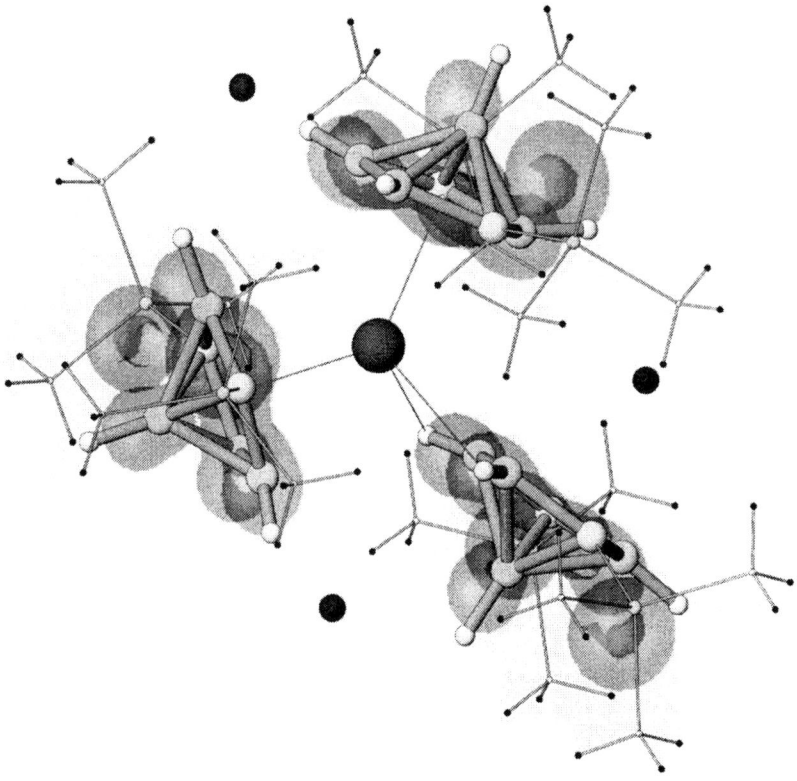

Figure 2. A perspective view of a lanthanacarborane complex containing three "carbons apart" carborane ligands about the central lanthanide metal.

An Oxide Ion-Encapsulating Tetralanthanide Tetrahedron Stabilized by Complexation with the "Carbons Apart" C_2B_4-Carborane Ligands

Previous work has shown that, unlike their larger cage (C_2B_9 and C_2B_{10}) analogues, the small-cage carboranes could form unusual oxo-lanthanacarboranes of the type $\{[\eta^5\text{-}1\text{-}Ln\text{-}2,3\text{-}(SiMe_3)_2\text{-}2,3\text{-}C_2B_4H_4]_3[(\mu\text{-}1\text{-}Li\text{-}2,3\text{-}(SiMe_3)_2\text{-}2,3\text{-}C_2B_4H_4)_3(\mu_3\text{-}OMe)]\text{-}[\mu\text{-}Li(THF)]_3(\mu_3\text{-}O)\}$ (Ln = Nd, Sm, Gd, Tb, Dy and Ho].[4b,26] As noted earlier, it was speculated that the reactions involve an initial formation of the respective half-sandwich lanthanacarborane, which then reacts with the remaining lithiacarborane precursors in the presence of degraded fragments of the THF solvent to yield the corresponding trinuclear oxo-Ln(III)-carborane clusters. It should be noted that only the carbons adjacent (2,3-C_2B_4) carboranes seem to show this sensitivity to solvent; the analogous carbons apart (2,4-C_2B_4) isomers,[7i,j] as well as the larger cage (C_2B_9) carboranes,[23b] gave only the simpler full- and half-sandwich lanthanacarboranes. The use of THF as an oxygen source makes it difficult to control the stoichiometry of the synthesis of oxo complexes, and, in addition, it introduces a number of other degradation products such as MeO$^-$ which can influence the reaction. These complexities led us to explore alternative methods for synthesizing oxo-lanthanacarboranes routinely. This investigation has resulted in the synthesis of the first oxide ion-encapsulating tetralanthanide tetrahedron, stabilized by complexation with the small-cage carbons apart C_2B_4-carborane ligands.[7m,27]

In this synthesis, anhydrous $LnCl_3$ was first treated with freshly distilled water under refluxing conditions in THF. The THF solution of *closo*-*exo*-5,6-Na(THF)$_2$-1-Na(THF)$_2$-2,4-(SiMe$_3$)$_2$-2,4-$C_2B_4H_4$ was then poured *in vacuo* onto the selective lanthanide chloride solution at –78 °C to give an overall carborane: $LnCl_3$: H_2O mole ratio of 5: 4: 1. After refluxing overnight, the new oxo-lanthanacarborane, $\{[\eta^5\text{-}1\text{-}Ln(THF)\text{-}2,4\text{-}(SiMe_3)_2\text{-}2,4\text{-}C_2B_4H_4]_4(\mu\text{-}Cl)_2(\mu_4\text{-}O)\}$, was obtained as a pale-yellow crystalline solid in 80-86 % yield (see **Scheme 5**).[7m,27] It is critical to the synthesis that the $LnCl_3/H_2O$ mixture must be refluxed until a homogeneous THF solution is obtained. If $LnCl_3$ and water are added to a solution of the carborane ligand without prior refluxing, the water decomposed the carborane; such decomposition occurred rapidly at room temperature. However, prior refluxing and a –78 °C reaction temperature produced the oxo-product in good yield. Because of the unprecedented nature of the synthetic route shown in **Scheme 5**, the structure of the oxo lanthacarborane was unambiguously determined by X-ray diffraction analysis.[7m,27] The molecular structure, depicted in **Figure 3**, shows that an oxide ion is tetrahedrally encapsulated by four lanthanide ions, each of which is stabilized by η^5-bonding to carbons apart C_2B_4-carborane cages. Each

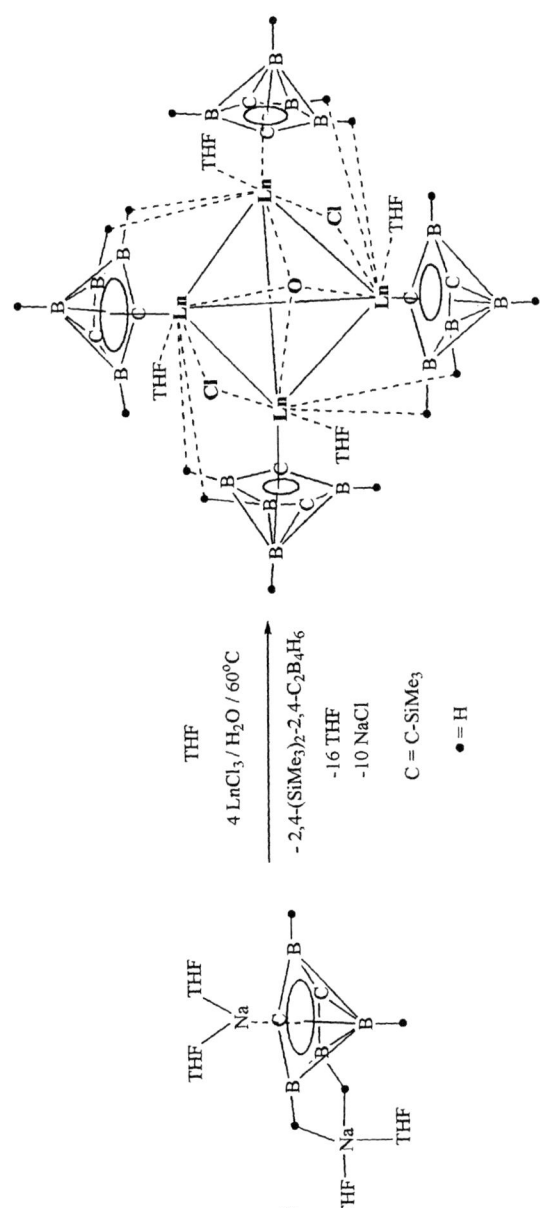

Scheme 5. Syntheses of oxo-lanthanacarborane clusters

lanthanide ion is also coordinated, *via* two Ln-H-B bridges, to a neighboring C_2B_4 cage. The O-metal distances are 2.369±0.007 Å, 2.299±0.005 Å, 2.283±0.001 Å, 2.244±0.008 Å for Nd, Tb, Ho and Lu-carborane complexes, respectively. The corresponding metal-oxygen-metal bond angles are: Ln(1)-O-Ln(2) = 107.48(3)°, 107.12(2)°, 106.21(2)°, 106.69(3)°; Ln(1)-O-Ln(1A) = 114.3(5)°, 115.3(4)°, 116.2(2)°, 115.5(6)°; and Ln(1)-O-Ln(2A) = 107.38(3)°, 106.89(2)°, 106.31(2)°, 106.56(3)°, respectively. These angles suggest a slightly distorted tetrahedral arrangement of the metals about the oxygen. Each lanthanide metal is associated with an η^5-bonded C_2B_4 cage, with the average Ln-cage centroid distances of 2.433±0.001 Å, 2.355±0.004 Å, 2.324±0.007Å, 2.279±0.003 Å for Nd, Tb, Ho and Lu-carborane complexes, respectively. These Ln-cage centroid distances are quite close to those found in the lanthanide complexes in other C_2B_4-carborane ligand systems.[4,7-9,16] The structures also show that each diagonal pair of Ln metals is linked by a bridged chlorine atom, to give the $[(C_2B_4Ln)_4Cl_2O]$ core. Both the Ln-O and the Ln-Cnt bond distances decrease in the order Nd > Tb > Ho > Lu which is the same order as the decrease in the ionic radii of the Ln^{3+} ions,[28] suggesting an essentially ionic metal-ligand interaction. This same trend has been noted in the corresponding trinuclear Ln-carborane complexes,[4b] lanthanacarboranes containing two or three carborane cages,[7i,j,16] and in the lanthanocenes.[29] The average metal-metal distances are 3.826±0.001 Å, 3.707±0.001 Å, 3.653±0.001 Å, 3.598±0.001 Å for lanthanacarborane complexes of Nd, Tb, Ho and Lu, respectively. Such long distances indicate that there are no direct metal-metal interactions in these structures.

Variations in the stoichiometric molar ratios of the reactants failed to produce any other lanthanacarborane complexes.

Half-Sandwich Halolanthanacarboranes of The Carbons Adjacent C_2B_4-Carborane Ligand Syatems: Important Synthons for New Polyhedral Cage Constructs.

The propensity of the C_2B_4 cages to form oxide encapsulated lanthacarboranes complicates any comparative studies of the reactivities of the lanthanacarboranes in the small cage system. What is needed in order to systematically study those factors that affect the reactivities and stabilities of the small C_2B_4-cage system is the development of a simple, direct method of synthesizing the half sandwich (2,3-C_2B_4) lanthanacarboranes. The reaction of *nido*-1-Na(THF)-2,3-(SiMe$_3$)$_2$-2,3-$C_2B_4H_5$[30] with anhydrous LnX_3 in a molar ratio of 2:1, in dry THF at 65 °C, produced the compounds, [*closo*-1(X)-1,1-(THF)$_2$-2,3-(SiMe$_3$)$_2$-1-Ln(η^5-2,3-$C_2B_4H_4$)]$_2$ (Ln = Ce, X = Br; Ln = Gd, X = Cl;

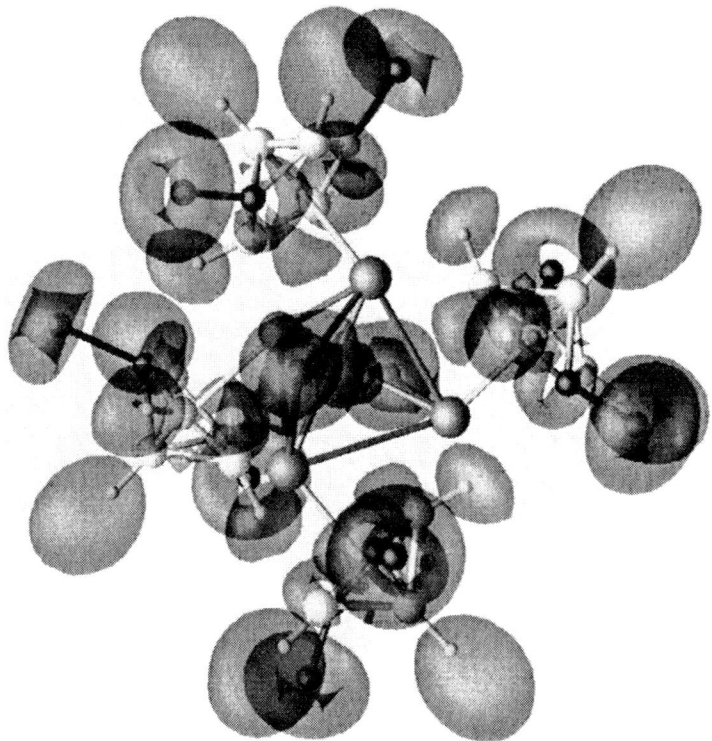

Figure 3. Electron-density map of a typical oxo-lanthanacarborane cluster as found in its X-ray diffraction analysis.

Ln = Lu, X = Cl), in yields of 92, 71, and 70%, respectively, along with one mole equivalent of the neutral *nido*-2,3-(SiMe$_3$)$_2$-2,3-C$_2$B$_4$H$_6$ (see **Scheme 6**).[31] The initial solubility of the LnX$_3$ in THF is quite critical in determining the

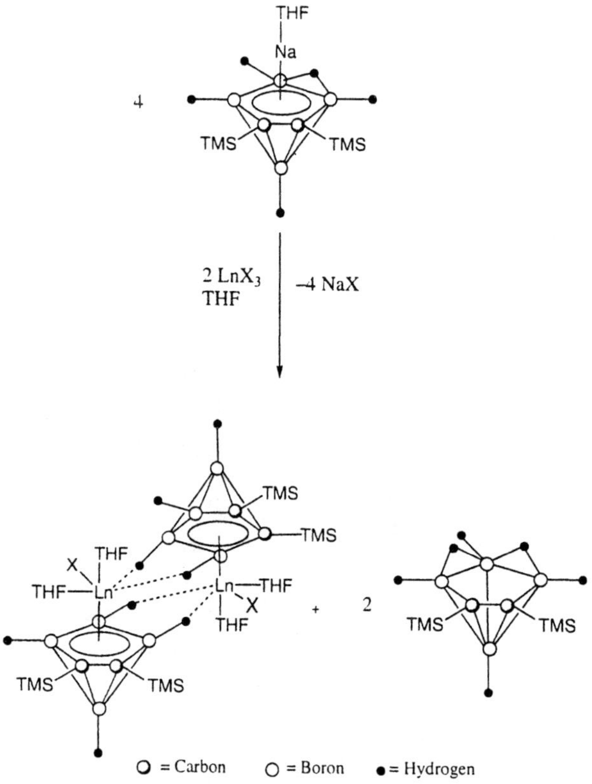

Scheme 6. Syntheses of half-sandwich halolanthanacarborane complexes

yields. The cerium reagent, CeCl$_3$, which has a low solubility in THF, gave extremely low yields of product, however, the more soluble CeBr$_3$ led to the ceracarborane in 92% yield. The more soluble GdCl$_3$ and LuCl$_3$ gave satisfactory yields of their respective metallacarboranes. The reaction clearly involves the transfer of a proton from a metallated carborane to an extra carborane mono-anion, however, it is not known whether this proton transfer occurs before, during or after metal coordination. The mechanism could also involve the initial formation of a bridged metal complex, followed by an intramolecular proton transfer and release of the formed neutral carborane by-product. Irrespective of the mechanism, the method covers the syntheses involving lanthanides at the start, in the middle and at the end of the period.

Therefore, it should be generally applicable to all lanthanides, and most likely the actinides. The method also has the advantage that it protects against the complications of THF decomposition found in other synthetic methods.[4b,17,26]

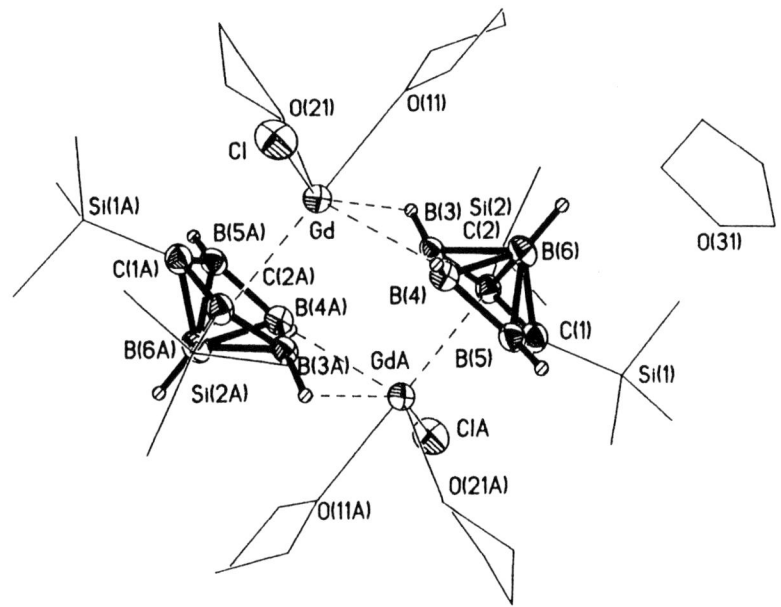

Figure 4. Crystal structure of the half-sandwich chlorogadolinacarborane dimer.

Because of the potential of this method to be a general one, the structure of the chlorogadolinacarborane complex was unambiguously determined by single crystal X-ray diffraction.[31] Figure 4 shows the compound to be a dimer of two oppositely oriented $closo$-1(X)-1,1-(THF)$_2$-2,3-(SiMe$_3$)$_2$-1-Ln(η^5-2,3-C$_2$B$_4$H$_4$) half-sandwich complexes. The carboranes are η^5-bonded to the metal, with an average Gd-ring atom distance of 2.733±0.007 Å, with no systematic variances attributable to the presence of bridged hydrogens. The Gd-(C$_2$B$_3$ centroid) distance of 2.394 Å is slightly greater than the Gd-centroid distances of 2.340 Å and 2.360 Å found in the full-sandwich, 2,2',4,4'-(SiMe$_3$)$_4$-3.5',6'-[(m-H)$_3$Na(THF)$_3$-1,1'-$commo$-Gd(2,4-C$_2$B$_4$H$_4$)$_2$,[16] but essentially the same as the 2.391±0.009 Å found for the average Gd-centroid distance in the trinuclear compound, {η^5-1-Gd-2,3-(SiMe$_3$)$_2$-2,3-C$_2$B$_4$H$_4$]$_3$[(μ-1-Li-2,3-(SiMe$_3$)$_2$-2,3-C$_2$B$_4$H$_4$)$_3$(μ_3-OMe)][μ-Li(THF)]$_3$(μ_3-O)}.[4b]

The resulting half-sandwich metallacarboranes have reactive, metal-bound halides that allow the study of the further reactivities of these compounds.

Acknowledgments: This work was supported by grants from the National Science Foundation (CHE-0241319), the donors of the Petroleum Research Fund, administered by the American Chemical Society, The Robert A. Welch Foundation (N-1322 to JAM) and Northern Illinois University through a Presidential Research Professorship. N.S.H. gratefully acknowledges the Forschungspreis der Alexander von Humboldt-Stiftung.

References

1. For general references see: (a) *Comprehensive Organometallic Chemistry II;* Abel, E. W.; Stone, F. G. A.; Wilkinson, G., Eds.; Elsevier Science Ltd.: Oxford, **1995**; Volume 1, Chapters 6–9. (b) Hosmane, N. S.; Maguire, J. A. In *Electron Deficient Boron and Carbon Clusters;* Olah, G. A., Wade, K., Williams, R. E., Eds; Wiley: New York, **1991**; Chapter 9, pp. 215–235. (c) Hosmane, N. S.; Maguire, J. A. *Adv. Organomet. Chem.* **1990**, *30*, 99–150. (d) Hosmane, N. S. *Pure and Appl. Chem.* **1991**, *63*, 375–378.
2. (a) Saxena, A. K.; Maguire, J. A.; Hosmane, N. S. *Chem. Rev.* **1997**, *97*, 2421–2462 and references therein. (b) Hosmane, N. S.; Lu. K.-J.; Zhang, H.; Maguire, J. A. *Organometallics* **1997**, *16*, 5163–5170. (c) Hosmane, N. S.; Yang, J.; Lu, K.-J.; Zhang, H.; Siriwardane, U.; Islam, M. S.; Thomas, J. L.C.; Maguire, J. A. *Organometallics* **1998**, *17*, 2784–2796.
3. (a) Hosmane, N. S.; Wang, Y.; Zhang, H.; Lu, K.-J.; Maguire, J. A.; Gray, T. G.; Brooks, K. A.; Waldhör, E.; Kaim, W.; Kremer, R. K. *Organometallics* **1997**, *16*, 1365–1377. (b) Zhang, H.; Wang, Y.; Saxena, A. K.; Oki. A. R.; Maguire, J. A.; Hosmane, N. S. *Organometallics* **1993**, *12*, 3933–3944. (c) Hosmane, N. S.; Wang, Y.; Zhang, H.; Maguire, J. A.; Waldhör, E.; Kaim, W. *Organometallics* **1993**, *12*, 3785–3787. (d) Saxena, A. K.; Hosmane, N. S. *Chem. Rev.* **1993**, *93*, 1081–1124. (e) Hosmane, N. S.; Wang, Y.; Zhang, H.; Maguire, J. A.; Waldhör, E.; Kaim, W.; Binder, H.; Kremer, R. K. *Organometallics* **1994**, *13*, 4156–4158. (f) Oki, A. R.; Zhang, H., Hosmane, N. S.; Ro, H.; Hatfield, W. E. *J. Am. Chem. Soc.* **1991**, *113*, 8531–8532. (g) Zhang, H.; Jia, L.; Hosmane, N. S. *Acta Crystallogr., Cryst. Struct. Commun.* **1993**, *C49*, 453–456. (h) Colacot, T. J.; Jia, L.; Zhang, H.; Siriwardane, U.; Maguire, J. A.; Wang, Y.; Brooks, K. A.; Weiss, V. P.; Hosmane, N. S. *Organometallics* **1995**, *14*, 1365–1376. (i) Hosmane, N. S.; Zhu, D.; Zhang, H.; Oki, A. R.; Maguire, J. A. *Organometallics* **1998**, *17*, 3196–3203.
4. (a) Hosmane, N. S.; Wang, Y.; Oki, A. R.; Zhang, H.; Zhu, D.; McDonald, E. M.; Maguire, J. A. *Phosphorus, Sulfur, and Silicon* **1994**, *93-94*, 253–256. (b) Hosmane, N. S.; Wang, Y.; Oki, A. R.; Zhang, H.; Maguire, J. A. *Organometallics* **1996**, *15*, 626–638. (c) Hosmane, N.S.; Wang. Y.; Zhang, H.; Maguire, J. A.; McInnis, M.; Gray, T. G.; Collins, J. D.; Kremer, R. K.; Binder, H.; Waldhör, E.;

Kaim, W. *Organometallics* **1996**, *15,* 1006–1013. (d) Zhang, H.; Oki, A. R.; Wang, Y.; Maguire, J. A.; Hosmane, N. S. *Acta Crystallogr., Cryst. Struct. Commun.* **1995**, *C51,* 635–638. (e) Hosmane, N. S.; Wang, Y.; Zhang, H.; Oki, A. R.; Maguire, J. A.; Waldhör, E.; Kaim, W.; Binder, H.; Kremer, R. K. *Organometallics* **1995**, *14,* 1101–1103. (f) Zhang, H.; Wang, Y.; Maguire, J. A.; Hosmane, N. S. *Acta Crystallogr., Cryst. Struct. Commun.* **1996**, *C52,* 640–643. (g) Zhang, H.; Wang, Y.; Maguire, J. A.; Hosmane, N. S. *Acta Crystallogr., Cryst. Struct. Commun.* **1996**, *C52,* 8–11. (h) Hosmane, N. S.; Oki, A. R.; Zhang, H. *Inorg. Chem. Commun.* **1998**, *1,* 101–104.
5. Hosmane, N. S.; Sirmokadam, N. N.; Mollenhauer, M. N. *J. Organomet. Chem.* **1985**, *279,* 359.
6. (a) Adams, L.; Hosmane, S. N.; Eklund, J. E.; Wang, J.; Hosmane, N. S. *J. Am. Chem. Soc.* **2002**, *124,* 7292–7293. (b) Adams, L.; Tomlinson, S.; Wang, J.; Hosmane, S. N.; Maguire, J. A.; Hosmane, N. S. *Inorg. Chem. Commun.* **2002**, *5,* 765-767.
7. (a) Hosmane, N. S.; Jia, L.; Zhang, H.; Bausch, J. W.; Prakash, G. K. S.; Williams, R. E.; Onak, T. P. *Inorg. Chem.* **1991**, *30,* 3793–3795. (b) Ezhova, M. B.; Zhang, H.; Maguire, J. A.; Hosmane, N. S. *J. Organomet. Chem.* **1998**, *550,* 409–422. (c) Hosmane, N. S.; Zhu, D.; McDonald, J. E.; Zhang, H.; Maguire, J. A.; Gray, T. G.; Helfert, S. C. *Organometallics* **1998**, *17,* 1426–1437. (d) Hosmane, N. S.; Zhu, D.; Zhang, H.; Oki, A. R.; Maguire, J. A. *Organometallics* **1998**, *17,* 3196–3203. (e) Zheng, C.; Wang, J.-Q.; Maguire, J. A.; Hosmane, N. S. *Main Group Met. Chem.* **1999**, *22,* 361–366. (f) Hosmane, N. S. *J. Organomet. Chem.* **1999**, *581,* 13–27. (g) Hosmane, N. S. *Current Sci.* **2000**, *78,* 475–486. (h) Rana, G.; Maguire, J. A.; Hosmane, S. N.; Hosmane, N. S. *Main Group Met. Chem.* **2000**, *23,* 527–547. (i) Hosmane, N. S.; Li, S.-J.; Zheng, C.; Maguire, J. A. *Inorg. Chem. Commun.* **2001**, *4,* 104–107. (j) Wang, J.; Li, S.-J.; Zheng, C.; Maguire, J. A.; Hosmane, N. S. *Organometallics* **2002**, *21,* 3314–3316. (k) Wang, J.; Li, S.-J.; Zheng, C.; Maguire, J. A.; Hosmane, N. S. *Inorg. Chem. Commun.* **2002**, *5,* 602–605. (l) Wang, J.; Li, S.; Zheng, C.; Maguire, J. A.; Hosmane, N. S. *Organometallics* **2002**, *21,* 5149-5151. (m) Wang, J.; Li, S. –J.; Zheng, C.; Hosmane, N. S.; Maguire, J. A.; Roesky, H. W.; Cummins, C. C.; Kaim, W. *Organometallics,* **2003**, *22,* 4390-4392.
8. Hosmane, N. S. *Pure Appl. Chem.,* **2003**, *75,* 1219-1229.
9. Hosmane, N. S.; Maguire, J. A *Euro. J. Inorg. Chem.,* **2003**, *2003,* 3989-3999.
10. (a) Zhu, Y.; Vyakaranam, K.; Maguire, J. A.; Quintana, W.; Teixidor, F.; Viñas, C.; Hosmane, N. S. *Inorg. Chem. Commun.* **2001**, *4,* 486–489. (b) Zhu, Y.; Maguire, J. A.; Hosmane, N. S. *Inorg. Chem. Commun.* **2002**, *5,* 296–299.
11. Teixidor, F.; Viñas, C.; Benakki, R.; Kivekäs, R.; Sillanpää, R. A *Inorg. Chem.* **1997**, *36,* 1719-1723.

12. Wang, J.; Zheng, C.; Maguire, J. A.; Hosmane, N. S. *Organometallics,* **2003,** *22,* 4839-4841.
13. Pu, W.; Wang, J. Ye, L.; Mu, Y.; Yang, G.; Fan, Y. *Acta Cryst.* **1999,** *C55,* 728-729.
14. Chen, Y.-X.; Fu, P.-F.; Stern, C. L.; Marks, T. J. *Organometallics* **1997,** 16, 5958-5963.
15. (a) Stevens, J. C.; Timmers, F. J.; Wilson, D. R.; Schmidt, G. F.; Nickias, P. N.; Rosen, R. K.; Knight, G. W.; Lai, S. Eur. Patent Appl. Ep 416 815-A2, 1991 (Dow Chemical Co.). (b) Carpenette, D. W.; Kloppenburg, L.; Kupec, J. T.; Peterson, J. L. *Organometallics* **1996,** *15,* 1572-1581.
16. Wang, J.; Li, S. –J.; Zheng, C.; Maguire, J. A.; Sarkar, B.; Kaim, W.; Hosmane, N. S. *Organometallics,* **2003,** *22,* 4334-4342.
17. (a) Oki, A. R.; Zhang, H.; Hosmane, N. S., *Angew. Chem. Int. Ed. Engl.,* **1992,** *31,* 432-434. (b) Hosmane, N. S.; Maguire, J. A., *J. Cluster Sci.,* **1993,** *4,* 297-349. (c) Hosmane, N. S.; Wang, Y.; Oki, A. R.; Zhang, H.; Zhu, D.; McDonald, E. M.; Maguire, J. A., *Phosphorus, Sulfur, and Silicon,* **1994,** *93-94,* 253-256. (d) Zhang, H.; Oki, A. R.; Wang, Y.; Maguire, J. A.; Hosmane, N. S., *Acta Crystallogr., Cryst. Struct. Commun.,* **1995,** *C51,* 635-638.
18. Hosmane, N. S.; Wang, Y.; Zhang, H.; Oki, A. R.; Maguire, J. A.; Waldhör, E.; Kaim, W.; Binder, H.; Kremer, R. K., *Organometallics,* **1995,** *14,* 1101-1103.
19. Zhang, H.; Wang, Y.; Maguire, J. A.; Hosmane, N. S., *Acta Crystallogr., Cryst. Struct. Commun.,* **1996,** *C52,* 8-11.
20. Hosmane, N. S.; Wang, Y.; Oki, A. R.; Zhang, H.; Maguire, J. A., *Organometallics,* **1996,** *15,* 626-638.
21. (a) Jung, M. E.; Blum, R. B. *Tetrahedron Letters* **1977,** 3791. (b) Kamata, K.; Terashima, M. *Heterocycles* **1980,** *14,* 205. (c) Schumann, H.; Palamidis, E.; Loebel, J. *J. Organomet. Chem.* **1990,** *384,* C49-52.
22. Evans, W. J.; Grate, J. W.; Bloom, I.; Hunter, W. E.; Atwood, J. L. *J. Am. Chem. Soc.* **1985,** *107,* 405-409.
23. (a) Fronczek, F. R.; Halstead, G. W.; Raymond, K. N. *J. Am. Chem. Soc.* **1977,** *99,* 1769-1775. (b) Manning, M. J.; Knobler, C. B.; Hawthorne, M. F., *J. Am. Chem. Soc.,* **1988,** *110,* 4458-4459. (c) Manning, M. J.; Knobler, C. B.; Khattar, R.; Hawthorne, M. F. *Inorg. Chem.,* **1991,** *30,* 2009.
24. (a) Evans, W. J.; Davis, B. L. *Chem. Rev.* **2002,** *102,* 2119-2136. (b) Evans, W. J.; Seibel, C. A.; Ziller, J. W. *J. Am. Chem. Soc.* **1998,** *120,* 6745-6752.
25. Evans, W.J.; Davis, B. L.; Ziller, J. W. *Inorg. Chem.* **2001,** *40,* 6341-6348.
26. Zheng, C.; Hosmane, N. S.; Zhang, H.; Zhu, D.; Maguire, J. A. *Internet. J. Chem.,* **1999,** *2,* 10 (URL: http://www.ijc.com/articles/1999v2/10/).

27. Wang, J.; Li, S. -J.; Zheng, C.; Li, A.; Hosmane, N. S.; Maguire, J. A.; Roesky, H. W.; Cummins, C. C.; Kaim, W. *Organometallics,* **2004**, submitted for publication.
28. Shannon, R. D. *Acta Crystallogr., Sect. A* **1976**, *32*, 751.
29. (a) Schumann, H.; Genthe, W.; Bruncks, N. *Angew.Chem.* **1981**, *93*, 126; *Angew. Chem., Int. Ed. Engl.* **1981**, *20*, 119. (b) Schumann, H.; Genthe, W.; Bruncks, N.; Pickardt, J. *Organometallics* **1982**, *1*, 1194. (c) Maginn, R. W.; Manastyrskj, S.; Dudeck, M. *J. Am. Chem. Soc.* **1963**, *85*, 672. (d) Evens, W. J.; Wayda, A. L.; Hunter, W. E.; Atwood, J. L. *Chem. Commun.* **1981**, 292-293.
30. Hosmane, N. S.; Saxena, A. K.; Barreto, R. D.; Zhang, H.; Maguire, J. A.; Jia, L.; Wang, Y.; Oki, A. R.; Grover, K. V.; Whitten, S. J.; Dawson, K.; Tolle, M. A.; Sirriwardane, U.; Demissie, T.; Fagner, J. S. *Organometallics* **1993**, *12*, 3001-3014
31. Wang, J.; Zheng, C.; Arikatla, G.; Li, A.; Maguire, J. A.; Hosmane, N. S. *Inorg. Chem. Commun.,* submitted for publication.

Material, Polymers, and Other Applications

Chapter 22

Polyhedral Boranes in the Nanoworld

M. Frederick Hawthorne, Omar K. Farha, Richard Julius, Ling Ma, Satish S. Jalisatgi, Tiejun Li, and Michael J. Bayer

Department of Chemistry and Biochemistry, University of California at Los Angeles, Los Angeles, CA 90095

Aromatic polyhedral boranes and carboranes lend themselves to the construction of nanoesque structures suited for unique applications. These include cyclic, water-soluble hosts having a lipophilic core; self-assembling nanorods and bilayer membranes using amphiphilic carborane derivatives; and the huge family of functional structures obtainable from the $[closo\text{-}B_{12}(OH)_{12}]^{2-}$ by derivatization of the hydroxyl groups. The latter species are potentially useful for molecular delivery devices important to biomedicine and material science. New developments in each of these areas of interest will be presented.

As previously pointed out,[1,2] of all the elements of the periodic table, only neighboring carbon and boron share the properties of wide-spread self-bonding (catenation) and the support of electron-delocalized structures based upon these catenated frameworks. Carbon catenation, of course, leads to the immense field of organic chemistry. Boron catenation provides the *nido-*, *arachno-*, and *hypho-*boranes, which may be considered the borane equivalents of aliphatic hydrocarbons, and discrete families of *closo-*borane derivatives which bear a formal resemblance to aromatic hydrocarbons, heterocycles and metallocenes. Aside from these analogs, boron and carbon chemistries are important to each other through their extravagant ability to mix in ways not available to other element pairs. This history of merging chemistries continues to offer new opportunities now and into the foreseeable future.

Recently, nanotechnology has received a great amount of attention due to the promise of its potential applications. Much of this attention has been focused upon carbon-based nanoparticles: nanotubes, nanoscrolls and most famously, buckminsterfullerene.[3] Integral to the success of these structures is carbon's aforementioned catenation ability. In this chapter, we demonstrate how boron's ability to catenate and mix with carbon in the form of polyhedral borane derivatives and carboranes allows the creation of novel nanoparticles, as well.

Commercially Available Boranes, Carboranes and Metallacarboranes

Boronhydride chemistry underwent a tremendous expansion in the 1950's. Extraordinarily stable $[closo\text{-}B_{12}H_{12}]^{2-}$ (**1**) and $[closo\text{-}B_{10}H_{10}]^{2-}$ (**2**) anions were discovered over 40 years ago.[4,5] Continued expansion in the field occurred with the synthesis of *closo*-carborane derivatives, $[closo\text{-}CB_{11}H_{12}]^{-}$ (**3**) ; the three isomeric dicarbon carboranes, $closo\text{-}1,2\text{-}C_2B_{10}H_{12}$ (**4**), $closo\text{-}1,7\text{-}C_2B_{10}H_{12}$ (**5**) and $closo\text{-}1,12\text{-}C_2B_{10}H_{12}$ (**6**), commonly known as *ortho*, *meta*, and *para*-carboranes.[6] The derivative chemistry of all of these molecules has been well-established,[7,8] with much new chemistry continuing to be discovered at an ever-expanding rate.[9,10]

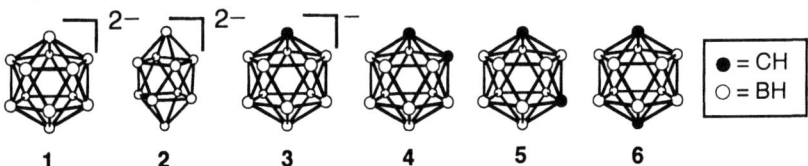

The icosahedral metallacarboranes are very stable metal complexes in which the metal may be considered to be coordinated to the open C_2B_3 pentagonal face of a dicarbollide ion ($nido\text{-}C_2B_9H_{11}^{2-}$), which is analogous to a cyclopentadienyl anion.[11] Due to a number of factors, metallacarboranes are even more stable than their cyclopentadienyl analogs. A wide variety of structures are available, in which both the metal and associated ligands may be varied in numerous ways, as shown in the representative structures **7**, **8** and **9**.[12]

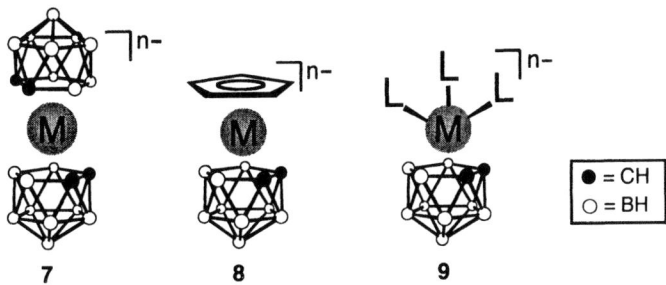

Permethylation of Icosahedral Borane Derivatives

Substitution of the vertices of icosahedral borane derivatives provides a route to the confluence of polyhedral borane and hydrocarbon chemistries, combining the unique chemical properties of the boranes, such as their extreme kinetic stability and three-dimensional aromaticity, with the synthetic arsenal of organic chemistry.[13] Methylation of all vertices of the icosahedral structure leads to the permethylated species, of which the dodecamethyl-1,12-dicarba-*closo*-dodecaborate, *closo*-1,12-C$_2$B$_{10}$(CH$_3$)$_{12}$, (**10**)[14] dodecamethyl-1-carba-*closo*-dodecaborate(1-), [*closo*-CB$_{11}$(CH$_3$)$_{12}$]$^-$ (**11**)[15] and dodecamethyl-*closo*-dodecaborate (2-), [*closo*-B$_{12}$(CH$_3$)$_{12}$]$^{2-}$ (**12**)[13] species are all known.

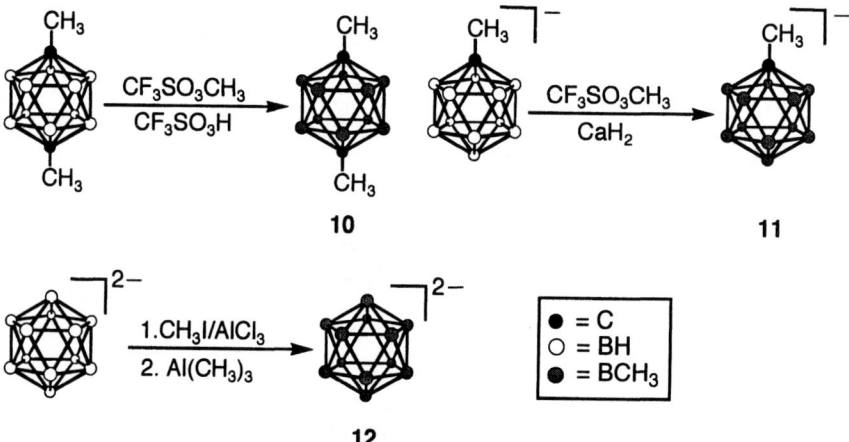

The resulting permethylated species have increased hydrophobicity and van der Waals diameters. These properties are analogous to those of buckminsterfullerene, as shown below, while maintaining the great chemical versatility found in borane chemistry.

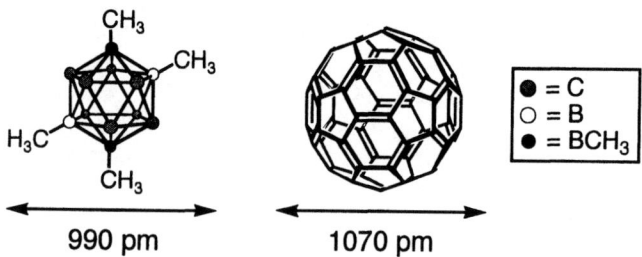

Perhydroxylation of Icosahedral Borane Derivatives

Icosahedral boranes and carboranes may also be hydroxylated to produce $[closo\text{-}B_{12}(OH)_{12}]^{2-}$ (13)[16], $[closo\text{-}CHB_{11}(OH)_{11}]^-$ (14) and $closo\text{-}1,12\text{-}C_2H_2B_{10}(OH)_{10}$ (15)[17] resulting in a hydrophilic species having easily derivatized BOH vertices and aromatic properties.

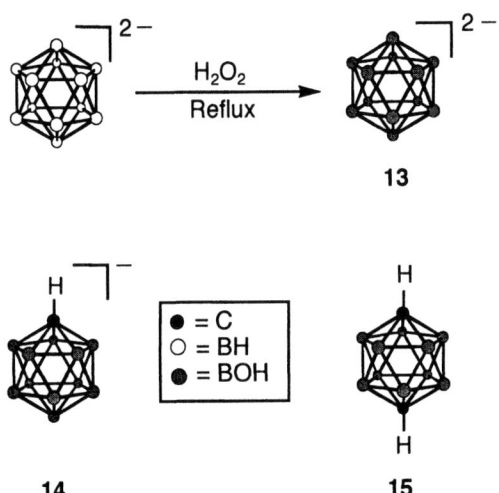

● = C
○ = BH
● = BOH

14   15

Further functionaliztion of $[closo\text{-}B_{12}(OH)_{12}]^{2-}$ was completed to obtain three different linker motifs, the ester[18], ether[19] and carbamate,[20] as shown below. Such derivatives are known as "closomers".

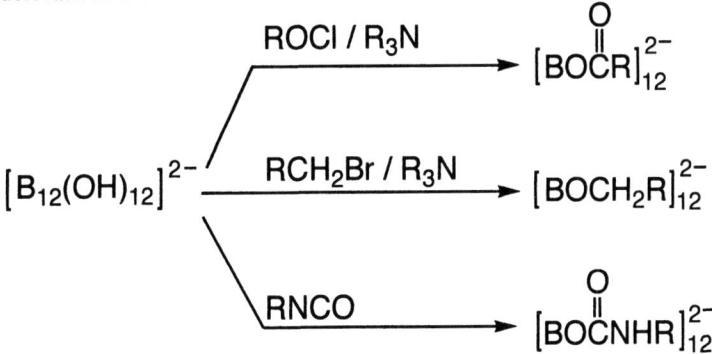

Current Applications of Boranes and Carboranes in Supramolecular Chemistry and Nanoscience

Stable Unilamellar Liposomes for Boron Neutron Capture Therapy

Liposomes, spherical phospholipid bilayer structures, have attracted a great deal of interest as drug delivery vehicles for cancer therapy, including Boron Neutron Capture Therapy (BNCT). Preferential uptake by cancer cells as well as reduced drug toxicity due to drug isolation support this interest. Compounds **16**[21] and **17**[22] have both been incorporated into phospholipids bilayers and used to encapsulate $Na_3[1-(2'-B_{10}H_9)-2-NH_3B_{10}H_8]$ in the aqueous inner core of liposomes with high boron content, a necessity for effective BNCT. Species **17** has been synthesized. This compound has the advantage of not requiring the addition of distearoylphosphocholine to form the liposome bilayer, resulting in a higher overall boron content.

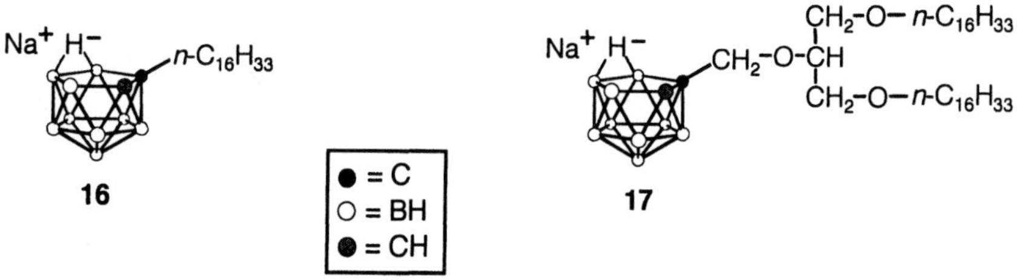

● = C
○ = BH
■ = CH

16: Na^+ H^- carborane-n-$C_{16}H_{33}$

17: Na^+ H^- carborane-CH_2-O-$CH(CH_2$-O-n-$C_{16}H_{33})_2$

Liposomes containing **16**:cholesterol:distearoylphosphocholine in a 1:3:3 molar ratio form 100 nm diameter liposomes which contain 2.5 wt % boron, while liposomes using **17**:cholesterol in a 1:1 molar ratio produce 40 nm liposomes with an increased 8.8 wt. % boron content. This significantly amplifies the efficiency of boron delivery to tumor cells.

Self-Assembly of Rod Structures

It has been found that linear and amphiphilic derivatives of C_{60}, which have a distinct hydrophobicity and hydrophilicity balance, could self-assemble into rod-like structures.[23] Recognizing the importance of the fact that the size, shape and hydrophobicity of permethylated carboranes resemble that of C_{60} and thus they may be regarded as surrogates, it was thought that an analogous permethylated carborane assembly could be formed. Employing the following scheme, rods were formed containing the proposed repeating structure shown below.[24]

317

The stability of the proposed repeating structure is based upon NH---Cl hydrogen bonding as well as hydrophobic interactions between the permethylated carborane cages. As shown by TEM, the rods were approx. 300 nm in diameter and greater than 70 micrometers in length.

Transmission Electron Microscope Image of Rods   *Optical Microscope Image of Rods in Water Suspension*

Carboracycles:Hydrophilic Support of Encapsulated Hydrophobic Species

Interest in the development of new supramolecular chemistry and its use in molecular recognition in solvent systems that range from organic to aqueous led to the development to new water-soluble carborane-containing structures.[25] The hydrophilic outer layer encapsulates a hydrophobic interior capable of solvating a hydrophobic guest.

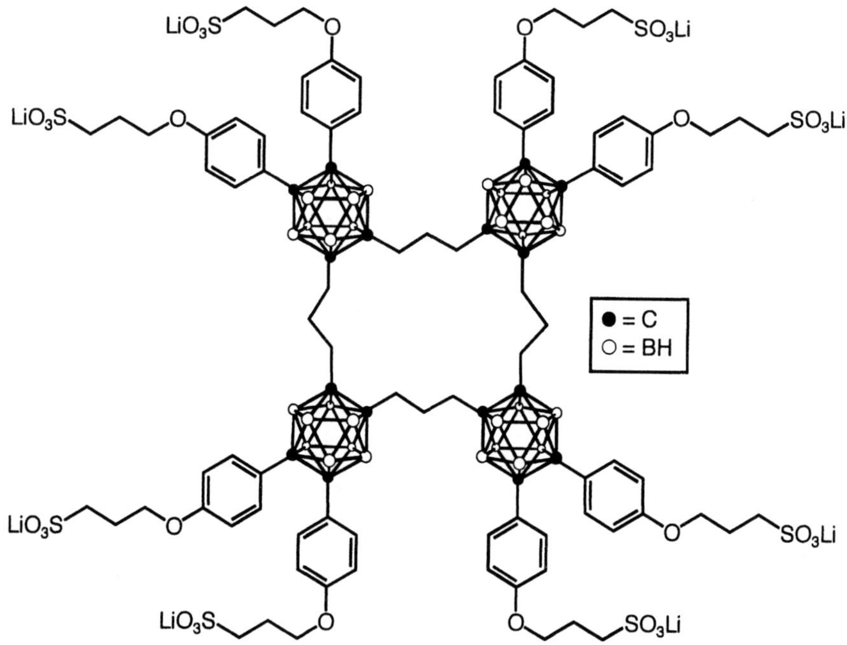

Closomers: A New Motif in Molecular Architecture

Extensive substitution of a *closo*-polyhedral borane or carborane surface by precisely constructed polyatomic moieties forms closomers. Closomers may have several types of substituents, such as branched (dendritic) or linear (oligomeric) chains with or without functional chain substituents.[18, 19]

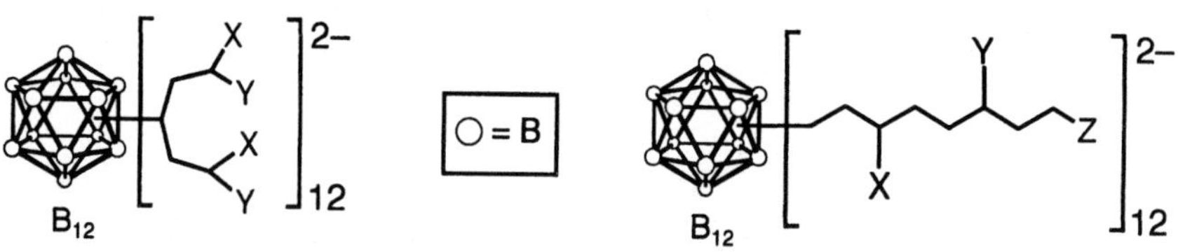

Dendritic Closomer

Oligomeric Closomer

Closomeric structures provide camouflaged, multifunctional modules of variable size, shape, charge, hydrophobicity and other properties designed to accomplish specific functions important to such fields as biomedicine and material science.

Jahn-Teller Distortion Accompanying the Two-Electron Oxidation of [Closo-$B_{12}(OCH_2Ph)_{12}]^{2-}$

Ether-linked closomers are unique in their susceptibility to oxidation, as shown below. The oxidation of the ether closomers have been shown to cause a Jahn-Teller distortion and a change in the point group of the molecule from I_h to D_{3d}.[19]

$[closo\text{-}B_{12}(OCH_2Ph)_{12}]^{2-}$ $\xrightleftharpoons[e^-]{-e^-}$ $[hypercloso\text{-}B_{12}(OCH_2Ph)_{12}]^{\bullet -}$

I_h 26 e$^-$ $\qquad\qquad$ Approx. I_h 25 e$^-$

$[hypercloso\text{-}B_{12}(OCH_2Ph)_{12}]^{\bullet -}$ $\xrightleftharpoons[e^-]{-e^-}$ $[hypercloso\text{-}B_{12}(OCH_2Ph)_{12}]$

Approx. I_h 25 e$^-$ $\qquad\qquad$ D_{3d} 24 e$^-$

An additional feature of ether-linked closomers is the tunability of their redox potentials made possible by changing their cage substituents.[26]

Closomers in Drug Delivery

The ability to conjugate up to twelve copies of pharmaceutical ligands to the icosahedral dodecaborate (2-) cage makes closomers attractive drug delivery vehicles which can result in an increase of drug released at the therapeutic site. Ampicillin, an inexpensive prototype pharmaceutical, has been used in proof of principle studies.[27]

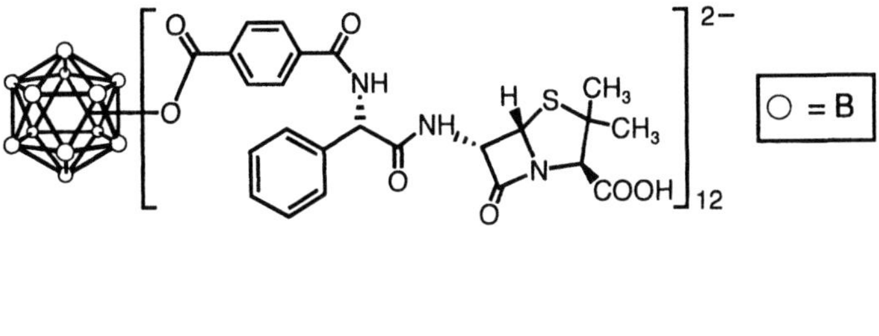

Closomers of High Boron Content for Use in Boron Neutron Capture Therapy

Derivatives of the icosahedral dodecaborate (2-) core surrounded by twelve pendant carborane groups have been synthesized using ester and ether linkages.[28,24] When the associated pendant carborane groups are degraded to the corresponding anionic *nido* species, water solubility greatly increases. The resulting boron-rich and water-soluble molecules are very attractive BNCT target candidates.

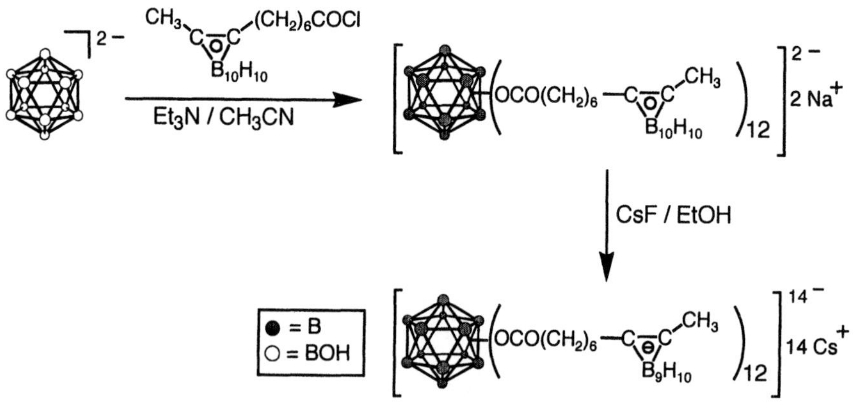

Vertex Differentiation for the Targeting of Specific Sites

The enhancement of the number of contrast agents per cell is desirable for greater image resolution in MRI.[29] One way to achieve this is through directed substitution of cell-targeting ligands on the icosahedral dodecaborate cage. This allows the precise control of size, shape and placement of functional groups, which is desirable for many life science applications. The reaction scheme shown below will generate a unique molecule that contains two different functional motifs in the ratio of 1:10.[30] This could lead to specific cell targeting of large payloads for MRI, drug delivery, radioimaging, chemotherapy of cancer and many other applications.

Water-soluble Carborane Scaffold for Closomer Chemistry

The exploratory study of *closo*-B-perhydroxylated-p-carborane shown below demonstrates that its C-H vertices can be utilized as platforms for substituents which control its solubility and reactivity. By placing sulfonate groups at the 1- and 12- positions of water-insoluble **18**, the methanol-soluble species **19** was formed and successfully converted to its water-soluble

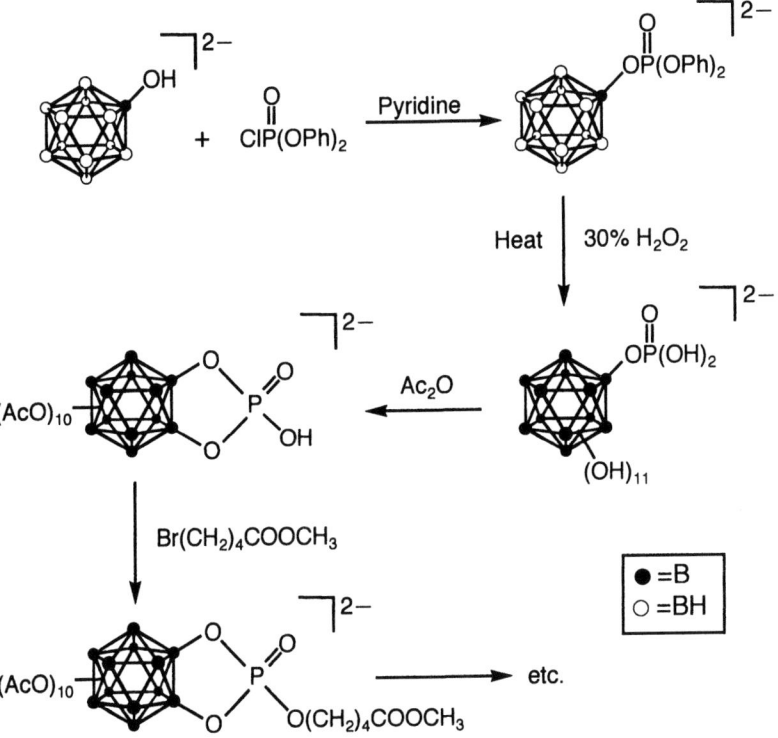

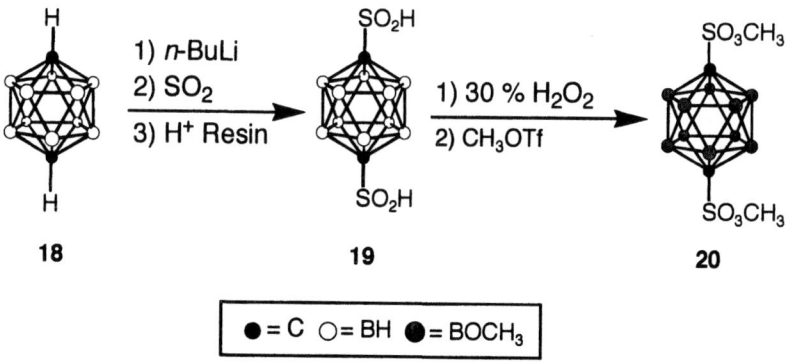

perhydroxylated derivative **20**.[31] The resulting species presents the advantages of facile derivitization of the BOH vertices and two distinguishable reactive sites in a hypothetical extended closomer structure.

A Rotary Molecular Motor Based on a Unique Nickelacarborane

The basis for the molecular device shown here has been known since the early 1970s. The following diagram shows the d^7 Ni(III) and d^6 Ni(IV) *commo*-bis-7,8-dicarbollyl metallacarboranes. The interconversion between these two geometries occurs when the oxidation state of nickel is changed and provides the basis for controlling the movements of this nanodevice.[32,33]

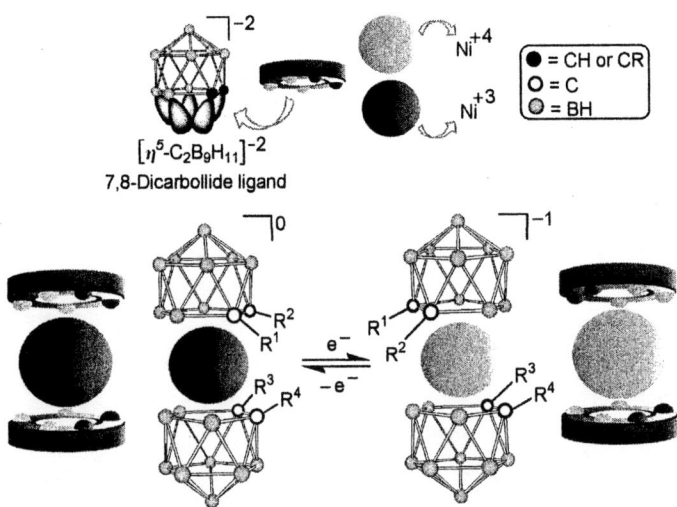

(Reproduced with permission from Reference 33. Copyright 2004 American Association for the Advancement for the Advancement of Science)

This rotation could allow for the control of nanovalves and switches, reversible exposure of catalytic sites and other applications where selective surface exposure is necessary.

References:

1. Hawthorne, M. F., Proceedings of the Ninth International Meeting on Boron Chemistry, Advances in Boron Chemistry, Special Publication No. 201, the Royal Society of Chemistry, **1996**, 261
2. Hawthorne, M. F., Proceedings of the Tenth International Conference on Boron Chemistry, Contemporary Boron Chemistry, the Royal Society of Chemistry, **2000**, 197
3. Köhler, T., Ilgner, J., Mietke, S. *Vakuum in Forschung und Praxis* **2003**, *15*, 292-297
4. Pitochelli, A. R., Hawthorne, M. F. *J. Am. Chem. Soc.* **1960**, *82*, 3228.
5. Hawthorne, M. F. *J. Am. Chem. Soc.* **1959**, *81*, 5836.
6. Green, J., Mayes, N., US Patent 3265726, **1966**.
7. Grimes, R. N., *Carboranes*; Academic Press: New York, **1970**
8. Valliant, J. F., Guenther, K. J., King, A. S., Morel, P., Schaffer, P., Sogbein, O. O., Stephenson, K. A., *Coor. Chem. Rev.* **2002**, *232*, 173-230.
9. King, B. T., Zharov, I, Michl, J. *Chem. Innov.* **2001**, *31(12)*, 23-31.
10. Stasko, D., Reed, C. A. *J. Am. Chem. Soc.* **2002**, *124(7)*, 1148-1149.
11. Hawthorne, M. F., Young, D. C., Wegner, P. A. *J. Am. Chem. Soc.* **1965**, *87*, 1818.
12. Hosmane, N. S., Saxena, A. K. *Chem. Rev.* **1993**, *93*, 1081-1124.
13. Peymann, T., Knobler, C. B., Khan, S. I., Hawthorne, M. F. *Inorg. Chem.* **2001**, *40*, 1291.
14. Jiang, W., Knobler, C. B., Curtis, C. E., Mortimer, M., Hawthorne, M. F. *Angew. Chem.* **1995**, *34*, 1332.
15. King, B. T., Janousek, Z., Gruener, B., Trammell, M., Noll, B. C., Michl, J. *J. Am. Chem. Soc.* **1996**, *118(13)*, 3313-14.
16. Bayer, M. J., Hawthorne, M. F. *Inorg. Chem.* **2004**. *43*, 2018-2020.
17. Peymann, T., Herzog, A., Knobler, C. B., Hawthorne, M. F. *Angew. Chem.* **1999**, *111*, 1129-1132; *Angew. Chem., Int. Ed.* **1999**, *38*, 1062-1064.
18. Maderna, A., Knobler, C. B., Hawthorne, M. F. *Angew. Chem. Int. Ed.* **2001**, *40*, 1662-1664.
19. Peymann, T., Knobler, C. B., Khan, S. I., Hawthorne, M. F. *Angew. Chem. Int. Ed.* **2001**, *40*, 1664-1667.
20. Li, T., Hawthorne, M. F. **2004** Unpublished Results
21. Shelly, K., Feakes, D. A., Hawthorne, M. F., Schmidt, P. G., Krisch, T. A., Bauer, W. F. *Proc. Natl. Acad. Sciences.* **1992**, *89(19)*, 9039-43.

22. Li, T., Thomas, J., Hawthorne, M. F., *Abstracts of Papers*, 225th National Meeting of the American Chemical Society, New Orleans, LA, March 23-27, **2003**; American Chemical Society: Washington, DC
23. Cassell, A. M., Asplund, C. L., Tour, J. M. *Angew. Chem. Int. Ed.* **1999**, *38*, 2403-2405.
24. Ma, L., Hawthorne, M. F. **2004** Unpublished Results
25. Bayer, M. J., Herzog, A., Diaz, M., Harakas, G. A., Lee, H., Knobler, C. B., Hawthorne, M. F. *Chem. Eur. J.* **2003**, *9*, 2732-2744.
26. Farha, O. K., Hawthorne, M. F. **2004**, Unpublished Results
27. Jalisatgi, S. S., Hawthorne, M. F. **2004**, Unpublished Results
28. Thomas, J., Hawthorne, M. F. *Chem. Commun.* **2001**, 1884-1885.
29. Artemov, D. *J. Cell. Mol. Med.* **2003**, *90*, 518-524.
30. Bayer, M. J., Hawthorne, M. F. **2004**, Unpublished Results
31. Herzog, A., Knobler, C. B., Hawthorne, M. F. *J. Am. Chem. Soc.* **2001**, *123*, 12791-12797.
32. Warren, Jr., L. F., Hawthorne, M. F. *J. Am. Chem. Soc.* **1970**, *92*. 1157-1173.
33. Hawthorne, M. F., Zink, J. I., Skelton, J. M., Bayer, M. J., Liu, C., Livshits, E., Baer, R., Neuhauser, D. *Science*, **2004**, 1849-1851.

Chapter 23

Synthesis and Reactivity of (Silylamino)- and (Silylanilino)phosphines

Robert H. Neilson, Pradeep Devulapalli, Bethany K. Jackson,
Andrew R. Neilson, Sahrah Parveen, and Bin Wang

Department of Chemistry, Texas Christian University,
Fort Worth, TX 76129

Phosphorus compounds that contain silicon-nitrogen functional groups have a very rich derivative chemistry. The wide variety of reactions that can occur at phosphorus in combination with facile cleavage of the Si-N bond makes them useful precursors to many types of cyclic and polymeric P-N systems. We report here on (a) the synthesis and reactivity of (silyl*amino*)phosphines, including some new systems that contain 4-aryl functional groups on phosphorus, and (b) the synthesis and characterization of two new types of (silyl*anilino*)phosphines that contain either a 4-aryl functional group on nitrogen or a phenylene ring between nitrogen and phosphorus. These anilino derivatives are potential precursors to poly(phenylenephosphazenes), a new class of inorganic-organic hybrid polymers in which phosphazene ($R_2P=N$) and phenylene (C_6H_4) units alternate along the backbone.

The preparative chemistry of compounds containing the Si-N-P linkage has been extensively developed over the last three decades (*1*). These studies have led to a number of synthetically useful reagents and methods in phosphorus-nitrogen chemistry. Depending on the types of substituents present on the three main group elements, and on the oxidation state and coordination number at phosphorus, such compounds are useful precursors to cyclic and polymeric poly(phosphazenes), $[R_2P=N]_n$, as well as various low-coordinate phosphorus systems containing P=N or P=C (p-p)π bonds.

In this paper, we will (a) review the chronological development of an important series of Si-N-P compounds, the (silylamino)phosphines, and (b) briefly summarize our recent results in this area including the synthesis of some new 4-aryl functionalized derivatives as well as some related (silylanilino)phosphines.

(Silylamino)phosphines

As reported by Cowley et al (*2*), some of the first (silylamino)phosphines, were prepared by treating phosphorus(III) halides with lithium bis(trimethylsilyl)amide (eq 1). At that time, these novel products were mainly of interest in relation to certain bonding and structure questions, including the determination of P-N bond rotation barriers by dynamic NMR measurements.

$$(Me_3Si)_2NLi \xrightarrow{PX_3 \text{ or } R_2PCl} \begin{array}{c} Me_3Si \\ \diagdown \\ N-P \\ \diagup \diagdown \\ Me_3Si R \end{array} \begin{array}{c} R \\ \\ \\ R \end{array} \quad (1)$$

$$R = F, Cl, CF_3, Me$$

For the preparation of organo-substituted derivatives (e.g., R = alkyl), however, this method is severely limited by the practical difficulties in preparing, storing, and handling large quantities of hazardous reagents like Me_2PCl. Thus, we were led to search for alternative synthetic routes to these systems.

The most useful synthetic entry to Si-N-P systems is the Wilburn method (*3*) (eqs 2 - 4) in which phosphorus halides are successively treated with silylamides and organometallic reagents. This is generally a convenient, one-pot synthesis of a wide variety of both symmetrical (**1**) and unsymmetrical (**2** and **3**) diorgano substituted (silylamino)phosphines. In cases with R groups more sterically demanding than methyl or ethyl, it is possible to isolate the mono-substituted chlorophosphines (**4**) (*4*).

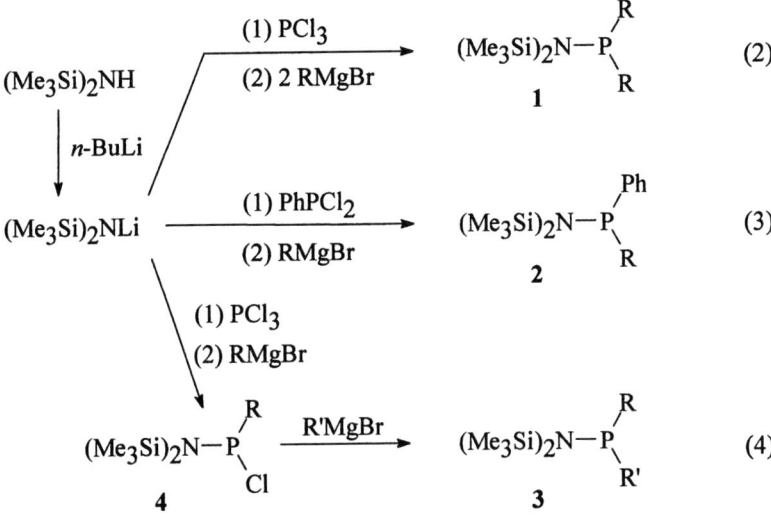

The (silylamino)phosphines (**1 - 3**) are then routinely converted to N-silylphosphoranimines (**5**) (eq 5) by sequential oxidative halogenation and substitution reactions (*5*). Subsequent thermolysis of these N-silylphosphoranimines (**5**) readily affords poly(alkyl/arylphosphazenes) (**6**) (eq 6) via a condensation polymerization process (*6*). This general method compliments other synthetic routes to poly(phosphazenes) (*7*) and significantly extends the overall scope of phosphazene chemistry.

R, R' = alkyl/aryl; R" = CH_2CF_3, Ph

The Wilburn method has proven to be very general over the years and variations of it have been used to prepare numerous types of Si-N-P derivatives, a few of which are reviewed here. For example, the use of bulky organolithium reagents, followed by dehydrohalogenation (eq 7), is a simple route to stable, 2-coordinate phosphines (**7**) containing P=C double bonds (8).

$$(Me_3Si)_2NLi \xrightarrow{PCl_3} \left[(Me_3Si)_2N-P\begin{matrix}Cl\\Cl\end{matrix} \right]$$

$$\downarrow Me_3SiC(E)H-Li \qquad (7)$$

$$\begin{matrix}Me_3Si\\Me_3Si\end{matrix}N-P\begin{matrix}\\C-SiMe_3\\E\end{matrix} \xleftarrow[-HCl]{base} \begin{matrix}Me_3Si\\Me_3Si\end{matrix}N-P\begin{matrix}Cl\\C\begin{matrix}H\\SiMe_3\end{matrix}\\E\end{matrix}$$

7: E = H, SiMe$_3$

By using vinyl Grignard reagents or ethynyl lithium reagents in the Wilburn synthesis, (silylamino)phosphines containing C=C (**8**) or C≡C (**9**) substituents have also been prepared and characterized (9). Subsequently, these compounds are readily converted to phosphazene precursors as shown in equations 5 and 6 above. Somewhat surprisingly, no adverse side reactions involving the multiple bonds are observed in such derivative chemistry.

$$(Me_3Si)_2N-P\begin{matrix}CH=CH_2\\R\end{matrix} \qquad (Me_3Si)_2N-P\begin{matrix}C\equiv C-R\\R'\end{matrix}$$

8: R = Me, Ph, CH=CH$_2$

9: R = SiMe$_3$, n-Bu, CH$_2$OCH$_3$
R' = Ph, i-Pr, CH$_2$SiMe$_3$, C≡CR

Other important variations of the Wilburn method include the preparation of fluorocarbon derivatives such as the trifluoromethyl compounds (**10**) and the trifluorovinyl ether substituted systems (**11**). Synthesis of the former (10) employed the novel CF$_3$ transfer reagent, [(Et$_2$N)$_3$PBr]$^+$(CF$_3$)$^-$, first reported by Ruppert (11), to afford the direct P-CF$_3$ linkage. In the latter case, the trifluorovinyl ether group was introduced using an organolithium reagent obtained from the p-bromo-trifluorovinyl aryl ether, CF$_2$=CFO-C$_6$H$_4$-Br, via metal-halogen exchange (12).

(Me₃Si)₂N—P(R)(CF₃)

10: R = Ph, n-Pr

(Me₃Si)₂N—P(R)(C₆H₄-O-CF=CF₂)

11: R = Cl, Br, Me

The substituents on silicon can also be varied in the Wilburn synthesis of (silylamino)phosphines. For example, by using unsymmetrical disilylamine reagents, t-BuR$_2$SiN(H)SiMe$_3$ (R = Me, Ph), in the first step, we have recently prepared a series of (silylamino)phosphines (**12**) containing sterically bulky silyl groups on nitrogen (*13*). Subsequent derivative chemistry (e.g., bromination as in equation 5) involved selective cleavage of the smaller Me$_3$Si group from nitrogen. Moreover, compounds containing the t-BuPh$_2$Si group are crystalline solids and several of them have been structurally characterized by single-crystal X-ray diffraction, e.g., the dimethylphosphine derivative (**12**: R = Ph; R' = R" = Me) shown in Figure 1 (*13*). Interestingly, both the P-N bond lengths and the R-P-R angles decrease as the electron withdrawing effect of the phosphorus substituents increases (CH$_3$ < Cl < F).

t-BuR₂Si, Me₃Si — N—P(R')(R")

12: R = Me, Ph
R' = R" = Me, F
R' = Ph, R" = Cl

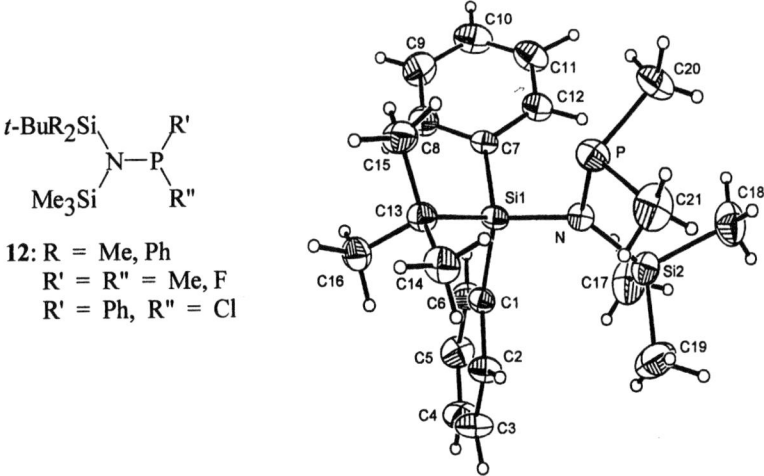

Figure 1. ORTEP representation of compound 12 (R = Ph; R' = R" = Me). Thermal ellipsoids are drawn at the 50% probability level. (Reproduced from reference 13. Copyright 2002 American Chemical Society.)

In a more recent modification of the Wilburn method, treatment of the intermediate dichloro(disilylamino)phosphine with 4-aryl substituted Grignard reagents gave the expected 4-aryl(chloro)phosphines (eq 8). Subsequent addition of MeMgBr afforded the desired alkyl/arylphosphines (**13**) in high yields as distillable liquids (*14*).

$$\begin{array}{c} (Me_3Si)_2N\text{-}PCl_2 \xrightarrow{XC_6H_4MgBr} (Me_3Si)_2N\text{-}P(Cl)(C_6H_4X) \xrightarrow{MeMgBr} (Me_3Si)_2N\text{-}P(Me)(C_6H_4X) \end{array} \quad (8)$$

13: X = Cl, Br, OMe, CF$_3$

By employing synthetic methodology described above (eqs 5 and 6), the (silylamino)phosphines (**13**) can be readily converted to a variety of new N-silylphosphoranimines (**5**) and poly(phosphazenes) (**6**) that contain 4-aryl functional groups. Moreover, as reported recently for some related systems (*15*), treatment of the N-silylphosphoranimines derived from these phosphines (**13**) with alcohols (e.g., CF$_3$CH$_2$OH) affords the corresponding cyclic trimers [4-X-C$_6$H$_4$(Me)P=N]$_3$ as mixtures of *cis* and *trans* isomers. All of the (silylamino)phosphines (**13**), as well as the polymeric and cyclic phosphazenes obtained from them, have been fully characterized by NMR spectroscopy and elemental analysis. Further details of the synthesis, characterization, and reactivity of these novel 4-aryl derivatives will be reported elsewhere.

(Silylanilino)phosphines

We are also currently interested in the possibility of preparing new types of inorganic-organic hybrid polymers including phenylenephosphazenes (**14**). Although currently unknown, such systems are likely to combine the varied and useful chemical and physical properties of phosphazene polymers, [R$_2$P=N]$_n$, with the electronic properties of poly(phenylene), [-C$_6$H$_4$-]$_n$.

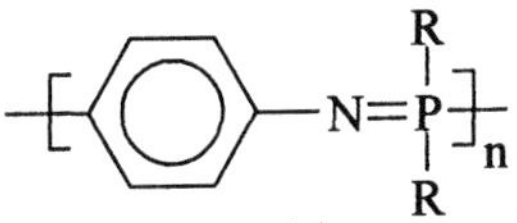

14

By analogy to our condensation polymerization route to poly(phosphazenes), we are studying various phosphorus compounds that contain reactive Si-N functional groups as possible starting materials for the eventual synthesis of polymers like **14**. Specifically, two types of (silyl*anilino*)phosphines in which the desired phenylene moiety is initially incorporated either on nitrogen (**15**) or between nitrogen and phosphorus (**16**).

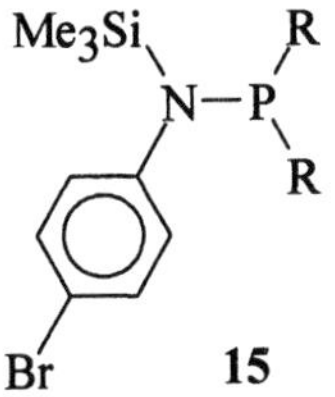

15 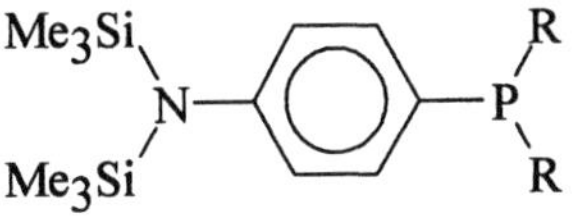   **16**

The first examples of both of these types of compounds are accessible from the N-silyl derivatives of *p*-bromoaniline. Dehydrohalogenation in the presence of triethylamine affords the mono-silyl derivative **17** (eq 9) that can be easily converted to the disilylaniline derivative **18** (eq 10).

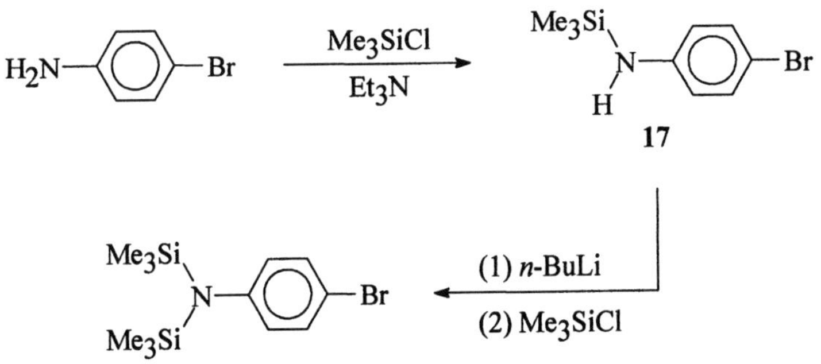

(9)

(10)

The mono-silylaniline **17** can be used as the starting silylamine in the Wilburn synthesis and, thus, is readily converted to the corresponding (silylanilino)phosphine (**19**) in ca. 60 % yield. The scope of this reaction, with the possibility of preparing many other compounds analogous to **19**, is currently being investigated.

$$\underset{17}{\underset{\text{Br}}{\text{Me}_3\text{Si}}\text{N-H}} \xrightarrow[\substack{(2)\ \text{PhPCl}_2 \\ (3)\ \text{MeMgBr}}]{(1)\ n\text{-BuLi}} \underset{19}{\underset{\text{Br}}{\text{Me}_3\text{Si}}\text{N-P}\underset{\text{Me}}{\overset{\text{Ph}}{}}} \quad (11)$$

Since it does not bear a reactive N-H group, the disilylaniline compound **18** is found to undergo facile metal-halogen exchange upon treatment with *n*-BuLi (eq 12) under mild conditions, producing the organo-lithium reagent in solution. Subsequent addition of various mono-chlorophosphines (eq 13) affords the novel disilylanilinophosphines **20** in moderate to high yields as thermally stable, distillable liquids. This process, however, is somewhat limited by the difficulties in preparing, storing, and/or handling large quantities of the necessary chlorophosphines. Consequently, we are also exploring alternative synthetic routes to phosphorus reagents such as Ph(Me)PCl.

$$\underset{18}{(\text{Me}_3\text{Si})_2\text{N-C}_6\text{H}_4\text{-Br}} \xrightarrow[\substack{\text{Et}_2\text{O} \\ 0\ °\text{C}}]{n\text{-BuLi}} \left\{ (\text{Me}_3\text{Si})_2\text{N-C}_6\text{H}_4\text{-Li} \right\} \quad (12)$$

$$(\text{Me}_3\text{Si})_2\text{N-C}_6\text{H}_4\text{-P}\underset{\text{R'}}{\overset{\text{R}}{}} \xleftarrow{\text{Cl-P}\underset{\text{R'}}{\overset{\text{R}}{}}} \quad (13)$$

20: R = R' = OCH$_2$CF$_3$, NMe$_2$, Ph
R = Ph, R' = OCH$_2$CF$_3$, Me

The new (silylanilino)phosphines **19** and **20** were obtained in moderate to good yields as colorless, distillable liquids and were fully characterized by multinuclear NMR spectroscopy and elemental analysis. Depending on the substituents at phosphorus, their P^{31} NMR chemical shifts range from 168 ppm (R = R' = OCH$_2$CF$_3$) to - 6.4 ppm (R = R' = Ph).

We are also currently investigating the derivative chemistry of (silylanilino)phosphines such as **19** and **20**, including various oxidative halogenation reactions. The overall goal of this work is to prepare appropriate phosphorus(V) derivatives that may serve as useful precursors to the first examples of the new phenylenephosphazene polymers (14). Consequently, full details of the synthesis, characterization, and reactivity of these novel (silylanilino)phosphine derivatives will be reported in due course.

Acknowledgments

This paper is dedicated to Professor Alan H. Cowley on the occasion of his 70th birthday and to the memory of the late Dr. James C. Wilburn for his pioneering work in Si-N-P chemistry in the labs of both RHN and AHC. The authors thank the Robert A. Welch Foundation and the TCU Research Fund for generous financial support of this work.

References

1. Neilson, R. H. in *Encyclopedia of Inorganic Chemistry, Vol 6*, 3180; King, R. B., Ed.; John Wiley & Sons; Chichester, England, **1994**.
2. (a) Neilson, R. H.; Lee, R. C.-Y.; Cowley, A. H. *Inorg. Chem.* **1977**, *16*, 1455. (b) Neilson, R. H.; Lee, R. C.-Y.; Cowley, A. H. *J. Am. Chem. Soc.* **1975**, *97*, 5302.
3. (a) Wilburn, J. C.; Neilson, R. H. *Inorg. Chem.* **1977**, *16*, 2519. (b) Neilson, R. H.; Wisian-Neilson, P. *Inorg. Chem.* **1982**, *21*, 3568.
4. See for example: (a) O'Neal, H. R.; Neilson, R. H. *Inorg. Chem.* **1983**, *22*, 814. (b) Klaehn, J. R.; Neilson, R. H. *Inorg. Chem.* **2002**, *41*, 5859.
5. (a) Wisian-Neilson, P.; Neilson, R. H. *Inorg. Chem.* **1980**, *19*, 1875. (b) Wisian-Neilson, P.; Ford, R. R.; Goodman, M. A.; Li, B.-L.; Roy, A. K.; Wettermark, U. G.; Neilson, R. H. *Inorg. Chem.* **1984**, *23*, 2063. (c) Ford, R. R.; Neilson, R. H. *Polyhedron* **1986**, *5*, 643.
6. (a) Wisian-Neilson, P.; Neilson, R. H. *J. Am. Chem. Soc.* **1980**, *102*, 2848. (b) Neilson, R. H.; Hani, R.; Wisian-Neilson, P.; Meister, J. J.; Roy, A. K.; Hagnauer, G. L. *Macromolecules*, **1987**, *20*, 910. (c) Neilson, R. H.; Wisian-Neilson, P. *Chem. Rev.* **1988**, *88*, 541. (d) Wisian-Neilson, P.; Neilson, R. H. *Inorg. Synth.* **1989**, *25*, 69.
7. (a) Allcock, H. R. *Phosphorus-Nitrogen Compounds*; Academic: New York, **1972**. (b) Mark, J. E.; Allcock, H. R.; West, R. *Inorganic Polymers*; Prentice Hall: New Jersey, **1992**. (c) Wisian-Neilson, P. in *Encyclopedia of Inorganic Chemistry, Vol. 6*, 3371; King, R. B., Ed.; John Wiley & Sons: Chichester, England, **1994**. (d) Allcock, H. R. *Chem. Eng. News* **1985**, *63(11)*, 22. (e) Allcock, H. R.; Nelson, J. M.; Reeves, S. D.; Honeyman, C. H.; Manners, I. *Macromolecules* **1997**, *30*, 50. (f) Matyjaszewski, K.;

Dauth, J.; Montague, R. A.; Reddick, C.; White, M. L. *J. Am. Chem. Soc.* **1990**, *112*, 6721. (g) D'Halluin, G.; De Jaeger, R.; Chambretie, J. P.; Potin, P. *Macromolecules* **1992**, *25*, 1254.
8. (a) Neilson, R. H. *Inorg. Chem.* **1981**, *20*, 1679. (b) Thoma, R. J.; Prieto, C. A.; Neilson, R. H. *Inorg. Chem.* **1988**, *27*, 784.
9. (a) Waters, K. R.; Neilson, R. H. *Phosphorus and Sulfur*, **1988**, *39*, 189. (b) Neilson, R. H.; Kucera, W. R. *J. Organometallic Chem.* **2002**, *646*, 223.
10. Karthikeyan, S.; Neilson, R. H. *Inorg. Chem.* **1999**, *38*, 2079.
11. Ruppert, I.; Schlich, K.; Volbach, W. *Tetrahedron Lett.* **1984**, *25*, 2195.
12. Ji, J.; Narayan-Sarathy, S.; Neilson, R. H.; Oxley, J. D.; Babb, D.A.; Rondan, N. G.; Smith, D. W., Jr. *Organometallics* **1998**, *17*, 783.
13. Samuel, R. C.; Kashyap, R. P.; Krawiec, M.; Watson, W. H.; Neilson, R. H. *Inorg. Chem.* **2002**, 7113.
14. (a) Wang, B.; Klaehn, J. R.; Neilson, R. H. *Phosphorus, Sulfur, and Silicon*, **2004**, *179*, 821. (b) B. Wang, Ph. D. Dissertation, Texas Christian University (2000).
15. Wisian-Neilson, P.; Johnson, R. S.; Zhang, H.; Jung, J.-H.; Neilson, R. H.; Ji, J.; Watson, W.; Krawiec, M. *Inorg. Chem.* **2002**, *41*, 4775.

Chapter 24

Cyclic and Polymeric Alkyl/Aryl Phosphazenes

Patty Wisian-Neilson, June-Ho Jung, and Srinagesh K. Potluri

Department of Chemistry, Southern Methodist University, Dallas, TX 75275

Main group compounds containing the Si-N-P linkage are vital precursors to both cyclic and polymeric phosphazenes containing P-C bonded substituents, including those with nongeminal substitution at phosphorus, e.g., [Me(Ph)PN]$_n$. Modification of both the polymers and the cyclics to incorporate a variety of functional groups affords a diverse set of new polymers and basket-shaped cyclic molecules. When heated, the cyclic trimers [Me(Ph)PN]$_3$ exist in equilibrium with all four isomers of the nongeminal tetramers [Me(Ph)PN]$_4$. All six isomers have been characterized by X-ray crystallography. Several of these cyclic and polymeric phosphazenes were found to stabilize metal nanoparticles.

Introduction

Phosphazenes are main group compounds in which alternating phosphorus and nitrogen atoms form either cyclic or polymeric structures. These P(V) compounds also have two substituents at each phosphorus, which greatly influence the properties of either the ring or polymer. In fact, literally hundreds of cyclic and polymeric compounds have been made. (*1, 2*) Most have been prepared by nucleophilic substitution reactions on either the trimer, [Cl$_2$PN]$_3$, or polymer, [Cl$_2$PN]$_n$, the latter of which is usually prepared by thermal ring opening of the trimer. (*1*) There are, however, some limitations to the nucleophilic substitution method. In the case of cyclic compounds there is a strong tendency for geminal substitution which precludes easy access to nongeminal geometric isomers of the cyclic products. For the polymers, nucleophilic substitution of the preformed dichloro polymer has successfully produced literally hundreds of amino, alkoxy, and aryloxy substituted compounds, but the process has failed to give high molecular-weight, fully substituted *alkyl* or *aryl* substituted polymers. (*2*)

In 1980, the first high molecular weight, di*alkyl* phosphazene polymer, [Me$_2$PN]$_n$, was prepared by condensation polymerization, which, at that time, was an unusual approach to any inorganic polymer system. (*3*) This process involved the loss of a substituted silane from an N-silylphosphoranimine (Scheme 1, *a*) and has been applied to the synthesis of a variety of polymers that include, **1** and **3**, (*2*, *4*) copolymers, **2**, (*5*) and a novel poly(phospholenazene), **4**, (*6*) all of which contain only P-C bonded substituents at phosphorus (Chart 1). Since then, other condensation polymerization methods have been used to prepare a variety of polyphosphazenes, including poly(dichlorophosphazene), [Cl$_2$PN]$_n$. (*7, 8, 9)*

Scheme 1

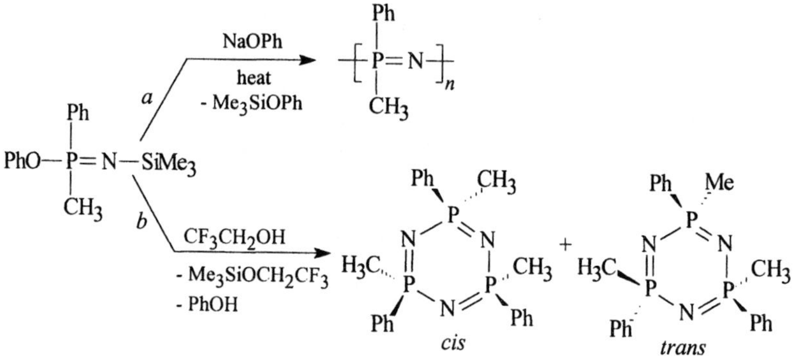

Chart 1

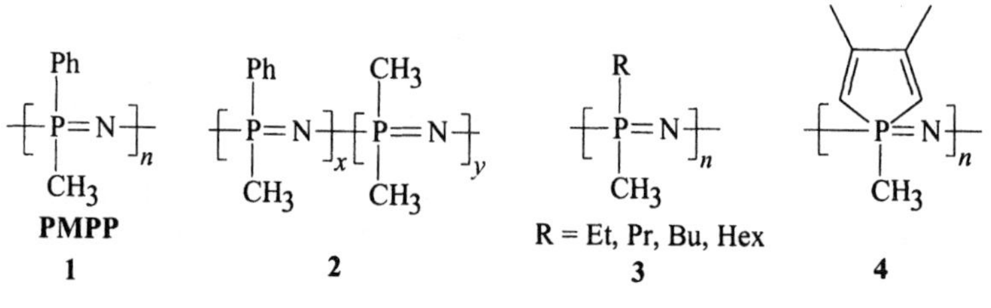

More recently, the same N-silylphosphoranimine polymer precursors have also been used to prepare nongeminally substituted cyclic phosphazenes (Scheme 1, *b*), (*10*) as discussed later in this paper.

Polymeric Phosphazenes

Poly(methylphenylphosphazene), PMPP, $[Me(Ph)PN]_n$, is the best studied of the poly(alky/arylphosphazenes) that are formed by condensation polymerization. It is not only soluble in many common organic solvents, but also possesses three sites of reactivity that affect this polymer's properties and its derivative chemistry.

Lewis acid-base reactions

electrophilic aromatic substitution

deprotonation-substitution

PMPP

First, a portion of the phenyl groups undergo at least one type of electrophilic aromatic substitution reaction, i.e., nitration under rigorous reaction conditions consisting of nitric and sulfuric acid to give the nitrated polymer **5** where R = NO_2. Many other aromatic substitution reactions are hindered by the deactivating effect of the PN backbone and by complications involving other reactive sites. Subsequent reduction converts the nitro group to an NH_2 group that can then be converted to amides. Purification of the nitrated polymers was difficult due to the high basicity at the lone pair on the nitrogen in the polymer backbone. (*11*) This second reactive site is readily protonated and forms strong complexes with many metals. In one study, silver and lithium cations were found to coordinate with the backbone nitrogen, **6**, causing decreased solubility in many organic solvents and increased glass transition temperatures, T_g. (*12*)

5
R = NO_2, NH_2, N=NAr, N(H)C(=O)R

6
R = Me, Ph
M^+ = Ag^+, Li^+

The third, and most useful reactive site in PMPP, is the methyl group which is readily deprotonated to form a polymeric anion intermediate, **7**, (eq 1) that reacts with electrophiles. In this manner, a variety of functional groups have been attached to the phosphazene backbone via a P-C linkage, e.g., E = Me, $CH_2CH=CH_2$, $SiMe_3$, $SiMe_2H$, $SiMe_2(CH_2)_3CN$, $SiMe_2CH=CH_2$, $SiMe_2(CH_2)_nCH_3$, (*13*) $SiMe_2(CH_2)_2CF_2)_nCF_3$, (*14*) C(OH)RR', (*15, 16*) C(OOCR")RR', (*17*) C(=O)R, (*18*) COOH, COOLi, (*19*) SR, and PR_2. The R groups in these functionalities include simple alkyl and aryl groups, thiophene, and ferrocene. In addition to the fact that many of the new functional groups provide sites for further reactions, they also impart a variety of different properties. For example, the surface hydrophobicity of the polymers is increased by incorporation of long alkyl groups as noted by higher contact angles (e.g., 90 to 100 ° versus ca. 73 ° for the parent polymer PMPP) and further increased if these side chains are changed to fluoroalkyl groups (contact angle θ >110 °) (*14, 15*). The gas permeabilities are also enhanced by the incorporation of simple silyl groups as well as fluoroalkyl silyl groups (*20*). Finally, the T_gs decreased with increasing length of the side-groups until the chains exceeded about 10 to 12 carbons. Deprotonation-substitution reactions have also been used to modify dialkyl polymers, [(*n*-hexyl)(Me)PN]$_n$ and [(*n*-butyl)(Me)PN]$_n$ (*21*) resulting in a series of polymers with T_gs below room temperature.

$$\underset{\mathbf{1}}{\left[\begin{array}{c}\text{Ph}\\\text{P}=\text{N}\\\text{CH}_3\end{array}\right]_n} \xrightarrow{n\text{-BuLi}}_{\text{THF}} \underset{\mathbf{7}}{\left[\begin{array}{c}\text{Ph}\\\text{P}=\text{N}\\\text{CH}_3\end{array}\right]_x \left[\begin{array}{c}\text{Ph}\\\text{P}=\text{N}\\\text{CH}_2^-\text{Li}^+\end{array}\right]_y} \quad (1)$$

$$\downarrow \text{electrophile}$$

$$\left[\begin{array}{c}\text{Ph}\\\text{P}=\text{N}\\\text{CH}_3\end{array}\right]_x \left[\begin{array}{c}\text{Ph}\\\text{P}=\text{N}\\\text{CH}_2\text{E}\end{array}\right]_y$$

The polymer anion **7** has also been used to initiate anionic addition polymerization of styrene and ring opening polymerization of hexamethylcyclotrisiloxane, $(Me_2SiO)_3$, to give the grafted copolymers, **6** and **7**. (*22, 23*). More recent efforts to combine the properties of polyphosphazenes and other organic and inorganic polymers via grafting have involved atom transfer radical polymerization (ATRP) which requires initial incorporation of a bromoalkyl group to initiate polymerization. (*24*) Anionic ring opening of lactones, ferrocenylsilane, and silacyclobutane has also been investigated. (*25*) In addition, PMMA grafts were prepared by anionic polymerization of methyl methacrylate after modification of **7** with diphenylethylene to give a sterically hindered polymer anion. (26)

$$\left[\begin{array}{c} Ph \\ | \\ P=N \\ | \\ CH_3 \end{array} \right]_x \left[\begin{array}{c} Ph \\ | \\ P=N \\ | \\ CH_2 \\ | \\ [CH_2-CH]_z-H \\ | \\ Ph \end{array} \right]_y \qquad \left[\begin{array}{c} Ph \\ | \\ P=N \\ | \\ CH_3 \end{array} \right]_x \left[\begin{array}{c} Ph \\ | \\ P=N \\ | \\ CH_2 \\ | \\ [Me_2Si-O]_z-SiMe_3 \end{array} \right]_y$$

8   **9**

More recently, PMPP was used to stabilize metal nanoparticles, presumably through interactions of the lone pair of electrons on the nitrogen atoms in the polymer backbone. (27) These nanocomposites were generated by reduction of $HAuCl_4$ in THF or toluene solutions of PMPP. The metal nanoparticles, which were approximately 5 to 8 nm in diameter as shown in Figure 1, display the characteristic gold nanoparticle surface plasmon absorption at ca. 540 nm, resist aggregation below 37 °C, the T_g of PMPP, for at least a year, and show some aggregation in solution over 4 to 7 days. Thermal gravimetric analysis (TGA) of the polymer composites indicated that the metal nanoparticles catalyzed the thermal degradation of the polymer. Other studies, still in progress in our lab, demonstrate that nanoparticle composites of derivatives of PMPP and several metals are readily formed and at least one of these nanocomposite catalyzes the hydrogenation of olefins. (28) Studies are also underway to use grafts and blends of PMPP with organic polymers to segregate metal nanoparticles into phosphazene domains. (29)

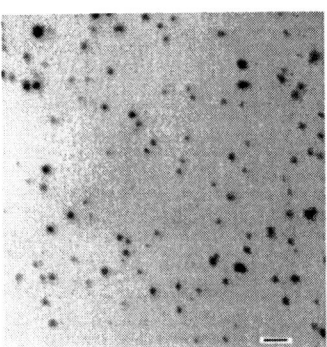

Figure 1. TEM image of 40:1 PMPP-gold nanoparticle composite. The scale bar is 20 nm. (Reproduced from reference 27. Copyright 2001 American Chemical Society.)

Cyclic Phosphazenes

As noted in Scheme 1*b*, cyclic phosphazenes can be made from the same N-silylphosphoranimine precursors as the polymers. (*10*) In this case, the phosphoranimine is treated with a nonequivalent amount of CF_3CH_2OH at room temperature in the absence of solvent. The known cyclic trimer, $[Me_2PN]_3$ and both the *cis* and *trans* isomers of $[(Ph)(Me)PN]_3$ were prepared in this manner. More recently, the geometric isomers of other alkyl/aryl cyclic phosphazenes have also been prepared in our labs. (*30*) A possible pathway in this synthesis is the initial cleavage of the Si-N bond to give an unstable H-N phosphoranimine which readily eliminates PhOH. This is consistent with the observation of both $Me_3SiOCH_3CF_3$ and PhOH in the reaction. (*10*)

The major advantages of this synthetic approach is the ability to make nongeminally substituted phosphazenes and to have access to both geometric isomers Nucleophilic substitution of halogens on $[Cl_2PN]_3$ generally does not afford nongeminal compounds. For $[(Ph)(Me)PN]_3$, the geometric isomers were separated by column chromatography and characterized by NMR spectroscopy, elemental analysis, and X-ray crystallography. The ratio of *trans* to *cis* isomers formed in this process is ca. 3:1. Perhaps the most interesting feature of the *cis* isomer is the basketlike shape (Figure 2) formed by the three phenyl groups on one side of the almost planar PN six-membered ring. (*10*)

Like the polymer PMPP, both isomers of the cyclic phosphazenes $[(Ph)(Me)PN]_3$, undergo simple derivatization via deprotonation-substitution reactions. (eq 2) This was first verified by quenching the anion formed by treatment with *n*-BuLi with the simple electrophile MeI. (*31*) In both cases, it was clear that the PN ring remained intact since there was no evidence for isomerization. The new ethylphenyl cyclics were characterized by spectroscopy and X-ray diffraction studies. Although the disubstituted derivative was formed and isolated when less than three equivalents of *n*-BuLi and MeI were used, it was not possible to detect monosubstituted derivatives. Since this work was published, a large range of functional groups, both hydrophobic and hydrophilic, have been attached by this approach producing molecules with

(2)

specific controlled architecture, a broad range of reactivity, and varied properties, including water solubility. (*30*) Subsequent reactions further extend the potential applications of these compounds to include ligands for stereospecific metal catalysts, molecular sensors, and self-assembly of supramolecular systems. Preliminary work also demonstrates that several of the new cyclic phosphazene trimers stabilize metal nanoparticles.

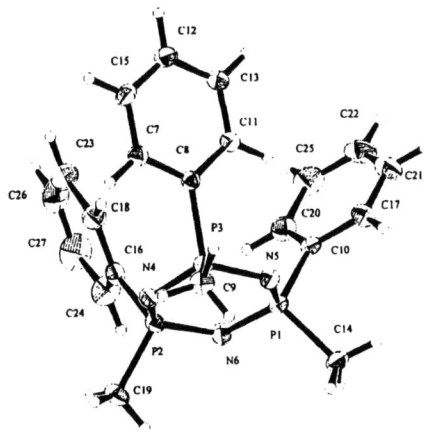

*Figure 2. Thermal ellipsoid plot of **6a**, cis-(Me(Ph)P=N)$_3$ (40% probability ellipsoids for non-hydrogen atoms are shown) [Reproduced from reference 10. Copyright 2002 American Chemical Society.]*

Heating either pure samples of the *cis* or *trans* isomers or mixtures of the isomers ultimately results in a ring opening reaction that produces mixtures of cyclic trimers and tetramers. (eq 3) (*32*) After less than 4 days at 220 °C predominantly cyclic trimers were observed with less than 5% tetramers, but this increased to about 10% tetramers after 11 days. At 250 °C, however, up to 70% of the equilibrium mixture was tetramers after 11 days. NMR spectroscopy indicated that both isomers of the trimer and all four isomers of the tetramer were present in the mixture, but there was no evidence for the formation of polymers. A similar ring equilibrium exists for [Me$_2$PN]$_n$ and [Ph$_2$PN]$_n$. (*33*) Heating the pure samples of the tetramers at 250 °C for several days also produced all six isomers of the trimer and tetramer. If either trimers or tetramers were heated in sealed ampoules at temperatures greater than 280 °C, degradation began to occur yielding significant quantities of insoluble black solids. (*32*)

By analogy to the positions of the aromatic rings in the isomers of calix-4-arene, these isomers have been named as cone, partial cone, 1,2-alternate, and 1,3-alternate as shown in Chart 2.

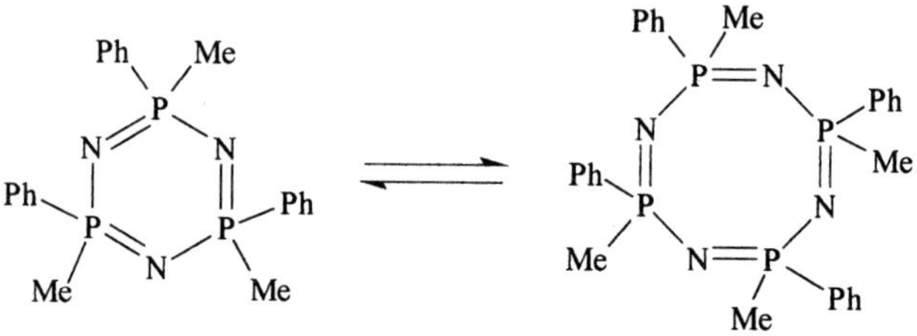

(3)

Chart 2

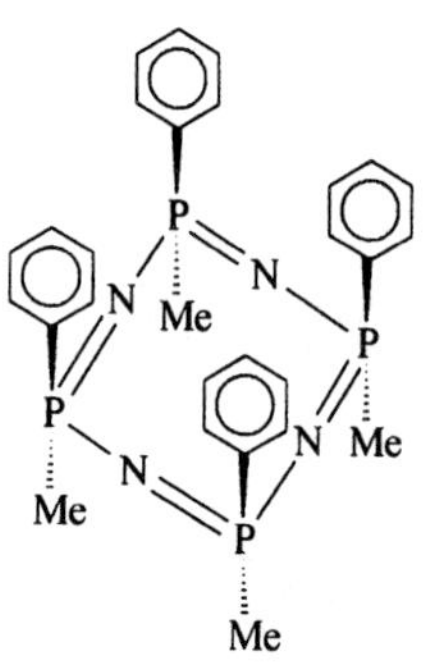

cone

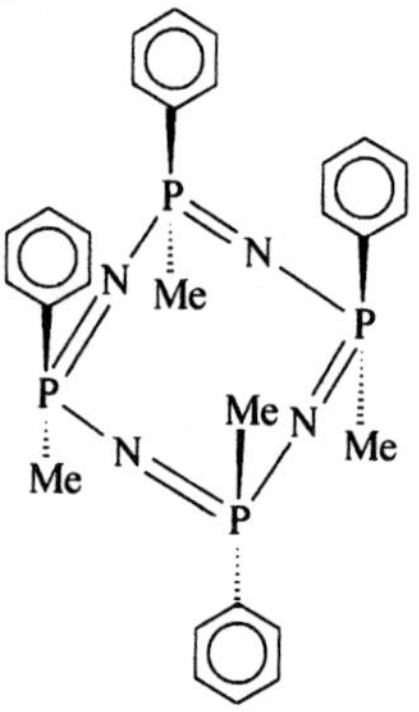

partial cone

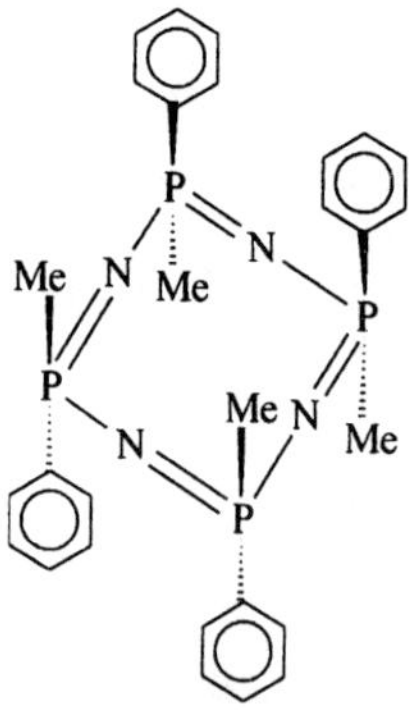

1,2-alternate

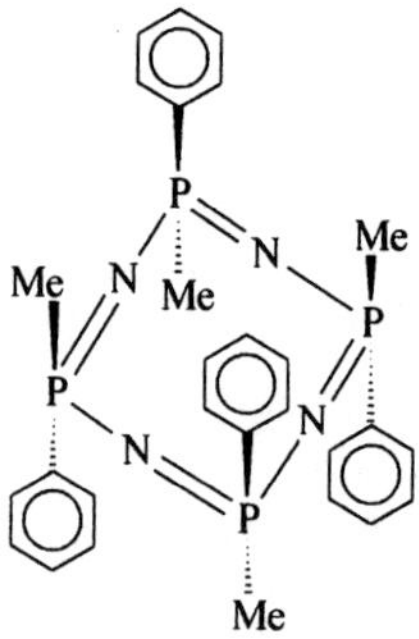

1,3-alternate

All six isomers were isolated by a series of column chromatography separations and by solubility differences. (Scheme 2) Once separated each isomer was fully characterized by spectroscopy and by X-ray crystallography. This was the first fully characterized set of all four isomers of a nongeminally substituted phosphazene tetramer. No similar set of isomers has been isolated for the isoelectronic siloxane analog.

Scheme 2

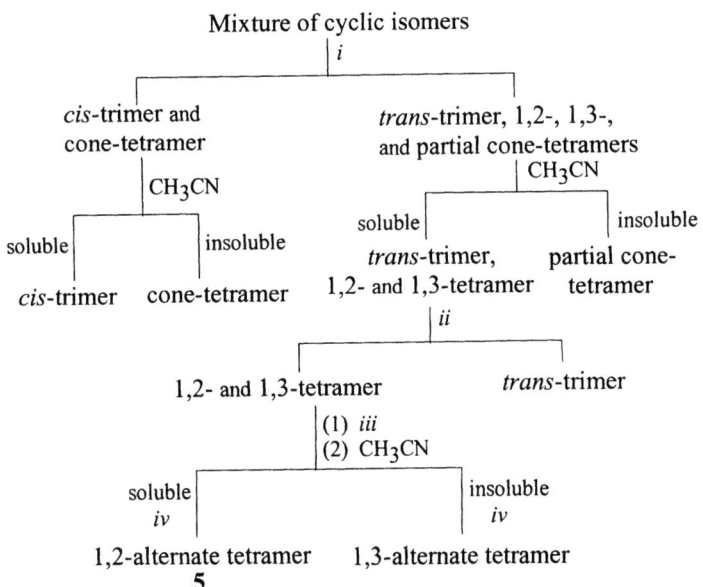

(i) column chromatography (ethyl acetate), (ii) column chromatography (ethyl acetate:hexanes = 1:1), (iii) HCl gas bubbled into ether solution to give white solids, (iv) extracted from 1.5 M aq. KOH solution with CH_2Cl_2. [Reproduced from reference 32. Copyright 2003 American Chemical Society.]

In general, the bond lengths in all the new nongeminal tetramer isomers are similar to closely related cyclic trimers and tetramers, $(Me_2PN)_3$, (*34*) $(Ph_2PN)_3$, (*35*) cis- and trans-[Ph(Me)PN]$_3$, (*10*) cis- and trans-[Ph(Et)PN]$_3$, (*31*) and $(Me_2PN)_4$. (*36*) The phosphazene ring in all four isomers is non-planar, and, the PN ring is essentially a twisted saddle for all but the 1,2-alternate isomer which has a chair conformation. (*32*)

In summary, both the alkyl/aryl substituted cyclic and polymeric phosphazenes are readily accessible from condensation reactions of N-silylphosphoranimines. These phosphazenes significantly extend the range of properties and reactivity of this family of main group inorganic rings and polymers. Moreover, simple deprotonation-substitution reactions on methyl substituents afford a large number of new polymers with unique properties and cyclic compounds with specific shape and controlled functionality on one side of a planar ring. In addition, the enhanced basicity of the nitrogen atoms that results from electron releasing groups at phosphorus, results in strong interactions with Lewis acids, including simple metal cations, H^+, and metal nanoparticles. Metal nanoparticle composites formed from the polymers show promise in catalysis of reactions such as hydrogenation of olefins. The rigid shape of the nongeminally substituted cyclic phosphazenes which allows for amphiphilic substitution on opposites sides of the ring could provide new self-assembling or molecular recognition systems.

Acknowledgements

This paper is dedicated to Alan H. Cowley on the occasion of his 70th birthday. The authors gratefully acknowledge the generous financial support of the Robert A. Welch Foundation and the National Science Foundation.

References

1. (a) Allcock, H. R. *Phosphorus-Nitrogen Compounds;* Academic Press, New York: 1972. (b) Allcock, H. R. *Chem. Rev.* **1972**, *72*, 315 – 356. (c) Allen, C. W. *Chem. Rev.* **1991**, *91*, 119–135.
2. See for example: (a) Neilson, R. H.; Wisian-Neilson, P. *Chem. Rev.* **1988**, *88*, 541–562. (b) Mark, J. E.; Allcock, H. R.; West, R. *Inorganic Polymers*; Prentice-Hall: New Jersey, 1992. (c) Wisian-Neilson, P. *Encyclopedia of Inorganic Chemistry*, King, R. B., Ed.; Wiley: England, 1994, Vol. 7, 3371–3403.
3. Wisian-Neilson, P.; Neilson, R. H. *J. Am. Chem. Soc.* **1980**, *102*, 2848.
4. Neilson, R. H.; Hani, R.; Wisian-Neilson, P.; Meister, J. J.; Roy, A. K.; Hagnauer, G. L. *Macromolecules* **1987**, *20*, 910–916.
5. Jinkerson, D. L., Ph. D. Dissertation, Texas Christian University, 1989.
6. Gruneich, J. A.; Wisian-Neilson, P. *Macromolecules* **1996**, *29*, 5511–5512.
7. D'Hallum, G.; De Jaeger, R.; Chambrette, J. P.; Potin, P. *Macromolecules* **1992**, *25*, 1254–1258.

8. Montague; R. A.; Matyjaszewski, K. *J. Am. Chem. Soc.* **1990**, *112*, 6721–6723.
9. Allcock, H. R.; Crane, C. A.; Morrissey, C. T.; Nelson, J. M.; Reeves, S. D.; Honeyman, C. H.; Manners, I. *Macromolecules* **1996**, *29*, 7740–7747.
10. Wisian-Neilson, P.; Johnson, R. S.; Zhang, H.; Jung, J.-H.; Neilson, R. H.; Ji, J.; Watson, W. H.; Krawiec, M. *Inorg. Chem.* **2002**, *41*, 4775–4779.
11. Wisian-Neilson, P.; Bahadur, M.; Iriarte, J. M.; Wood, C. E.; Ford, R. R. *Macromolecules* **1994**, *27*, 4471–4476.
12. Wisian-Neilson, P.; Garcia-Alonso, F. J. *Macromolecules* **1993**, 26, 7156–7160.
13. Wisian-Neilson, P.; Ford, R. R.; Neilson, R. H.; Roy, A. K. *Macromolecules* **1986**, *19*, 2089–2091.
14. Wisian-Neilson, P.; Bailey, L.; Bahadur, M. *Macromolecules* **1994**, *27*, 7713–7717.
15. Wisian-Neilson, P.; Xu, G.-F; Wang, T. *Macromolecules* **1995**, *28*, 8657–8661.
16. Wisian-Neilson, P.; Ford, R. R. *Macromolecules* **1989**, *22*, 72–75.
17. Wisian-Neilson, P; Huang, L.; Islam, M. Q.; Crane, R. A. *Polymer* **1994**, *35*, 4985–4989.
18. Wisian-Neilson, P.; Zhang, C. *Macromolecules* **1998**, *31*, 9084–9086.
19. Wisian-Neilson, P.; Islam, M. S.; Ganapathiappan, S.; Scott, D. L.; Raghuveer, K. S.; Ford, R. R. *Macromolecules*, **1989**, *22*, 4382–4384.
20. Wisian-Neilson, P.; Xu, G.-F *Macromolecules* **1996**, *29*, 3457–3461.
21. Wisian-Neilson, P.; Koch, K. A.; Zhang, C. *Macromolecules* **1998**, *31*, 1808–1813.
22. Wisian-Neilson, P.; Schaefer, M. A. *Macromolecules* **1989**, *22*, 2003–2007.
23. Wisian-Neilson, P.; Islam, M. S. *Macromolecules* **1989**, *22*, 2026–2028.
24. Cambre, J. N.; Wisian-Neilson, P. *Polym. Prepr., (Am. Chem. Soc. Div. Polym. Chem.)* **2003**, *44(1)*, 919–920.
25. Potluri, S. K.; Wisian-Neilson, P. *Polym. Prepr., (Am. Chem. Soc. Div. Polym. Chem.)* **2003**, *44(1)*, 925.
26. Garcia-Alonso, F. J.; Wisian-Neilson, P., unpublished results.
27. Walker, C. H.; St. John, J. V.; Wisian-Neilson, P. *J. Am. Chem. Soc* **2001** *123*, 3846–3847.
28. Jung, J.-H.; Tomlinson, R. E.; Truong, H., Wisian-Neilson, P., unpublished results.
29. Jung, J.-H.; Wisian-Neilson, P., unpublished results.
30. Jung, J.-H.; Potluri, S. K.; Wisian-Neilson, P., unpublished results.
31. Jung, J.-H.; Zhang, H.; Wisian-Neilson, P. *Inorg. Chem.* **2002**, *41*, 6720–6725. .

32. Jung, J.-H.; Pomeroy, J. C.; Zhang, H.; Wisian-Neilson, P. *J. Am. Chem. Soc.* **2003**, *125*, 15537–15542.
33. (a) Allcock, H. R.; McDonnell, G. S.; Desorcie, J. L. *Inorg. Chem.* **1990**, *29*, 3839–3844. (b) Allcock, H. R.; Patterson, D. B. *Inorg. Chem.* **1977**, *16*, 197–200.
34. Oakley, R. T.; Paddock, N. L.; Rettig, S. J.; Trotter, J. *Can. J. Chem.* **1977**, *55*, 4206–4210.
35. Ahmed, F. R.; Singh, P.; Barnes, W. H. *Acta Crystallogr., Sect. B 25* **1969**, 316–328.
36. Dougill, M. W. *J. Chem. Soc., London* **1961**, 5471–5479.

Chapter 25

Giant Dendrimer Construction: Hydroboration versus Hydrosilylation as a Growth Strategy

Jaime Ruiz[1], Gustavo Lafuente[1], Sylvia Marcen[1], Catia Ornelas[1], Sylvain Lazare[2], Eric Cloutet[3], Jean-Claude Blais[4], and Didier Astruc[1,*]

[1]Nanosciences and Catalysis Group, LCOO, UMR CNRS N°5802, University Bordeaux I, 33405 Talence Cedex, France
[2]LCPM, UMR CNRS N°5803, University Bordeaux I, 33405 Talence Cedex, France
[3]LCPO, University Bordeaux I, 33405 Talence Cedex, France
[4]LCSOB, UMR CNRS N°7613, University Paris VI, 75252 Paris, France

Giant dendrimers were constructed using generations consisting in two reactions: hydroelementation followed by a nucleophilic reaction of a phenoltriallyl brick. After having synthesized a third-generation dendrimer using hydroboration with disiamylborane, our group switched to Karsted-catalyzed hydrosilylation with dimethylchloromethylsilane. This successfully led us to the 9th generation (> 10^5 tethers) without being marred by the bulk problem at the dendrimer periphery. This construction was achieved far beyond the de Gennes "dense-packing" limit, indicating that the extremities of the tethers must back fold towards the core. Thus, the construction must be limited by the volume rather than surface.

© 2006 American Chemical Society

Introduction

Dendrimers have numerous potential applications in biology, catalysis and materials science.[1-15] Although numerous syntheses of large dendrimers were available,[8-10] none with well-characterized compounds existed before this work[16a] with numbers of tethers larger than those defined by the de Gennes "dense-packing" limit involving steric congestion at the periphery.[11] According to this theory, the space A_Z available on the dendrimer for each terminal group at the dendrimer periphery being given by:

$$A_Z = A_D/N_Z \; \alpha \; r^2/N_c N_b G$$

in which A_D is the total surface of the periphery, N_Z the number of termini, r the radius of the dendrimer sphere, N_c the number of core branches, N_b the multiplication number at each focal point and G the generation number at each focal point and G the generation number. De Gennes' dense-packing (limit) generation follows:

$$G_l = a \, (\ln P + b)$$

P: length of the branch-cell segment; a, b: parameters that depend on the type of dendrimer (a = 2.88 and b = 1.5 for Tomalia's PAMAM dendrimers[1,8]). Several groups who reported large dendrimers did not proceed beyond this limit, implying that the termini do not back fold towards the core.[8-10] Several theoreticians suggested, however, that the steric congestion does not intervene at the periphery due to back folding that fills the inner cavities,[13,14] and a number of studies have been carried out on the back-folding problem.[15] Thus, the problem of the validity of the de Gennes "dense-packing" limit was open.

Results and Discussion

We wished to synthesize large dendrimers and check the validity of the de Gennes "dense-packing" limit when the termini are small and non-polar (polar and hydrogen-bonded termini can interact with one another which may prevent back folding). Therefore, we designed dendritic cores and dendronic building blocks using the polyfunctionalization of arenes mediated by the CpFe$^+$ group. The basis of the functionalization of arenes using the perallylation of the methyl groups in benzylic positions is the large increase of acidity of these benzylic protons when the arene is coordinated to the CpFe$^+$ fragment (for instance decrease of the pKa of C_6Me_6 from 33 to 29 in DMSO).[16b] Another advantage of this strategy is that it is possible to mix excess of KOH and allylbromide to selectively perfunctionalize polymethylaromatics under ambient or sub-ambient conditions.[16c]

For instance, mesitylene was nona-allylated which is a powerful way to provide a nona-branched dendritic core. The application of this perallylation reaction to *p*.ethoxytoluene was a challenge because the application of the multifunctionalization to functional aromatics would open up applications to the functionalization of surfaces, nanoparticles, dendrimers, polymers, etc. Thus, the finding of the one-pot transformation of its $CpFe^+$ complex to the iron-free phenoltriallyl dendron was gratifying and turned out to be a useful breakthrough for our ongoing dendrimer syntheses as well as for those of other branched nanomaterials.[16d] The dendritic growth was based on the hydroelementation of polyallyl dendrimers followed by the nucleophilic substitution of a polymesylate or polyiodo dendrimer by the phenoltriallyl dendron. In this way, the polyallyl dendrimer of the next generation is obtained in which the number of branches is multiplied by three.[17]

Thus, the choice of hydroelementation was crucial. We first started the dendrimer construction using the hydroboration reaction with disiamylborane.[16a] This construction starting from ferrocene is represented on Scheme 1. This synthesis was pursued until the third generation polyallyl dendrimer containing a theoretical number of 243 allyl branches that is represented on Chart 1.

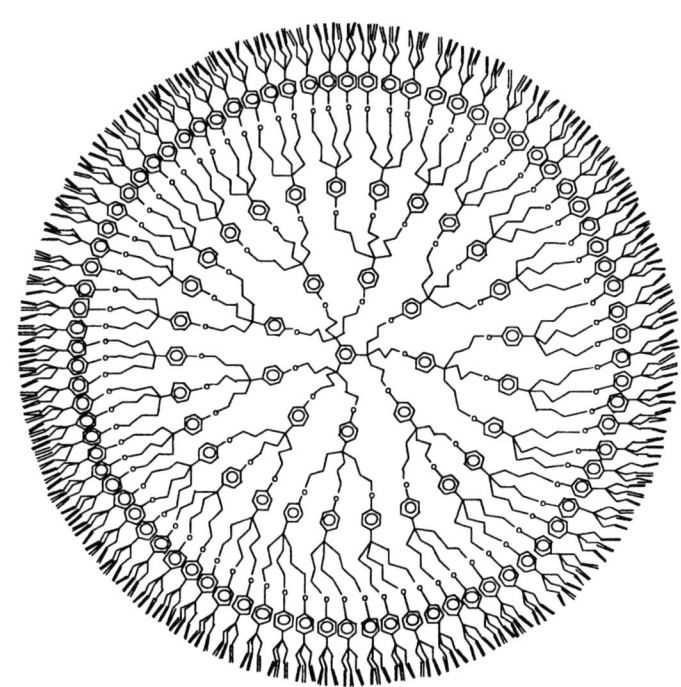

Chart 1

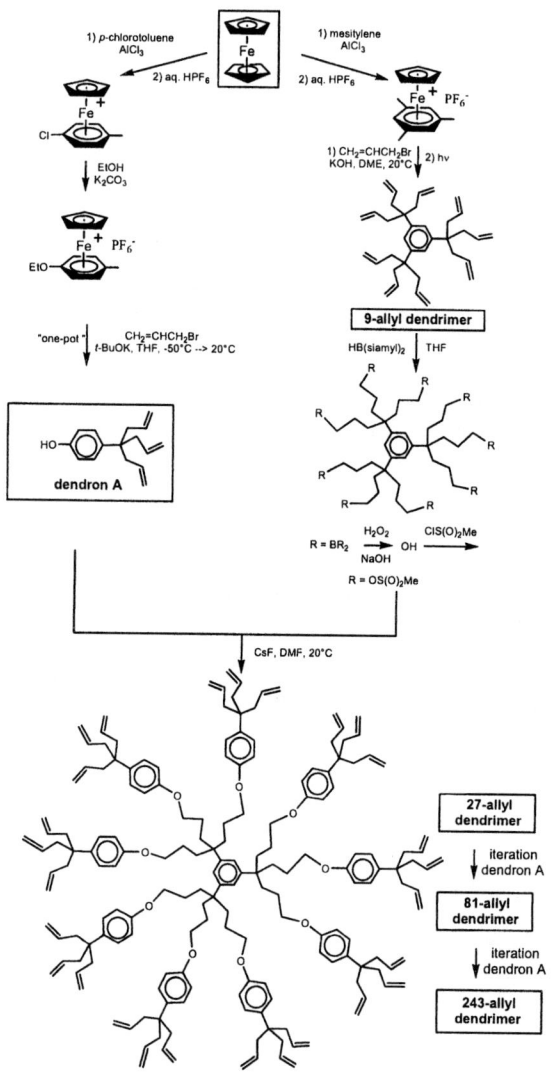

Scheme 1: Overall dendrimer construction starting from ferrocene. The nona-allyl core synthesized by CpFe$^+$ induced nonaallylation of mesitylene followed by decomplexation using visible light (right column) can be considered as G_0 and thus the 27-allyl derivative, represented below, as G_1(abbreviation: G_1-27).

The MALDI TOF mass spectra showed the dominant molecular peaks for the G_1-27-allyl and G_2-81 allyl dendrimers, but the mass spectrum of G_3-243 allyl could not be obtained, although the ^1H and ^{13}C seemingly showed a correct structure. It is probable that the number of defects were very large, however at the third generation. Also the rate of the hydroboration was much lower for the third generation than for the two first ones. It is probable that the steric bulk of the borane kept the terminal groups at the periphery, and that this dendrimer periphery was highly congested. Therefore, at this point, we decided to switch to the less sterically constraining hydrosilylation series using chloromethyldimethylsilane already used by Seyferth for dendrimer growth.[18] This hydrosilylation, carried out at 20-40°C using the Karsted catalyst, turned out to be very regioselective on the terminal carbon atoms, so that the expected isomer was the only one observed by ^1H, ^{13}C and ^{29}Si NMR.[16b] It is probable that the bulk provided by the dendrimer is at the origin of this selectivity. The ^{29}Si spectra were particularly simple. The hydrosilylation of the polyallyl dendrimers gave a compound that showed two ^{29}Si NMR signals at 0.73 ppm and 3.8 ppm, and the polyolefin dendrimers showed only one ^{29}Si NMR signal at 3.8 ppm. The dendrimers remained pentane soluble until the ninth generation and the ^1H and ^{29}Si NMR spectra are shown in Figure 1 and 2. The construction based on this hydrosilylation is represented on Scheme 2. At this time, the MALDI TOF mass spectra also showed only the molecular peak for G_1-27 allyl and a dominant molecular peak for the G_2-81 allyl dendrimers (Figure 3), but a minor molecular peak was also found for G_3-243 allyl shown in Figure 4.

Moreover, it was possible to continue the construction and to follow all the reactions by ^1H and ^{29}Si NMR until the 9th generation with a theoretical number of branches of 177407. The dendrimers were characterized by a number of techniques besides multinuclear NMR indicated above. The MALDI TOF of G_4-729 shows a massif vanishing near the molecular mass that is not observed as a single peak. The size exclusion chromatograms (SEC) were obtained with low polydispersities (between 1.00 and 1.02) from G_1-27 allyl to G_5-2187 allyl (see Figure 5). Above this size, the globular shape does not allow to overtake molecular weights larger than 10^5. Some aggregation[19] could be observed by SEC (and other techniques, *vide infra*) when the dendrimer generation increases, and are clear in the SEC of G_4-729 and G_5-2187. The dendrimers being organic, they cannot be seen by transmission electron microscopy, because the samples are lacking heavy atoms. By vaporizing OsO_4 onto the sample in a well-ventilated hood [CAUTION: OsO_4 is toxic], osmametallocycles are formed with the double bounds at the dendrimer periphery, and the single dendrimers can be observed by TEM. For instance, G_4-729 was observed in this way, and the diameter matches that obtained using the molecular model (Figure 6). For the largest dendrimer, G_9-177407 allyl, the hydrosilylation could still be carried out and the nucleophilic substitution of chloride by iodide using NaI produced the polyiodo dendrimer that is visible by TEM due to the heavy iodine atoms. A diameter of 13 nm was found in this way for G_9-177407 allyl (Figure 7).

Perhaps the most useful microscopy technique, however, was atomic force microscopy (AFM), that allowed to measure the heights of the dendrimers for all the generations from G_1-27-allyl to G_9-177407 allyl (Figure 8)

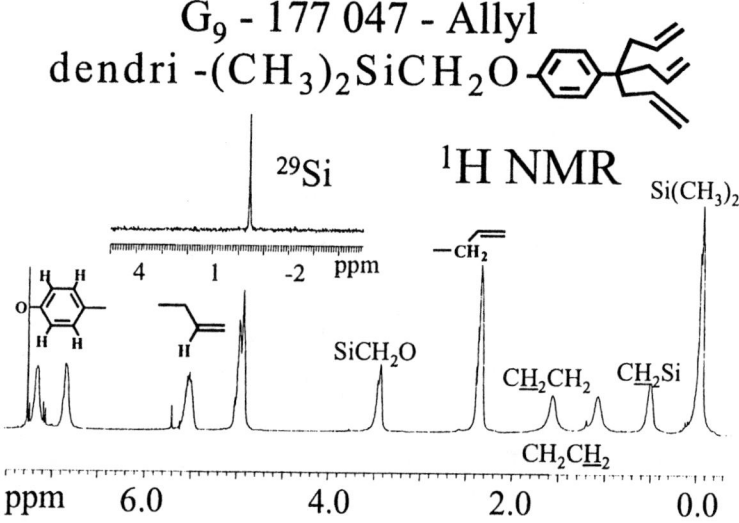

Figure 1: 400 MHz ^{1}H and ^{29}Si NMR spectra of the G_9-177047-allyl dendrimer which very clearly show the signals of the surface phenoxy, allyl and methylene protons and the single ^{29}Si peak. The large peak at 1.25 ppm and the multiplet signal at 0.85 ppm progressively appear beyond G_5 and indicate the presence of dendrimer encapsulated alkanes found as impurities in the commercial xylene solution of the Karstedt catalyst used for the hydrosilylation reactions.

353

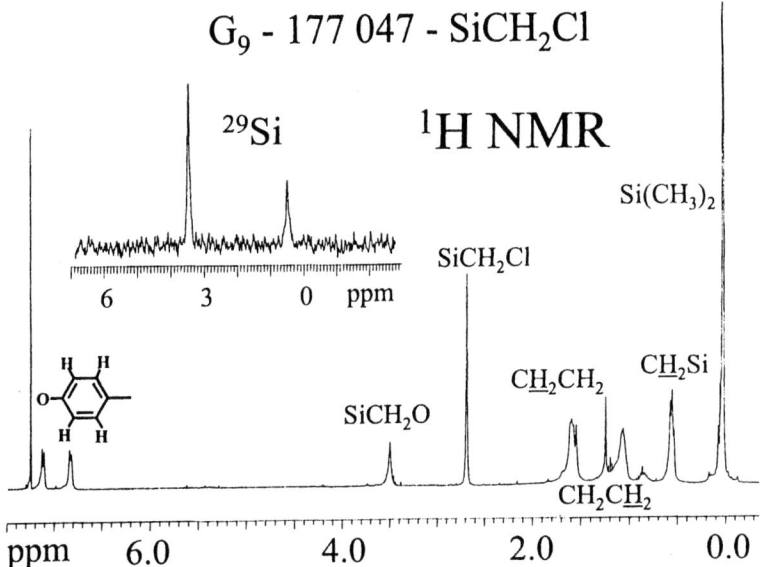

Figure 2: ^1H and ^{29}Si NMR spectra of G_9-177147-$SiCH_2Cl$ resulting from the catalyzed hydrosilylation of G_9-177147-allyl (see text) followed by reaction with NaI at 80°C in acetone to provide exchange of chloro by iodo. The signals corresponding to small amounts of ether (which could not be removed under vacuum) are also observed on this ^1H NMR spectrum. See the HRTEM of this compound in Figure 7.

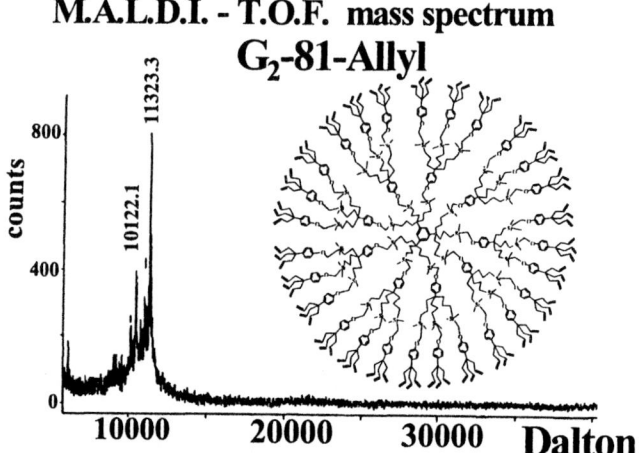

Figure 3: MALDI TOF mass spectrum of G_2-81-allyl-dendrimer. The major peak is the molecular peak (M = 11299; M+Na = 11322). The minor peak at 10122 represents the same dendrimer in which one triallylphenol unit is missing possibly because of the isomerization of a double bond during the catalyzed hydrosilylation reaction (see text)

G_3 - 243 - Allyl

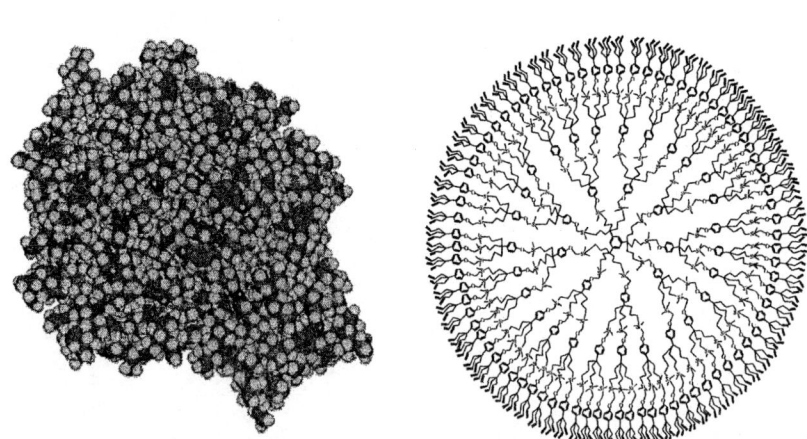

Figure 4: G_3-243-allyl dendrimer (right, see Scheme 1 for the synthesis) and its molecular model (left). The molecular peak is observed in its MALDI TOF mass spectrum which yields a polydispersity value of 1.02 (see also Figure 5 for SEC and polydispersity data).

355

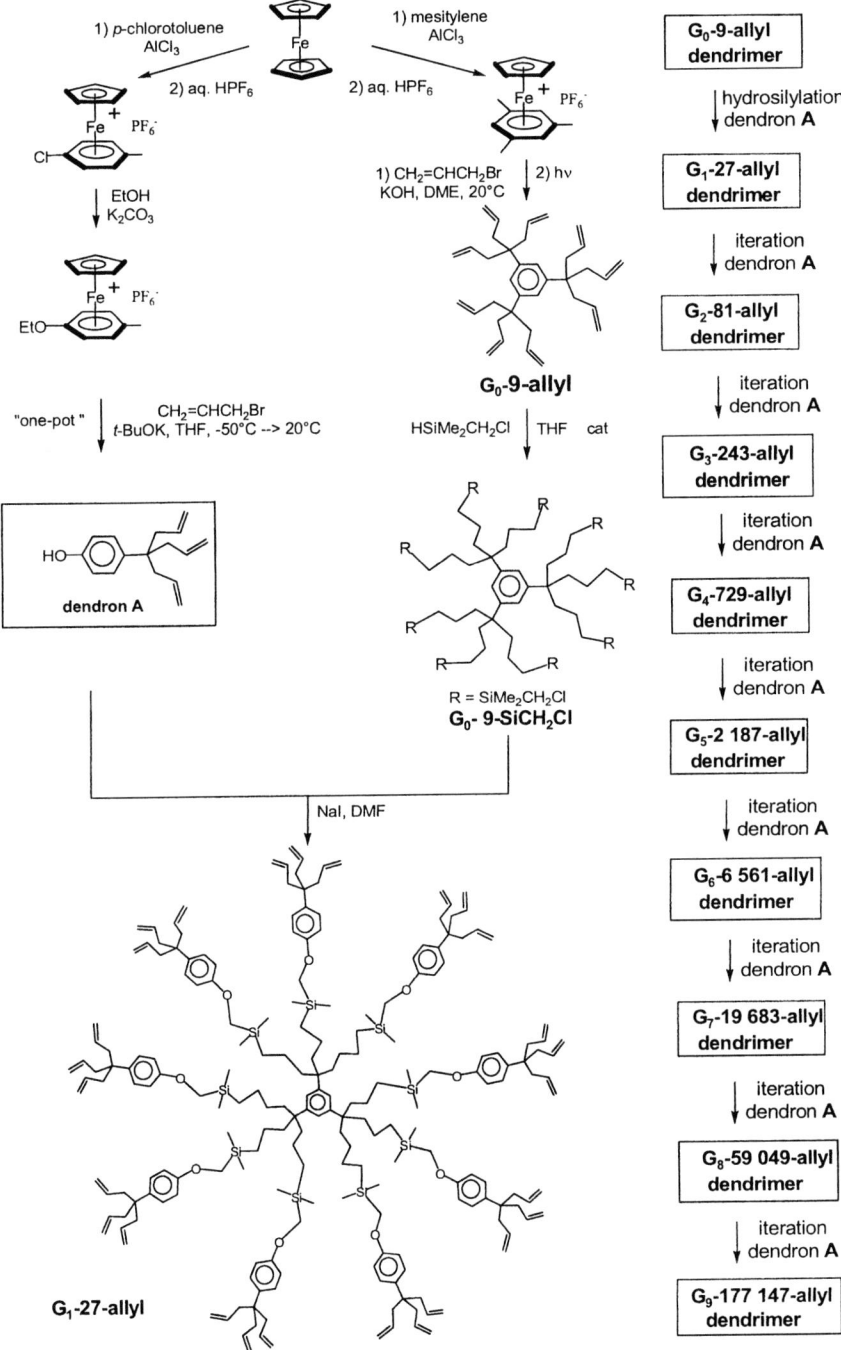

Scheme 2

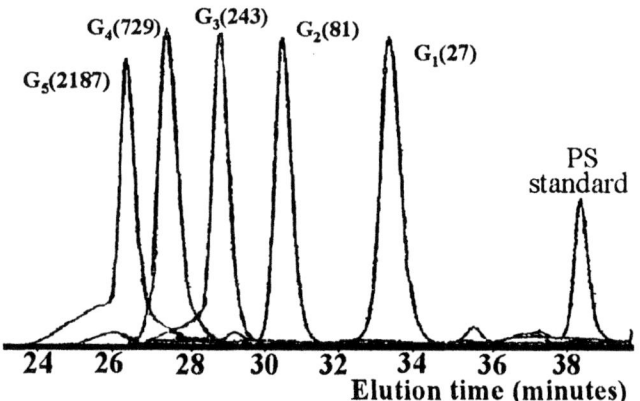

Figure 5: Size exclusion chromatograms (SEC) of all the dendrimers from G_1 to G_5. The generations and numbers of allyl groups are noted above each chromatogram on the figure. The polydispersities obtained from these SEC's have values between 1.00 and 1.02 up to G_5. Above this generation, the SEC could not be obtained so far because the dendrimer size overtakes 10^6 uma and the globular shape requires much larger pores than linear polymers (SEC for dendrimers overtaking 10^6 uma have, to our knowledge, not yet been recorded). The presence of the minor peak at high masses is not yet clarified given the purity observed by NMR, but it could be due to oxidatively induced cationic polymerization in the SEC column

G_4 - 729 - allyl

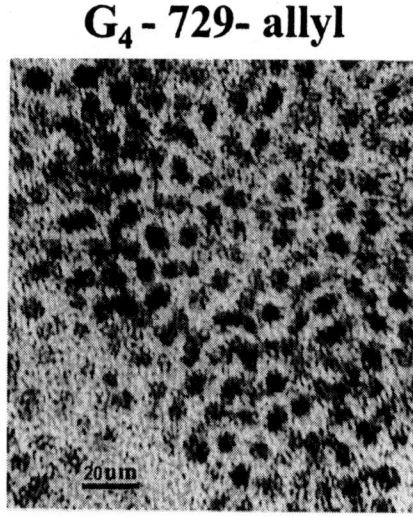

Figure 6: HRTEM picture of G_4-729 coated with osmium oxide vapor and recorded on amorphous graphite support. Osmium oxide forms a five-membered glycolate metallocycle by addition of the terminal double bonds of the branches onto oxo ligands[36] (the proportion of reacted terminal double bonds is unknown)

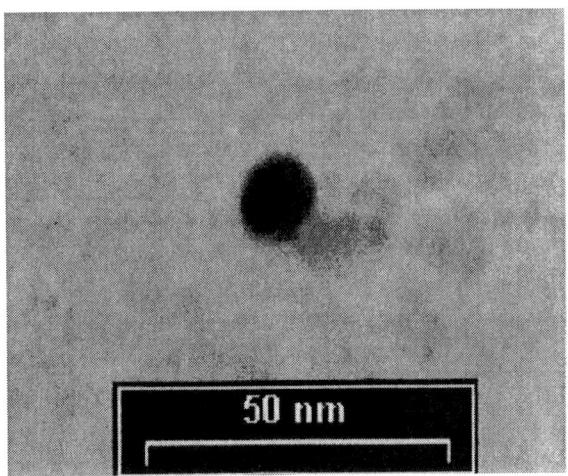

Figure 7: HRTEM picture of G_9-177047-$SiCH_2I$ recorded on amorphous graphite support showing the globular shape with a diameter of 13 nm (compare Table 1 and Figure 8) lower than that calculated (16.7 nm) for tethers with maximum extension. See the ^1H and ^{29}Si NMR spectra of this compound on Figure 2.

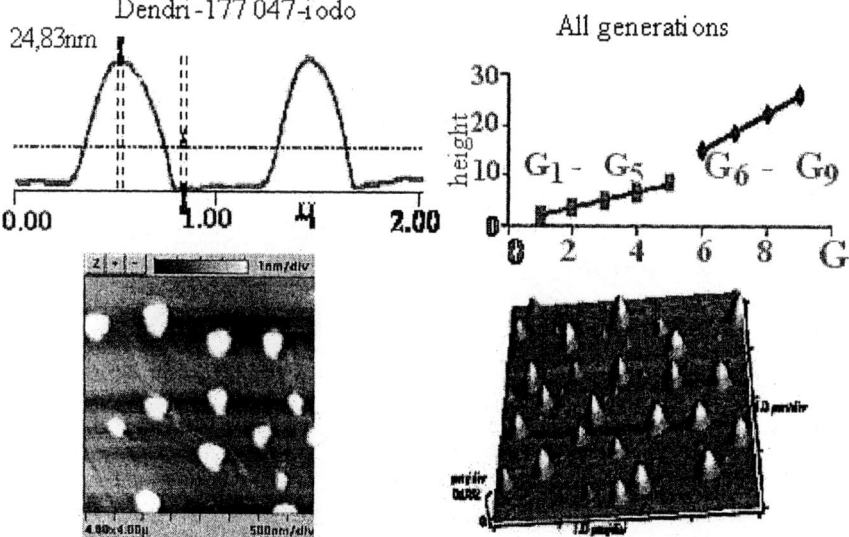

Figure 8: AFM images of G_9-177047- $SiCH_2I$ on graphite HOPG support shown in two dimensions (left) and three dimensions (right). The diameters are 100 nm in average, and the height is 25 nm (top) indicating the aggregation of about 100 dendrimers in a double layer (compare Table 1 and Figure 5). AFM was operated in the tapping-mode with a resonance frequency of around 200 MHz. The tip has a radius of 10 nm. The top right graph shows the progression of the height of the polyallyl dendrimers determined by AFM from the first to the ninth generation (G_1-G_9) involving monolayers from G_1 to G_5 and bilayers from G_6 to G_9

Monolayers were found for the first 4 generations, then it appeared that a double layer was observed up to the last one, with a height of 25 nm for G_9-177407 allyl. Flattening and aggregation of dendrimers recorded by AFM have already been observed in other cases.[19] The heights of the monolayers are slightly lower than those calculated until G_5. Finally, we believe that, although the number of defects increases with generations, the total number of termini in G_9-177407 allyl is larger than 10^5.

Although the reaction time increases with generations, it is necessary that the number of back folding methylene termini increase when the dendrimers become larger than the size corresponding to the "dense-packing limit". We have estimated that this number is around 6000 in our series.[16a] In the largest dendrimers, the proportion of methylene termini back folding must be of the order of 90%, although this should be viewed as a dynamic phenomenon. This means that the construction is limited by the volume, not by the surface. The limit of generation is thus found much later than the de Gennes "dense-packing limit" if the termini are small enough and flexible This finding is indeed true only for dendrimers with small flexible termini. When the size of the termini is larger, the termini do not back fold as was confirmed by calculation,[14] and then the periphery can be severely congested at a much earlier generation.[2]

Conclusion

The largest known well-characterized dendrimers have been synthesized with up to $>10^5$ termini for G_9, far beyond the de Gennes "dense-packing" limit, upon multiplying the number of allyl branches by three at each generation. The dendritic core and dendronic bricks were synthesized by $CpFe^+$ polyfunctionalization of arenes and covalently assembled into these robust dendrimers. The synthetic strategies involved hydroelementation of the allyl groups for which the hydroboration using disiamylborane and the hydrosilylation using chloromethyldimethylsilane catalyzed by the Karsted catalyst were compared. Whereas the former one encountered steric problems at the third generation, this was not the case for the latter one that allowed growing up to the 9th generation. The constructions were monitored by ^1H, ^{13}C and ^{29}Si NMR, MALDI TOF mass spectroscopy, SEC, TEM and AFM. Consequently, the small and flexible methylene termini must back fold towards the core (filling the intra-dendrimer cavities) to take the feasibility of such a large construction into account.

Acknowledgement

This article is dedicated to our distinguished friend Professor Alan, H. Cowley, F. R. S. and Dr H. C. for his guiding spirit and outstanding contributions to main-group chemistry at the occasion of his 70th birthday. Financial support from the Institut Universitaire de France (IUF), the Centre

National de la Recherche Scientifique (CNRS) and the Universities Bordeaux I and Paris VI is gratefully acknowledged.

References
(1) (a) Tomalia, D. A.; Taylor, A. N., Goddard III, W. A. *Angew. Chem. Int. Ed. Engl.* **1990**, *29*, 138-175; (b) Fréchet, J. M. J. *Science*, **1994**, *263*, 1710-1715; (c) Ardoin, N.; Astruc, D. *Bull. Soc. Chim. Fr.* **1995**, *132*, 875-909.
(2) (a) Newkome, G. R.; Moorefield, C. N.; Vögtle, F. *Dendrimers and Dendrons: Concepts, Syntheses and Applications*, VCH, Weinheim, 2001; (b) *Advances in Dendritic Molecules*, Newkome, G. Ed.; JAI, Greenwich, CT, Vol 1-5, 1994-1998 and 2002; (c) *Dendrimers, Top. Curr. Chem.*; Vögtle, F. Ed., Springer Verlag, Berlin Vol 1, 1997 and Vol 2, 2000; (d) *Dendrimers and Other Dendritic Polymers*, Tomalia, D.; Fréchet, J. M. J. Eds., Wiley-VCH, New York, 2002; (e) *Dendrimers and Nanosciences*, Astruc, D. Guest Ed., *C. R. Chimie*, Elsevier, Paris, **2003**, *6* (Vol. 8-10).
(3) (a) Zeng, F.; Zimmerman, S. C. *Chem. Rev.* **1997**, *97*, 1681-1712; (b) Matthews, O. A.; Shipway, A. N.; Stoddart, J. F. *Prog. Polym. Sci.* **1998**, *23*, 1-56.
(4) Venturi, M.; Serroni, S.; Juris, A.; Campagna, S.; Balzani, V. In *Dendrimers, Topics Curr. Chem.* Springer-Verlag, Berlin, Ed. Vögtle, F. 1998, Vol. *197*, pp. 193-228.
(5) (a) Bosman, A. W.; Janssen, H. M.; Meijer, *Chem. Rev.* **1999**, *99*, 1665-1688; (b) Sijbesma, R. P.; Meijer, E. W. *Curr. Opin. Coll. Int. Sci.* **1999**, *4*, 24-34; (c) Weener, J.-W.; Baars, M. W. P. L.; Meijer, E. W. *In* ref. 2d, pp.387-424.
(6) (a) Hecht, S.; Fréchet, J. M. J. *Angew. Chem. Int. Ed.* 2001, 40, 74-91; (b) Fréchet, J. M. J. *Chem. Rev.*, **2001**, *101*, 3819-3867; (b) Crooks, R. M.; Zhao, M.; Sun, L.; Chechik, V.; Yeung, L. K. *Acc. Chem. Res.* **2001**, *34*, 181-190.
(7) Tomalia, D. A.; Baker, H.; Dewald, J. R.; Hall, M.; Kallos, G.; Martin, S.; Roeck, J.; Ryder, J.; Smith, P., *Macromolecules* **1986**, *19*, 2466-2470.
(8) van der Made, A. W.; van Leeuwen, P. W. N. M.; de Wilde, J. C.; Brandes, R. A. C. *Adv. Mater.* **1993**, *5*, 466-468.
(9) Caminade, A. M.; Majoral, J. P. In *Dendrimers* Ed. Vögtle, F. Springer, Berlin, *Topics Cur. Chem.* 1998, pp. 79-124.
(10) Sournies, F.; Crasnier, F.; Graffeuil, M.; Faucher, J.-P.; Lahana, R.; Labarre, M. C.; Labarre, J.-F. *Angew. Chem. Int. Ed. Engl.* **1995**, *34*, 578-581.
(11) de Gennes, P.-G.; Hervet, H. *J. Phys. Lett.* **1983**, *44*, 351-360.
(12) Percec, V.; Johansson, G.; Ungar, G.; Zhou, J. P. *J. Am. Chem. Soc.* **1996**, *18*, 9855-9866.

(13) (a) Lescanec, R. L.; Muthukumar, M. *Macromolecules*, **1990**, *23*, 2280-2288; (b) Mansfield, M. L.; Klushin, L. I. *Macromolecules* **1993**, *26*, 4262-4268; (c) Boris, D.; Rubinstein, M. *Macromolecules*, **1996**, *29*, 7251-7260.

(14) Naidoo, K. J.; Hughes, S. J.; Moss, J. R., *Macromolecules* **1999**, *32*, 331-341.

(15) (a) Mourey, T. H.; Turner, S. R.; Rubinstein, M.; Fréchet, J. M. J.; Hawker, C. J.; Wooley, K. L. *Macromolecules* **1992**, *25*, 2401-2406; (b) Wooley, K. L.; Klug, C. A.; Tasaki, K.; Schaefer, J. *J. Am. Chem. Soc.* **1997**, *119*, 53-58; (c) Meltzer, A. D.; Tirrell, D. A.; Jones, A. A.; Inglefield, P. T.; (d) Hedstrand, D. M.; Tomalia, D. A. *Macromolecules* **1992**, *25*, 4549-4552; (e) De Backer, S.; Prinzie, Y.; Verheijen, W.; Smet, M.; Desmedt, K.; Dehaen, W.; De Schryver, F. C. *J. Phys. Chem. A* **1998**, *102*, 5451-5455.

(16) (a) Ruiz, J.; Lafuente, G.; Marcen, S.; Ornelas, C.; Lazare, S.; Cloutet, E.; Blais, J.-C.; Astruc, D. *J. Am. Chem. Soc.* **2003**, *125*, 7250-7257; (b) Trujillo, H. A.; Casado, C. M.; Ruiz, J.; Astruc, D. *J. Am. Chem. Soc.* **1999**, *121*, 5674-5686; (c) Sartor, V.; Djakovitch, L.; Fillaut, J.-L.; Moulines, F.; Neveu, F.; Marvaud, V.; Guittard, J.; Blais, J.-C.; Astruc, D. *J. Am. Chem. Soc.* **1999**, *121*, 2929-2930; (d) Daniel, M.-C.; Ruiz, J.; Nlate, S.; Blais, J.-C.; Astruc, D. *J. Am. Chem. Soc.* **2003**, *125*, 2617-2628.

(17) Newkome, G. R.; Yao, Z.; Baker, G. R.; Gupta, V. K. *J. Org. Chem.* **1985**, *50*, 2003-2004.

(18) Krsda, S. W.; Seyferth, D. *J. Am. Chem. Soc.* **1998**, *120*, 3604-3612.

(19) (a) Hierleman, A.; Campbell, J. K.; Crooks, A. M.; Rico, A. J. *J. Am. Chem. Soc.* **1998**, *120*, 5323-5324; (b) Li, J.; Swanson, D. R.; Qin, D.; Brothers, H. M.; Piehler, L. T.; Tomalia, D.; Meier, J. *Langmuir* **1999**, *15*, 7347-7350; (c) Tsukruk, V. V.; Rinderspacher, F.; Bliznyuk, V. N. *Langmuir* **1997**, *13*, 2171-2176; (d) Sheiko, S. S.; Gauthier, M.; Möller, M. *Macromolecules* **1997**, *30*, 2343-2349; (e) Díaz, D. J.; Storrier, G. D.; Bernhard, S.; Takada, K.; Abruña, H. D. *Langmuir* **1999**, *15*, 7351-7354; (f) Sano, M.; Okamura, J.; Ikeda, A., Shinkai, S. A. *Langmuir* **2001**, *17*, 1807-1810.

Chapter 26

Electrical Properties of Boron Nanowires

Carolyn Jones Otten[1], Dawei Wang[2], Jia G. Lu[2,*],
and William E. Buhro[1,*]

[1]Department of Chemistry and Center for Materials Innovation,
Washington University, St. Louis, MO 63130
[2]Department of Chemical Engineering and Materials Science, and
Department of Electrical Engineering and Computer Science,
University of California at Irvine, Irvine, CA 92697

One-dimensional boron nanostructures are grown by catalyzed CVD, and found to be dense nanowires rather than hollow nanotubes. Electrical-transport studies on individual boron nanowires establish them to be p-type semiconductors ($\sigma = 10^{-2}$ $(\Omega\ cm)^{-1}$) with low carrier mobility ($\mu_H = 10^{-3}$ cm^2/Vs). These characteristics are similar to those of bulk boron. The carrier mobility is considerably below those of other nanowire and nanotube semiconductors.

Introduction

The gains in computing speed and power realized over the past decades have been achieved by advances in microelectronics fabrication. Specifically, the shrinking feature sizes of transistors and attendant circuitry have allowed ever increasing numbers of devices to be incorporated into a microchip (1). The rate of advance is captured in Moore's Law, which states that the number of transistors per chip doubles every 1.5 years, and to a rough first approximation

Moore's Law has held true for over 30 years (2). Commercial microchips are now available with feature sizes of 90 nm and more than 100M transistors per chip (3). The continued validity of Moore's Law is, however, very much in doubt as the photolithographic-patterning technology used in microelectronics fabrication appears to be reaching its minimum size limit (2), a dimension comparable to the wavelength of the UV photons employed.

Furthermore, to support higher device densities, the power used by individual components must be decreased. This requires a thinning of the gate oxide layer to maintain the current response of the transistor at lower gate voltages (4). Currently the oxide layer can be made as thin as seven SiO_2 formula units. The international roadmap for semiconductors projects that by 2012 the gate-oxide thickness must be reduced to five SiO_2 units as device sizes continue to miniaturize. Such scaling will then have reached its fundamental physical limit, as the quantum-mechanical tunneling of electrons is known to cause electrical breakdown of the SiO_2 gate insulator at thicknesses below five formula units (4). Thus, as device dimensions shrink further into the nanometer regime and microelectronics give way to nanoelectronics, continued progress will require either that we accept slower incremental improvements in the existing technologies (1), likely at rates considerably below that predicted by Moore's Law, or that we identify quite different construction and fabrication technologies.

Much basic-research effort is now invested in developing new, bottom-up fabrication strategies for molecular electronics and nanoelectronics (5, 6). One-dimensional nanostructures such as single-wall carbon nanotubes (7, 8) and semiconductor nanowires (5, 9) are receiving particular attention. Field-effect transistors (FETs) (10), single-electron transistors (SETs) (11), p–n diodes (12), rectifiers (13), logic circuits (14), address decoders (15), and other devices have now been constructed either with single nanowires or in crossed-nanowire arrays.

The fabrication of nanoelectronic circuitry by the bottom-up assembly of nanowires would ideally require two types of nanowire: (1) semiconducting to provide the device functions, and (2) metallically conducting to serve as the device interconnects. Carbon nanotubes can be either semiconducting or metallically conducting, depending on diameter and chirality (wall orientation) (16). Unfortunately, these electrical and structural characteristics are not yet controlled by synthesis. Synthetic mixtures contain both semiconducting and metallic carbon nanotubes, which are not readily sorted and separated. One might imagine that nanowires composed of metallic elements would be the ideal nanoscale interconnect. However, metal nanowires with diameters smaller than ~60 nm are highly susceptible to electromigration thinning and failure induced by local heating (17). Semiconducting nanowires composed of conventional semiconductors such as silicon may also be susceptible to failure due to local heating. Therefore, the ideal conducting and semiconducting nanowires may not yet have been discovered.

We suggest that ideal nanoelectronic components should be composed of refractory, chemically stable, covalently bonded elements or compounds. Furthermore, the nanowires should have electrical properties that do not depend strongly on size or crystallographic orientation, and are therefore uniform from wire to wire. Nanostructures are known to have lower melting points and higher chemical reactivities than bulk materials of the same composition, and thus nanowires of particularly refractory, chemically stable compositions should best withstand the effects of local heating induced by current. Covalent bonding should afford protection against electromigration (8). Consequently, we are investigating boron and metal-boride nanotubes and nanowires, expecting to achieve both semiconducting and metallic character, thermal and chemical stability, and high mechanical strength.

Although elemental boron is a semiconductor, theoretical studies have predicted that boron nanotubes should be stable (18, 19) and highly electrically conductive irrespective of diameter or wall orientation (18). Additionally, boron is a refractory, chemically stable element that produces stiff and strong fibers (20). Metal borides are also refractory and chemically stable (20), and often have low resistivities comparable to those of their constituent pure metals (see Table I). In fact, metal borides can be metallically conducting (Table I), semiconducting (such as BP (21)), and even superconducting (MgB$_2$ (22)). Therefore the boron and metal boride family of materials might provide ideal nanowires for both the metallically conducting and semiconducting components sought for nanoelectronics. Ultimately, we intend to survey broadly the electrical properties of this nanowire family.

Table I. Room-temperature Resistivities of Some Metals and Metal Diborides

Metal (M)	Resistivity of M ($\mu\Omega$ cm)[a]	Resistivity of MB$_2$ ($\mu\Omega$ cm)
Al	2.7	31[b]
Ti	42	9[b]
Mg	4.5	9.6[c]

[a]*CRC Handbook of Chemistry and Physics;* Lide, D. R., Ed.; CRC Press: Boca Raton, 2002; pp. 12-45, 12-47.

[b]*Boron and Refractory Borides;* Matkovich, V. I., Ed.; Springer Verlag: Berlin, 1977; p. 401.

[c]Ref. 22.

The present report addresses the electrical properties of boron nanowires. At the outset, we hoped to determine if boron-nanowire synthesis would afford hollow nanotubes, dense nanowires, or both. As noted above, boron nanotubes should exhibit metallic conductivity, whereas dense nanowires should be semiconducting. If dense, semiconducting nanowires were obtained, we intended to determine if they might be potentially useful components for nanoelectronics.

As described below, we indeed obtained dense, semiconducting nanowires rather than hollow boron nanotubes. We found that the carrier mobilities in these boron nanowires were considerably smaller than those of other semiconducting nanowires, suggesting that elemental-boron nanowires do not have ideal potential for use in nanoelectronic devices. The origins of the low mobilities and the implications for future research are described. The results included herein have appeared previously (23, 24), although some new analysis and discussion are provided.

Results and Discussion

Synthesis

Crystalline boron nanowires were synthesized by a catalyzed-CVD method at 1100 °C, using diborane as the boron precursor and NiB as the growth catalyst (22). The boron nanowires grew from the surface of a liquid nickel boride layer formed upon the alumina substrate under the growth conditions. An SEM image of the boron nanowires is shown in Figure 1a.

The nanowires had diameters ranging from 20–200 nm with a mean diameter of ca. 60 nm, and lengths of 1–10 μm (23). Elemental analysis by EDXS and PEELS in the TEM detected boron and also a small amount carbon, ascribed to adventitious surface deposits. TEM and electron diffraction revealed that the one-dimensional nanostructures were dense nanowhiskers, not hollow nanotubes, containing a twinning plane along the crystallographic c axis. A TEM image of a boron nanowire with discernable twinning boundary is shown in Figure 1b.

Nanodiffraction characterized the crystal structure as orthorhombic with $a = 9.4$ Å, $b = 7.1$ Å, and $c = 5.4$ Å, which does not correspond to any known form of boron. However, according to previous studies boron has shown a propensity to form boron-rich compounds by incorporating other elements into its crystal structure at levels potentially below our detection limits (25) ; thus we do not presently assert that the nanowires constitute a new allotrope of boron.

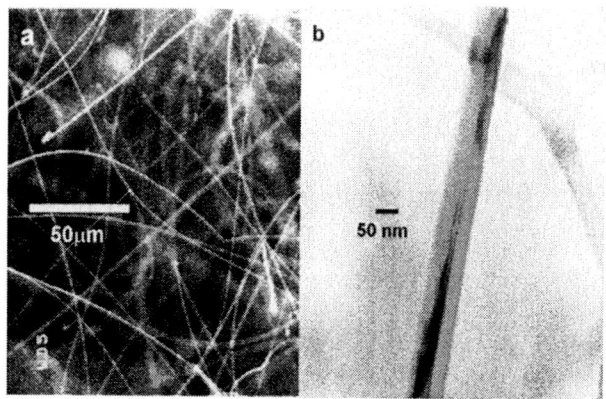

Figure 1. (a) An SEM image of boron nanowires; (b) a TEM image of an individual nanowire showing the twinning plane along the wire axis.

Device Fabrication

The nanowires were removed from the nickel boride slag by sonication in 1-octanol. The grey suspension was allowed to settle for ca. 1 hour to allow the larger pieces to fall to the bottom of the vessel. The remaining gray suspension was pippetted into test tubes and centrifuged for several minutes to induce further settling. The "supernatant" was separated from the packed solid by pipetting and one drop was placed on a clean silicon wafer for imaging. This cycle was repeated until individual nanowires were identifiable by SEM on the silicon wafer, affording a stock nanowire suspension for subsequent use.

Devices were constructed by pipetting drops of the nanowire suspension onto a heavily doped p-type Si wafer capped with a 500 nm oxide layer (24). Nanowire coordinates were calculated from predefined alignment marks present on the silicon wafer, which enabled controlled placement of the contact electrodes during fabrication. A bi-layer of electron beam resist (380 nm MMA/MMA bottom, 170 nm PMMA top) was spin coated onto the silicon wafer and baked for 15 minutes at 180 °C. Electrodes were deposited on top of individual nanowires by electron-beam lithography, followed by metal deposition and liftoff. Electrical transport measurements were done using an Agilent 4156C semiconductor parametric analyzer, using the degeneratively doped silicon wafer as a back gate (24). A diagram of a boron nanowire device is shown in Figure 2.

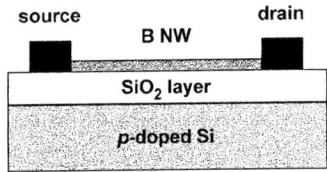

Figure 2. Diagram of a boron-nanowire device constructed for electrical-transport measurements.

Electrical Transport

Figure 3 shows I–V data collected from an individual boron-nanowire device fabricated with Ti/Au electrodes (24). Negligible current was measured at low applied bias (source-drain) voltages, which was indicative of resistance gap in the device, evident in the region near 0 V. After annealing the device briefly at 400 °C in an Ar atmosphere, the current increased by a factor of 1.5 and the resistance gap was reduced. However, the response was still non-ohmic, which precluded us from determining the inherent nanowire conductivity due to the considerable contribution from the contact resistances of the electrode-nanowire junctions.

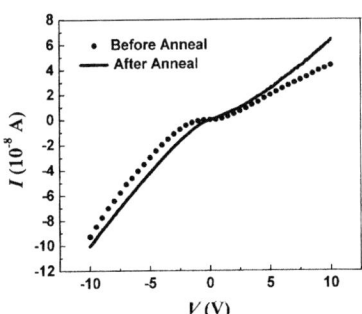

Figure 3. I–V curves of a boron-nanowire device with Ti/Au contacts, before and after annealing, giving evidence of non-ohmic (Schottky barrier) contacts. Adapted with permission from reference 24. Copyright 2003 AIP.

Figure 4 shows the electrical response of an annealed boron-nanowire device fabricated with Ni/Au contacts (24). The I–V measurements were collected on three segments of a boron nanowire having different lengths. For each segment, the I–V response was linear at low applied biases, which indicated that the device exhibited ohmic behavior.

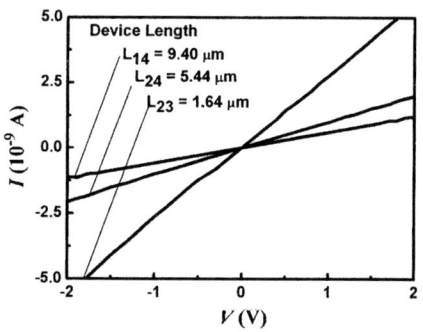

Figure 4. I–V data generated from different length segments of a boron-nanowire device having four variously spaced Ni/Au electrodes (1–4). Adapted with permission from reference 24. Copyright 2003 AIP.

Figure 5 plots the resistance (R) vs. length (L) data extracted from Figure 4. The R vs. L data were linear, consistent with Ohm's Law, and intersected the origin, indicating that the contact resistances were negligible. Thus the electrical resistance of the device reflected the inherent nanowire conductivity.

The conductivity of the boron nanowire was calculated from the slope of the line in Figure 5 and found to be on the order of 10^{-2} $(\Omega$ cm$)^{-1}$ in the annealed device, as compared to 10^{-3} $(\Omega$ cm$)^{-1}$ in the unannealed device (data not shown). These values are higher than the literature value for bulk β- rhombohedral boron $(10^{-6}$ $(\Omega$ cm$)^{-1})$, and slightly outside the range of values measured for other forms of elemental boron $(10^{-4}$–10^{-7} $(\Omega$ cm$)^{-1})$ (26). The conductivities of bulk boron carbides $B_{1-x}C_x$ ($x = 0.1$ to 0.2) (27) and carbon-doped bulk boron (28) are in the range of 10^{0}–10^{2} $(\Omega$ cm$)^{-1}$, which is higher than that of the boron nanowires. That the boron nanowires exhibited lower conductivity than those of boron-carbide or carbon-doped boron phases provided further evidence that the trace carbon detected in the boron nanowires (see above) was primarily a surface deposit rather than a solute species distributed throughout the volume of the nanowires.

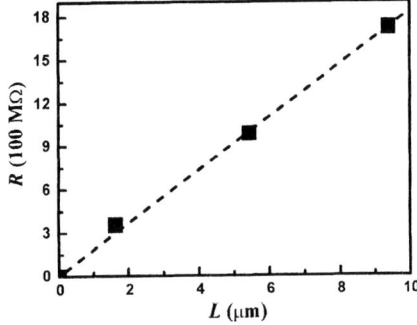

Figure 5. Resistance vs. length in a boron-nanowire device. Successful extrapolation of the linear fit through the origin established that the contact resistances were negligible. Adapted with permission from reference 24. Copyright 2003 AIP.

The conductivities of the boron nanowires, elemental boron, and boron carbides and carbon-doped boron are graphically compared to the conductivities of common insulators, semiconductors, and metals in Figure 6. The boron-nanowire conductivities fall in the center of the semiconductor regime. Thus the dense boron nanowires have semiconducting character similar to that of bulk boron, rather than the metallic character predicted for boron nanotubes.

Electrical-transport measurements from a boron-nanowire device with ohmic, Ni/Au contacts showed that the boron nanowire exhibited a gate-dependent I–V response characteristic of a p-type semiconductor (24). In these experiments various gate voltages were applied using the silicon substrate as a third electrode (a back gate), capacitively coupled to the nanowire through the insulating SiO_2 layer. At each gate voltage, the current through the device was determined as a function of the bias (source-drain) voltage, as above. Figure 7 reveals that the slope of the I–V data was greater at negative gate voltages, indicating hole conduction and thus p-type character.

The simplified band diagrams in Figure 8 demonstrate how gate voltage influences the conductance of a p-type nanowire device, in which the majority carriers are holes. As the semiconducting nanowire makes contact with the metallic electrodes, its valence band bends upward at the contacts to equilibrate the Fermi levels of the nanowire and the metallic electrodes (Figure 8a). The holes in the boron nanowire rise to the top of the valence band. A positive gate voltage (Figure 8b) pulls the valence-band edge downward in the nanowire, causing the equilibrium hole concentration in the nanowire to decrease as holes

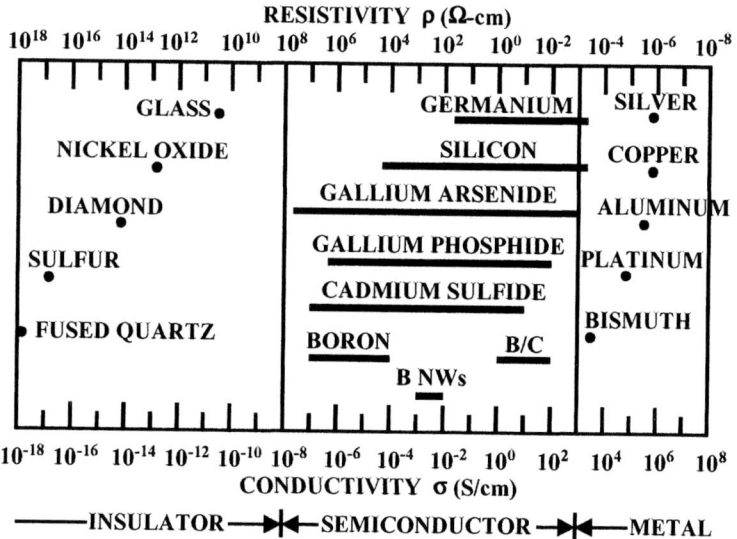

Figure 6. Classification of common materials by conductivity. Bulk boron, boron nanowires, and boron carbides are included for comparison. Adapted from Ng, K. K. Complete Guide to Semiconductor Devices, 2nd ed., with permission of the author. Reprinted with permission of John Wiley & Sons, Inc. Copyright 2002 Wiley.

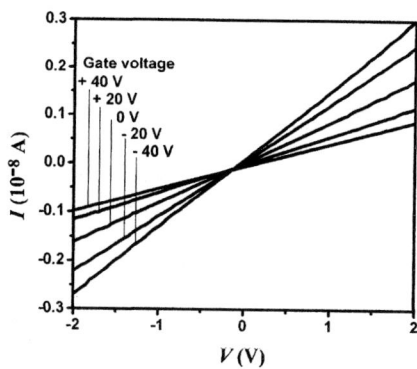

Figure 7. I–V data from a boron-nanowire device with Ni/Au contacts at various gate voltages, which indicate p-type character. Adapted with permission from reference 24. Copyright 2003 AIP.

are driven into the electrodes. The lower concentration of charge carriers results in a lower conductance through the device. However, a negative gate voltage (Figure 8c) pushes the valence-band edge upwards, increasing the equilibrium hole concentration in the nanowire and therefore increasing the conductance through the device. Thus, the data shown in Figure 7 indicate that the majority charge carriers are holes, and therefore that the nanowire is a *p*-type semiconductor.

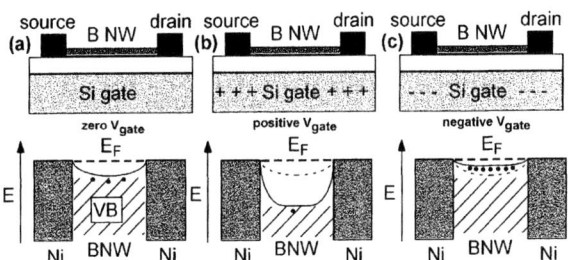

Figure 8. *Diagrams to analyze the effect of gate voltage on the conductance of the p-type semiconducting boron nanowire, for (a)* $V_g = 0$, *(b)* $V_g > 0$, *and (c)* $V_g < 0$. *The corresponding band diagram is shown below the device for each case.*

The *p*-type behavior of the boron nanowire is consistent with the properties of bulk boron, which is an intrinsic *p*-type semiconductor due to a high density of acceptor states at 0.2 eV above the valence band edge (*29*). The preponderance of acceptor states is most likely due to intrinsic structural defects and Jahn-Teller distortions in its atypical icosahedral-cluster-based crystalline lattice (*30*).

The gate-dependent behavior observed in the ohmic boron-nanowire device afforded a measurement of the transconductance; that is, the ratio of change in source-drain current to change in gate voltage (dI/dV_g). The transconductance evaluates the gain or degree of signal amplification that the device is capable of. The transconductance plot in Figure 9 was collected by varying gate voltage and measuring the current through the device, at a constant bias (source-drain) voltage of 10 V (*24*). The carrier mobility in the nanowire was calculated from the transconductance, specifically from the slope of the curve at $V_g \leq 10$ V, and was found to be on the order of 10^{-3} cm^2/Vs (*24*). Table II lists the carrier mobilities of other one-dimensional nanotubes and nanowires, which reveals that the carrier mobility of the boron nanowire is several orders of magnitude lower in comparison. This is unfortunate, as the speed and sensitivity of a semiconductor device increase with increasing carrier mobility.

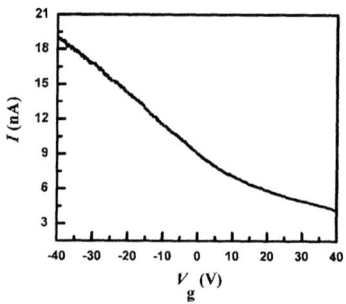

Figure 9. An I – V_g curve showing the transconductance of a boron-nanowire device at a constant 10 V source-drain bias. Adapted with permission from reference 24. Copyright 2003 AIP.

Table II. Comparison of mobilities for different one-dimensional nanostructures

Material	Mobility cm^2/Vs
Boron nanowire[a]	10^{-3}
Intrinsic silicon nanowire[b]	5.9×10^{-3}
p-silicon nanowire[b]	3.17
GaN nanowire[c]	150-650
SnO_2 nanowire[d]	40
In_2O_3 nanowire[e]	98.1 (air)
	46 (vacuum)
Single-walled carbon nanotube d = 1.6 nm[f]	20 (holes)
Single-walled carbon nanotube d = 4 nm[g]	220 (holes)
	150 (electrons)

[a]Ref. 24.

[b]Cui, Y.; Duan, X.; Hu, J.; Lieber, C. M. *J. Phys. Chem. B* **2000**, 104, 5213-5216.

[c]Huang, Y.; Duan, X.; Cui, Y.; Lieber, C. *Nano Lett.* **2002**, 2, 101-104.

[d]Liu, Z.; Zhang, D.; Han, S.; Li, C.; Tang, T.; Hin, W.; Liu, X.; Lei, B.; Zhou, C. *Adv. Mater.* **2003**, 15, 1754-1757.

[e]Zhang, D.; Li, C.; Han, S.; Liu, X.; Tang, T.; Jin, W.; Zhou, C. *Appl. Phys. Lett.* **2003**, 82, 112-114.

[f]Martel, R.; Schmidt, T.; Shea, H. R.; Hertel, T.; Avouris, Ph. *Appl. Phys. Lett.* **1998**, 73, 2447-2449.

[g]Javey, A.; Shim, M.; Dai, H. *Appl. Phys. Lett.* **2002**, 80, 1064-1066.

Interestingly, the carrier mobility we measured for the boron nanowire agrees well with the previously measured values for bulk boron, which are in the range of 10^{-2}–10^{-8} cm^2/V s (31). The low mobilities in bulk boron are likely due to its complex crystal structure. Carrier mobility is limited by scattering, or the mean free path of the charge carrier. The non-close-packed crystal structures of elemental boron, which are based on the assembly of icosahedral B$_{12}$ units, provide favorable sites for impurity dopant atoms that may scatter carriers (32). Phonons (quantized lattice vibrations) also scatter carriers, and the complex crystal structures of boron produce a wide spectrum of phonon energies for carriers to interact with (33). Intrinsic structural defects in the complex structures induce a high density of trapping centers for carrier localization (34), which induce a hopping-type conduction $(35, 36)$ that is detrimental to carrier mobility. Thus, the low carrier mobility exhibited by the boron nanowire is apparently inherited from bulk boron, and presumably has similar origins. Elemental boron phases containing the B$_{12}$ structural unit, with the possible exception of α-rhombohedral boron (37), are not good semiconductors for potential device applications.

Summary

We have found the electrical properties of dense boron nanowires to be very similar to those of bulk boron; both are low-mobility *p*-type semiconductors. The nanowires studied here had diameters of 75–80 nm, which are fairly large. Property differences between boron nanowires and bulk boron are more likely to emerge in nanowires of much smaller diameter (ca. 2–20 nm). Indeed, hollow boron nanotubes (3-nm diameters), which are predicted to be metallically conductive (18), have recently been prepared by a template-based strategy (38). Unfortunately, their sensitivity to electron-beam damage in the TEM has precluded a full characterization. The electrical properties of boron nanotubes remain unknown.

Our studies suggest that elemental-boron nanowires are unlikely to serve as ideal components for either the interconnect (conducting) or device (semiconducting) functions in nanoelectronics. Consequently, we are now directing our efforts towards nanowires of metal borides such as TiB$_2$, AlB$_2$, and MgB$_2$, which should exhibit good metallic conductivities, and the semiconductor BP, which should exhibit good carrier mobility comparable to the better values shown in Table II (*39*).

Finally, we note that until a method for the massively parallel integration and interconnection of nanowire and nanotube components is developed, bottom-up nanoelectronics as discussed here will perhaps remain an interesting area for scientific research, but will not significantly impact technology nor perpetuate Moore's Law.

Acknowledgment

This work was supported by an NSF Nanoscale Interdisciplinary Research Team grant, No. EEC-0210120.

References

1. Hutcheson, G. D. *Scientific American* **2004**, *290(4)*, 76-83.
2. Normile, D. *Science* **2001**, *293*, 787.
3. *Intel Microprocessor Quick Reference Guide*, URL http://www.intel.com/pressroom/kits/quickrefyr.htm.
4. Schulz, M. *Nature* **1999**, *399*, 729-730.
5. Lieber, C. M. *Scientific American* **2001**, *285(3)*, 58-64.
6. Whitesides, G. M.; Love, J. C. *Scientific American* **2001**, *285(3)*, 39-47.
7. Avouris, Ph. *Acc. Chem. Res.* **2002**, *35*, 1026-1034.
8. McEuen, P. L.; Park, J.-Y. *MRS Bull.* **2004**, *29(4)*, 272-275.
9. Lieber, C. M. *MRS Bull.* **2003**, *28(7)*, 486-491.
10. Tans, S. J.; Verschureren, A. R. M.; Dekker, C. *Nature* **1998**, *393*, 49-52.
11. Tans, S. J.; Devoret, M. H.; Dai, H.; Thess, A.; Smalley, R. E.; Geerlings, L. J.; Dekker, C. *Nature* **1997**, *386*, 474-477.
12. Duan, X.; Huang, Y.; Cui, Y.; Wang, J., Lieber, C. M. *Nature* **2001**, *409*, 66-69.
13. Fuhrer, M. S.; Nygård, J.; Shih, L.; Forero, M.; Yoon, Y.-G.; Mazzoni, M. S. C.; Choi, H. J.; Ihm, J.; Louie, S. G.; Zettl, A.; McEuen, P. L. *Science* **2000**, *288*, 494-497.
14. Derycke, V.; Martel, R.; Appenzeller, J.; Avouris, Ph. *Nano Lett.* **2001**, *1*, 453-456.
15. Zhong, Z.; Wang, D.; Cui, Y.; Bockrath, M. C.; Lieber, C. M. *Science* **2003**, *302*, 1377-1379.
16. Dekker, C. *Physics Today* **1999 (May)**, 22-28.
17. Durkan, C.; Schneider, M. A.; Welland, M. E. *J. Appl. Phys.* **1999**, *86*, 1280-1286.

18. Boustani, I.; Quandt, A.; Hernández, E.; Rubio, A. *J. Chem. Phys.* **1999**, *110*, 3176-3185.
19. Gindulyte, A.; Lipscomb, W. N.; Massa, L. *Inorg. Chem.* **1998**, *37*, 6544-6545.
20. *Boron and Refractory Borides*; Matkovich, V. I., Ed.; Springer Verlag: Berlin, 1977.
21. Kumashiro, Y. *J. Mater. Res.* **1990**, *5*, 2933-2947.
22. Canfield, P. C.; Finnemore, D. K.; Bud'ko, S. L.; Ostenson, J. E.; Lapertot, G.; Cunningham, C. E.; Petrovic, C. *Phys. Rev. Lett.* **2001**, *86*, 2423-2426.
23. Otten, C. J.; Lourie, O. R.; Yu, M.-F.; Cowley, J. M.; Dyer, M. J.; Ruoff, R. S.; Buhro, W. E. *J. Am. Chem. Soc.* **2002**, *124*, 4564-4565.
24. Wang, D.; Lu, J. G.; Otten, C. J.; Buhro, W. E. *Appl. Phys. Lett.* **2003**, *83*, 5280-5282.
25. Ploog, K. *J. Electrochem. Soc.* **1974**, *121*, 846-848.
26. *Landolt-Börnstein Numerical Data and Functional Relationships, New Series*; Hellwege, K.-H., Ed.; Springer-Verlag: Berlin, 1983; Vol. III/17e, pp 16-18.
27. Wood, C.; Emin, D. *Phys. Rev. B* **1984**, 29, 4582-4587.
28. Reference 26, p 272.
29. Werheit, H.; de Groot, K.; Malkemper, W.; Lundström, T. *J. Less-Common Met.* **1981**, 82, 163.
30. Werheit, H.; Schmechel, R. *J. Solid State Chem.* **2000**, 154, 61.
31. Reference 26, p 11.
32. Greenwood, N. N.; Earnshaw, A. *Chemistry of the Elements*, 2nd ed.; Butterworth-Heinemann: Oxford, 1997, pp 139-144.
33. Werheit, H.; Kuhlmann, U. *Solid State Commun.* **1993**, *88*, 421-425.
34. Schmechel, R.; Werheit, H. *J. Solid State Chem.* **2000**, *154*, 61-67.
35. *Boron Preparation and Properties*; Niemyski, T., Ed.; Polish Scientific Publishers: Warszawa, Poland, 1970, Vol. 3, p 18.
36. Reference 20, p 52.
37. Golikova, O. A. *Chemtronics* **1991**, *5*, 3-9.
38. Ciuparu, D.; Klie, R. F.; Zhu, Y.; Pfefferle, L. *J. Phys. Chem. B* **2004**, *108*, 3967-3969.
39. Shohno, K.; Takigawa, M.; Nakada, T. *J. Cryst. Growth* **1974**, *24-25*, 193-196.

Chapter 27

Molecular Design of Precursors for the Chemical Vapor Deposition of Group 13 Chalcogenides

Claire J. Carmalt, Emily S. Peters, Simon J. King,
John D. Mileham, and Derek A. Tocher

Department of Chemistry, University College London,
20 Gordon Street, London WC1H 0AJ, United Kingdom

A range of aluminum and gallium alkoxide and thiolate complexes have been synthesized and characterized. These complexes are of interest as precursors to group 13 oxides and sulfides. Thermal gravimetric analysis was used to study the decomposition of some of the compounds. The formation of thin films of alumina and gallium oxide on glass substrates was achieved via low pressure chemical vapor deposition. The films were analyzed by SEM/EDAX, powder XRD, XPS and band gap measurements.

Introduction

Traditional synthetic routes to many materials involve high temperatures and/or the use of toxic and pyrophoric reagents. The formation of a range of materials, including metal nitrides, sulfides and phosphides *via* single-source chemical vapor deposition (CVD) has been of interest ([1–3]). Single-source precursors offer several advantages over traditional routes. These include lower deposition temperatures, reduced contamination from carbon and removal of the use of toxic/pyrophoric reagents. In general, a successful single-source precursor should contain the elements of the desired material (e.g. Ga and O for Ga_2O_3), possess easily removable ligands and exhibit sufficient volatility.

Alumina thin films offer, in principle, great potential in a number of applications including as wear resistant coatings, insulating layers, optical filters and corrosion protective coatings (*4*). The conventional CVD method for preparing alumina thin films involves the hydrolysis of $AlCl_3$ with a mixture of H_2 and CO_2 at 700–1000 °C (*5*). The formation of alumina films from other precursors, such as the pyrophoric $AlMe_3$ have also been described (*6*). However, highly volatile and non-pyrophoric precursors are preferred for the deposition of alumina films in order to make the process safer. The versatility of metal alkoxides as precursors to metal oxides *via* CVD processes is widely recognized; they are easy to prepare and purify and are intrinsically non-corrosive so can be stored almost indefinitely in a dry atmosphere (*7*). Metal alkoxides are ideal for use in low pressure (LP)CVD as a result of their high volatility. Despite the interest in alumina there are few reports of the use of organoaluminum alkoxides for the deposition of Al_2O_3 thin films (*8*).

Gallium oxide (Ga_2O_3) is considered to be one of the most ideal materials for application as thin film gas sensors at high temperature (*9*). Gallium oxide is an electrical insulator at room temperature and semiconducting above 400 °C; it is chemically and thermally stable. As well as detecting oxygen gas at temperatures above 900 °C, a gallium oxide thin film operates as a surface-control-type sensor to reducing gases below 900 °C (*10*). Therefore, it may be possible to switch the function of the sensor with temperature (*11*). CVD is the most practical method for preparing thin films for large-scale applications. CVD has a number of advantages over other deposition techniques as it offers the potential for good film uniformity and composition control, large area growth and excellent step coverage. Despite the interest in gallium oxide for thin film gas sensor applications there have been very few reports using CVD to deposit Ga_2O_3 films. The precursors [Ga(hfac)$_3$] (hfac = hexafluoroacetoacetonate), [Ga(OCH(CF$_3$)$_2$)$_3$(HNMe$_2$)] and [Ga(O^{t}Bu)$_3$]$_2$ have been used to deposit amorphous Ga_2O_3 thin films at ~ 450 °C (*12, 13, 14*). Group 13 sulfides (e.g. Al_2S_3) are also of interest and are reported to have potential applications as alternatives to II/VI materials for photovoltaic and optoelectronic devices (*15*).

The synthesis and characterization of a range of potential precursors to aluminum and gallium oxides and sulfides are described herein. The decomposition of some of the compounds has been studied using thermal gravimetric analysis and the formation of thin films of alumina and gallium oxide is described.

Synthesis of Molecular Precursors

A requisite for metal alkoxides to be precursors in CVD processes are that they should be sufficiently volatile and sublimation should occur at as low a temperature as possible. The presence of hydride ligands in place of alkoxide groups, e.g. [AlH(OR)$_2$]$_n$ *versus* [Al(OR)$_3$]$_n$, can increase the volatility of the precursor. Another method for increasing the volatility of metal alkoxides is to use the 'donor functionalization' concept (*16*). Using this concept a ligand type can be developed which unifies the advantages of both steric demands and σ-donor stabilization. The reaction of alane with alcohols and thiols and the formation of aluminum and gallium 'donor functionalized' alkoxides are described below.

Reactivity of Alane with Alcohols and Thiols

The reaction between [$AlH_3(OEt_2)$] (generated *in situ* from $LiAlH_4$ and $AlCl_3$ (*17*)) and 1 equivalent of *i*-PrOH in diethyl ether at room temperature resulted in the isolation of colorless crystalline **1** (*18*), as shown in Scheme 1.

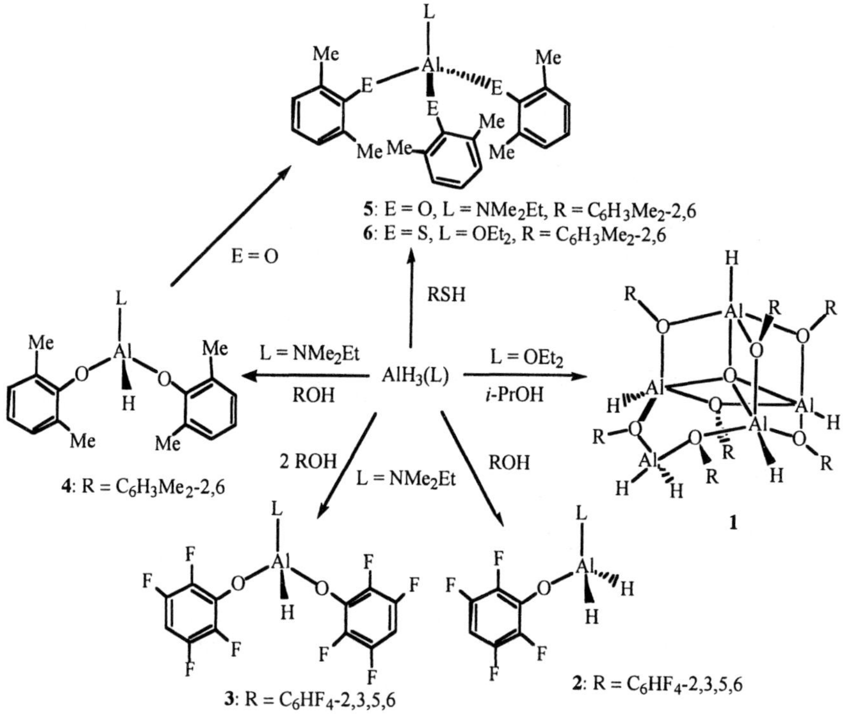

5: E = O, L = NMe2Et, R = C6H3Me2-2,6
6: E = S, L = OEt2, R = C6H3Me2-2,6
E = O
RSH
L = NMe2Et
ROH
AlH3(L)
L = OEt2
i-PrOH
2 ROH
L = NMe2Et
ROH
4: R = C6H3Me2-2,6
1
3: R = C6HF4-2,3,5,6
2: R = C6HF4-2,3,5,6

Scheme 1

An X-ray structure determination showed that the pentanuclear complex [$Al_5(\mu_4$-O)(μ-O-*i*-Pr)$_7H_6$] (**1**) had formed (Figure 1). The formation of compound **1** could be the result of hydrolysis, however oxoalkoxides have been isolated previously, even where exceptional precautions against adventitious hydrolysis were taken. For example, triisopropoxides of a number of metals are actually pentanuclear oxoaggregates of formula [$M_5(\mu_5$-O)(μ_3-O-*i*-Pr)$_4$(μ-O-*i*-Pr)$_4$(O-*i*-

Pr)₅] (e.g. M = In, Sc, Yb) *(19, 20)*. The central, four-coordinate, oxygen atom in 1 adopts a distorted square pyramidal geometry, with Al(4) occupying the apical position, but with the basal site opposite to Al(3) vacant. The Al(2)–O–Al(2A) angle is 154.09(15)° whilst the remaining Al–O–Al angles are in the range 97.88(8) to 99.02(8)°. The Al–O(isopropoxide) distances involving Al(2), Al(3) and Al(4) are typical (1.843(2)–1.858(2) Å), whereas those to Al(1) are significantly shorter by 0.04 Å; conversely, the Al–O bonds to the central oxygen atom are 0.04 Å longer with that to the "apical" Al(4) being the longest.

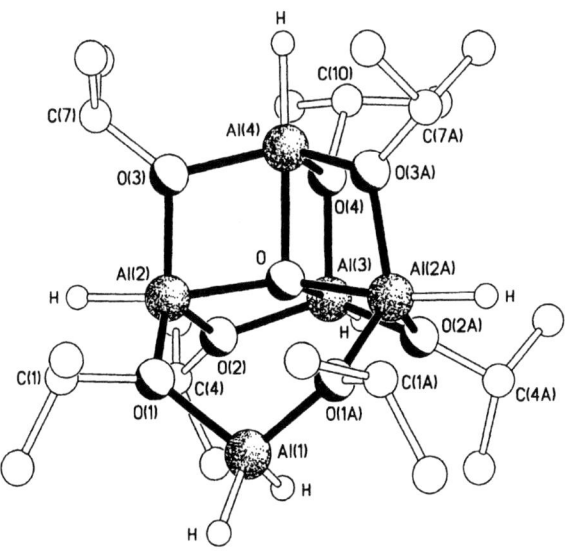

Figure 1. Molecular structure of the C_s symmetric complex 1. (Reproduced by permission of The Royal Society of Chemistry (RSC) on behalf of the Centre National de la Recherche Scientifique (CNRS). Copyright 2002.)

The reaction of [AlH₃(NMe₂Et)] and one equivalent of ROH (R = C₆HF₄-2,3,5,6) in diethyl ether resulted in the formation of the dihydrido complex [H₂Al(OR)(NMe₂Et)] (**2**) (Scheme 1). The ¹H NMR spectrum of **2** shows the expected 1:1 ratio of amine:alkoxide peaks. The presence of the hydride group was confirmed by the FT-IR spectra of **2**, which showed characteristic absorption and deformation bands at 1897 and 772 cm⁻¹, respectively *(21)*. Based on the presence of the intense band at 1897 cm⁻¹ in the IR spectrum, the structure of **2** is most probably monomeric (Scheme 1) *(22)*.

The monohydrido complexes [HAl(C₆HF₄-2,3,5,6)₂(NMe₂Et)] (**3**) and [HAl(OC₆H₃Me₂-2,6)₂(NMe₂Et)] (**4**) were isolated from the reaction of [AlH₃(NMe₂Et)] and two equivalents of ROH (R = C₆HF₄-2,3,5,6 or C₆H₃Me₂-

2,6) in diethyl ether (Scheme 1). The ^{1}H NMR spectra of **3** exhibits broad peaks suggesting that a fluxional process is occurring in solution at room temperature. The 1:2 ratio of amine:alkoxide peaks is observed as expected and the FT-IR shows characteristic Al–H bands (1896 and 776 cm^{-1}). Analytical and spectroscopic data confirmed the formation of compound **4** (FT-IR 1843 and 760 cm^{-1}). The observed sharp band in the IR spectra of **3** and **4** are consistent with monomeric structures (Scheme 1) (*22*).

Crystals of (**5**) were obtained on recrystallization of **4**, and from the 1:3 reaction of [AlH$_3$(NMe$_2$Et)] with 2,6-Me$_2$C$_6$H$_3$OH. Analytical and spectroscopic data for **5** were consistent with the formation of the triphenoxide [Al(OC$_6$H$_3$Me$_2$-2,6)$_3$(NMe$_2$Et)]. The X-ray structure of **5** (Figure 2) shows the coordination of the central aluminum to be a flattened tetrahedral, the N–Al–O angles ranging between 102.49(6) and 107.56(6)°, whereas those for O–Al–O are between 111.73(6) and 116.36(6)°. The Al–O distances are unexceptional, ranging between 1.7007(12) and 1.7172(12) Å [the Al–N distance is 1.9736(14) Å]. The structure of **5** is similar to related compounds, such as the ketone adduct [Al(OC$_6$H$_2$-*t*-Bu$_2$-2,6-Me-4)(O=C(C$_5$H$_9$)-*t*-Bu-4)] (*21*).

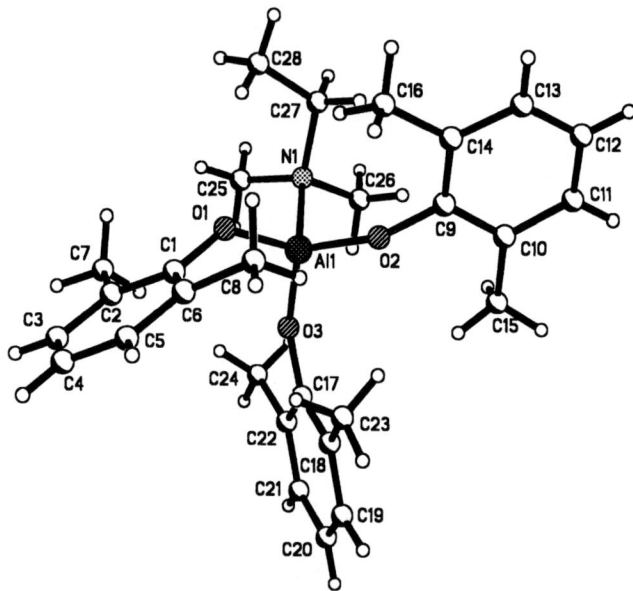

*Figure 2. The Molecular Structure of **5***.

The reaction of [AlH$_3$(OEt$_2$)] and one equivalent of 2,6-Me$_2$C$_6$H$_3$SH in diethyl ether at room temperature afforded colorless crystals of [Al(SC$_6$H$_3$Me$_2$-2,6)$_3$(OEt$_2$)] (**6**) (*23*). Analytical and spectroscopic data for **6** were consistent with the formation of [Al(SC$_6$H$_3$Me$_2$-2,6)$_3$(OEt$_2$)] rather than that of the expected product [H$_2$Al(SC$_6$H$_3$Me$_2$-2,6)]$_n$. The X-ray structure of **6** is shown in Figure 3. The structure, ignoring the two ethyl groups, has pseudo C_3 symmetry about the O–Al direction creating a propeller-like geometry. The three 2,6-dimethylphenyl ring systems are oriented approximately orthogonally (77, 82 and 96°) to the S$_3$ plane. The "all down" geometry observed here is in contrast to the "one up, two down" conformation seen in compound **5**. The geometry at aluminum is best described as flattened tetrahedral, the metal center being displaced towards the "basal" S$_3$ face, lying only 0.43 Å out of this plane in the direction of the oxygen atom. The angles subtended at aluminum are in the range 97.34(7)–120.81(5)°, and the sum of the three S–Al–S angles is 349°. The Al–S distances are unexceptional, ranging between 2.227(1) and 2.238(1) Å [the Al–O distance is 1.907(2) Å]. The propeller-like geometry results in one of the methyl groups of each 2,6-dimethylphenyl ring system being directed into the face of its nearest neighbor aryl ring, the shortest H···π distance being 2.78 Å.

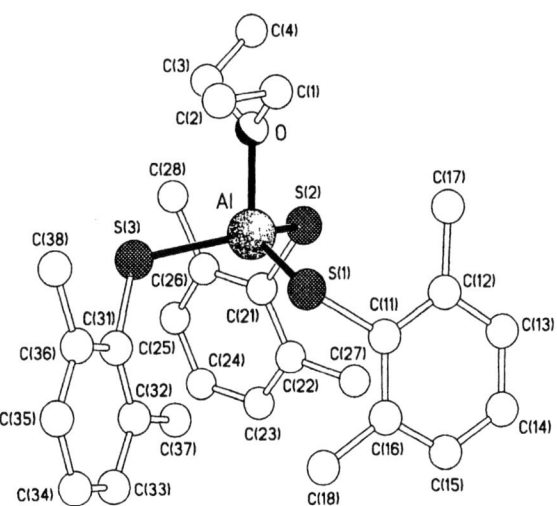

Figure 3. Molecular Structure of 6. (Reproduced with permission from reference 23. Copyright 2003 Elsevier.)

In order to obtain compound **6**, the reaction mixture formed between LiAlH$_4$ and AlCl$_3$ to produce [AlH$_3$(OEt$_2$)] was allowed to stir for 30 min before the addition of 2,6-Me$_2$C$_6$H$_3$SH. However, if 2,6-Me$_2$C$_6$H$_3$SH is added immediately to the slurry of LiAlH$_4$ and AlCl$_3$ in diethyl ether, colorless crystals of a new compound [Li(OEt$_2$)$_3$][Al(SC$_6$H$_3$Me$_2$-2,6)$_4$] (**7**) are obtained (*21*). The formation of **7** is the result of incomplete reaction of aluminum(III) chloride with LiAlH$_4$ before the addition of the thiol. Spectroscopic data shows only the formation of **7** rather than a mixture of products. A single crystal X-ray diffraction analysis showed the anion in **7** to be the S_4-symmetric homoleptic complex shown in Figure 4. The geometry at aluminum is distorted tetrahedral with unique angles of 104.85(3) and 119.16(6)°, reflecting a slight flattening of the tetrahedron about the S_4 axis [which bisects the S–Al–S(0A) angle], and an Al–S distance of 2.253(1) Å. The $[(Et_2O)_4Li]^+$ cation is disordered.

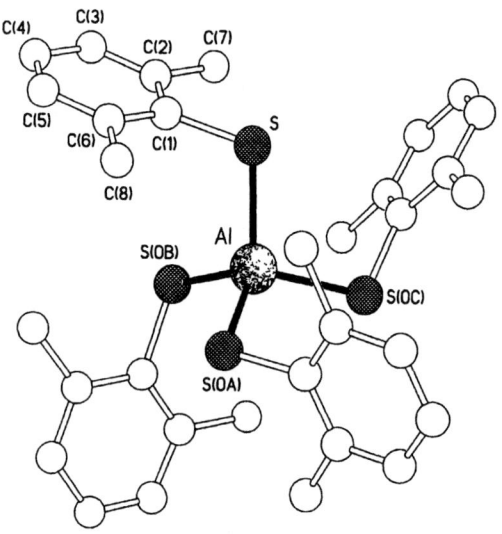

Figure 4. The structure of the homoleptic, S_4-symmetric anion 7. (Reproduced with permission from reference 23. Copyright 2003 Elsevier.)

Decomposition Studies

Thermal gravimetric analysis (TGA) of **3** (10 °C/min from 20 to 500 °C; under N_2) showed a total weight loss of 78%, which is less than the 88% weight loss expected if decomposition to Al_2O_3 has occurred. The TGA of **4** also showed that incomplete decomposition has occurred as the total weight loss (77%) is less than that expected (85%). Three weight losses were observed in the TGA of **3** and **4**, the first of which (19% and 23% respectively) corresponds to the loss of NMe_2Et (calculated as 17% and 21% respectively). The decomposition properties of $[Al(SC_6H_3Me_2-2,6)_3(OEt_2)]$ (**6**) were also studied by TGA (Figure 5). The decomposition of **6** is clean and shows a total weight loss of 73%. This is in good agreement with the calculated value for decomposition of **6** to Al_2S_3 (71%). Two distinct weight losses are observed, the first (19%) corresponds to the loss of solvent from **6** (20%). The second (54%) corresponds to the formation of Al_2S_3 from $[Al(SC_6H_3Me_2-2,6)_3]$ (52%). One possible decomposition mechanism involves the formation of R_2S and Al_2S_3.

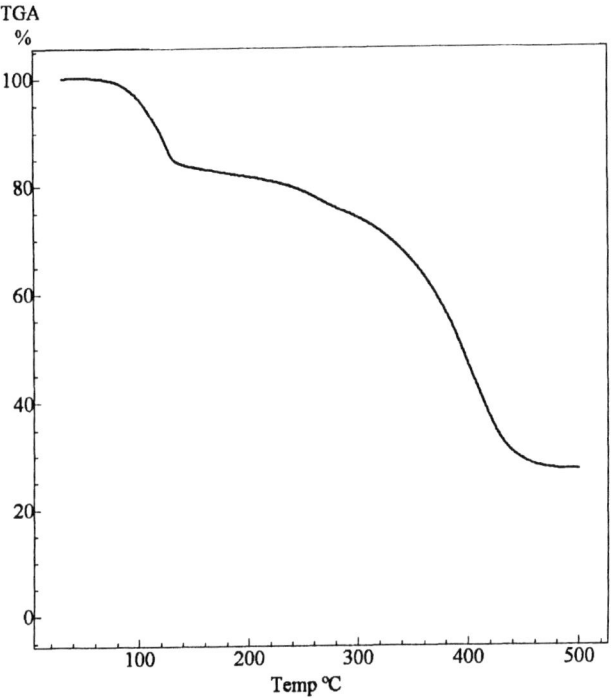

Figure 5. TGA of compound 6.

Donor-Functionalized Aluminum and Gallium Alkoxides

The reaction of Et_3M (M = Al or Ga) with one equivalent of $HOCH_2CH_2NMe_2$ resulted in the formation of white crystalline solids (compounds **8** and **9**), according to Scheme 2. Analytical and spectroscopic data were consistent with the formation of the monoalkoxides $[Et_2Al(OCH_2CH_2NMe_2)]_2$ (**8**) and $[Et_2Ga(OCH_2CH_2NMe_2)]_2$ (**9**). On the basis of mass spectroscopic data these compounds are thought to be dimeric *via* alkoxy bridges in the solid-state. By TGA, compound **9** gave a mass loss of 71% with a clean decomposition. This sample showed the best deposition characteristics. Therefore LPCVD using compounds **8** and **9** were studied.

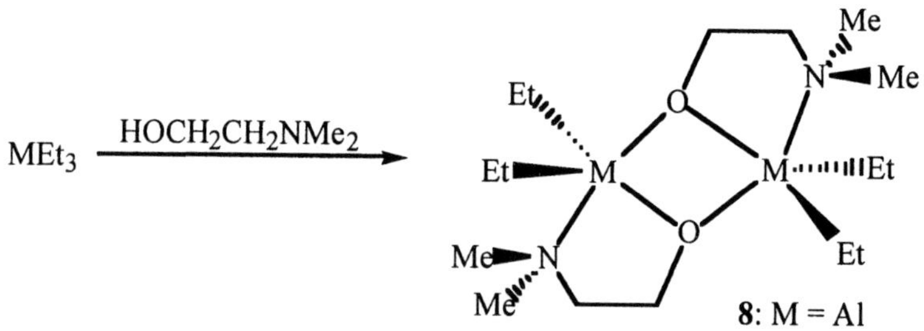

MEt₃ $\xrightarrow{\text{HOCH}_2\text{CH}_2\text{NMe}_2}$

8: M = Al
9: M = Ga

Scheme 2

Thin-Film Growth

LPCVD of **8** and **9** were investigated, the details of which are described in the Experimental section. Both complexes formed gray/white colored films on glass under LPCVD conditions at 500 °C. SEM analysis of the films produced from **8** showed that these films consist of agglomerates of approximately 0.5 μm. The EDAX analysis showed the presence of aluminum and oxygen in the films. Significant breakthrough of the coating to the underlying glass was observed and so accurate quantitative analysis by EDAX was difficult. X-ray photoelectron spectroscopy (XPS) of the film revealed the presence of aluminum and oxygen. The Al 2p binding energy shift of 74.69 eV and the O 2p shift of 532.79 eV are in good agreement with previous literature measurements (Al 2p, 74.7 eV; O 2p 531.6 eV) (*24*). The films showed negligible carbon content by both EDAX and XPS (0–1%). Overall films grown from **8** have an approximate composition of Al$_2$O$_3$. The films were amorphous according to powder XRD.

SEM analysis of the films produced from **9** showed the formation of spherical particles of approximately 0.5–2 μm in size. EDAX analysis showed the presence of gallium and oxygen in the films. However, the films were amorphous according to powder XRD, typical for Ga$_2$O$_3$ films grown at this temperature (*13*). Therefore, the films were annealed at 900 °C for 24 hr, resulting in crystalline β-Ga$_2$O$_3$ films as shown by powder XRD (Figure 6). X-ray photoelectron spectroscopy (XPS) of the film revealed the presence of gallium, oxygen and carbon. However, the carbon concentration was significantly reduced after the first etch, indicating that the carbon contamination was graphitic. The Ga 2p binding energy shift was 1119.2 eV and the O 2p shift was at 532.8 eV for films grown from compound **9**. The XPS peak profiles are consistent with a single environment for gallium and oxygen. The binding energy shifts are in agreement with previous literature values (*14*). Overall films grown from **9** have an approximate composition of Ga$_2$O$_3$. Conducting a Tauc plot of the UV/Visible data indicated that Ga$_2$O$_3$ prepared by

LPCVD had an indirect bandgap of 4.6 eV, comparable to other literature values for Ga_2O_3 of 4.2–4.9 eV (14).

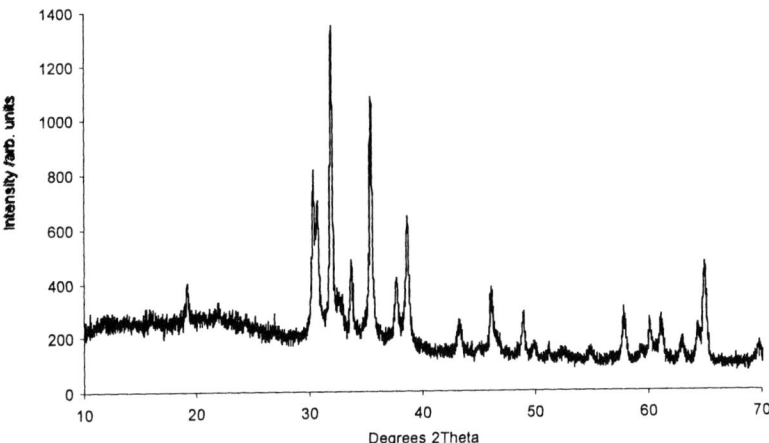

Figure 6. *X-ray powder diffraction pattern of the annealed Ga_2O_3 film grown from compound 9.*

Experimental

General Procedures

All manipulations were performed under a dry, oxygen-free dinitrogen atmosphere using Schlenk techniques or in a MBraun Unilab glove box. All solvents were distilled from appropriate drying agents prior to use (sodium - toluene and hexanes; sodium/benzophenone for diethyl ether). All other reagents were procured from Aldrich and used without further purification. Microanalytical data were obtained at University College London (UCL). The synthesis of compounds 1, 5 and 6 were described in references 18 and 23.

Physical Measurements

NMR spectra were recorded on Brüker AMX400 spectrometers at UCL. The NMR spectra are referenced to CD_2Cl_2, which was degassed and dried prior to use; ^{1}H and ^{13}C chemical shifts are reported relative to SiMe$_4$ (0.00 ppm). Mass spectra (CI) were run on a micromass ZABSE instrument, and IR spectra on a Nicolet 205 instrument. TGA of the compounds were obtained from the Thermal Methods Laboratory at Birkbeck College. EDXA/SEM results were obtained on a Hitachi S570 instrument using the KEVEX system. Glancing angle X-ray diffraction analysis of the films was obtained using a Siemens D5000 diffractometer in reflection mode (1.5° incidence) using CuK$_\alpha$ (λ = 15406 Å) radiation. X-ray photoelectron spectra were recorded using a VG ESCALAB 220i XL instrument using focused (300 mm spot) monochromatic Al Kα radiation at a pass energy of 20 eV. Scans were acquired with steps of 50 meV. A flood gun was used to control charging and the binding energies were referenced to an adventitious C 1s peak at 284.8 eV. Depth profile measurements were obtained by using argon beam sputtering.

Synthesis of 2

A solution of HOC$_6$HF$_4$-2,3,5,6 (0.63 g, 3.77 mmol) in diethyl ether (10 mL) was added dropwise to [AlH$_3$(NMe$_2$Et)] (0.5 mL, 3.77 mmol) in diethyl ether (10 mL) at room temperature. The evolution of hydrogen was observed and the mixture was allowed to stir for 1 hour. The volume of the solution was reduced under vacuum to *ca.* 5 mL. Cooling to -20 °C afforded a white powder of **2** (0.58 g, 79%), mp 42 °C. ^{1}H NMR (CD$_2$Cl$_2$): δ 1.26 (t, *J* = 14 Hz, 3H, NCH$_2$C*H$_3$*), 2.60 (s, 6H, N(C*H$_3$*)$_2$), 2.97 (q, *J* = 25 Hz, 2H, NC*H$_2$*CH$_3$), 6.53 (spt, *J* = 21 Hz, 1H, C$_6$F$_4$*H*), Al*H* not detected. ^{13}C{^{1}H} NMR (CD$_2$Cl$_2$): δ 8.8 (s, NCH$_2$*C*H$_3$), 43.6 (s, N(*C*H$_3$)$_2$), 94.5 (s, N*C*H$_2$CH$_3$), 139.3 (s, *p*-O*C*$_6$F$_4$H), 141.7 (s, *m*-O*C*$_6$F$_4$H), 145.6 (s, *o*-O*C*$_6$F$_4$H), 148.0 (s, *ipso*-O*C*$_6$F$_4$H). IR (nujol, cm^{-1}): 3079 w, 1897 s, 1650 s, 1515 s, 1261 s, 1169 s, 1108 s br, 1038 w, 1020 w, 944 s, 925 w, 820 w, 805 m, 772 m, 738 m, 722 m, 687 w, 570 w, 480 w, 428w.

Synthesis of 3

Compound **3** was prepared in a similar manner to **2** but using HOC$_6$HF$_4$-2,3,5,6 (1.25 g, 7.54 mmol) and [AlH$_3$(NMe$_2$Et)] (0.5 mL, 3.77 mmol). Cooling to -20 °C afforded a white powder of **3** (1.01 g, 75%), mp 46-48 °C. ^{1}H NMR (CD$_2$Cl$_2$): δ 1.34 (t, *J* = 14 Hz, 3H, NCH$_2$C*H$_3$*), 2.72 (s, 6H, N(C*H$_3$*)$_2$), 3.07 (d br, *J* = 11 Hz, 2H, NC*H$_2$*CH$_3$), 6.53 (s br, 2H, C$_6$F$_4$*H*), Al*H* not detected. ^{13}C{^{1}H} NMR (CD$_2$Cl$_2$): δ 8.8 (s, NCH$_2$*C*H$_3$), 43.0 (s, N(*C*H$_3$)$_2$), 94.7 (s, N*C*H$_2$CH$_3$), 95.4 (s, N*C*H$_2$CH$_3$), 139.4 (s, *p*-O*C*$_6$F$_4$H), 141.7 (s, *m*-O*C*$_6$F$_4$H), 145.6 (s, *o*-O*C*$_6$F$_4$H), 148.0 (s, *ipso*-O*C*$_6$F$_4$H). IR (nujol, cm^{-1}): 3075 w, 1896 s, 1651 s, 1505 s br, 1280 w, 1261 m, 1168 s, 1109 s br, 1037 m, 993 w, 940 vs, 821 w, 806 w, 773 w, 727 w, 702 w, 686 m, 644 m, 568 m, 538 m, 531 m, 517 w, 481 s, 458 w.

Synthesis of 4/5

Compound **4** was prepared in a similar manner to **2** but using 2,6-Me$_2$C$_6$H$_3$OH (1.84 g, 15.1 mmol) and [AlH$_3$(NMe$_2$Et)] (1.0 mL, 7.54 mmol). Cooling to -20 °C afforded colourless crystals of **4** (1.24 g, 48% yield), mp 58-60 °C. Data for **4**: Anal. Calc. for C$_{20}$H$_{30}$NO$_2$Al: C, 69.94; H, 8.81; N, 4.08. Found: C, 68.72; H, 8.24; N, 4.15. ^{1}H NMR (CD$_2$Cl$_2$): δ 1.39 (t, J = 10 Hz, 3H, NCH$_2$CH_3), 2.28 (s, 12H, OC$_6$H$_3$(CH_3)$_2$), 2.75 (s, 6H, N(CH_3)$_2$), 3.15 (q, J = 20 Hz, 2H, NCH_2CH$_3$), 6.75 (t, J = 17 Hz, 2H, p-OC$_6$$H_3$(CH$_3$)$_2$), 7.01 (d, J = 8 Hz, 4H, m-OC$_6$$H_3$(CH$_3$)$_2$). ^{13}C{^{1}H} NMR (CD$_2Cl_2$): δ 8.4 (s, NCH$_2$$CH_3$), 18.1 (s, OC$_6H_3$($CH_3$)$_2$), 18.3 (s, N(CH$_3$)$_2$), 43.3 (s, N$CH_2CH_3$), 118.4 (s, p-OC$_6H_3$(CH$_3$)$_2$), 127.0 (s, p-O$C$$_6H_3$(CH$_3$)$_2$), 128.6 (s, o-O$C$$_6H_3$(CH$_3$)$_2$), 155.9 (s, *ipso*-OC$_6H_3$(CH$_3$)$_2$). IR (nujol, cm^{-1}): 1843 s, 1594 s, 1429 w, 1281 s, 1237 s, 1182 m, 1093 s, 1036 m, 997 m, 919 s, 897 s, 760 s, 749 s, 679 s br, 532 m, 402 w. Compound **5** was prepared either from the recrystallization of **4** in diethy ether or *via* the route described above for **4** but using a 1:3 ratio of [AlH$_3$(NMe$_2$Et)]: 2,6-Me$_2$C$_6$H$_3$OH. Data for **5**: ^{1}H NMR (CD$_2$Cl$_2$): δ 1.28 (t, J = 10 Hz, 3H, NCH$_2$CH_3), 2.12 (s, 18H, OC$_6$H$_3$(CH_3)$_2$), 2.75 (s, 6H, N(CH_3)$_2$), 3.24 (q, J = 20 Hz, 2H, NCH_2CH$_3$), 6.61 (t, J = 17 Hz, 3H,p- OC$_6$$H_3$(CH$_3$)$_2$), 6.87 (d, J = 8 Hz, 6H, m- OC$_6$$H_3$(C$H_3$)$_2$). ^{13}C{^{1}H} NMR (CD$_2Cl_2$): δ 7.5 (s, NCH$_2$$CH_3$), 17.9 (s, OC$_6H_3$(CH$_3$)$_2$ and N(CH$_3$)$_2$), 44.1 (s, NCH$_2$CH$_3$), 118.6 (s, p-OC$_6$H$_3$(CH$_3$)$_2$), 127.0 (s, p-O$C$$_6H_3$(CH$_3$)$_2$), 128.8 (s, o-O$C$$_6H_3$(CH$_3$)$_2$), 155.2 (s, *ipso*-OC$_6H_3$(CH$_3$)$_2$). IR (nujol, cm^{-1}): 1589 s, 1562 s, 1475 s, 1430 m, 1279 s, 1234 s, 1175 m, 1092 s, 1029 m, 979 s, 933 s, 851 s, 786 s, 654 s, 609 s, 550 m.

Synthesis of 8

HOCH$_2$CH$_2$NMe$_2$ (2.34 g, 26.3 mmol) was added dropwise to a solution of [Et$_3$Al] (3.00 g, 26.3 mmol) in toluene (30 mL) at –78 °C with stirring. The reaction mixture was allowed to warm slowly to room temperature and stirred for a further 2 h to yield a colourless solution. Removal of the solvent *in vacuo* afforded a white solid. The solid was redissolved in hexane (~5 mL) and cooled to –20 °C. Compound **8** was obtained as white crystals after several days at this temperature (4.56 g, yield 100%). Anal. Calc. for C$_8$H$_{20}$NOAl: C, 55.47; H, 11.64; N, 8.06. Found: C, 54.31; H, 11.88; N, 8.90. ^{1}H NMR (CD$_2$Cl$_2$): δ -0.30 (quartet, J = 8.1 Hz, 4H, AlCH_2CH$_3$), 1.00 (t, quartet, J = 8.1 Hz, 6H, AlCH$_2$CH_3), 2.30 (s, 6H, NCH_3), 2.54 (t, J = 5.4 Hz, 2H, OCH$_2$CH_2N), 3.70 (t, J = 5.4 Hz, 2H, OCH_2CH$_2$N, 2H). IR (nujol, cm^{-1}): 2852 vs, 2794 m, 1464 vs, 1416 m, 1378 s, 1355 m, 1272 m, 1185 w, 11096 vs, 1073 w, 1031 m, 988 m, 904 m, 791 m, 674 m, 635 m, 575 m, 517 m, 443 m, 408 m. Mass Spec (CI): (m/z) 346 ([M]), 317 ([M]-Et), 173 (Et$_2$Al(OCH$_2$CH$_2$NMe$_2$), 144 (EtAl(OCH$_2$CH$_2$NMe$_2$)), 72 (CH$_2$CH$_2$NMe$_2$).

Synthesis of 9

Compound **9** was prepared in a similar manner to **8** but using HOCH$_2$CH$_2$NMe$_2$ (0.64 mL, 6.37 mmol), [Et$_3$Ga] (0.95 mL, 6.37 mmol) and hexane rather than toluene. Compound **9** was obtained as a pale white microcrystalline solid after several days at –20 °C (1.14 g, yield 83%). Anal. Calc. for C$_8$H$_{20}$NOGa: C, 44.49; H, 9.33; N, 6.49. Found: C, 44.38; H, 9.40; N, 6.59. ^{1}H NMR (CD$_2$Cl$_2$): δ 0.28 (quartet, *J* = 8.1 Hz, 4H, GaC*H*$_2$CH$_3$,), 1.08 (t, J = 8.1 Hz, 6H, GaCH$_2$C*H*$_3$), 2.24 (s, 6H, NC*H*$_3$), 2.40 (t, J = 5.4 Hz, 2H, OCH$_2$C*H*$_2$N), 3.67 (t, J = 5.4 Hz, 2H, OC*H*$_2$CH$_2$N). ^{13}C{^{1}H} NMR (CD$_2$Cl$_2$): δ 3.8 (GaC*H*$_2$CH$_3$), 10.6 (GaCH$_2$*C*H$_3$), 45.2 (N*C*H$_3$), 58.9 (OCH$_2$*C*H$_2$N), 61.8 (O*C*H$_2$). IR (nujol, cm^{-1}): 2856 vs, 2789 m, 1460 vs, 1420 m, 1379 s, 1356 m, 1273 m, 1186 w, 1101 vs, 1072 w, 1036 m, 999 m, 954 m, 932 w, 893 m, 785 m, 629 m, 554 m, 503 m, 435 m. Mass Spec (CI): (m/z) 433/432 ([M]), 403 ([M]-Et), 344 ([M]-(OCH$_2$CH$_2$NMe$_2$)), 216 (Et$_2$Ga(OCH$_2$CH$_2$NMe$_2$), 186 (EtGa(OCH$_2$CH$_2$NMe$_2$)), 127 (Et$_2$Ga), 72 (CH$_2$CH$_2$NMe$_2$).

Low pressure CVD experiments

0.5 g of compound was loaded into the sealed end of a glass tube (400 mm length x 16 mm diameter) in the glovebox. Glass substrates (100 mm x 7 mm x 1 mm) were placed along the inside of the tube. The tube was then put in a furnace such that 30 cm was inside the furnace and the end containing the sample protruded by 4 cm. The tube was heated to a temperature of 600°C under vacuum (0.1 Torr). The tube was slowly drawn into the furnace until the sample started to melt. It was held at that point until all of the precursor had decomposed and then the furnace was allowed to cool to room temperature. All the compounds produced white/gray films deposited on the glass substrates. The films were analyzed by EDAX/SEM, band gap measurements, XRD and XPS.

X-ray structure determination

Crystal data for **5** has been deposited in the Cambridge Structural Database (CCDC 239164).

Acknowledgements

We are grateful to Epichem Ltd for [AlH$_3$(NMe$_2$Et)], EPSRC for studentships (ESP, JDM) and the Nuffield foundation for an undergraduate research bursary (SJK). We would like to dedicate this article to Professor A. H. Cowley on the occasion of his 70th birthday.

389

References

1. Carmalt, C. J.; Mileham, J. D.; White, A. J. P.; Williams, D. J. *Dalton Trans.* **2003**, 4255.
2. Carmalt, C. J.; Newport, A. C.; Parkin, I. P.; Mountford, P.; Sealey A. J.; Dubberley, S. R. *J. Mater. Chem.* **2003**, *13*, 84.
3. Blackman, C.; Carmalt, C. J.; O'Neill, S.; Parkin, I. P.; Apostilco, L.; Molloy, K. C.; White, A. J. P.; Williams, D. J. *J. Chem. Soc., Dalton Trans*, **2002**, 2702.
4. Blittersdorf, S.; Bahlawane, N.; Kohse-Höinghaus, K.; Atakan, B.; Müller, J. *Chem. Vap. Deposition* **2003**, *9*, 194.
5. Fredriksson, E.; Carlsson, J. -O. *Surf. Coat. Technol.* **1993**, *56*, 165.
6. Fredriksson, E.; Carlsson, J. -O. *Chem. Vap. Deposition* **1993**, *1*, 333.
7. Rees, Jr., W. S. *CVD of Nonmetals*; VCH, Weinheim 1996.
8. Barreca, D.; Battiston, G. A.; Gerbasi, R.; Tondello, E. *J. Mater. Chem.* **2000**, *10*, 2127.
9. Fleischer, M.; Höllbauer, L.; Meixner, H. *Sensors and Actuators B* **1994**, *18-19*, 119.
10. Ogita, M.; Yuasa, S.; Kobayashi, K.; Yamada, Y.; Nakanishi, Y.; Hatanaka, Y. *Appl. Surf. Sci.* **2003**, *212*, 397.
11. Ogita, M.; Saika, N.; Nakanishi, Y.; Hatanaka, Y. *Appl. Surf. Sci.* **1999**, *142*, 188.
12. Battiston, G. A.; Gerbasi. R.; Porchia, M.; Bertoncello, R.; Caccavale, F. *Thin Solid Films* **1996**, *279*, 115.
13. Valet, M.; Hoffman, D. M. *Chem. Mater.* **2001**, *13*, 2135.
14. Mînea, L.; Suh, S.; Bott, S. G.; Liu, J. -R.; Chu, W. -K.; Hoffman, D. M. *J. Mater. Chem.* **1999**, *9*, 929.
15. Lazell, M.; O'Brien, P.; Otway, D. J.; Park, J. -H. *J. Chem. Soc., Dalton Trans.* **2000**, 4479.
16. Herrmann, W. A. *Angew. Chem. Int. Ed. Engl.* **1995**, *34*, 2187.
17. Nöth, H.; Suchy, H. *Z. Anorg. Allg. Chem.* **1968**, *358*, 44.
18. Carmalt, C. J.; Mileham, J. D.; White, A. J. P. ; Williams, D. J. *New J. Chem.* **2002**, *26*, 902.
19. Bradley, D. C.; Chudzynska, H.; Frigo, D. M.; Hursthouse, M. B.; Mazid, M. A. *J. Chem. Soc., Chem. Commun.* **1988**, 1258
20. Hubert-Pfalzgraf, L. G. *New. J. Chem.* **1995**, *19*, 727.
21. Veith, M.; Faber, S.; Wolfanger, H.; Huch, V. *Chem. Ber.* **1996**, *129*, 381.
22. Healy, M. D.; Mason, M. R.; Gravelle, P. W.; Bott, S. G.; Barron, A. R. *J. Chem. Soc., Dalton Trans.* **1993**, 441.
23. Carmalt, C. J.; Mileham, J. D.; White, A. J. P. ; Williams, D. J. *Polyhedron*, **2003**, *22*, 2655.
24. George, V. C. *Thin Solid Films* **1988**, *165*, 163.

Chapter 28

Phosphate Ester Cleavage with Binuclear Boron Chelates

Amitabha Mitra and David A. Atwood*

Department of Chemistry, University of Kentucky, Lexington, KY 40506-0055

This chapter reviews the application of binuclear boron halide compounds of the type $L[BX_2]_2$ (L= Salen ligands, X= Cl, Br) in the dealkylation of organophosphate esters. These boron compounds can cleave a series of phosphates at room temperature in good yield. The reaction is catalytic in presence of excess BBr_3 and is thought to proceed through the formation of a phosphate coordinated boron cation.

Introduction

Organophosphate esters are a commercially important class of compounds. They are used in flame retardants, plasticizers, selective extraction of metal compounds, hydraulic fluids and lubricants (1, 2). The well known esters of orthophosphoric acid can be classified into phosphate monoesters, diesters and triesters according to the number of ester groups present (Figure 1). Mono- and di-esters have ionizable hydrogen atoms that can be replaced by cations. Triesters are covalent compounds and they do not exist in nature. The first laboratory synthesis of an organophosphate ester was reported in 1820 (3). However, the first commercial use of organophosphate esters was developed during the 1930s for potential use in nerve agents (1).

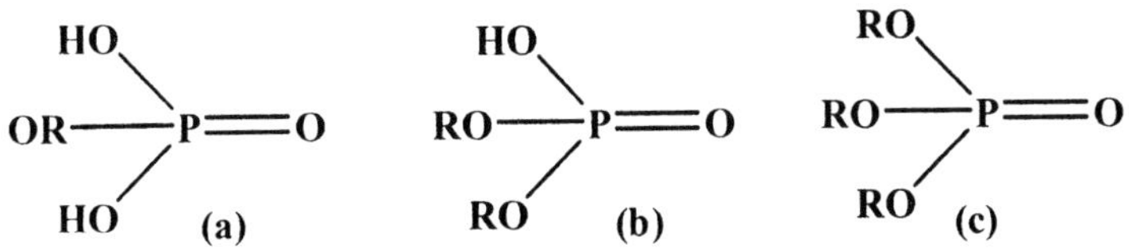

391

```
HO                  HO                  RO
  \                   \                   \
OR——P=O           RO——P=O           RO——P=O
  /                   /                   /
HO    (a)         RO    (b)         RO    (c)
```

Figure 1. Phosphate esters: monoester (a), diester (b) and triester (c)

Phosphate esters are very important in biological systems *(4)*. They participate in storage and transfer of genetic information, carry chemical energy and regulate the activity of enzymes and signaling molecules in the cell. Examples of biological molecules having phosphate ester bonds include DNA and RNA, and ATP (Figure 2). Also, many coenzymes are esters of phosphoric acid.

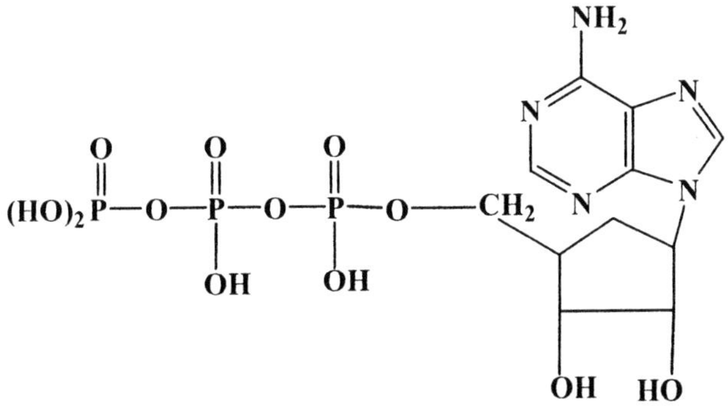

```
                               NH₂
                                |
                          N≡≡≡≡≡N
         O    O    O     ‖      ‖
         ‖    ‖    ‖     N      N
(HO)₂P—O—P—O—P—O——CH₂
              |    |
              OH   OH
                              OH  HO
```

Figure 2. Phosphate ester bond in ATP

Organophosphate esters are active components of chemical warfare agents and pesticides. They can enter the body by inhalation, through contact with the skin or orally. They function by irreversibly phosphorylating the serine hydroxy group in the enzyme acetylcholinesterase *(5-7)*. Acetylcholinesterase is responsible for the hydrolysis of the neurotransmitter acetylcholine. Loss of the activity of the enzyme acetylcholinesterase causes buildup of acetylcholine in the synapse and as a result the neural signal to the muscle is not terminated which leads to muscular paralysis and death. This forms the basis of the extreme neurotoxicity of nerve agents like VX and Sarin and the pesticides chloropyrifos and paraoxons (Figure 3). Today, organophosphate pesticides are the most widely used type of pesticides and have replaced organochlorine pesticides due to their increased environmental instability compared to that of organochlorine compounds. Despite their toxicity, the use of organophosphate pesticides has been increasing and is predicted to continue to incease in the future because of the lack of suitable substitutes *(8)*. Decontamination of nerve agents is required in battle fields, laboratories, storage and destruction sites. Long-lived pesticides pose a threat when they spread beyond their intended application. All of these compounds possess a P-O-C linkage in their structure. The cleavage of the P-O-C bond has been targeted as a means of deactivating these toxic agents *(9, 10)*.

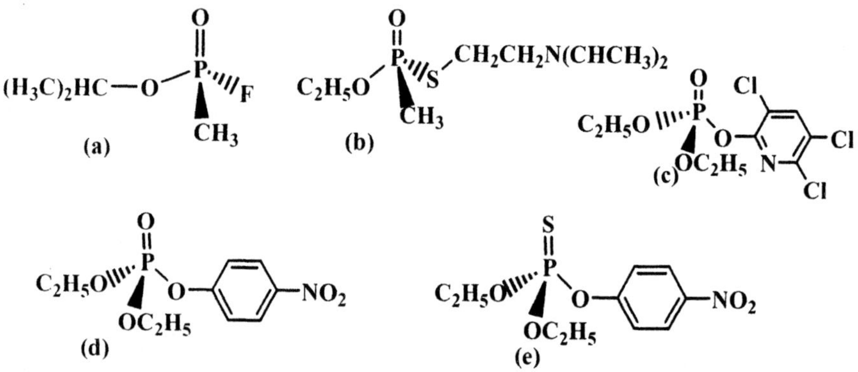

Figure 3. Structures of some organophosphate nerve gases and pesticides: (a) Sarin Gas, (b) VX gas, (c) Chloropyrifos, (d) Paraoxon and (e) Parathion

Reagents that have been used for the decontamination of nerve gases *(9, 10)* include bleach, that oxidizes the nerve agent to less toxic inorganic phosphates, and alkali, that hydrolyzes the P-O or P-S bond. One disadvantage of using bleach is that a large excess is required. Also the active chlorine content of bleach solutions decreases with time. Moreover, bleach is indiscriminately corrosive to any surface or compound it comes into contact with. Base hydrolysis also has some limitations such as the requirement of large quantities of base to maintain a high pH level. Also VX has low solubility in alkaline solution and reacts with base very slowly. Catalytic hydrolyses involving metal ions *(10-13)* and enzymes *(10, 14-18)* have been proposed but so far only limited success has been achieved.

Here we review the use of chelated binuclear boron halide compounds to cleave the P-O-C bond in organophosphate esters. This review also includes biological cleavage of phosphate esters and cleavage by models of biological systems.

Phosphate ester cleavage in biology

In biology, the hydrolytic cleavage of phosphate esters is catalyzed by a number of enzymes. These reactions have been the subject of much discussion *(19-23)* and can be divided into two types: associative and dissociative. A dissociative path proceeds through a hydrated metaphosphate ion PO_3^-, and an associative path goes through a five coordinate phosphorus intermediate. Currently it is thought that in solution the hydrolysis of di- and triesters follows a more associative-like pathway, whereas, that of monoesters proceeds through a dissociative mechanism *(24-26)*.

The enzymes catalyzing the hydrolytic cleavage contain metal ions and use the functional groups of the amino acids and/or the Lewis acidity of the metal ions. A number of recent reviews have been published on this subject *(27, 28)*. These enzymes can be divided into several classes depending on the phosphate containing substrate. These are phosphate monoesterase, diesterase and triesterase. Important monoesterase enzymes are glucose-6-phosphatase, alkaline phosphatase and purple acid phosphatase. Phosphate diesterases are exemplified by P1 nuclease, phospholipase C and 3'-5' exonuclease from the Klenow fragment of DNA polymerase I *(29)*.

Alkaline phosphatase

Alkaline phosphatase (AP) *(27, 28, 30-33)* is probably the most widely studied enzyme of metallohydrolases. These enzymes show low substrate specificity and optimal activity in alkaline conditions (pH7.5) and are found in prokaryotes and eukaryotes. The most extensively studied of these enzymes is that from *E. coli*. This is a homodimeric protein (94kDa) that contains two Zn^{2+} ions and one Mg^{2+} ion in each active site. The hydrolysis proceeds through a phosphorylated intermediate formed by the nucleophilic attack of Ser-102 on the phosphate monoester. This intermediate is subsequently attacked by an adjacent Zn-OH group (Figure 4) *(34)*. A retention of configuration at the phosphorus center is observed. The two Zn^{2+} ions are important for the activity of AP, but the Mg^{2+} ion plays only an ancillary role. One or two of the Zn^{2+} ions can be substituted by Mg^{2+}, Mn^{2+} or Co^{2+} *(33)* but this results in a considerable reduction in activity.

Purple Acid Phosphatase

Purple acid phosphatases (PAPs) *(35-38)* are a group of phosphomonoesterases that catalyze phosphate ester hydrolysis under acidic condition (low pH optima at 4.9-6.0), resist inhibition by tartrate and exhibit a characteristic intense purple color. They contain two irons or one iron and another dipositive cation centers (Figure 5) in the active form. Most of the enzymes having Fe^{III}-Fe^{II} centers are isolated from bovine spleen *(39)* or porcine uterine fluid *(40)*. PAPs isolated from kidney beans *(41)* generally contain one iron and one zinc or manganese.

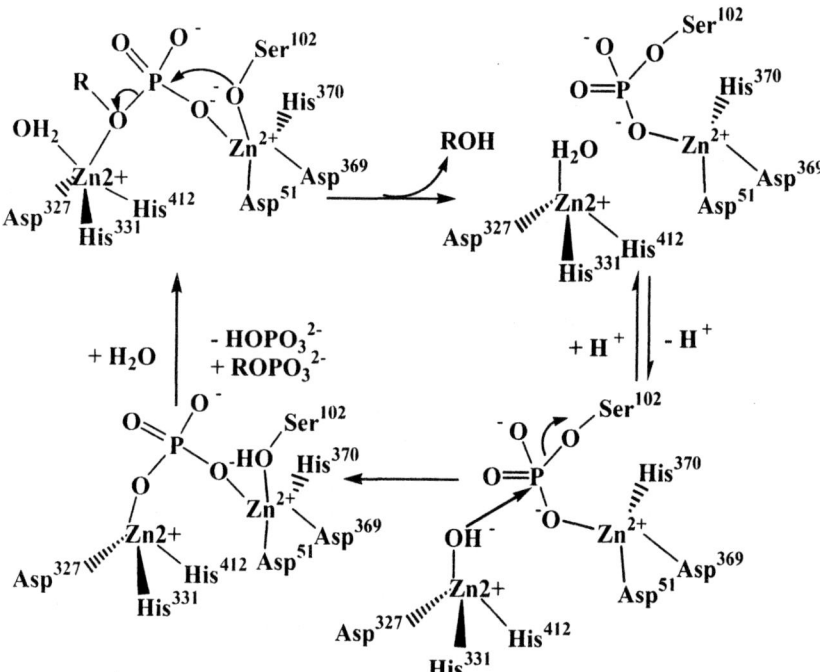

Figure 4. Proposed mechanism for phosphate ester cleavage by the active site of enzyme Alkaline Phosphatase.

P1 Nuclease

P1 nuclease *(46)* is a phosphodiesterase that preferentially cleaves the bond between 3'-hydroxy and and 5'-phosphate group of single stranded RNA and DNA. It requires the presence of three zinc ions *(47)* per molecule and the hydrolysis causes the inversion of the configuration at phosphorus center *(48)*.

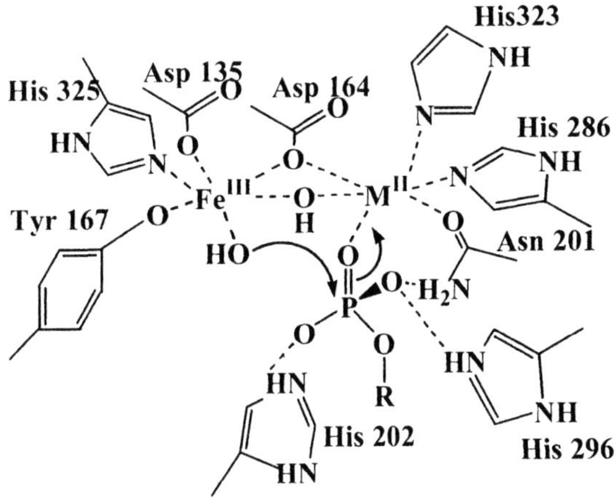

Figure 5. Binding mode and attack of Fe^{III} bound nucleophile for proposed mechanism of phosphate ester hydrolysis by PAP (42-45) (redrawn from reference 27).

Phospholipase C

Phospholipase C enzymes catalyze the hydrolysis of various phospholipids, for example, phosphatidylinositol and phosphatidylcholine. The PLC *(49, 50)* from soil bacterium *B. Cereus* contain two tightly bound Zn(II) ions per molecule, whereas, a third more weakly bound zinc(II) ion was revealed in its X-ray crystal structure *(51)*. Other metal ions like cobalt, magnesium and manganese can replace zinc in some PLC enzymes, however this results in reduced activity and altered substrate specificity *(52-54)*.

Phosphate triesterase

Phosphate triesterases (PTEs) *(55)* catalyze the hydrolysis of organophosphate triesters. The triesters do not occur in nature but they are released into nature by the use of pesticides and insecticides. The best characterized phosphotriesterase is the enzyme from soil bacteria *Pseudomonas diminuta* which is a monomeric metalloprotein that hydrolyzes P-O, P-F, P-CN, and P-S bonds between the phosphorus center and the leaving group *(56-58)*. The native enzyme contains two Zn^{2+} ions in the active site but they can be substituted with Cd^{2+}, Co^{2+}, Ni^{2+}, or Mn^{2+} without affecting catalytic activity *(59)*.

Phosphate ester cleavage using models of biological catalysts

To elucidate the role of the metal ions in the hydrolysis of phosphate esters extensive studies have been conducted during the last several years *(29, 60, 61)*. Besides kinetic and structural studies of the original enzymes, several well-defined model compounds have been developed to understand the process of the cleavage. Besides shedding light on the mechanism of phosphate ester cleavage, studies of metal containing model compounds are also relevant to the synthesis of artificial metallohydrolases. These model compounds use d- and f-block and alkaline earth metals such as cobalt *(62, 63)*, zinc *(64-68)*, copper *(69-71)* and lanthanides *(72-80)*. These systems include mono-, bi- (Figure 6) and trinuclear metal compounds.

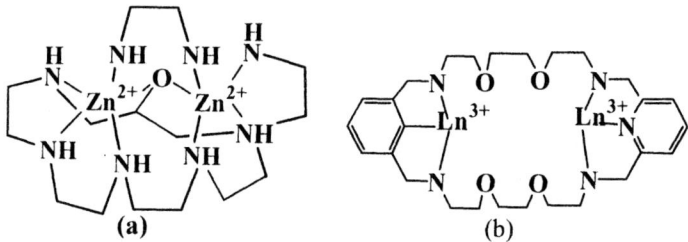

Figure 6. Examples of binuclear zinc (a) and lanthanide (b) model compounds.

For most of these models the binuclear compounds have been found to be much more active than the mononuclear counterparts. For example, the hydrolysis rate of bis(*p*-nitrophenyl) phosphate (BNP) increases almost ten fold by the binuclear $[Zn_2L1(OH)_2]^+$ (**L1**= [30]aneN_6O_4] complex with respect to the mononuclear **L2**-Zn-OH^+ (**L2**=[15]aneN_3O_2) *(81)*. The higher reactivity with binuclear systems is attributed to the double Lewis acidic activation from the coordination of two phosphoryl oxygen atoms with two metal centers as shown for a peroxide bridged lanthanide binuclear complex in Figure 7.

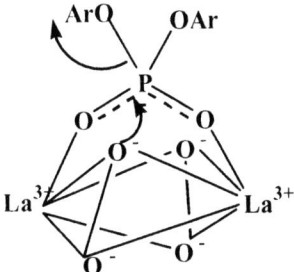

Figure 7. Proposed mechanism for phosphate ester cleavage by a peroxide bridged binuclear lanthanide compound (79, 80).

Phosphate ester cleavage using chelated boron compounds

Chelated boron compounds

One advantage of chelated boron complexes over simple boron Lewis acidic compounds like BCl_3 is the possibility of tuning their structure by varying the chelate ring. Chelated mononuclear (1:1 ligand to metal) boron compounds have been known for a long time. Some of the recent examples of these includes complexes of β-diketonates *(82)*, β-diketiminate *(83)*, substituted anthracene *(84)*, polysaccharides*(85)*, L-cysteine *(86)*, nitroles *(87, 88)*, tropolonates *(89)* and salilcylaldimines *(90)*.

Recently binuclear boron compounds *(91-100)* have attracted a lot of attention due to their potential use as two-point Lewis acids. Most of these binuclear compounds are derived from Salen ligands (Figure 8). Salen ligands are a class of Schiff bases *(101)* which are formed from the condensation of one mole of salicylaldehyde with two moles of an amine. These are tetradentate dianionic ligands with two coordinative and two covalent sites. They are similar to porphyrins but are less expensive and easier to synthesize. They provide a wide range of options in terms of varying the steric and electronic properties around the metal center by tuning the backbone of the ligand. Moreover, their solubility can be varied by incorporating appropriate substituent groups on the phenyl ring. Salen(^{t}Bu) ligands are more appealing than unsubstituted Salen(H_2) ligands because of their better solubility in organic solvents and hence better scope of characterizing their solution state properties. Salen ligands have been used to prepare both mono- and binuclear group 13 complexes *(102)*.

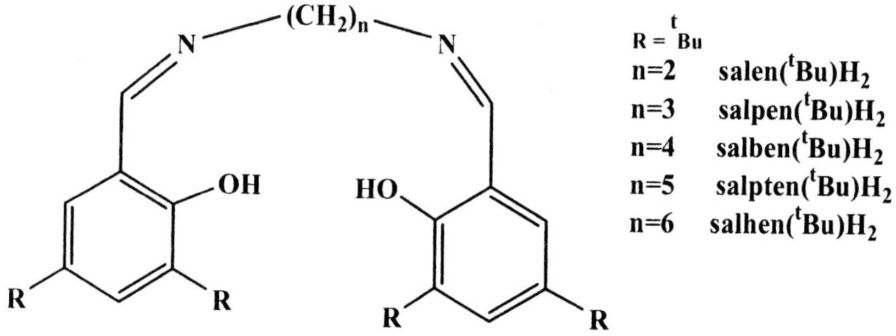

```
         ─(CH₂)ₙ─
      N            N                 R = ᵗBu
                                     n=2    salen(ᵗBu)H₂
                                     n=3    salpen(ᵗBu)H₂
        OH      HO                   n=4    salben(ᵗBu)H₂
                                     n=5    salpten(ᵗBu)H₂
                                     n=6    salhen(ᵗBu)H₂
  R        R   R       R
```

Figure 8. Salen(ᵗBu) ligands

Binuclear boron alkoxides and borosiloxides

Binuclear boron alkoxides *(93)* can be prepared by the alcohol elimination reaction between SalenH₂ and corresponding alkyl borates (Figure 9(a)). The resulting compounds show interesting extended structures due to intra- and intermolecular hydrogen bonding of the alkoxide oxygens with the imine protons.

These binuclear boron compounds form readily even in the presence of excess ligand. This is in contrast to aluminum that forms both mono- and binuclear complex depending on the metal to ligand ratio. This is due to the smaller size of boron and its preference for a four-coordinate tetrahedral geometry. Similar binuclear boron compounds can be prepared by the condensation of boronic acids with the Salen ligands *(97)*. These binuclear boron alkoxides can further be derivatized to bimetallic boron siloxides (Figure 9(b)) *(95)*.

Binuclear boron halides

Although the bimetallic boron alkoxides and siloxides are interesting from a fundamental standpoint, they are not ideal for use as precursors to two-point Lewis acids. One step towards this goal was the synthesis of binuclear boron bromides and chlorides of the type L[BX₂]₂ *(98-100)* opening the possibility of creating cations through salt elimination reactions. Binuclear boron chlorides can be prepared by combining the corresponding

Salen(^{t}Bu)[B(OMe$_2$)$_2$] reagents with BCl$_3$ in stoichiometric ratio at room temperature in almost quantitative yield (Figure 10).

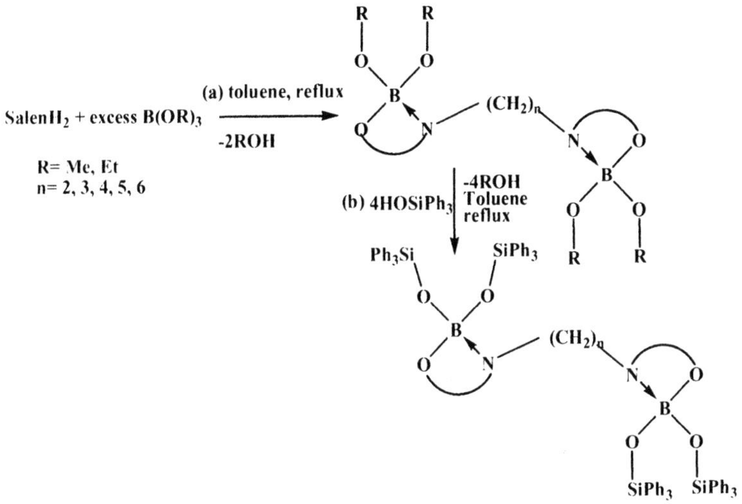

Figure 9. Synthesis of binuclear boron alkoxides (a) and siloxides (b).

L[B(OMe)$_2$]$_2$ $\xrightarrow[\text{-4/3 B(OMe)}_3]{\text{4/3 BX}_3}$ L[BX$_2$]$_2$

L= Salen(^{t}Bu) ligands   toluene, RT
X= Cl, Br

Figure 10. Synthesis of binuclear boron halides

Similarly, binuclear boron bromides can be prepared using BBr$_3$ instead of BCl$_3$. However, while L[BCl$_2$]$_2$ remains unaffected by the presence of excess BCl$_3$, the addition of excess BBr$_3$ to L[BBr$_2$]$_2$ results in the cleavage of one B-Br bond and the formation of L[BBr$_2$(BBr)$^+$]BBr$_4^-$ *(98)*. This was the first indication of the lability of B-Br bonds in these compounds. It was further evident by the formation of a cation upon the addition of THF to L[BBr$_2$]$_2$ *(98)*. In L[BX$_2$]$_2$ (X= Br, Cl) compounds the boron atoms are in a distorted tetrahedral geometry and trans to one another as evident from the crystal structure (Figure 11).

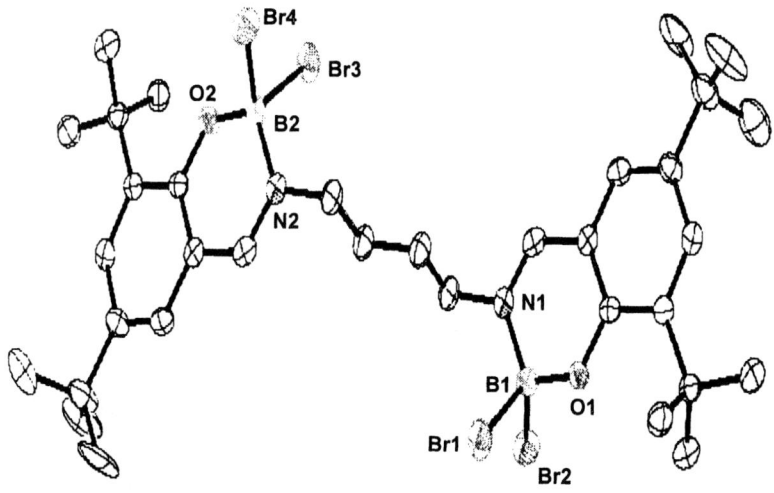

Figure 11. Crystal structure of salben(ᵗBu)[BBr₂]₂ (Reproduced from reference 98. Copyright 2001 American Chemical Society.)

In addition to binuclear boron tetrahalides of the type $L[BX_2]_2$, Salen phenyl boron bromide compounds $L[B(p\text{-tolyl})Br]_2$ *(100)* can be prepared from Salen (ᵗBu)[(p-tolyl)B(OMe)]₂ and BBr_3 (Figure 12).

Phosphate dealkylation with boron halide chelates

The binuclear boron halide chelate compounds can dealkylate a wide range of phosphates at ambient temperature *(98-100)* (Table I, II and III). For example, compound salpen(ᵗBu)[BBr₂]₂ dealkylates $(MeO)_3P(O)$ by 89% and $(^nBuO)_3P(O)$ by 99% in only 30 minutes *(100)*. This is significant considering the fact that BCl_3 or BBr_3 is ineffective for phosphate dealkylation. In the dealkylation reaction MeCl or MeBr is produced along with a yet unidentified phosphate material. The reaction can be monitored by comparing the ¹H NMR peak integration of methyl halide to that of the original phosphate.

The active species in this type of reaction is thought to be a cation formed through the heterolytic cleavage of a B-X bond. The cation then coordinates to the phosphate thereby activating the ester carbon (Figure 13).

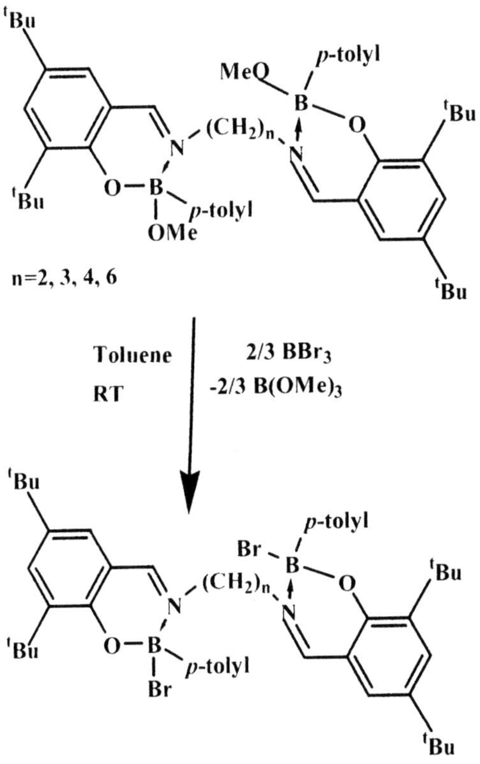

Figure 12. Synthesis of Salen[B(^{p}tolyl)Br]$_2$.

Table I. Dealkylation of trimethyl phosphate with Salen boron chloride compounds

L[BCl$_2$]$_2$	% conversion, 30min	% conversion, 24h
Salen(^{t}Bu)[BCl$_2$]$_2$	14	45
Salpen(^{t}Bu)[BCl$_2$]$_2$	20	32
Salben(^{t}Bu)[BCl$_2$]$_2$	11	53
Salpten(^{t}Bu)[BCl$_2$]$_2$	42	62
Salhen(^{t}Bu)[BCl$_2$]$_2$	7	47

SOURCE: Adapted from Reference 99.

Table II. % dealkylation of various phosphates with Salen boron bromide compounds

Compound	Salen(^{t}Bu)[BBr$_2$]$_2$	Salpen(^{t}Bu)[BBr$_2$]$_2$	Salben(^{t}Bu)[BBr$_2$]$_2$	Salhen(^{t}Bu)[BBr$_2$]$_2$
Carbons in backbone of the ligand	2	3	4	6
Phosphate				
(MeO)$_3$P(O)	76	89	90	81
(^{n}BuO)$_3$P(O)	42	99	77	69
(^{n}PentO)$_3$P(O)		98		
(MeO)$_2$P(O)H		85		
(MeO)$_2$P(O)Me	61	98	87	47
(^{i}PrO)$_2$P(O)H		63		
(PhO)$_2$((2-Et)HexO)P(O)	48	71	64	88
(Me$_3$SiO)$_3$P(O)	88	98	90	79
(PhO)$_3$P(O)	0	0	0	0

NOTE: Reaction time: 30 minutes.
SOURCE: Adapted from Reference 100.

Table III. % dealkylation of various phosphates with Salen phenyl boron bromide compounds

Compound L[(Ptolyl)BBr]$_2$	L= Salen(^{t}Bu)	L= Salpen(^{t}Bu)	L= Salben(^{t}Bu)	L= Salhen(^{t}Bu)
Carbons in backbone of the ligand	2	3	4	6
Phosphate				
(MeO)$_3$P(O)	45	52	92	80
(EtO)$_3$P(O)	40	50	71	56
(^{n}BuO)$_3$P(O)	31	58	70	47
(MeO)$_2$P(O)H	59	67	50	50
(MeO)$_2$P(O)Me	47	50	50	50
(^{i}PrO)$_2$P(O)H	42	32	37	26
(PhO)$_2$((2-Et)HexO)P(O)	12	46	52	42
(Me$_3$SiO)$_3$P(O)	43	48	69	89
(PhO)$_3$P(O)	0	0	0	0

NOTE: Reaction time: 30 minutes.
SOURCE: Adapted from Reference 100.

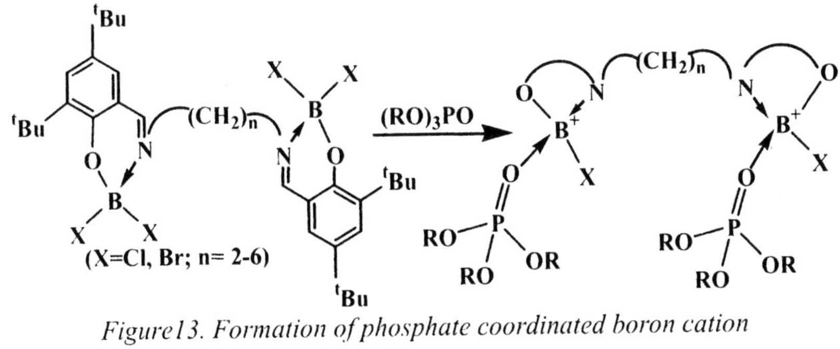

Figure 13. Formation of phosphate coordinated boron cation

This is supported by the easy formation of this type of cation by the simple addition of a Lewis base to $L[BX_2]_2$ *(98)*. When THF was added to a solution of salpen(^{t}Bu)[BBr$_2$]$_2$ in CDCl$_3$ in a NMR tube the ^{11}B NMR shifted downfield by 4.85 ppm, indicating the displacement of Br by THF and formation of a cation. The coordination of the P=O bond to the boron atom makes the P-O-R bonds susceptible to nucleophilic attack by the halide anion (Figure 14). A similar type of activation of an alkyl group is found during the dealkylation of phosphoramidates in strongly acidic medium *(103)* and also during the dealkylation of phosphonates with boron tribromide *(104)*. All of the alkoxy bonds are cleaved in the process to produce RX and an as yet uncharacterized boron phosphate material.

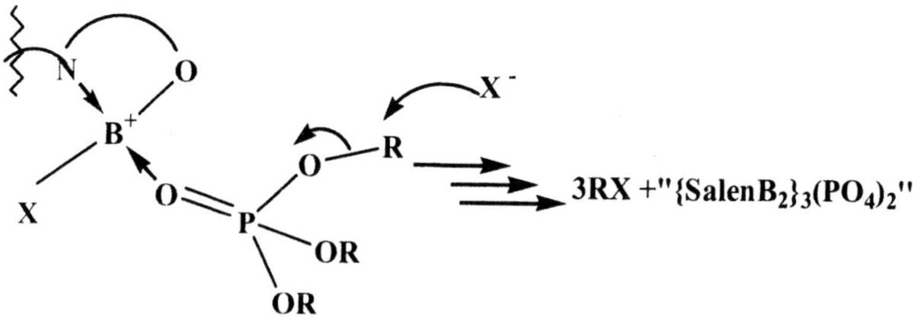

Figure 14. Attack of P-O-R bond by halide anion

The process could be made catalytic by conducting the reaction in the presence of excess BBr$_3$ *(98, 100)*. In general, bromide compounds were found to be more active compared to chloride analogues, which is in accordance with the weaker B-Br bond strength by comparison to B-Cl. It was interesting to note that none of the chelate compounds could dealkylate (ArO)$_3$P(O).

It was also observed that Salen compounds with longer backbones were more efficient than the shorter backbones probably due to unfavorable interaction between two boron centers. Also, the dealkylation activity of the Salen boron bromide compounds decreases with the branched phosphates such as (PhO)$_2$(2-Et)hexylO)P(O). It was also found that one mononuclear boron chelate complex, LBX$_2$ (L= ^{t}Bu Sal(^{t}Bu), X= Br) could dealkylate trimethyl phosphate at 92% conversion *(100)*. Thus, it seems that the presence of two boron centers may not be necessary for the dealkylation to occur. If this is really the case then this will be in contrast with biological phosphate ester cleavage by enzymes and many binuclear d-block and lanthanide compounds where the presence of two metal centers increases the rate. Also, with these compounds, the cleavage involves the direct attack of a nucleophile (water or hydroxide) to the phosphorus center which is not the case with boron chelates where the attack seems to be on the ester carbon of the phosphate.

Conclusion and future studies

Binuclear boron halide chelates can dealkylate a series of organic phosphates at room temperature. They can be considered as promising candidates for decontamination of nerve gas agents and pesticides. However, the scope, limitations, and full utility of the possibilities have not been determined. For instance, an in depth study of a series of mononuclear boron chelates would tell us convincingly whether two Lewis acidic sites are necessary, or whether a single Lewis acidic site is sufficient. This should include kinetic studies to compare the relative rates for bi- and mononuclear chelates. Once this has been determined it will be useful to know if other group 13 elements, such as aluminum, could also be employed in the reaction. The high activity of boron halide Schiff base chelate complexes towards phosphate dealkylation demands a similar study with aluminum chelates (for example Salen(^{t}Bu)AlCl) since aluminum centers in these compounds may also act as Lewis acids and thus facilitate the initial coordination of Lewis basic oxygen of phosphate esters. Both neutral and cationic Schiff base complexes of aluminum have been found to catalyze the polymerization of oxiranes, a reaction that has been attributed to the Lewis acidic property of the aluminum center *(105, 106)*. Lewis acids such as $AlCl_3$ *(107)* and $CyBCl_2$ *(108, 109)* can dealkylate phosphates and phosphonates but only with long reaction times (12 hours to 2 days) and elevated temperatures (70°C or more). Compounds with Al-Br bonds such as Salen(^{t}Bu)AlBr are expected to be more active compared to the chloride analogues due to the greater lability of the bromide. Since these bromide compounds are not known so far, their synthesis and application to phosphate dealkylaion studies might be of interest.

References

1. Quinn, L. D. In *A Guide to Organophosphorus Chemistry;* Wiley Interscience: 2000; 153.
2. Corbridge, D. E. C.; *Phosphorus: An Outline of its Chemistry, Biochemistry and Technology;* Elsevier: Amsterdam, Netherlands, 1995.
3. Lassaigne, F. *Justus Liebigs Ann. Chiim.* **1820**, *3*, 294.
4. Westheimer, F. H. *Science.* **1987**, *235*, 1173-1178.
5. Chambers, J. E.; Oppenheimer, S. F. *Toxicol. Sci.* **2004**, *77*, 185-187.
6. *Organophosphates, Chemistry, Fate, and Effects.;* Chambers, J. E.; Levi, P. E., Eds.; Academic Press: San Diego, 1992.
7. Burgen, A. S. V. *Br. J. Pharmacol.* **1949**, *4*, 219-228.
8. Walker, B., Jr. ; Nidiry, J. *Inhal.Toxicol.* **2002**, *14*, 975-990.
9. Yang, Y.-C. *Acc. Chem. Res.* **1999**, *32*, 109-115.

10. Yang, Y. C.;Baker, J. A.; Ward, J. R. *Chem. Rev.* **1992**, *92*, 1729-1743.
11. Wagner-Jauregg, T.;Hackley, B. E., Jr.;Lies, T. A.; Owens, O. O.; Proper, R. *J. Am. Chem. Soc.* **1955**, *77*, 922-929.
12. Ward, J. R.; Hovanec, J. W.; Albizo, J. M.; Szafraniec, L. L.; Beaudry, W. T. *J. Fluorine Chem.* **1991**, *51*, 277-282.
13. Gustafson, R. L.; Chaberek, S. J.; Martell, A. E. *J. Am. Chem. Soc.* **1963**, *67*, 576-582.
14. Sethunathan, N.; Yoshida, T. *Can. J. Microbiol.* **1973**, *19*, 873-875.
15. Racke, K. D.; Coats, J. R. *J. Agric. Food Chem.* **1987**, *35*, 94-99.
16. Siddaramappa, R.; Rajaram, K. P.; Sethunathan, N. *Appl. Microbiol.* **1973**, *26*, 846-847.
17. Harper, L. L.; McDaniel, C. S.; Miller, C. E.; Wild, J. R. *Appl. Environ. Microbiol.* **1988**, *54*, 2586-2589.
18. DeFrank, J. J. In *Applications of Enzymes Biochemistry;* Kelly, J. W. and Baldwin, T. O., Plenum Press: New York, 1991; 165-180
19. Thatcher, G. R. J.; Kluger, R. *Adv. Phys. Org. Chem.* **1989**, *25*, 99.
20. Florian, J.; Warshel, A. *J. Phys. Chem. B.* **1998**, *102*, 719-734.
21. Futatsugi, N.; Hata, M.;Hoshino, T.; Tsuda, M. *Biophys. J.* **1999**, *77*, 3287-3292.
22. Bianciotto, M.; Barthelat, J.-C.; Vigroux, A. *J. Am. Chem. Soc.* **2002**, *124*, 7573-7587.
23. Cavalli, A.; Carloni, P. *J. Am. Chem. Soc.* **2002**, *124*, 3763-3768.
24. Aqvist, J.; Kolmodin, K.; Florian, J. *Chem. Biol.* **1999**, *6*, R71-80.
25. Admiraal, S. J.; Herschlag, D. *Chem. Biol.* **1995**, *2*, 729-739.
26. Maegley, K. A.; Admiraal, S. J.; Herschlag, D. *Proc. Natl. Acad. Sci. USA.* **1996**, *93*, 8160-8166.
27. Straeter, N.; Lipscomb, W. N.; Klabunde, T.; Krebs, B. *Angew. Chem. Int. Ed. Engl.* **1996**, *35*, 2025-2055.
28. Wilcox, D. E. *Chem. Rev.* **1996**, *96*, 2435-2458.
29. Chin, J. *Curr. Opin. Chem. Biol.* **1997**, *1*, 514-521.
30. McComb, R. B.; Bowers, G. N., Jr.; Posen, S.; *Alkaline Phosphatase;* Plenum Press: N. Y., 1979.
31. Coleman, J. E.; Gettins, P. *Metal Ions in Biology.* **1983**, *5*, 219-252.
32. Coleman, J. E.; Gettins, P. *Advances in Enzymology and Related Areas of Molecular Biology.* **1983**, *55*, 381-452.
33. Coleman, J. E. *Annu. Rev. Biophys. Biomol. Struct.* **1992**, *21*, 441-483.
34. Bazzicalupi, C.; Bencini, A.; Berni, E.;Bianchi, A.; Fedi, V.;Fusi, V.; Giorgi, C.; Paoletti, P.; Valtancoli, B. *Inorg. Chem.* **1999**, *38*, 4115-4122.
35. Antanaitis, B. C. *Adv. Inorg. Biochem.* **1983**, *5*, 111-136.
36. Doi, K.; Antanaitis, B. C.; Aisen, P. *Struct. Bonding.* **1988**, *70*, 1-26.
37. Que, L., Jr.; True, A. E. *Prog. Inorg. Chem.* **1990**, *38*, 97-200.
38. Vincent, J. B.; Olivier-Lilley, G. L. *Chem. Rev.* **1990**, *90*, 1447-1467.

39. Campbell, H. D.; Zerner, B. *Biochem. Biophys. Res. Commun.* **1973**, *54*, 1498-1503.
40. Chen, T. T.; Bazer, F. W.; Cetorelli, J. J.; Pollard, W. E.; Roberts, R. M. *J. Biol. Chem.* **1973**, *248*, 8560-8566.
41. Beck, J. L.; McConachie, L. A.; Summors, A. C.; Arnold, W. N.; De Jersey, J. Z., Burt. *Biochim. Biophys. Acta.* **1986**, *869*, 61-68.
42. Strater, N.; Klabunde, T.;Tucker, P.;Witzel, H.; Krebs, B. *Science.* **1995**, *268*, 1489-1492.
43. Klabunde, T.; Straeter, N.; Froehlich, R.; Witzel, H.; Krebs, B. *J. Mol. Biol.* **1996**, *259*, 737-748.
44. Dietrich, M.; Muenstermann, D.; Suerbaum, H.; Witzel, H. *Eur. J. Biochem.* **1991**, *199*, 105-113.
45. Aquino, M. A. S.; Lim, J. S.; Sykes, G. *J. Chem. Soc. Dalton Trans.* **1992**, 2135-2136.
46. Fraser, M. J.; Low, R. L. In *Nucleases;* Linn, S., Lloyd, R. S. and Roberts, R. J., Cold Spring Harbor Laboratory Press: NY, 1993; 171-207
47. Fujimoto, M.; Kuninaka, A.; Yoshino, H. *Agric. Biol. Chem.* **1975**, *39*, 1991-1997.
48. Potter, B. V. L.; Connolly, B. A.; Eckstein, F. *Biochemistry.* **1983**, *22*, 1369-1377.
49. Ottolenghi, A. C. *Biochim. Biophys. Acta.* **1965**, *106*, 510-518.
50. Little, C. *Methods Enzymol.* **1981**, *71*, 725-730.
51. Hough, E.; Hansen, L. K.; Birknes, B.; Jynge, K.; Hansen, S.; Hordvik, A.; Little, C.; Dodson, E.; Derewenda, Z. *Nature.* **1989**, *338*, 357-360.
52. Little, C. *Acta Chem. Scand.* **1981**, *B35*, 39-44.
53. Little, C.; Aakre, S. E.; Rumsby, M. G.; Gwarsha, K. *Biochem. J.* **1982**, *203*, 799-801.
54. Otnaess, A. B. *FEBS Lett.* **1980**, *114*, 202-204.
55. Caldwell, S. R.; Raushel, F. M. *Appl. Biochem. Biotechnol.* **1991**, *31*, 59-73.
56. Dumas, D. P.; Caldwell, S. R.;Wild, J. R.; Raushel, F. M. *J. Biol. Chem.* **1989**, *264*, 19659-19665.
57. Lai, K.; Stolowich, N. J.; Wild, J. R. *Arch. Biochem. Biophys.* **1995**, *318*, 59-64.
58. Dumas, D. P.; Durst, H. D.; Landis, W. G.;Raushel, F. M.; Wild, J. R. *Arch. Biochem. Biophys.* **1990**, *277*, 155-159.
59. Omburo, G. A.; Kuo, J. M.; Mullins, L. S.; Raushel, F. M. *J. Biol. Chem.* **1992**, *267*, 13278-13283.
60. Bashkin, J. K. *Curr. Opin. Chem. Biol.* **1999**, *3*, 752-758.
61. Blasko, A.; Bruice, T. C. *Acc. Chem. Res.* **1999**, *32*, 475-484.
62. Jones, D. R.; Lindoy, L. F.; Sargeson, A. M. *J. Am. Chem. Soc.* **1984**, *106*, 7807-7819.
63. Vance, D. H.; Czarnik, A. W. *J. Am. Chem. Soc.* **1993**, *115*, 12165-12166.

64. Kaminskaia, N. V.; He, C.; Lippard, S. J. *Inorg. Chem.* **2000**, *39*, 3365-3373.
65. Yamami, M.; Furutachi, H.; Yokoyama, T.; Okawa, H. *Inorg. Chem.* **1998**, *37*, 6832-6838.
66. Yamami, M.; Furutachi, H.; Yokoyama, T.; Okawa, H. *Chem. Lett.* **1998**, 211-212.
67. Chapman, W. H., Jr.; Breslow, R. *J. Am. Chem. Soc.* **1995**, *117*, 5462-5469.
68. Molenveld, P.; Kapsabelis, S.; Engbersen, J. F. J.; Reinhoudt, D. N. *J. Am. Chem. Soc.* **1997**, *119*, 2948-2949.
69. McCue, K. P.; Morrow, J. R. *Inorg. Chem.* **1999**, *38*, 6136-6142.
70. Scrimin, P.; Ghirlanda, G.; Tecilla, P.; Moss, R. A. *Langmuir.* **1996**, *12*, 6235-6241.
71. Gajda, T.; Duepre, Y.; Toeroek, I.; Harmer, J.; Schweiger, A. S., Juergen; ; Kuppert, D.; Hegetschweiler, K. *Inorg. Chem.* **2001**, *40*, 4918-4927.
72. Morrow, J. R.; Buttrey, L. A.; Berback, K. A. *Inorg. Chem.* **1992**, *31*, 16-20.
73. Morrow, J. R.; Amin, S.;Lake, C. H.; Churchill, M. R. *Inorg. Chem.* **1993**, *32*, 4566-4572.
74. Morrow, J. R.; Chin, K. O. A. *Inorg. Chem.* **1993**, *32*, 3357-3361.
75. Morrow, J. R.; Buttrey, L. A.; Shelton, V. M.; Berback, K. A. *J. Am. Chem. Soc.* **1992**, *114*, 1903-1905.
76. Amin, S.; Morrow, J. R.; Lake, C. H.; Churchill, M. R. *Angew. Chem. Int. Ed. Engl.* **1994**, *33*, 773-775.
77. Schneider, H. J.; Rammo, J.; Hettich, R. *Angew. Chem. Int. Ed. Engl.* **1993**, *32*, 1716-1719.
78. Breslow, R.; Huang, D. L. *Proc. Natl. Acad. Sci. USA.* **1991**, *88*, 4080-4083.
79. Takasaki, B. K.; Chin, J. *J. Am. Chem. Soc.* **1993**, *115*, 9337-9338.
80. Takasaki, B. K.; Chin, J. *J. Am. Chem. Soc.* **1995**, *117*, 8582-8585.
81. Bazzicalupi, C.; Bencini, A.; Bianchi, A.; Fusi, V.; Giorgi, C.; Paoletti, P.; Valtancoli, B.; Zanchi, D. *Inorg. Chem.* **1997**, *36*, 2784-2790.
82. Reutov, V. A.; Gukhman, E. V.; Kafitulova, E. *Russ. J. Gen. Chem.* **2003**, *73*, 1441-1444.
83. Qian, B.; Baek, S. W.; Smith, M. R., III. *Polyhedron.* **1999**, *18*, 2405-2414.
84. Yamashita, M.; Kamura, K.; Yamamoto, Y.; Akiba, K.-Y. *chem. Eur. J.* **2002**, *8*, 2976-2979.
85. Miyazaki, Y.; Yoshimura, K.; Miura, Y.; Sakashita, H. I., Katsutoshi. *Polyhedron.* **2003**, *22*, 909-916.
86. Gonzalez, A.; Granell, J.; Piniella, J. F.; Alvarez-Larena, A. *Tetrahedron.* **1998**, *54*, 13313-13322.
87. Kliegel, W.; Metge, J.; Rettig, S. J.; Trotter, J. *Can. J. Chem.* **1998**, *76*, 1082-1092.

88. Kliegel, W.; Metge, J.; Rettig, S. J.; Trotter, J. *Can. J. Chem.* **1998**, *76*, 389-399.
89. Hopfl, H.; Hernandez, N. P.; Lima, S. R.; Santillan, R. F., Norberto. *Heteroatom Chem.* **1998**, *9*, 359-368.
90. Wei, P.; Atwood, D. A. *Inorg. Chem.* **1998**, *37*, 4934-4938.
91. Hohaus, E. *Fresenius Z. Anal. Chem.* **1983**, *315*, 696-699.
92. Atwood, D. A.; Jegier, J. A.; Remington, M. P.; Rutherford, D. *Aust. J. Chem.* **1996**, *49*, 1333-1338.
93. Wei, P.; Atwood, D. A. *Inorg. Chem.* **1997**, *36*, 4060-4065.
94. Wei, P.; Atwood, D. *Chem. Commun.* **1997**, 1427-1428.
95. Wei, P.; Keizer, T.; Atwood, D. A. *Inorg. Chem.* **1999**, *38*, 3914-3918.
96. Sanchez, M.; Sanchez, O.; Hopfl, H.; Ochoa, M.-E.; Castillo, D.; Farfan, N.; Rojas-Lima, S. *J. Organomet. Chem.* **2004**, *689*, 811-822.
97. Sanchez, M.; Hopfl, H.; Ochoa, M. E.; Farfan, N.; Santillan, R.; Rojas, S. *Inorg. Chem.* **2001**, *40*, 6405-6412.
98. Keizer, T. S.; De Pue, L. J.; Parkin, S.; Atwood, D. A. *J. Am. Chem. Soc.* **2002**, *124*, 1864-1865.
99. Keizer, T. S.; De Pue, L. J.; Parkin, S.; Atwood, D. A. *Can. J. Chem.* **2002**, *80*, 1463-1468.
100. Keizer, T. S.; De Pue, L. J.; Parkin, S.; Atwood, D. A. *J. Organomet. Chem.* **2003**, *666*, 103-109.
101. Schiff, H. *Ann.* **1864**, *Suppl. 3*, 343.
102. Atwood, D. A.; Harvey, M. J. *Chem. Rev.* **2001**, *101*, 37-52.
103. Modro, T. A. *J.Chem. Soc. Chem. Commun.* **1980**, *5*, 201-202.
104. Gauvry, N.; Mortier, J. *Synthesis.* **2001**, 553-554.
105. Munoz-Hernandez, M.-A.; McKee, M. L.; Keizer, T. S.; Yearwood, B. C.; Atwood, D. A. *J. Chem. Soc., Dalton Trans.* **2002**, *3*, 410-414.
106. Jegier, J. A.; Munoz-Hernandez, M.-A.; Atwood, D. A. *J. Chem. Soc., Dalton Trans.* **1999**, *15*, 2583-2587.
107. Pinkas, J.; Wessel, H.; Yang, Y.; Montero, M. L.; Noltemeyer, M.; Froeba, M.; Roesky, H. W. *Inorg. Chem.* **1998**, *37*, 2450-2457.
108. Mortier, J.; Gridnev, I. D.; Fortineau, A.-D. *Org. Lett.* **1999**, *1*, 981-984.
109. Mortier, J.; Gridnev, I. D.; Guenot, P. *Organometallics.* **2000**, *19*, 4266-4275.

Chapter 29

Synthesis and Characterization of Divalent Main Group Diamides and Reactions with CO_2

Yongjun Tang[1], Ana M. Felix[1], Virginia W. Manner[1],
Lev N. Zakharov[2], Arnold L. Rheingold[2], Bahram Moasser[3],
and Richard A. Kemp[1,4,*]

[1]Department of Chemistry, University of Mexico, MSC03 2060,
Albuquerque, NM 87131
[2]Department of Chemistry and Biochemistry, University of California at San Diego, 9500 Gillman Drive, La Jolla, CA 92093
[3]General Electric Global Research Center, Building CEB, Room 134, One Research Circle, Niskayuna, NY 12309
[4]Sandia National Laboratories, Advanced Materials Laboratory,
1001 University Boulevard, SE, Albuquerque, NM 87106

It is known that bulky Sn and Ge N-silylamides insert CO_2 to form either silyl isocyanides or silyl carbodiimides, albeit relatively slowly. As part of a research effort to eventually prepare ^{11}C-labelled radiopharmaceuticals derived from $^{11}CO_2$, we have been interested in expanding the scope of this reaction by investigating other species that may react more rapidly. We have synthesized and structurally characterized a large variety of new diamides based on other metals such as Mg, Ca, Ba, and Zn, as well as new divalent Sn species with several new sterically-demanding, trimethylsilyl-containing ligands. Reactions of these species with CO_2 are discussed.

The insertion of carbon dioxide into well-defined organometallic species in order to prepare more valuable organic molecules has been an active subject of study over the past several years (*1*). Goals of these previous investigations have ranged from detailed mechanistic studies of how CO_2 interacts with metals to synthetic schemes that use CO_2 as reactant and/or solvent. While most of the previous work has focused on the interactions of CO_2 with transition metals, more recently the insertion of CO_2 into main group element bonds has been of interest. More specifically insertion of CO_2 into main group element–nitrogen bonds to form carbamates has been under investigation, which has culminated in a recent review (*2*). Sita and coworkers have very recently studied the insertion of CO_2 into divalent Sn and Ge bis(bistrimethylsilylamides) (Figure 1) (*3*). They discovered that under relatively mild conditions CO_2 would insert into

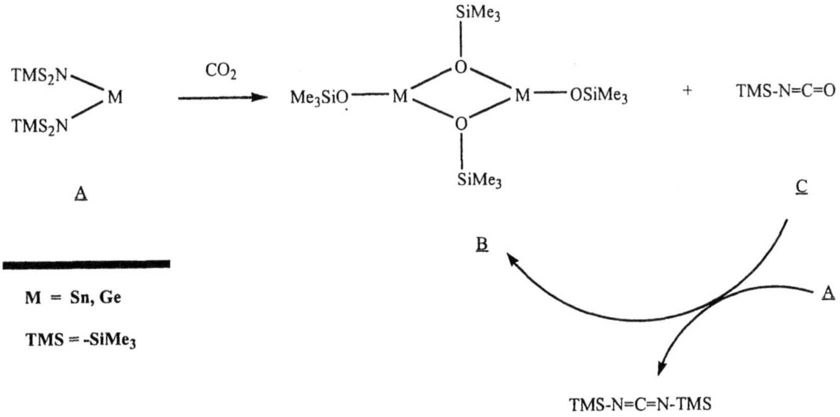

M = Sn, Ge

TMS = -SiMe$_3$

Figure 1. Overall reaction scheme to generate trimethylsilylisocyanate and bis(trimethylsilyl)carbodiimide from CO_2 and Group 14 divalent amides.[3]

these Sn or Ge amide bonds to form *in situ* carbamates, and these carbamates would subsequently extrude either trimethylsilylisocyanate or bis(trimethylsilyl)carbodiimide. We were intrigued with the possibility that by using these divalent metal amides we might be able to use radio-labeled $^{11}CO_2$ as a reagent for preparing ^{11}C-carbamate or urea-containing radiopharmaceuticals that could be useful in diversifying the options for positron emission tomography (PET) (*4*). However, the rates of CO_2 insertion observed by Sita, while impressive, were not nearly fast enough for our purposes. Since the half-life of ^{11}C is 20.3 minutes (*5*) we are limited to 3-4 half lives (or approximately 60-80 minutes) to accomplish the chemistry we need to do in order to possibly make radiopharmaceuticals effectively. Thus, we were interested in reactions of CO_2 with other main group amides that would mimic the reaction chemistry seen by Sita but at a <u>significantly faster rate</u> than that observed with the subvalent Sn or Ge amides. Additionally, we were interested in new ligands that with Sn or other Group 14 congeners might increase reaction rates beyond those seen earlier by Sita.

Syntheses of Divalent Metal Amides

Our initial investigations have involved the synthesis and characterization of a variety of divalent metal amides with both known and new silyl-containing ligands. Towards this end we have prepared and structurally characterized new metal amides based on Mg, Ca, Ba, Zn, Hg, and Sn.

Magnesium-Based Diamides

Interactions of CO_2 with magnesium compounds are well-known, ranging as far back as the preparation of organic acids from Grignard reagents (6). More relevant recent research comes from the group of Chang (7) who examined the insertions of CO_2 into Mg and mixed Mg-Al species to generate carbamato-type complexes of Mg. Of much interest to us was the observation that the insertions of CO_2 were much more rapid than those seen by Sita earlier; however, the complexes formed in the Chang work did not lose trimethylsilylisocyanate spontaneously since his work was done with alkyl or aryl amides. We thus set forth to synthesize a range of $Mg(NRR')_2$ complexes, where at least one of the R groups was a trimethylsilyl (TMS) group.

We prepared the magnesium diamides by either of two routes commonly used to prepare metal amides (8). The first method involved alkane elimination upon direct reaction of dibutylmagnesium with the desired silylamine. The second route was a metathesis reaction, whereby LiX (or KX) was eliminated from the interaction of lithium (or potassium) amide with $MgBr_2$. In both of these routes solvents were found to play non-innocent roles, much as they do in the preparation of related metal alkoxides (9). Figure 2 shows the types of ligands used (**1a/1b**, benzyl; **2**, adamantyl; **3**, i-propyl; **4a/4b**, mesityl; **5**, *t*-butyldimethylsilyl-; **6**, *t*-butyldiphenylsilyl-; **7**, cyclohexyl; **8**, 4-adamantyl-2,6-di-*i*-propylphenyl-) along with the range of magnesium coordination geometries seen in these preparations. While detailed discussions of the preparations and X-ray structures will be published elsewhere, several salient points can be made here. Generally, the [(TMS)(R)N]- ligands are insufficient in steric bulk to keep solvent molecules from entering the coordination sphere of the Mg atom. The majority of the structures are consistent with a Mg atom containing two solvent molecules in addition to the two Mg-N bonds, leading to a pseudo-tetrahedral geometry around Mg. In two cases we see a Mg atom with only one solvent ligand coordinated (the bulky R groups adamantyl **2** and *t*-butyldimethylsilyl **5**) and in one case we see <u>no</u> solvent molecules attached to Mg (the extremely bulky R group *t*-butyldiphenylsilyl **6**). The ligand used to prepare **8** deserves some comment as well. We were attempting to prepare the bulky amine (adamantyl)(2,6-di-i-propylphenyl)NH by reacting adamantyl bromide with 2,6-di-i-propylphenylamine at 240 °C. Rather than forming the desired product we obtained instead the new compound formed by attack of the adamantyl bromide at the *para*-position of the aromatic ring to liberate HBr and couple the rings.

Table 1. Magnesium Disilylamides [Mg(N(TMS)R)$_2$] Prepared and Characterized

R Group	Solvent	Resulting Formula	X-Ray Structure Comments
benzyl (PhCH$_2$)	a) HMPA b) 4-DMAP	Mg(N(TMS)R)$_2$[Solv]$_2$ (1)	Pseudo-tetrahedral Mg
adamantyl	ether	Mg(N(TMS)R)$_2$[Et$_2$O] (2)	Tri-coordinate Mg
i-Propyl	4-DMAP	Mg(N(TMS)R)$_2$[DMAP]$_2$ (3)	Pseudo-tetrahedral Mg
2,4,6-trimethylphenyl (mesityl)	a) THF b) pyr	Mg(N(TMS)R)$_2$[Solv]$_2$ (4)	Pseudo-tetrahedral Mg
(t-Butyl)Me$_2$Si-	pyr	Mg(N(TMS)R)$_2$[pyr] (5)	Tri-coordinate Mg
(t-Butyl)Ph$_2$Si-	ether	Mg(N(TMS)R)$_2$ (6)	Monomeric, Di-coordinate Mg
cyclohexyl	4-DMAP	Mg(N(TMS)R)$_2$[DMAP]$_2$ (7)	Pseudo-tetrahedral Mg
(adamantyl)(2,6-di-i-propylphenyl)	THF	Mg(N(TMS)R)$_2$[THF]$_2$ (8)	Pseudo-tetrahedral Mg

Solvents: HMPA - hexamethylphosphoramide; 4-DMAP - 4-dimethylaminopyridine; pyr - pyridine; THF - tetrahydrofuran; ether - diethyl ether

All of these magnesium species have been subjected to reaction with CO_2. In each case, simple bubbling of CO_2 through a pentane solution of the Mg diamide at room temperature led to an extremely rapid reaction (<5 minutes). These reaction rates are similar to those seen by Chang (7), and much faster than those seen by Sita (60 psig CO_2 pressure, 1 hour reaction time) (3). We are in the early stages of identifying products, but it appears at this time that the carbamates presumably formed by the insertion of CO_2 do not appear to liberate either trimethylsilylisocyanate or bis(trimethylsilyl)carbodiimide. This is somewhat similar to the alkyl/arylamide results seen by Chang earlier (7). However, when we take a solution of the known compound $Mg[N(TMS_2)]_2$ (10) and bubble CO_2 through it we do observe a species with only one TMS-resonance in the ^1H or ^{13}C NMR, but as yet the exact identity is not known.

We also prepared a non-silylated Mg complex, $[Mg(NCy_2)_2]_4$ (Cy = cyclohexyl) (11) and again bubbled CO_2 into a pentane solution. Again, the reaction was almost instantaneous at room temperature. After workup and isolation of the product in the presence of HMPA we structurally characterized a dinuclear Mg species, $[Mg_2(O_2CNCy_2)_4](HMPA)$ 9, with four carbamato groups and containing a previously unknown Mg-carbamate bonding mode. Chang had previously identified the possible modes of bonding of carbamates to mononuclear or dinuclear Mg species (Figure 2) and had commented quite

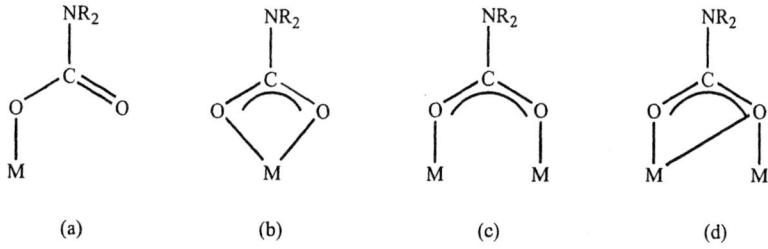

Figure 2. Possible bonding modes for carbamato ligands to either a mononuclear or dinuclear metal atom system: (a) η^1; (b) η^2; (c) μ_2-η^2; (d) μ_2-η^3.

specifically on the <u>absence</u> of a terminal chelating mode to Mg (mode b), the only mode with no literature precedent (7). He attributed the nonexistence of mode b to the instability caused in the strained ring structure by excess electron density when two highly-electron rich oxygen atoms (O,O) are used as donor atoms to Mg. Further evidence for this position can be found from the related, but less electron donating, systems utilizing N and S ligands. Mode b <u>has</u> been found in the analogous carbodiimides (N,N), isothiocyanates (N,S), and carbon disulfide (S,S) systems, presumably due to the lowering of electron density on Mg using these less electron donating ligands. However, we have now found in the X-ray structure of **9** that mode *b* can indeed form in carbamato (O,O) complexes of Mg. The structure of **9** is shown in Figure 3. Immediately obvious are two features – the first is that there is one HMPA solvent molecule bound <u>only</u> to Mg(1), thus leaving Mg(2) solvent-free. Second, there is a carbamato group attached solely to Mg(2) via previously unknown bonding mode *b*. We feel these features are related. The lack of any coordinated solvent to Mg(2) should allow for more electron density to be donated from the chelating carbamato group without disruption of the 4-membered ring. In previous examples where only modes *a*, *c*, and *d* were formed, the Mg atoms in the structures were coordinatively saturated with electron density donated from either solvents or alkoxide ligands. In the case of **9**, this is not the situation.

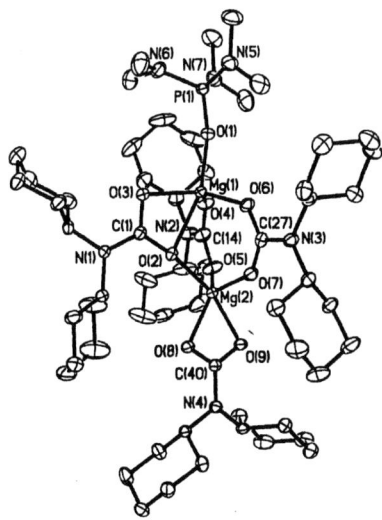

Figure 3. Thermal ellipsoid plot of 9 showing terminal bidentate bonding mode b.

Calcium- and Barium-Based Disilylamides

The number of previously known and well-characterized Ca disilylamides is quite small (*12,13*). We have prepared and structurally-characterized Ca(N(TMS)(SiMe$_2$t-Bu)$_2$[solv]$_2$ with a variety of coordinating solvents such as HMPA, 4-DMAP, THF, and pyr (Figure 4). Interestingly, despite the donation of electron density by the coordinated solvents to the electropositive Ca centers, the X-ray structures of all of the Ca(NRR')$_2$(solv)$_2$ complexes showed several

t-BuMe₂Si\\NH/Me₃Si →KH→ [t-BuMe₂Si\\N-K-N/SiMe₂t-Bu / Me₃Si\\K/SiMe₃] →CaI₂/Solvent→ t-BuMe₂Si\\N-Ca-N/SiMe₂t-Bu / Me₃Si / SiMe₃ (with L above and below Ca)

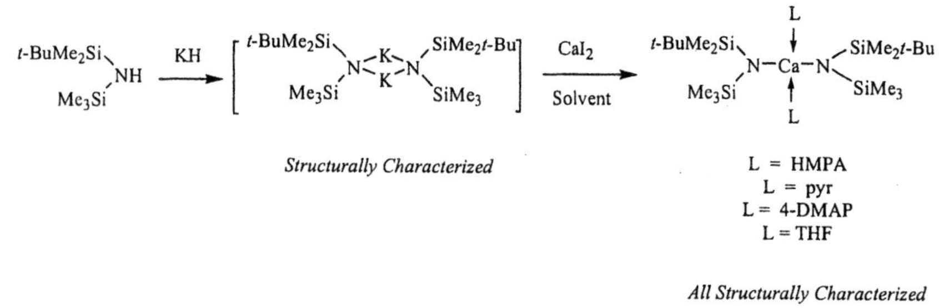

Structurally Characterized

L = HMPA
L = pyr
L = 4-DMAP
L = THF

All Structurally Characterized

Figure 4. Overall route to silylamido-substituted calcium complexes.

extremely close interactions between the Ca atom and the H atoms on the methyl groups on either the TMS- or t-BuMe₂Si- groups. This shows clearly the electron-deficient nature of the Ca atoms in these species even with two donating ligands attached. As in the Mg cases, these Ca amides react essentially instantaneously with CO_2; however, the identities of the species formed have not yet been identified definitively by X-ray crystallography.

As was the case with Ca, the number of known Ba silylamides is quite limited; relevant examples consist of the parent Ba[N(TMS)₂]₂ (*12*) and the related bulky amide Ba[N(TMS)(2,6-di-i-Pr₂C₆H₃)]₂ (*13*). In this study we prepared the parent Ba[N(TMS)₂]₂ and new compound Ba[N(TMS)(t-Bu)]₂(THF)₂ **10** by transmetalation reactions using the Hg diamide generated *in situ*. While we have not investigated as yet the CO_2 insertion chemistry of these Ba amides, we have shown the inherent reactivity of both of these Ba compounds towards H₂S. Both compounds with H₂S form crystalline BaS at room temperature and atmospheric pressure. BaS is currently produced at high temperatures via carbon reduction of BaSO₄ (*14*) The BaS formed was characterized and found to be pure by elemental analysis as well as crystalline by powder X-ray diffraction analysis.

Zinc- and Mercury-Based Disilylamides

We have also prepared and characterized new solvent-free Zn and Hg complexes that contain extremely bulky silylamides as ligands. These compounds include M[N(TMS)(SiPh₂t-Bu)]₂ and M[N(TMS)(adam)]₂ (adam = adamantyl), M = Zn, Hg. These compounds all adapt essentially linear N-Zn-N or N-Hg-N backbones, with the backbone angles ranging from 168 to 179°. Structures of the Zn compounds are shown in Figure 5 – the structures of the corresponding Hg compounds are essentially identical and are thus not shown. We find that the Zn compounds are very reactive to CO_2, very much like the Mg compounds we have prepared earlier. However, the product(s) of the presumed insertion of CO_2 are proving difficult to crystallize. We note as well that the

CO_2 reactions with the Zn compounds appear to be even more exothermic than in the Mg cases. However, quite surprisingly the Hg compounds appear to be totally unreactive at 25 °C and one atmosphere pressure of CO_2. Under these conditions only unreacted starting materials are obtained. The less-reactive Hg compounds will require more forcing conditions (if at all) to effect this reaction.

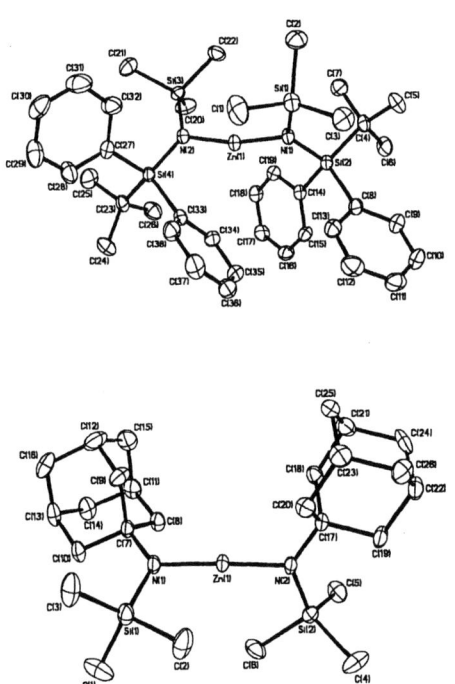

Figure 5. Thermal ellipsoid plots for the bulky Zn disilylamides $Zn[(TMS)(SiPh_2t-Bu)]_2$ (top) and $Zn[N(TMS)(adam)]_2$ (bottom).

Alkyl Zinc Amides

We have also investigated the interactions of CO_2 with alkyl zinc amides (2,15). When $MeZnNR_2$ (R = i-propyl, i-butyl, piperidino) is allowed to react with atmospheric pressure CO_2 at 25 °C, an immediate insertion reaction takes place to afford a Zn_4 carbamato complex (Figure 6). This Zn tetramer formally has CO_2 molecules inserted into each Zn-NR_2 bond. In each of the three cases the initially formed tetramer can be isolated and identified by standard methods

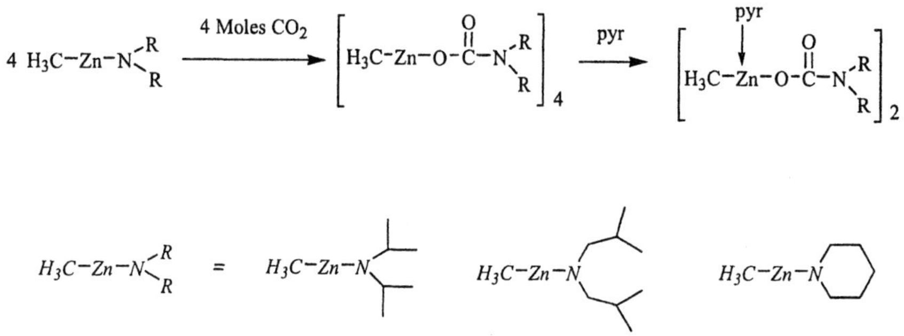

$$4\ H_3C-Zn-N{\overset{R}{\underset{R}{\diagdown}}} \xrightarrow{4\ Moles\ CO_2} \left[H_3C-Zn-O-\overset{\overset{O}{\|}}{C}-N{\overset{R}{\underset{R}{\diagdown}}} \right]_4 \xrightarrow{pyr} \left[H_3C-Zn(pyr)-O-\overset{\overset{O}{\|}}{C}-N{\overset{R}{\underset{R}{\diagdown}}} \right]_2$$

$$H_3C-Zn-N{\overset{R}{\underset{R}{\diagdown}}} = H_3C-Zn-N{\diagdown\!\!\diagup} \quad H_3C-Zn-N{\diagdown\!\!\diagup} \quad H_3C-Zn-N\bigcirc$$

Figure 6. Insertion reactions of CO_2 into Me-Zn-NR_2 Species.

such as NMR and elemental analysis, and as well as by single-crystal X-ray diffraction. All three tetramers have similar structures and the carbamato groups all bind via bonding mode *d* discussed earlier. Each of the $-CO_2$ fragments bind to three Zn atoms – one oxygen atom of the pair binds directly to one Zn atom, while the other oxygen atom bridges two additional Zn atoms. When a strongly coordinating solvent like pyridine is added to the tetramer the cluster is broken apart and a pyridine-containing Zn dimer is subsequently formed. We have analyzed these Zn dimers by typical spectroscopic techniques as well as X-ray crystallography. Again, the solid-state structures of all three dimeric compounds are similar. The carbamato groups now bridge the two Zn atoms by bonding as in mode *c*, resulting ultimately in a puckered 8-membered ring. Figure 7 shows the thermal ellipsoid plots of the tetrameric structure (shown for the *i*-propyl compound, minus the methyl and pyridine bonded to Zn as well as the *i*-propyl groups in order to better see the core) and the dimer structure (shown for the *i*-butyl compound with all atoms shown in order to more clearly see the puckered nature of the ring.)

Sn(II) Compounds with New Silylamine Ligands

As mentioned in the introductory section Sita had examined the addition of CO_2 to lower valent Sn(II) and Ge(II) species (*3*). He had seen significantly more active and facile additions using Sn rather than Ge. We were interested in whether changing the sterics on the silylamino ligand bound to Sn could also affect the rate. We have prepared and characterized several new stannylene complexes containing new silylamino ligands in place of $-N(TMS)_2$. In all of these cases, the group replacing one of the TMS groups is more sterically congested than the TMS group itself. These compounds are shown in Table 2. The N-Sn-N bond angles in all cases are between 104 to 109°, normal for a bis(amido)stannylene (*16*). We then subjected these bulky Sn amides to higher pressures of CO_2. Based on the work of Sita (*3*) we did not expect these Sn amides to be as active as the Group 2 or Group 12 compounds prepared earlier.

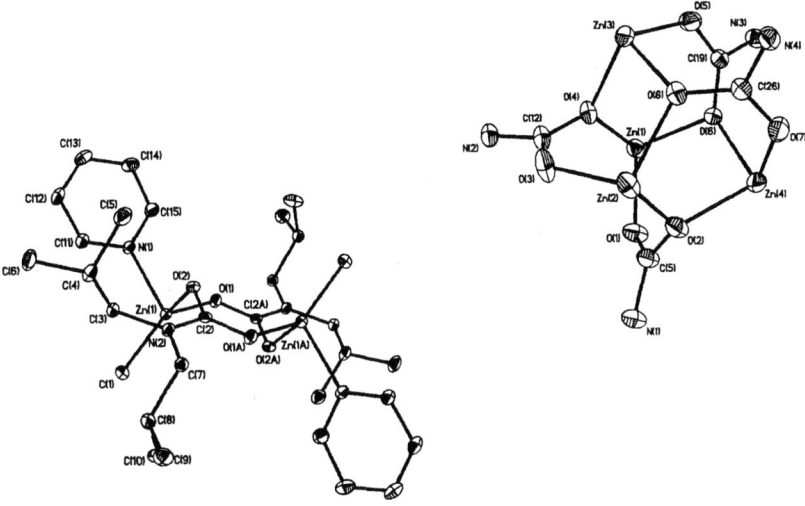

Figure 7. Thermal ellipsoid plots for the core of the Zn tetramer structure (top right) and the Zn dimer structure (bottom left).

Table 2. Tin Disilylamides Prepared and Characterized

R Group	Resulting Formula	X-Ray Structure?
adamantyl	Sn[N(TMS)(adamantyl)]$_2$	Yes
Ph$_3$Si-	Sn[N(TMS)(SiPh$_3$)]$_2$	Yes
t-BuMe$_2$Si-	Sn[N(TMS)(SiMe$_2t$-Bu)]$_2$	No
t-BuPh$_2$Si-	Sn[N(TMS)(SiPh$_2t$-Bu)]$_2$	Yes
mesityl	Sn[N(TMS)(mesityl)]$_2$	Yes

The results from these CO_2 reactions were quite interesting to us. When the rates of insertion are compared, it is found that the more sterically congested ligands give measurably faster CO_2 insertions. The data shown below indicate that both the time of reaction as well as the pressures required for complete CO_2 insertion can be significantly reduced relative to $Sn[N(TMS)_2]_2$. This may indicate that the relief of steric strain in the starting bis(amido) Sn complex may be accelerating the insertion reaction of CO_2 into these Sn compounds.

-N(TMS)$_2$	-N(TMS)(SiMe$_2$$t$-Bu)	-N(TMS)(SiPh$_2$$t$-Bu)
25 °C	25 °C	25 °C
60 psig CO_2	30 psig CO_2	30 psig CO_2
60 min	40 min	20 min

Conclusions

We have prepared and characterized a wide range of divalent metal amides in order to try to find compounds that are more active than subvalent Group 14 bis(amides). We have successfully found other metal systems that are significantly more active than Sn or Ge, but the chemistry in these more active systems in many cases does not mimic the Sn system by producing isocyanates or carbodiimides. We have also prepared new ligands on Sn that cause the CO_2 insertion reaction to be much more facile. While our goal is to eventually make this process catalytic in order to produce radio-labeled drugs, we find there is much interesting fundamental chemistry in the reactions of these metal silylamides with small molecules.

Acknowledgements

We gratefully acknowledge the funding provided by the National Science Foundation, Grant Number CHE-0213165. In addition, RAK would like to acknowledge the influence that his Ph.D. advisor and friend Professor Alan H. Cowley has had on his professional career. From the beginnings in graduate school 25 years ago up to the current time, Professor Cowley has always served as a role model of someone whose vocation is his avocation. His passion and enthusiasm for chemistry and discovery is inspiring, and I look back upon the time spent in his laboratory with incredible fondness and happiness. It is in honor of his many contributions to inorganic chemistry (and, of course, his 70th birthday) that we celebrate CowleyFest 2004, and wish for many more to come!

References

1. Halmann, M. M. *Chemical Fixation of Carbon Dioxide. Methods for Recycling CO_2 into Useful Products*; CRC Press: Boca Raton, FL, 1993.
2. Dell'Amica, D. B.; Calderazzo, F.; Labella, L.; Marchetti, F.; Pampaloni, G. *Chem. Rev.* **2003**, *103*, 3857.
3. (a) Sita, L. R.; Babcock, J. R.; Xi, R. *J. Am. Chem. Soc.*, **1996**, *118*, 10912. (b) Xi, R.; Sita, L. R. *Inorganica Chimica Acta*, **1998**, *270*, 118.
4. Phelps, M. E. *J. Nucl. Med.*, **2000**, *41*, 661.
5. Heath, R. L. In *Handbook of Chemistry and Physics, 57th Ed.*; Weast, R. C., Ed.; Chemical Rubber Company: Cleveland, OH, 1976; p B-272.
6. March, J., *Advanced Organic Chemistry: Reactions, Mechanisms, and Structure, 4th Edition*; John Wiley: New York, NY; 1992, pp 935-936.
7. (a) Chang, C.-C.; Srinivas, B.; Wu, M.-L.; Chiang, W.-H.; Chiang, M. Y., Hsiung, C.-S. *Organometallics* **1995**, *14*, 5150. (b) Chang, C.-C.; Ameerunisha, M. S. *Coord. Chem. Rev.* **1999**, *189*, 199. (c) Yang, K.-C.; Chang, C.-C.; Yeh, C.-S., Lee, G.-H.; Peng, S.-M. *Organometallics* **2001**, *20*, 126. (d) Yang, K.-C.; Chang, C.-C.; Yeh, C.-S.; Lee, G.-H.; Wang, Y. *Organometallics* **2002**, *21*, 1296.
8. Lappert, M. F., Sanger, A. R., Srivastava, R. C., Power, P. P. *Metal and Metalloid Amides: Synthesis, Structure, and Physical and Chemical Properties*; Ellis Horwood: Chichester UK; 1980.
9. Bradley, D. C., Mehrotra, R. C., Rothwell, I. P., Singh, A. *Alkoxo and Aryloxo Derivatives of Metals*; Academic Press: San Diego, CA; 2001.
10. Wannagat, U.; Kuckertz, H. *Angew. Chem.* **1963**, *75*, 95.
11. Olmstead, M. M.; Grigsby, W. J.; Chacon, D. R.; Hascall, T.; Power, P.P. *Inorganica Chimica Acta*, **1996**, *251*, 273.
12. Westerhausen, M. *Inorg. Chem.*, **1991**, *30*, 96.
13. Vargas, W.; Englich, U.; Ruhland-Senge, K. *Inorg. Chem.* **2002**, *41*, 5602.
14. Ahmed, Y. M. Z.; El-Toni, A.; Ibrahim, I. A.; Shalabi, I. M. E. H. *European Journal of Mineral Processing and Environmental Protection* **2002**, *2*, 103.
15. (a) Hursthouse, M. B.; Malik, M. A.; Motevalli, M.; O'Brien, P. *J. Chem. Soc., Chem. Commun.* **1991**, 1690.
16. Fjeldberg, T.; Hope, H.; Lappert, M. F.; Power, P. P.; Thorne, A. J. *J. Chem. Soc., Chem. Commun.* **1983**, 639.

Chapter 30

Organotin–Sulfur Intramolecular Interactions: An Overview of Current and Past Compounds and the Biological Implications of Sn---S Interactions

Teresita Munguia[1], Francisco Cervantes-Lee[1], László Párkányi[2], and Keith H. Pannell[1,*]

[1]Deparment of Chemistry, University of Texas at El Paso, El Paso, TX 79968
[2]Institute of Chemistry, Chemical Research Center, Hungarian Academy of Sciences, H–1525, P.O. Box 17, Budapest, Hungary

Lewis acid interactions of tin are of interest because it is often through intermolecular hypervalent mechanisms that organotin compounds interact with biological materials, resulting in their characteristic biocidal capabilities. We investigate the nature of the Sn---S intramolecular interactions in previously known and new organotin sulfides that may prove to be useful in modifying biocidal activity due to competition with intermolecular Sn---S biological interactions.

The restrictions on organotin antifouling paints, has lead to the search for materials that are as efficacious as the currently used formulations, but less toxic to the environment. It has been suggested that the biocidal properties of oragnotin compounds results from the ability of the tin atom to participate in intermolecular interactions, in particular with thiol groups in cystine type residues (1). For this reason, we believe it useful to investigate organotin compounds with a potential for *intra*molecular interactions with sulfur, so as to modify biocidal activity *via* competition between the intramolecular sulfur and biological sulfur groups, Figure 1.

Figure 1. Potential competition between sulfur containing biological residues and organotin compounds with intramolecular Sn---S interactions.

Biocidal Activity of Organotin Compounds

Organotin compounds have found application in a large array of societal uses. They are used as stabilizers for vinyl chloride resins, catalysts, wood preservatives, antifouling agents, agrochemicals, precursors for tin oxide films, and as potentially useful therapeutic drugs. (2-5). Each usage draws upon a basic property of organotin compounds; that is, the ability of tin to become hypervalent and contain coordination numbers larger than four. However, such widespread usage has led to accumulation in the environment and their use is to be severely limited.

The mammalian toxicity of organotin compounds has been well-reviewed (6). Of major concern is the recent finding by Whalen el al. that "normal" human blood samples contained concentrations varying from 64 to 155 ng/mL of tributyl tin, dibutyl tin and monobutyl tin (7). The same group has shown that tributyl tin can inhibit human natural killer lymphocyte function *in vitro* and since tributyl tin has been linked to neurotoxicity, hepatotoxicity, immunotoxiciy, and cutaneous toxicity in rats and mice (6) the overall finding is potentially significant.

While different organotin toxicities are well-documented, a complete and detailed understanding of the mechanism of cell-organotin interactions is not available (8). However, histidine and cysteine residues have been implicated in binding of trialkyltin compounds in cat and rat hemoglobin (8).

Cysteine in particular has the thiol group that can bind to organotin compounds, Figure 2.

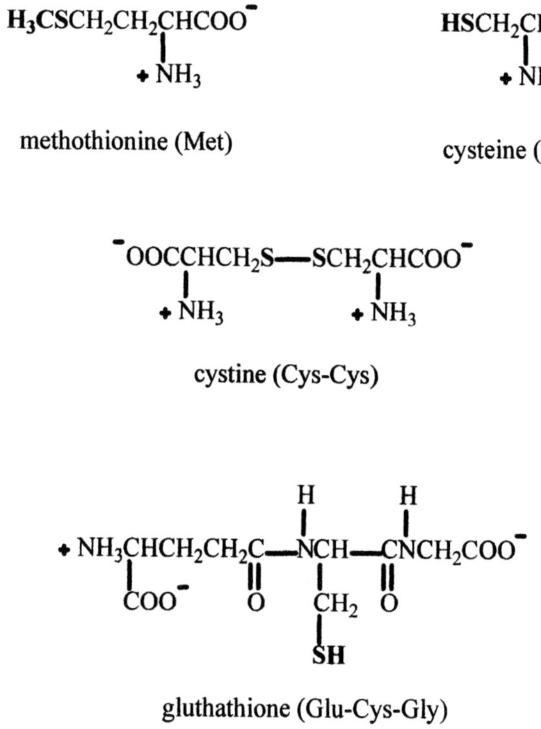

$$H_3CSCH_2CH_2\underset{\underset{+NH_3}{|}}{C}HCOO^-$$

methothionine (Met)

$$HSCH_2\underset{\underset{+NH_3}{|}}{C}HOO^-$$

cysteine (Cys)

$$^-OOCCHCH_2S\text{—}SCH_2CHCOO^-$$
$$\underset{+NH_3}{|} \qquad \underset{+NH_3}{|}$$

cystine (Cys-Cys)

$$^+NH_3\underset{\underset{COO^-}{|}}{C}HCH_2CH_2\underset{\underset{O}{\|}}{C}\text{—}\underset{\underset{CH_2}{|}}{\overset{\overset{H}{|}}{N}}CH\text{—}\underset{\underset{O}{\|}}{C}\overset{\overset{H}{|}}{N}CH_2COO^-$$
$$\underset{SH}{|}$$

gluthathione (Glu-Cys-Gly)

Figure 2. Biologically important sulfur-containing residues.

An excellent review (8) has summarized organotin amino acid and peptide interactions. Recent studies (1, 9) have illustrated a trigonal bipyramidal solution coordination geometry for di- and tri-organotin derivatives of sulfur-containing amino acids and peptides. The strong Sn---S intermolecular interactions apparently act as an anchoring site for the organotin molecules thereby permitting further coordination with neighboring N groups in the biological residue (1). Structure and activity of organotin compounds can be condensed into three main points (8, 9):

- The accessibility of coordination sites at tin.
- The ability to create stable (but not too stable) ligand---Sn interactions via Sn---S or Sn---N interactions.
- The capability of the ligand---Sn interaction to withstand hydrolysis.

An aim of our current research is the synthesis and biocidal evaluation of organotin compounds containing sulfur groups that can exhibit intramolecular Sn---S interactions. Here we present an overview of such known interactions including those exhibited by new *o*-thiomethylbenzyl derivatives from our laboratory.

Tin-Sulfur Intramolecular Interactions

A query of non-bonded contact distances between tin and sulfur was performed using the Cambridge Crystallographic Database (10). A maximum for the sum of the van der Waals radii for Sn and S of 4.0 Å was used. We further restricted our analysis to the organotin compounds with 1,1-ditholate family and simple alkyl-alkyl or alkyl-aryl sulfides.

Intramolecular Organotin 1,1-Dithiolate Interactions

The family of organotin derivatives of dithiocarbamates, xanthates, and dithiocarboxylates all have the same ligand arrangement of ^-S_2CY, where Y = aryl/alkyl, NR_2, and OR, respectively. Several reviews (11-13) have demonstrated the remarkable diversity in the structures of these organotin(IV) compounds; therefore our analysis of these compounds will be brief and only a few representative samples will be discussed.

Ligands such as 1,1-dithiolates can coordinate with organotin compounds in three ways, Figure 3. The 1,1-ditholate ligand can be isobidentate, in which both sulfur atoms bind equally with tin, anisobidentate, in which one covalent bond has formed and the other S interacts intramolecularly to form a secondary interaction, and monodentate, in which the secondary S---Sn interaction is "too long" to contribute to changes in the coordination geometry at the tin atom.

In addition, Tiekink (12, 14) found that many of the bis-(1,1-dithiolates) adopt one of four structural arrangements that are a combination of isobidentate, anisobidentate, and monodentate. The R groups can adopt a trans or cis configuration in the distorted octahedral geometry which the bis-(1,1-dithiolates) can form. A skew-trapezoidal bipyramidal geometry can also be formed in which the R groups lean over the longer Sn---S interactions and represents the most

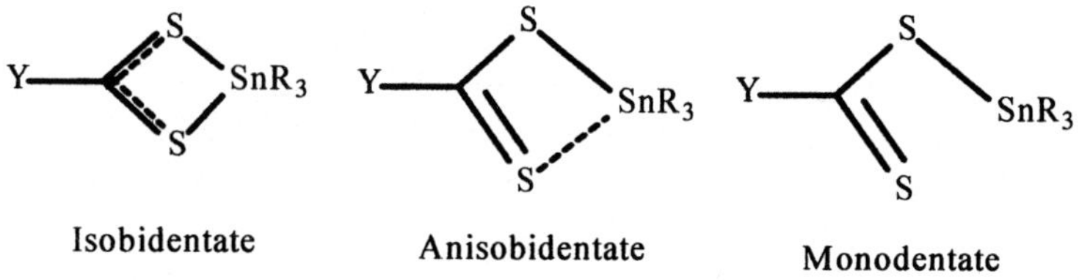

Isobidentate   Anisobidentate   Monodentate

Figure 3. Structural motifs of organotin 1,1-dithiolates

commonly found bonding motif. Finally, distorted trigonal bipyramidal geometry about tin due to the monodentate binding of a second dithiolate ligand can be observed.

Table 1 summarizes a selection of organotin-1,1-ditholate compounds with their relevant Sn---S distances. The great majority of organotin-1,1-dithiolates coordinate in an anisobidentate manner with two clearly distinguishable Sn---S atom separations. The complexes $Ph_2Sn(S_2CN(Et)C_6H_{10})_2$, $Ph_2Sn(S_2CN[C_6H_{10}]_2)_2$, and $Ph_2ClSnS_2C(C_5H_3NH_2)$ approximate isobidentate coordination observed by the lengthening of the S=C bond and the shortening of the S—C bond.

Table I. Selected Organotin 1,1-dithiolates.

	Sn—S (Å)	S→Sn (Å)	S—C (Å)	S=C (Å)	ref
$Me_3SnS_2CNMe_2$	2.47	3.16	1.80	1.70	(15)
	2.47	3.33	1.78	1.71	
	2.47	3.16	1.75	1.72	(16)
n-$BuPh_2SnS_2CNMe_2$	2.466	3.079	1.762	1.680	(17)
$Ph_3SnS_2CNC_4H_5$	2.468	3.106	1.776	1.702	(18)
$Ph_3SnS_2CN(Et)_2$	2.428	3.095	1.74	1.67	(19)
$Ph_3SnS_2CN(Me)n$-Bu	2.463	3.084	1.753	1.668	(20)
$Ph_3SnS_2C(o$-$tol)$	2.446	3.207	1.737	1.645	(21)
$Ph_2ClSnS_2C(C_5H_3NH_2)$	2.439	2.649	1.756	1.720	(22)
$(n$-$Bu)_2Sn(S_2CNC_4H_4O)_2$	2.525	3.001	1.740	1.694	(23)
$Ph_2Sn(S_2CN[Et]_2)_2$	2.428	3.095	1.74	1.67	(19)
$Ph_2Sn(S_2CN[C_6H_{10}]_2)_2$	2.575	2.687	1.720	1.735	(24)
	2.568	2.692	1.720	1.735	
$Ph_2Sn(S_2CN(Et)C_6H_{10})_2$	2.601	2.660	1.734	1.708	(24)
	2.571	2.735	1.741	1.717	
$Ph_2Sn(S_2COi$-$Pr)_2$	2.482	3.179	1.747	1.652	(25)
	2.500	3.067	1.733	1.663	
$Me_2Sn(S_2COMe)_2$	2.482	2.900	1.733	1.655	(14)
	2.538	---	1.744	1.609	
$PhMeSn(S_2COMe)_2$	2.498	3.019	1.736	1.644	(14)
	2.509	3.089	1.725	1.660	
$Ph_2Sn(S_2COMe)_2$	2.498	3.126	1.749	1.658	(14)
	2.502	3.042	1.737	1.652	
$PhSn(S_2C$-o-$Tol)_3$	2.594	2.813	1.701	1.663	(21)
	2.600	2.751	1.691	1.673	

It is also important to note that the compounds with the shortest S→Sn distances have S—C—S bite angles ranging from 115.1-116.9° and S—Sn—S angles ranging from 67.26-70.65°. The compounds with longer S→Sn distances have larger S—C—S angles (117-124.4°) and smaller S—Sn—S angles (60-64.64°) (14-25).

Intramolecular Organotin(IV)-Sulfide Interactions

We are interested in simple alkyl/aryl sulfide systems where the soft sulfide ligand can act as a weaker donor toward tin halides, which are considered hard acids (26). A simple way to describe and assess potential intramolecular tin-sulfur interactions is to monitor the progress from tetrahedral to trigonal bipyramidal geometry. Two primary methods for monitoring this transition is through Sn---S bond order (26-29) and inspection of the six bond angles of the inner tetrahedron (30, 31). Taken together, a fair assessment of the intramolecular tin-sulfur bond can be determined.

The general types of such organotin sulfides are illustrated in Figure 4. The first group involves 1,5-transannular Sn---S interactions of the 8 member ring systems containing tin and sulfur, Figure 4a (28, 29, 32-36), while the others are aryl-alkyl sulfides connected to tin *via* flexible alkyl chains with 1-3 methylene groups, Figure 4b (26, 37, 38), and systems synthesized in our laboratory that contain an ortho benzylsulfide ligand that enables sulfur to be in close enough proximity to the tin atom, Figure 4c (39).

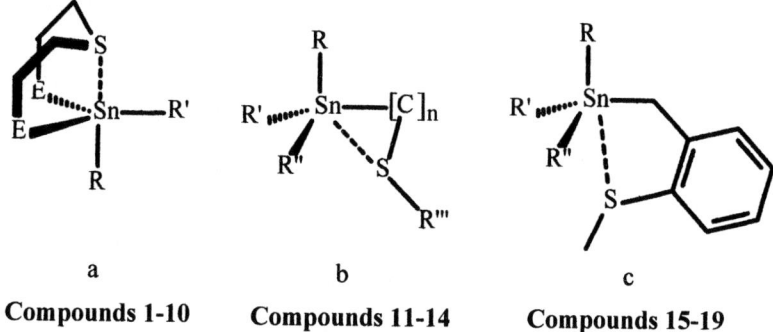

a
Compounds 1-10

b
Compounds 11-14

c
Compounds 15-19

Figure 4. a. Eight membered heterocycles of organotin compounds. b. Thioalkyl organotin compounds. c. (Ortho-thiomethyl)benzyl organotin compounds.

We have also systematically increased the Lewis acidity of the tin atom by removal of the phenyl groups and addition of chlorine, i.e. Ph_3Sn, Ph_2ClSn and $PhCl_2Sn$. Table II lists the various known compounds within this general group that contain Sn---S atom distances shorter than the sum of the van der Waals radii, i.e. 4.02 Å (40).

Table II. Compounds Containing Tin---Sulfur Intramolecular Interactions.

Compound (ref)	E	n	R	R'	R''	R'''
1 (32)	S		Cl	Cl		
2 (33)	S		Ph	Ph		
3 (28)	C		Cl	Cl		
4 (29)	S		Br	Br		
5 (29)	S		I	I		
6 (34)	S		Cl	Me		
7 (34)	S		Br	Me		
8 (34)	S		Me	Me		
9 (35)	S		$SCH_2CH_2SCH_2CH_2S$			
10 (36)	S		a	n-Bu		
11 (37)		1	c-Hex	c-Hex	c-Hex	p-ClC_6H_4
12 (38)		2	Ph	Ph	Ph	o-$NO_2C_6H_4$
13 (38)		2	Ph	Ph	Ph	2-NO_2-4-$CH_3C_6H_3$
14 (26)		3	Cl	Ph	Ph	o-Tol
15 (39)			Ph	Ph	Ph	
16 (41)			Cl	Ph	Ph	
17 (41)			Cl	Cl	Ph	
18 (41)			p-tBuPh	p-tBuPh	p-tBuPh	
19 (41)			Cl	p-tBuPh	p-tBuPh	

Note: Structural identification see Fig. 4; a: $S_2P(OCH_2C(Et_2)CH_2O)$; c-Hex = Cyclohexyl

The simplest and preferred way to analyze the Sn---S intramolecular interactions in these molecules is to determine their bond order (BO) and their "closeness" to tetrahedral or trigonal bipyramidal geometries. To determine the formal bond Sn---S order of the compounds we used the equation used by Dräger (26, 27): $[d(Sn—E)_{av}] + 1 - [d(Sn---E)] = BO$, where E is the element in question, in these particular examples either C, S or halogen, and $d(Sn—E)_{av}$ are the standard single bond distances according to O'Keeffe and Brese with corrections for electronegativity omitted (42). The distances used were Sn—S 2.4 Å, Sn—C 2.15 Å, Sn—Cl 2.36 Å, Sn—Br 2.50 Å, and Sn—I 2.71 Å. According to Dräger and co-workers (29), the Sn—I distance of O'Keeffe and

Brese is too long and the distance 2.71 Å determined from Mes₃SnI was used instead. Here we also use this distance to simplify structural comparisons. Table III lists the tin-sulfur distances, bond orders, angle sums.

Table III. Important Bond Distances, Angle Sums, and "Closeness" to a Trigonal Bipyramid.

	S→Sn Å	Sn—Rax Å	BO S→Sn	BO Sn—Rax	Σ_{eq} (deg)	Σ_{ax} (deg)	Σ_{eq}- Σ_{ax}
1	2.760	2.392	0.64	0.91	357.1	287.1	70
2	3.246	2.156	0.15	0.99	342.9	311.9	31
3	2.851	2.449	0.55	0.99	353.6	293.6	60
4	2.767	2.545	0.63	0.96	357.1	287.2	70
5	2.779	2.786	0.62	0.92	357.2	287.1	70
6	2.863	2.444	0.54	0.92	361.2	285.6	76
7	2.835	2.582	0.57	0.92	357.9	284.6	73
8	3.514	2.147	0	1	332.2	323.4	9
9	3.074	2.434	0.32	0.97	--	--	--
10	2.940	2.5379	0.46	0.86	354.7	291	64
11	3.29	2.22	0.11	0.93	332.6	324.2	8
	3.26	2.15	0.14	1	332.2	324.6	8
12	3.67	2.128	0	1	334.1	322.5	12
13	3.58	2.139	0	1	334.9	321.6	13
14	3.195	2.442	0.21	0.92	355.2	292.2	63
15	3.699	2.145	0	1	333.4	323.3	10
	3.829	2.170	0	1	325.7	331.1	5
16	3.062	2.4130	0.34	0.95	350.5	300.9	50
17	2.994	2.4042	0.41	0.96	349.4	299.8	50
18		2.143		1	335.8	320.4	15
19	3.351	2.3891	0.05	0.97	345.7	305.3	40

Note: **9** Bicapped tetrahedron; **10** 1,1-dithiolate interaction not considered here; **11** and **15** have two molecules in the asymmetric unit.

The transition from tetrahedral to trigonal bipyramidal geometry can be mapped by comparing the "equatorial angles" to the "axial" angles that arise as tin changes from 4 coordinate to 5; the sum of the equatorial angles change from 328.5° to 360° and the sum of the axial angles change from 328.5° to 270° in the transition of tetrahedral to trigonal bipyramidal, respectively.

Angles a—Sn—b, a—Sn—c, and a—Sn—d are the "axial" angles which give Σ_{ax}. The equatorial angles are b—Sn—c, c—Sn—d, and b—Sn—d and give Σ_{eq}. Simply taking the difference between the two sums, one can determine the geometry of a compound, i.e. the sum of pure tetrahedral will be 0° and the sum of pure trigonal bipyramidal will be 90°, so the resulting value in the range 0 – 90 reflects the proximity of the structure to the tbp form. Figure 5 shows this transition and the angles in question.

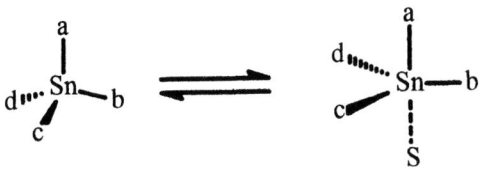

Figure 5. Transition from tetrahedral to trigonal bypyramidal.

Figure 6 maps the distortion from tetrahedral for each compound. It can be seen that long S---Sn distances give geometries that can be called distorted tetrahedral whereas short S---Sn distances begin to approximate trigonal bipyramidal geometry.

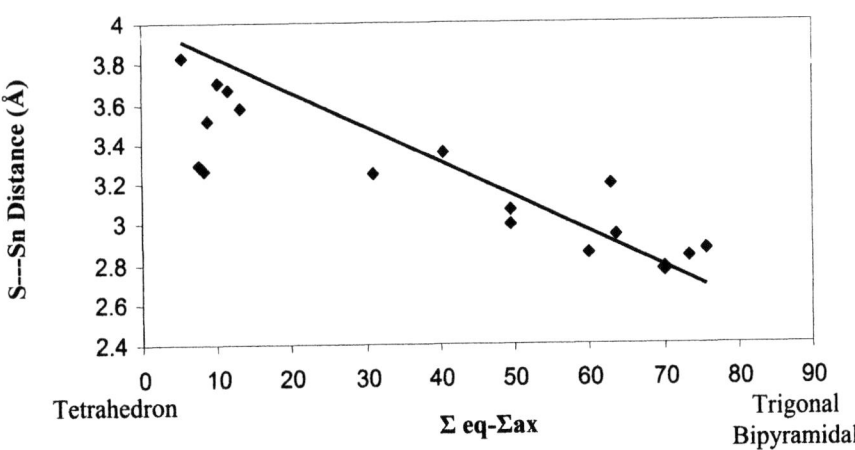

Figure 6. Transition from tetrahedral to trigonal bipyramidal geometry as a function of S..Sn distance.

The Sn---S bond order increases as the intranuclear distance decreases; however, there is no significant weakening of the Sn—R_{ax} bond upon complexation of tin by the sulfur. A compilation of the Sn---S and Sn—R_{ax} bond orders is presented in Figure 7. It is interesting to note that even for the compounds which contain one or more halogens, and therefore the largest BO for Sn---S (compounds **1** and **3-7**), the corresponding Sn—R_{ax} BO does not decrease very much if at all. Furthermore, even the strongest bonding interactions exhibit a BO of significantly less than unity.

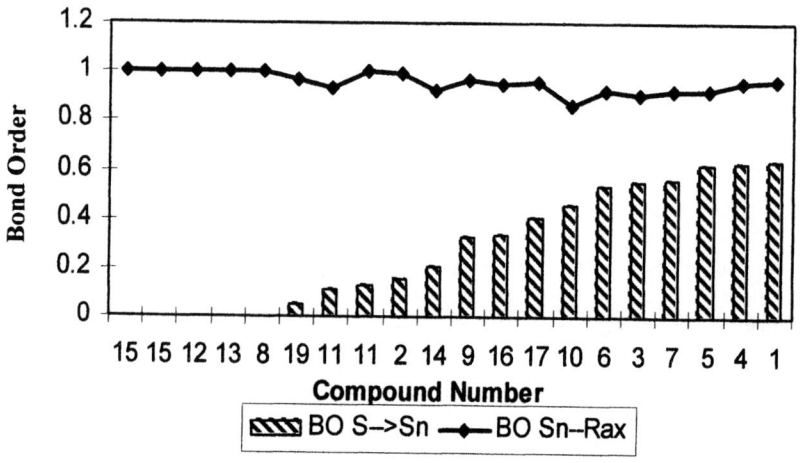

Figure 7. Comparison of Sn---S and Sn—R_{ax} bond orders.

Although compound **11** has Sn---S distances that at a glance look short, 3.26 and 3.39 Å, it has both low bond order (0.11 and 0.13) and very little distortion about the tin atom. The short intramolecular distance is due to their only being one C atom separating tin from sulfur. Normal C—S—C bond angles and normal Sn—C and C—S bond distances will place sulfur approximately 3.1-3.4 Å apart (37).

It is worth noting that the chlorinated organotin compounds, **16**, **17**, and **19** all have relatively short S---Sn distances and significant bond orders that result in clearly observable distortions at the tin atom. It is hoped that these compounds will exhibit modified biocidal activity by challenging the manner in which the central tin atom can interact with biologically available sulfur. Tests on such activity are presently in progress.

At above approximately 3.4 Å the bond order calculation, using the afore mentioned equation, begins to give negative values, which essentially means that the bond order is zero. Compounds **8, 13, 12,** and **15** all have Sn---S bond orders of zero. Figure 8 illustrates the "cut off" distance of 3.4 Å from which a bond order can be calculated.

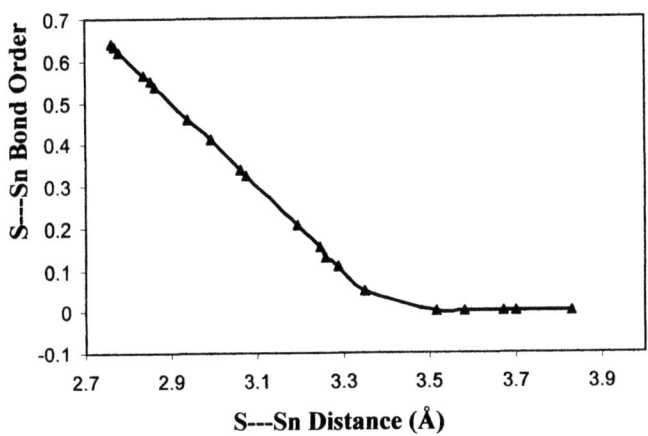

Figure 8. S-Sn bond order as a function of S-Sn intramolecular distance.

Each of these are triorganotin compounds, and by virtue of this, we can see how diminished the Lewis acidity of tin is in comparison to compounds such as **6, 14,** and **16** which have one organic group replaced by a chloride atom and have comparably larger bond orders and shorter Sn---S distances.

Concluding Remarks

Despite the simplicity of these compounds, they help provide an eloquent way to begin to understand how organotin species are able to interact with biologically important residues, namely sulfur containing residues. By analyzing what parameters are important for the coordination of tin, i.e. proximity of donor atoms and Lewis acidity of the tin center, we can perhaps bring forth some ideas into mechanisms by which these compounds work.

References

1. Gajda-Schrantz, K.; Jancso, A.; Pettinari, C.; Gajda, T. *J. Chem. Soc., Dalton Trans.* **2003**, 2919-2916.
2. Omae, I. *Applied Organometallic Chemistry* **2003**, *17*, 81-105.
3. Gielen, M. *Applied Organometallic Chemistry* **2002**, *16*, 481-494.
4. Saki, T.; Hiyama, Y.; Sato, Y. U.S. Patent 3,511,803, 1970.
5. Majumdar, K. K.; Kundu, A.; Das, I.; Roy, S. *Applied Organometallic Chemistry* **2000**, *14*, 79-85.
6. Snoeij, N. J.; Penninks, A. H.; Seinen, W. *Environmental Research* **1987**, *44*, 335-353.
7. Whalen, M. M.; Loganathan, B. G.; Kannan, K. *Environmental Research Section A* **1999**, *81*, 108-116.
8. Nath, M.; Pokharia, S.; Yadav, R. *Coordination Chemistry Reviews* **2001**, *215*, 99-149.
9. Nath, M.; Pokharia, S.; Eng, G.; Song, X.; Kumar, A. *Journal of Organometallic Chemistry* **2003**, *669*, 109-123.
10. Allen, F. H. *Acta Cryst.* **2002**, *B58*, 380-388.
11. Tiekink, E. R. T. *Applied Organometallic Chemistry* **1991**, *5*, 1.
12. Tiekink, E. R. T. *Main Group Metal Chemistry* **1992**, *15*, 161.
13. Tiekink, E. R. T. *Trends in Organomet. Chem.* **1994**, *1*, 71.
14. Mohamed-Ibrahim, M. I.; Chee, S. S.; Buntine, M. A.; Cox, M. J.; Tiekink, E. R. T. *Organometallics* **2000**, *19*, 5410-5415.
15. Sheldrick, G. M.; Sheldrick, W. S. *J. Chem. Soc., (A)* **1970**, 490-493.
16. Sheldrick, G. M.; Sheldrick, W. S.; Dalton, R. F.; Jones, K. *J. Chem. Soc., (A)* **1970**, 493-497.
17. Das, V. G. K.; Wei, C.; Sinn, E. *Journal of Organometallic Chemistry* **1985**, *290*, 291-299.
18. Holt, E. M.; Nasser, F. A. K.; Wilson, A.; Zuckerman, J. J. *Organometallics* **1985**, *4*, 2073-2080.
19. Hook, J. M.; Linahan, B. M.; Taylor, R. L.; Tiekink, E. R. T.; van Gorkom, L.; Webster, L. K. *Main Group Metal Chemistry* **1994**, *17*, 293-311.
20. Kana, A. T.; Hibbert, T. G.; Mahon, M. F.; Molloy, K. C.; Parkin, I. P.; Price, L. S. *Polyhedron* **2001**, *20*, 2989-2995.
21. Kato, S.; Tani, K.; Kitaoka, N.; Yamada, K.; Mifune, H. *Journal of Organometallic Chemistry* **2000**, *611*, 190-199.
22. Tarassoli, A.; Sedaghat, T.; Neumuller, B.; Ghassemzadeh, M. *Inorganica Chimica Acta* **2001**, *318*, 15-22.
23. Vrabel, V.; Kello, E. *Acta Cryst.* **1993**, *C49*, 873-875.

24. Hall, V. J.; Tiekink, E. R. T. *Main Group Metal Chemistry* **1995**, *18*, 611-620.
25. Donoghue, N.; Tiekink, E. R. T. *Journal of Organometallic Chemistry* **1991**, *420*, 179-184.
26. Cox, P. J.; Doidge-Harrison, S. M. S. V.; Nowell, I. W.; Howie, R. A.; Wardell, J. L.; Wigzell, J. M. *Acta Cryst.* **1990**, *C46*, 1015-1017.
27. Drager, M. *Z. anorg. allg. Chem.* **1976**, *423*, 53-66.
28. Jurkschat, K.; Schilling, J.; Mugge, C.; Tzschach, A. *Organometallics* **1988**, *7*, 38-46.
29. Kolb, U.; Beuter, M.; Drager, M. *Inorganic Chemistry* **1994**, *33*, 4522-4530.
30. Kolb, U.; Drager, M.; Jousseaume, B. *Organometallics* **1991**, *10*, 2737-2742.
31. Drager, M. *Journal of Organometallic Chemistry* **1983**, *251*, 209.
32. Drager, M.; Engler, R. *Chem. Ber.* **1975**, *108*, 17-25.
33. Drager, M.; Guttmann, H.-J. *Journal of Organometallic Chemistry* **1981**, *212*, 171-182.
34. Kolb, U.; Beuter, M.; Drager, M. *Organometallics* **1994**, *13*, 4413-4425.
35. Cea-Olivares, R.; Lomeli, V.; Hernandez-Ortega, S.; Haiduc, I. *Polyhedron* **1995**, *14*, 747-755.
36. Garcia y Garcia, P.; Cruz-Almanza, R.; Toscano, R.-A.; Cea-Olivares, R. *Journal of Organometallic Chemistry* **2000**, *598*, 160-166.
37. Cox, P. J.; Doidge-Harrison, S. M. S. V.; Nowell, I. W.; Howie, R. A.; Randall, A. P.; Wardell, J. L. *Inorganica Chimica Acta* **1990**, *172*,
38. Howie, R. A.; Wardell, J. L.; Zanetti, E.; Cox, P. J.; Doidge-Harrison, S. M. S. V. *Journal of Organometallic Chemistry* **1992**, *431*, 27-40.
39. Munguia, T.; Pavel, I. S.; Kapoor, R. N.; Cervantes-Lee, F.; Párkányi, L.; Pannell, K. H. *Canadian Journal of Chemistry* **2003**, *81*, 1388-1397.
40. Huheey, J. E.; Keiter, E. A.; Keiter, R. L. *Inorganic Chemistry: Principals of Structure and Reactivity*; 4th ed.; Harper Collins College Publishers: New York, NY, 1993; pp 292.
41. Munguia, T.; Kapoor, R. N.; Cervantes-Lee, F.; Pannell, K. H. *Unpublished Data*
42. O'Keeffe, M.; Brese, N. E. *J. Am. Chem. Soc.* **1991**, *113*, 3226-3229.

Author Index

Abernethy, Colin D., 252
Armstrong, A., 66
Astruc, Didier, 347
Atwood, David A., 390
Baber, R. Angharad, 137
Bayer, Michael J., 312
Bertrand, Guy, 81
Blais, Jean-Claude, 347
Boeré, R. T., 66
Borisenko, Konstantin B., 94
Bourg, Jean-Baptiste, 81
Bourget-Merle, Laurence, 192
Buhro, William E., 362
Burford, Neil, 280
Carmalt, Claire J., 376
Cervantes-Lee, Francisco, 422
Charmant, Jonathan P.H., 137
Cheng, Yanxiang, 192
Chivers, T., 66
Cloutet, Eric, 347
Clyburne, Jason A. C., 266
Cowley, Alan H., 2
Culver, John P., 252
Devulapalli, Pradeep, 325
Dodds, Christopher A., 252
Doyle, David J., 192
Dreissig, D., 152
Duesler, E. N., 152
Eichler, Jack F., 122
Ellis, Bobby D., 108
Farha, Omar K., 312
Felix, Ana M., 410
Gabbaï, François P., 208
Gandon, Vincent, 81
Gaspar, Peter P., 52
Gross, Michael L., 52

Habereder, T., 152
Hawthorne, M. Frederick, 312
Hinchley, Sarah L., 94
Hitchcock, Peter B., 192
Hosmane, Narayan S., 293
Ivanova, Diana, 52
Jackson, Bethany K., 325
Jalisatgi, Satish S., 312
Janik, J. F., 152
Jones, Richard A., 221
Julius, Richard, 312
Jung, June-Ho, 335
Just, Oliver, 122
Kemp, Richard A., 410
Khvostov, Alexei V., 192
King, Simon J., 376
Lafuente, Gustavo, 347
Lappert, Michael F., 192
Lattman, Michael, 237
Lazare, Sylvain, 347
Lesley, M. J. Gerald, 137
Li, Tiejun, 312
Liu, Xinping, 52
Lu, Jia G., 362
Ma, Ling, 312
Macdonald, Charles L. B., 108
Maguire, John A., 293
Manner, Virginia W., 410
Marcen, Sylvia, 347
Mileham, John D., 376
Mitra, Amitabha, 390
Moasser, Bahram, 410
Munguia, Teresita, 422
Neilson, Andrew R., 325
Neilson, Robert H., 325
Nogai, Stefan, 32

Norman, Nicholas C., 137
Nöth, H., 152
Ornelas, Catia, 347
Orpen, A. Guy, 137
Otten, Carolyn Jone, 362
Paine, R. T., 152
Pannell, Keith H., 422
Párkányi, László, 422
Parveen, Sahrah, 325
Peters, Emily S., 376
Potluri, Srinagesh K., 335
Powers, Philip P., 179
Präsang, Carsten, 81
Prell, James S., 52
Protchenko, Andrey V., 192
Ragogna, Paul J., 280
Ramnial, Taramatee, 266
Rankin, David W. H., 94
Read, David, 52
Rees, William S., Jr., 122
Rheingold, Arnold L., 410
Rodriguez, Amor, 81
Roesky, Herbert W., 20
Rossi, Jean, 137
Ruiz, Jaime, 347
Schmidbaur, Hubert, 32
Solé, Stéphane, 208
Spicer, Mark D., 252
Tang, Yongjun, 410
Tocher, Derek A., 376
Waheed, Abdul, 221
Wang, Bin, 325
Wang, Dawei, 362
Wang, Huadong, 208
Wang, Jianhui, 293
Wei, Xue-hong, 192
West, Robert, 166
Wiester, Michael, 221
Wisian-Neilson, Patty, 335
Yang, Xiaoping, 221
Zakharov, Lev N., 410
Zhang, Lilu, 221

Subject Index

A

Acyclic triphosphenium cations, oxidation of, 111, 112
Adamantane cages, Ga_4S_6, 33
Adenosine triphosphate (ATP), phosphate ester bond, 391
Al–H–C chemistry
  alternative routes to $LAl(OH)_2$, 22
  alumoxane with terminal hydroxides, 23
  hydrolysis of trimesitylaluminum and -gallium, 25
  minerals, 21
  molecular structure of $Me_3Ga \cdot OH_2 \cdot 2THF$, 25$f$
  natural compounds, 21
  organoaluminum diamide, 29
  organoaluminum dihydroxide, 21–23
  organoaluminum diselenol, 28
  organoaluminum dithiol, 26–28
  organoaluminum monohydroxide, 24–26
  trimeric alumoxane, 23
  X-ray crystal structure of $Mes_6Ga_6O_4(OH)_4 \cdot 4THF$, 26$f$
Alkali metals. See β-Diketiminates
Alkaline phosphatase, phosphate ester cleavage, 393, 394f
Alkene analogs
  addition of PMe_3 to Ar'(H)GeGe(H)Ar', 187f
  derivative of mixed valence isomer III, 182–184
  derivatives of singly bridged isomer IV, 186–189
  derivatives of trans-doubly bridged isomer V, 184–186
  heavier group 14 elements, 180–182
  isomeric forms of E_2H_4 (E=C–Pb), 180f
  relative energies and E–E bond lengths, 181f
  structure of Ar'(H)GeGe(H)Ar', 186f
  structure of bridged Ar'Sn(μ-Br)Sn(Ar')$CH_2C_6H_4$-4-Pr^i, 188f
  structure of divalent hydride Ar*Sn(μ-H)$_2$SnAr*, 185f
  trans-pyramidal configuration, 180f, 184, 186f
  unsymmetric stannylstannylene, 183–184
  X-ray crystal structure of unsymmetric Ar*SnSn(Me)$_2$Ar, 183f
Alkyl zinc amides, synthesis, 417–418
Alumina
  chemical vapor deposition for thin films, 377
  hydrated forms, 21
  polymorphs, 21
Aluminum, resistivity of, and diboride, 364t
Aluminum alkoxides, donor-functionalized, 383, 384
Antimony, Sb. See Pnictogens
Architecture, molecular. See Closomers
Aromatic compounds, silicon atom in ring, 171–172
Arsenic
  insertion into calix[4]arene, 244
  See also Pnictogens; Tetrakis(disyl)diarsine
Arsenic-arsenic double bonds, stable compounds, 6–7

439

Arsenidenes, cyclic oligomers, 3–4
Artificial tetrahedron atom (T-atom), Ga_4S_6 core, 33
Atomic force microscopy (ATM), heights of dendrimers, 351, 358f, 359
Atoms in Molecules (AIM) analysis, 1,8-diborylnaphthalenes, 215

B

Ball and spring model, dissociation, 102f
Barium-based disilylamides, synthesis, 415–416
Bayerite, aluminum trihydroxide, 21
Bicyclo[1.1.0] structure
  diradical versus, 88–90
  energy profile for inversion of bicyclo[1.1.0]butanes, 82f
  See also Diradicals
Bidentate Lewis acids. *See* 1,8-Diborylnaphthalenes
Binuclear boron chelates. *See* Phosphate esters
Binuclear zinc, phosphate ester cleavage, 296
Biocidal activity, organotin compounds, 423–425
Biological catalyst models, phosphate ester cleavage, 396
Biological systems, phosphate esters, 391
Biology, phosphate ester cleavage, 393–396
Bis(diphenylphosphino)ethane (dppe), chelating pnictogen cations, 110, 115–118
Bis(disyl)phosphido radical
  radical distribution curve for, 97f
  structure, 98f
  See also Tetrakis(disyl) diphosphine
Bismuth. *See* Pnictogens

Bis(tetramethylsilylamido)(di-isopropylamido)phosphido radical
  ligand shape and size, 105–106
  structure, 106f
Boehmite, hydrated alumina, 21
Bond angles
  crystals containing diborane compound [$B_2Cl_4(NHMe_2)_2$], 148t
  germanium- and lead-nitrogen heterocubanes, 128, 131t, 132t, 133t
  organotin-sulfur compounds, 430–431
Bond cleavage. *See* Phosphate esters
Bonding
  dimers of carbene-like fragments, 6–8
  phosphinidenes, 8
  pnictinidene complexes, 9
Bond length
  Al–OH of organoaluminum monohydroxide, 24
  crystals containing diborane compound [$B_2Cl_4(NHMe_2)_2$], 148t
  E–E for E=Ge or Sn, 181f
  Ge–Ge distance, 186f
  heavier group 14 element ethylene analogs, 181f
  Si=C bond, 170
  Si=Si bond, 174
  Sn–Sn distance, 183f, 188–189
Bond order, Sn---S intramolecular interactions, 429–430, 432–433
Bonds, multiple. *See* Silicon multiple bonds
Bond-stretch isomerism
  concept, 90
  See also Diradicals
Borane complex, imidazol-2-ylidene, exhibiting intermolecular dihydrogen bonding, 275–277
Boranes
  commercially available, 313

current applications, 316–317
See also Polyhedral boranes
Boron
  catenation, 312–313
  clusters, 12–14
  lone pair orbitals, 12f
  singlet and triplet states of four-valence-electron group 13 fragments, 11–12
  See also Carboranes; Diborane(4) compounds; Polyhedral boranes
Boron and phosphorus
  stable singlet diradical, 83–85
  See also Diradicals
Boron chelates
  attack of P–O–R bond by halide anion, 404f
  binuclear boron alkoxides and borosiloxides, 398, 399f
  binuclear boron halides, 398–400, 401f
  dealkylation of trimethyl phosphate with Salen boron chloride compounds, 402t
  formation of phosphate coordinated boron cation, 403f
  future studies, 405
  percent dealkylation of various phosphates with Salen boron bromide compounds, 402t
  percent dealkylation of various phosphates with Salen phenyl boron bromide compounds, 403t
  phosphate dealkylation with boron halide chelates, 400, 404
  phosphate ester cleavage using, 397, 398f
  See also Phosphate esters
Boron nanowires
  carrier mobility, 365, 373
  classification of common materials by conductivity, 370f
  comparison of mobilities for one-dimensional nanostructures, 372t
  conductivity, 368–369
  device fabrication, 366–367
  diagram of device for electrical-transport measurements, 367f
  effect of gate voltage on conductance of p-type semiconducting, 371f
  electrical transport, 367–373
  gate-dependent behavior, 371
  I–V curves of device with Ti/Au contacts before and after annealing, 367f
  I–V data for device with Ni/Au contacts at various gate voltages, 370f
  I–V data of annealed device with Ni/Au contacts, 368f
  p-type behavior, 371
  resistance vs. length, 368, 369f
  scanning electron microscopy (SEM) image, 366f
  synthesis, 365–366
  transconductance, 371, 372f
  transmission electron microscopy (TEM) image, 366f
Boron neutron capture therapy
  closomers of high boron content for use in, 320
  stable unilamellar liposomes for, 316
Boron-phosphorus ring and cage chemistry
  boron phosphide (BP) synthesis, 152
  cage assembly chemistry, 156, 158
  diborane(4) building blocks, 158, 161, 163
  1,2,3,4-diphosphadiboretane (tmpBPH)$_2$ as building unit for five and six vertex closo cage compounds, 161
  formation of H$_3$BP(SiMe$_3$)$_3$, 152–154
  molecular chemistry, 152

molecular structure of [(Me$_2$N)B–B(NMe$_2$)PH]$_2$ (12), 162f
molecular structure of [PhB(Cl)P(SiMe$_3$)$_2$]$_2$ (4), 157f
molecular structure of [(THF)(DME)Li]P$_2$[(i-Pr$_2$NB)$_2$PH]$_2$, 160f
molecular structure of (PhB)$_5$P$_5$(SiMe$_3$)$_4$ (5), 159f
molecular structure of PhB(Cl)$_2$P(SiMe$_3$)$_3$ (3), 157f
molecular structure of (tmpB)$_2$[PB(NMe$_2$)B(Cl)(NMe$_2$)]PH (13), 162f
molecular structure of {[(Me$_3$Sn)$_2$P](Me$_2$N)B$_2$ (11), 160f
molecular structures of [H$_2$BP(SiMe$_3$)$_2$]$_3$ (1), 155f
molecular structures of P[μ-H-B$_2$H$_2$]{(H$_2$B)$_2$P(SiMe$_3$)$_2$]$_2$}$_2$ (2), 155f
phosphine borane systems, 152–156
proposed assembly for cage molecules, 159
reactions of B$_2$H$_6$ and H$_3$B-base adducts with organyl phosphines, 154, 156
reactions of haloboranes with P(SiMe$_3$)$_3$, 154
Bridging pnictinidene complexes, bonding, 9
Bromotrichloromethane, reaction with stable singlet diradical, 87

C

C$_6$H$_7$Si$^+$. *See* Phenylsilyl cation Ph–SiH$_2^+$
Cage structures
molecular gallium sulfides, 44, 47
See also Boron-phosphorus ring and cage chemistry

Calcium-based disilylamides, synthesis, 415–416
Calicenes, silapentatriafulvalene, 171
Calixarenes
abbreviations and idealized conformations of calix[4]arenes, 239f
calix[4]arene framework supporting six-, five-, four-, and three-coordinate phosphorus atom, 238, 240f
calix[4]arenes, 238–246
calix[5]arenes, 248–249
catalytic applications, 246
combination of constraint and flexibility, 238
derivatives with two trivalent phosphorus atoms as bidentate ligands, 249, 250f
1,2-disubstituted calix[4]arenes using dimethylsilyl protecting group, 244, 245f
^{1}H NMR spectroscopy determining conformation, 244
insertion of Ar into calix[4]arene, 244
insertion of one and two bridging silyl groups into calix[5]arene, 247f, 248
insertion of tungsten and control of phosphorus/metal interaction, 248–249
monosubstituted calix[4]arenes, 244, 245f
partial cone and cone conformers of four-coordinate silicon calix[4]arene, 241
phosphorus calix[4]arene containing Ti–O bond, 246, 247f
phosphorus-containing calix[5]arene ligands, 249
synthesis of four- and five-coordinate silicon calix[4]arenes, 241f

tetraisopropyldisiloxane (TIPDS)
protecting group for 1,2-
disubstituted calix[4]arenes,
244, 245f
tricoordinate phosphorus insertion
into calix[5]arene, 248
X-ray crystal structures of
calix[4]arenes, 242f, 243f
Cancer therapy
liposome as drug delivery vehicles,
316
substitution on icosahedral
dodecaborate cage, 320, 321
Carbene analogs
charged six-valence electron, 53–59
See also Six-valence electron
species
Carbenes
main group analogues, 2–3
one electron oxidation, 273–275
synthesis of first imidazol-2-
ylidenes, 267
See also Imidazol-2-ylidenes; Main
group chemistry; N-heterocyclic
carbenes (NHCs)
Carbon dioxide, insertion into
organometalic species, 411–412
"Carbons apart"-carborane cages
lanthanacarborane complex, 299f
lanthanacarboranes with, 296–298
Carboracycles, encapsulation of
hydrophobic species, 317, 318
Carboranes
carboranyl-thiol-appended half-
sandwich titanocene, 294–295, 296f
commercially available, 313
current applications, 316–317
half-sandwich
halolanthanacarboranes of
carbons adjacent C_2B_4-carborane
ligand systems, 302, 304–305
lanthanacarboranes with "carbons
apart"-carborane cages, 296–298

ligands for various metals, 293–294
oxide ion-encapsulating
tetralanthanide tetrahedron with
"carbons-apart", 300–302
perhydroxylation, 315
water-soluble, scaffold for
closomer chemistry, 320, 322
See also Metallacarboranes
Carrier mobility, boron nanowires,
365, 373
Catalysis, calix[4]arene applications,
246
Catenation, boron, 312–313
Cationic low oxidation state. See
Pnictogens
Cell targeting, substitution on
icosahedral dodecaborate cage,
320, 321
Chalcogenides. See Group 13
chalcogenides
Chelation. See 1,8-
Diborylnaphthalenes
Chemical reduction, imidazolium ions,
272–273
Chemical vapor deposition (CVD)
alumina thin films, 377
gallium oxide (Ga_2O_3), 377
low pressure CVD, 384–385, 388
precursors, 377
single-source precursors, 376
See also Group 13 chalcogenides
Chemical warfare agents,
organophosphate esters, 391–392
Chloroform-d, reaction with stable
singlet diradical, 86
Chloropyrifos, organophosphate ester,
391–392
Cleavage of bonds. See Phosphate
esters
Closomers
dendritic or oligomeric, 318
description, 315
drug delivery, 319
high boron content for use in boron
neutron capture therapy, 320

Jahn–Teller distortion
  accompanying two-electron
  oxidation, 319
  motif in molecular architecture,
    318–319
  vertex differentiation for targeting
    specific sites, 320, 321
  water-soluble carborane scaffold
    for closomer chemistry, 320,
    322
  See also Polyhedral boranes
Clusters, group 13 fragments, 12–14
Collision-induced dissociation (CID)
  product structure, 56
  unreactive $C_6H_7Si^+$ from two
    sources, 61–62
Colorimetric fluoride ion sensor,
  bidentate boranes as, 211–213
Column chromatography, isolation of
  cyclic phosphazene isomers, 343
Complexes. See Lanthanides with
  unusual main group ligands
Conductivity
  catenation of singlet diradicals,
    82
  classification of common materials,
    370f
  See also Boron nanowires
Constraint, calixarenes, 238
Coordination chemistry of
  phosphorus(III)
  bifunctional amines tethering
    neutral phosphines, 286–287
  bifunctional diphosphine ligands,
    287–288
  coordination complexes, 281
  coordination complexes of
    diphenylphosphenium cation,
    285–289
  coordination complexes of
    phosphadiazonium cation, 281–
    284
  formation of
    (gallane)Ga→P(phosphine)
    complex, 289

homoatomic P→P coordinate
    bonding model, 286
  Lewis acceptor, 280–281
  ligand exchange on
    diphenylphosphenium cation,
    289
  NMR and P–(E) distances for
    complexes of $[Ph_2P]^+$ and
    related compounds, 287t
  NMR and structural parameters for
    complexes of phosphadiazonium
    cations and related compounds,
    282t
  ^{31}P NMR spectroscopy, 281
  P–P coordinate interactions, 284
  solid state structure of bicyclic
    di(phosphino)phosphonium
    cation, 288f
  solid state structure of cation in
    [Mes*NP(Bipy)][OTf], 284f
  solid state structure of cation in
    [Mes*NP(Pyr)][OTf], 283f
  solid state structure of cation in
    [Ph(Cl)P–PPh$_2$][GaCl$_4$], 285f
  solid state structure of cation in
    [Ph$_2$P-dpph-PPh$_2$][OTf], 286f
  structure of phosphadiazonium
    cation, 281
  structure of phosphenium cation,
    281
Cp$_2$Zr compounds, organoaluminum
  dithiol, 27–28
Cubanes. See Heterocubanes
Cyclic dications, oxidation of P(I)
  cations, 118–119
Cyclic phosphazenes
  cis and trans isomers of
    [(Ph)(Me)PN]$_3$, 340–341
  isolation of isomers by column
    chromatography, 343
  isomers of calix-4-arene, 341, 342
  [Me$_2$PN]$_3$, 340
  See also Phosphazenes
Cyclic triphosphenium cations,
  synthesis, 114–115

1,3-Cyclobutanediyls
  stable singlet diradical, 82
  See also Diradicals
Cyclotetraarsine, synthesis, 3
Cyclotetraphosphine, reaction with tetramethyl diarsine, 3–4
Cyclotriphosphine, synthesis, 3
Cysteine, sulfur-containing, 424f
Cystine, sulfur-containing, 424f

D

Dendrimers
  applications, 348
  atomic force microscopy (AFM) measuring height, 351, 358f
  construction based on hydrosilylation, 355
  de Gennes "dense–packing" limit, 348, 359
  diameter for G_9–177407 allyl, 351, 357f
  diameter of G_4–729 matching molecular model, 351, 356f
  flattening and aggregation, 359
  ^{1}H NMR spectra, 352f
  hydroelementation, 349
  hydrosilylation using Karsted catalyst, 351, 355
  MALDI TOF mass spectra, 351
  MALDI TOF mass spectra for G_2–81 and G_3–243 allyl dendrimers, 354f
  overall construction starting from ferrocene, 350
  reaction time with generations, 359
  ^{29}Si NMR spectra, 353f
  size exclusion chromatography (SEC) from G_1–27 allyl to G_5–2187 allyl, 351, 356f
  third generation polyallyl, 349
Dense packing limit

  steric congestion at periphery of dendrimers, 348, 359
  See also Dendrimers
Density functional theory (DFT), structure optimization of diborane, 211, 212f
Dialkyl phosphido radicals, dissociation of tetraalkyldiphosphine, 95
Diamide, organoaluminum, 29
Diarsine. See Tetrakis(disyl)diarsine
Diaspore, hydrated alumina, 21
1,3-Dibora-2,4-diphosphoniocyclobutane-1,3-diyls, ring closure into bicyclo[1.1.0] isomer, 88
Diborane(4) compounds
  ^{11}B nuclear magnetic resonance (NMR) chemical shift, 137
  condensed borinane species $B_4O_2(OH)_4$ (2), 137–137
  derivative of borinane, $B_4O_2(OH)_4\cdot2[NH_2Me_2]Cl$ (2a), 137
  hydrogen bonding interactions, 140, 148
  hydrolysis chemistry, 137–137, 140
  molecular structure of $[B_2Cl_4(NHMe_2)_2][2\text{-picH}]Cl$, 143$f$
  molecular structure of $[B_2Cl_4(NHMe_2)_2][4\text{-picH}]Cl$, 143$f$
  molecular structure of $[B_2Cl_4(NHMe_2)_2][NH_2Me_2]Cl$, 146$f$
  molecular structure of borinane ring and dimethyl ammonium chloride units (2a), 139f
  selected bond lengths and angles for crystals containing $[B_2Cl_4(NHMe_2)_2]$, 148t
  solid state structure of $[B_2Cl_4(NHMe_2)_2][2\text{-picH}]Cl$, 144$f$

solid state structure of
  $[B_2Cl_4(NHMe_2)_2][4\text{-picH}]Cl$,
  145f
solid state structure of
  $[B_2Cl_4(NHMe_2)_2][NH_2Me_2]Cl$,
  147f
solid state structures of (2), 141f
solid state structures of (2a), 142f
See also Boron-phosphorus ring
  and cage chemistry
1,8-Diborylnaphthalenes
  Atoms in Molecules (AIM)
    analysis, 215
  bidentate boranes as colorimetric
    fluoride ion sensor, 211–213
  bidentate Lewis acids, 208
  1,8-bis(dimethylboryl)naphthalene
    (1), 208–209
  1,8-bis(diphenylboryl)naphthalene
    (2) and isoelectronic
    dicarbocation (3^{2+}), 209–210
  Boys localized orbital in $[3\text{-}F]^+$,
    216f
  C–F–C bridged species, 214–218
  crystal structure of (2) and (3^{2+}),
    210f
  diborane (6) from 10-bromo-9-thia-
    10-boranthracene (4) and
    dimesityl-1,8-
    naphthalenediylborate (5), 211–
    213
  fluoride complexation of (6), 213
  formation of cation containing C–
    F→C bridge, 214–218
  optimization of (6) using density
    functional theory (DFT), 211,
    212f
  Ortep view of $[3\text{-}F]^+$, 215f
  Ortep view of borate anion
    complexed with fluoride [6·μ_2-
    F]⁻, 213f
  stable methylium cations, 209
  structure of $[3\text{-}F]^+$, 217–218
  variable ^{1}H NMR of $[3\text{-}F]^+$,
    217f

Dicyclopentenyl–Zr compounds,
  organoaluminum dithiol, 27–28
β-Diketiminates
  di- and tri-anionic, 203–204
  examples of metal, 194f
  experiments with alkali metal N,N'-
    bis(diisopropylphenyl)-β-
    diketiminates, 204–205
  Li and K, from
    $M[CH(SiMe_3)(SiMe_2OMe)]$,
    196
  ligands, 192–193
  lithium, from
    $[Li\{CH(SiMe_3)(SiMe(OMe)_2)\}]_∞$
    infinity and PhCN, 199
  lithium, from
    $Li[CH(SiMe_3)\{SiMe_{3-n}(OMe)_n\}]_∞$ (n=1 or 2), 196–197
  lithium and aluminum CH_2-bridged
    bis(β-diketiminate)s having
    diverse structures, 199, 202
  paramagnetic dilithium, and
    diamagnetic trilithium
    derivative, 203–204
  proposed pathways, 197, 198, 200
  reaction for formation of dilithio
    compound from CH_2-bridged
    bis(β-diketimine), 202
  reactions of $Li[CH(SiMe_3)_2]$ with
    PhCN, 195
  terminal or bridging mode for
    binding, 193
Dilead, triple-decker cation, 15, 16
Dilithiated tetrakisimidophosphate
  radicals
  electron paramagnetic resonance
    (EPR) characterization, 75–76
  structure and stability, 74–75
  See also Phosphorus-containing
    radicals
Dimethylammonium chloride co-
  crystals. See Diborane(4)
  compounds
Diphenyl diselenide, reaction with
  stable singlet diradical, 86

Diphenylphosphenium cation
  application of ligand exchange on,
    289
  bifunctional diphosphine ligands,
    287–288
  coordination complexes of, 285–
    289
  formation of
    (gallane)Ga→P(phosphine)
    complex, 289
  homoatomic P→P coordinate
    bonding model, 286
  NMR data and P–(E) distances for
    complexes of $[Ph_2P]^+$ and
    related compounds, 287t
  solid state structure of cation in
    $[Ph_2(Cl)P\text{-}PPh_2][GaCl_4]$, 285f
  solid state structure of cation in
    $[Ph_2P\text{-}dpph\text{-}PPh_2][OTf]$, 286f
  See also Coordination chemistry of
    phosphorus(III)
1,3-Diphosphaallyl radical
  electron paramagnetic resonance
    (EPR) characterization, 73
  preparation, 72–73
  structure and stability, 73–74
  See also Phosphorus-containing
    radicals
1,3-Diphosphacyclobutane-2,4-diyl
  and 1,3-diphosphabicyclo-
  [1.1.0]butane
    bond-stretch isomers, 90–91
    schematic of, and highest occupied
      molecular orbitals (HOMO),
      88f
Diphosphanyl radical
  decomposition and half-life, 70–71
  electron paramagnetic resonance
    (EPR) characterization, 71–72
  preparation, 70
  structure and stability, 72
  See also Phosphorus-containing
    radicals
Diradicals
  catenation of singlet, 82

concept of "bond-stretch
    isomerism", 90
  description, 81
  1,3-diphosphacyclobutane-2,4-diyl
    and 1,3-diphosphabicyclo-
    [1.1.0]butane, 90–91
  diradical versus bicyclo[1.1.0]
    structure, 88–90
  energy profile for inversion of
    bicyclo[1.1.0]butanes, 82f
  Hoffmann predictions of existence
    of two bond-stretch isomers, 90f
  inherent instability, 82
  molecular view of singlet 1,3-
    diradical 1a, 84f
  most stable singlet, 1,3-
    cyclobutanediyls, 82f
  reaction of 1a with
    bromotrichloromethane, 87
  reaction of 1a with $CDCl_3$, 86
  reaction of 1a with elemental
    selenium and PhSeSePh, 86
  reaction with trimethyltin hydride
    at room temperature, 86, 87f
  reactivity of stable singlet diradical
    1a, 85–88
  route to 1,2-diphosphinodiboranes,
    84
  schematic of bicyclic compounds,
    diradicals, and their highest
    occupied molecular orbitals
    (HOMO), 88f
  synthesis and structure of first
    stable diradical (1a), 83–85
  types of organic, 81–82
Discovery, multiply-bonded silicon
  compounds, 167
Diselenol, organoaluminum, 28
Disilenes
  example of stable, 6
  intermediates, 172–173
  tetramesityldisilene, 173–174
Disilyne polymers, silicon multiple
  bond compounds, 176
Dissociation

ball and spring model, 102f
comparing intrinsic bond, energies for diphosphines and disilanes, 104f
energetics of, of molecule X–X to two X radicals, 102f
energies for diphosphine and disilanes, 105t
tetraalkyldiphosphine to radicals, 95
See also Thermodynamics
Dithiol, organoaluminum, 26–28
1,1-Dithiolate
intramolecular organotin, interactions, 425–428
structural motifs of organotin, 426f
Divalent metal amides
alkyl zinc amides, 417–418
calcium- and barium-based disilylamides, 415–416
insertion of carbon dioxide, 411–412
magnesium-based diamides, 412–415
Sn(II) compounds with new silylamine ligands, 418–420
zinc- and mercury-based disilylamides, 416–417
Doubly bridged isomer, group 14 element alkene analogs, 184–186
Drug delivery
closomers in, 319
unilamellar liposomes, 316

E

Electrical transport. *See* Boron nanowires
Electrochemical reduction, imidazolium ions, 272–273
Electron-deficient species
reasons to study, 52–53

See also Four-valence electron species; Six-valence electron species
Electron deformation density (EDD), diphosphenes, 7–8
Electronic structure, pnictogen cations, 109
Electron paramagnetic resonance (EPR) spectroscopy
dilithiated tetrakisimidophosphate radicals, 75–76
1,3-diphosphaallyl radical, 73
diphosphanyl radical, 71–72
phosphaverdazyl radicals, 77–78
phosphinyl radicals, 68–69
technique for characterizing radicals, 67
Encapsulation, carboracycles for, of hydrophobic species, 317, 318
Erbium
mixed metal 3d-4f complexes with benzimidazole based ligand, 232f, 233
See also Lanthanides with unusual main group ligands
Ethylene analogs. *See* Alkene analogs
Europium
mixed metal 3d-4f complexes with benzimidazole based ligand, 232f, 233
See also Lanthanides with unusual main group ligands

F

Ferrocene
dendrimer construction starting from, 349, 350
See also Dendrimers
Ferrocenium salts, reaction of N-heterocyclic carbenes (NHCs) with, 275
Flexibility, calixarenes, 238

Fluoride ion complexation. *See* 1,8-Diborylnaphthalenes
Fluoride sensors
  bidentate boranes as colorimetric, 211–213
  molecular recognition units for design, 208–209
Four-valence electron species $C_6H_7Si^+$, 59–64
  collision-induced dissociation (CID) spectra of unreactive $C_6H_7Si^+$, 61, 62*t*
  effect of charge on philicity of, 64
  energy profile for reaction of HSi:$^+$ and benzene, 63*f*
  ion-molecule reactions forming $C_6H_7Si^+$ and suggested structures, 62*f*
  nature of product from HSi:$^+$ and benzene, 64
  phenylsilicon cations Ph–Si:$^+$ attack on pyridine, 59, 61*f*
  quadrupole ion trap mass spectrometry (QITMS) experiments, 61
  reactions of PhSi$^+$ with pyridine and HSi:$^+$ with diethylamine, 61*f*
  silanetriyl (silyne) cations and reactions, 60*f*
  silatropylium ion, 59
  silyne cation, 59
  supersilylenes, 59
  See also Six-valence electron species

G

Gallium
  calculated minimum geometry for $(Et_3P)GaHCl_2$, 35, 39*f*
  Ga_4S_6 and adamantane cages, 33
  gallium(III) sulfide, Ga_2S_3, 33
  gallium trihydride, trihydride/halide and trihalide complexes of pyridines, 38, 41, 44
  gallium trihydride, trihydride/halide and trihalide complexes of tertiary phosphines, 34–35, 38
  hydrolysis of trimesitylaluminum and -gallium, 25
  intermolecular contacts in lattice of $(Py)GaHCl_2$, 41, 42*f*
  molecular gallium sulfides with ring and cage structures, 44, 47
  molecular structure of $(3,5-Me_2-Py)_2GaH_3$, 41, 43*f*
  molecular structure of $(3,5-Me_2-Py)_2GaHCl_2$, 38, 42*f*
  molecular structure of $(3,5-Me_2-Py)(4-H_2-3,5-Me_2-Py)GaH_2$, 41, 43*f*
  molecular structure of $(3,5-Me2-Py)GaCl_3$, 44, 45*f*
  molecular structure of $(3,5-Me_2-Py)GaHCl_2$, 38, 40*f*
  molecular structure of (4-NC-Py)GaHCl$_2$, 38, 40*f*
  molecular structure of $[(3,5-Me_2-Py)GaCl_2]_2$, 44, 45*f*
  molecular structure of $[(3,5-Me_2-Py)GaSBr]_3$, 44, 46*f*
  molecular structure of $[(3,5-Me_2-Py)GaSCl]_3$, 44, 46*f*
  molecular structure of $[(4-Me_2N-Py)Ga_4S_6]$, 47, 49*f*
  molecular structure of $[CH_2Ph_2PGaHCl_2]_2$, 37*f*
  molecular structure of $(Cy_3P)GaHCl_2$, 36*f*
  molecular structure of $(Et_3P)Cl_2GaGaCl_2(PEt_2)$, 38, 39*f*
  molecular structure of $(Ph_3P)GaHCl_2$, 36*f*
  structure of dication $[(4-Me_2N-Py)_6Ga_4S_5]^{2+}$ in bromide salt, 47, 48*f*

superposition of (Ph$_3$P)GaHCl$_2$ in monoclinic and triclinic modifications, 37f
ternary compounds GaYX (Y = chalogen, X = halogen), 33
Gallium alkoxides, donor-functionalized, 383, 384
Gallium oxide, chemical vapor deposition for thin films, 377
Gallium sulfides
defect ZnS structures, 33
molecular, with ring and cage structures, 44, 47
Geometry, organotin-sulfur compounds, 430–431
Germanium. *See* Alkene analogs
Germanium bis(bistrimethylsilylamide), carbon dioxide insertion, 411
Germanium-nitrogen heterocubanes
crystal structure, 126f
interatomic distances and bond angles, 128, 131t
synthesis, 122, 125, 134–135
See also Heterocubanes
Giant dendrimers. *See* Dendrimers
Gibbsite, aluminum trihydroxide, 21
Gluthathione, sulfur-containing, 424f
Green chemistry, electrochemical and chemical reduction of imidazolium ions, 272–273
Group 13 chalcogenides
[Al5(μ_4-O)(μ-O-iPr)$_7$H$_6$] (1) pentanuclear complex, 378, 379f
[Al(OC$_6$H$_3$Me$_2$-2,6)$_3$(NMe$_2$Et)] (5), 380
[Al(SC$_6$H$_3$Me$_2$-2,6)$_3$(OEt$_2$)] (6), 380–381
chemical vapor deposition (CVD) for thin films, 377
deposition studies, 382–383
donor-functionalized aluminum and gallium alkoxides, 383, 384
experimental, 385–388

[Li(OEt$_2$)$_3$][Al(SC$_6$H$_3$Me$_2$-2,6)$_4$] (7), 381–382
low pressure CVD (LPCVD), 384–385
LPCVD experiments, 388
reactivity of alane with alcohols and thiols, 378–383
syntheses, 386–388
TGA of compound (6), 383f
thin-film growth, 384–385
X-ray powder diffraction pattern of annealed Ga$_2$O$_3$ film, 385f
X-ray structure determination, 388
Group 14 elements. *See* Alkene analogs
Group 15 elements. *See* Pnictogens
Group chemistry. *See* Main group chemistry

H

^{1}H nuclear magnetic resonance (NMR), conformation determination, 244
Half-sandwich halolanthanacarboranes, carbons adjacent C$_2$B$_4$-carborane ligand systems, 302, 304–305
Half-sandwich titanocene, carboranyl-thiol-appended, 294–295, 296f
Heavier elements, multiply-bonded silicon compounds, 175t
Heptachlorodigallate, salts of [Mes*NP]$^+$, 282–283
Heterocubanes
crystal structure of compound (1), 128, 130f
crystal structures of (2) and (3), 126f, 127f
experimental, 134–135
germanium species, 122
interatomic distances and bond angles for compounds 1–3, 128, 131t, 132t, 133t

lead-nitrogen species, 122
lithium dimer of compound (1), 128, 130f
metal-N-metal and N-metal-N angles in Group 14-nitrogen heterocubane series, 128, 129t
preparation of [Ge(μ_3-NSiMe$_3$)]$_4$ (2), 122, 125
preparation of [(Me$_3$Sn)(Me$_2$Sn)NLi·Et$_2$O]$_2$ (1), 122, 124
preparation of [Pb(μ_3-NSiMe$_3$)]$_4$ (3), 122, 125
Sn$_4$N$_4$, 122
trimethylsilyl incorporation into germanium- or lead-nitrogen cubanes, 122, 128
Hoffmann predictions, bond-stretch isomers, 90
Hydrocarbons, unsaturated, reaction of phosphenium ions with, 9–10
Hydroelementation, choice for dendrimer synthesis, 349
Hydrogen atom addition, imidazol-2-ylidenes, 268–269
Hydrogen bonding, imidazol-2-ylidene borane complex exhibiting, 275–277
Hydrogen bonding interactions, diborane(4) compounds, 140, 148
Hydrolysis, trimesitylaluminum and -gallium, 25
Hydrolysis chemistry, diborane(4) compounds, 137–137, 140
Hydrosilylation
dendrimer growth, 351, 355
See also Dendrimers
Hydroxides of aluminum
organoaluminum dihydroxide, 21–23
organoaluminum monohydroxide, 24–26

I

Icosahedral borane
perhydroxylation of derivatives, 315
permethylation of derivatives, 314
Imidazol-2-ylidenes
borane complex of, exhibiting intermolecular dihydrogen bonding, 275–277
electrochemical and chemical reduction of imidazolium ions, 272–273
estimated partial charges on carbine-borane adduct, 277f
experimental confirmation using muonium (Mu), 269–271
hydrogen atom addition to, 268–269
one electron oxidation of carbenes, 273–275
possible reactions with hydrogen atoms or muonium (Mu), 268
reaction with tetracyanoethylene (TCNE), 273–275
synthesis of ^{13}C labeled, 271
synthesis of borane complex, 275–276
X-ray structure of carbine-borane adduct, 276f
See also N-heterocyclic carbenes (NHCs)
Imidazolium ions, electrochemical and chemical reduction, 272–273
Indium, triple-decker cation, 15, 16
Interatomic distances, germanium- and lead-nitrogen heterocubanes, 128, 131t, 132t, 133t
Intermolecular dihydrogen bonding, imidazol-2-ylidene borane complex exhibiting, 275–277
Intramolecular interactions. See Organotin-sulfur intramolecular interactions

Isomerism, bond-stretch
concept, 90
See also Diradicals

J

Jahn–Teller distortion, two-electron oxidation of closomers, 319

K

Karsted catalyst, hydrosilylation for dendrimer growth, 351

L

Lanthanacarboranes
"carbons apart"-carborane cages bonding to Ln(III) metal, 296–298
half-sandwich halolanthanacarboranes of carbons adjacent C_2B_4-carborane ligand systems, 302, 304–305
perspective view of complex, 299f
Lanthanide elements. See Metallacarboranes
Lanthanide model compounds, phosphate ester cleavage, 396, 397f
Lanthanides with unusual main group ligands
absorption spectra of modified Schiff-base ligand (3), and complexes (5) and (6), 225f
complexes involving coordination of Zn_2 and $ZnEu_2$ groups, 227, 230f, 231f
conventional "salen" type Schiff-base ligand with -C≡CSiMe_3 units, 221f
fluorescence of Yb(III) ions, 226
ligands from 1,2-diaminobenzene, 228f
mechanism of formation of Tb_{10} complex (14), 235
mixed metal $3d$-$4f$ complexes with benzimidazole based ligand, 232f, 233
mixed metal $3d$-$4f$ complexes with vanillin based materials, 233–235
near infrared (NIR) luminescence of (6), 225f
photophysical properties of lanthanides, 225
Schiff-base ligands for stabilization of $3d$-$4f$ complexes, 223f
synthesis of Zn_3 complex (5) and conversion to Zn_2Yb_2 dimer (6), 223f
Tb_{10} complex, 233f, 234f, 235
"tetra-decker" complex $[Tb_3L_4(H_2O)_2]Cl$, 227, 229f
trinuclear complexes of Tb (15) and Er (16), 235f
unusual Zn_3CO_8 complex (2), 222f
X-ray structure of $[ZnLnL_3(NO_3)_3]$ (8) (Ln = Nd, Eu, Tb, and Yb), 227f
X-ray structure of $[ZnYbL_2(NO_3)_2(OAc)] \cdot C_6H_5OH$ (7), 226f
X-ray structures of (5) and (6), 224f, 225
X-ray structures of dinuclear examples (Cu(II)–Yb) (12) and Cu(II)–(Yb, Eu, Tb and Er) (13), 232f, 233
Zn_3CO_8 preparation, 221–222
Lead
derivatives of singly bridged isomer, 186–189
trans-doubly bridged alkene analog, 184–186
triple-decker cation, 15, 16
See also Alkene analogs

Lead-nitrogen heterocubanes
  crystal structure, 127f
  experimental, 134–135
  interatomic distances and bond
    angles, 128, 132t
  synthesis, 122, 125
  See also Heterocubanes
Lewis acceptors
  phosphorus(III) centers, 280–281
  See also Coordination chemistry of
    phosphorus(III)
Lewis acid-base complexes, group 13
  complexes, 14–15
Lewis acids. See 1,8-
  Diborylnaphthalenes
Ligands
  1,1-dithiolates, 425, 426f
  modified Schiff-base, 221
  Schiff-base, for stabilization of 3d-
    4f complexes, 223f
  shape and size of
    tetrakis(disyl)diphosphine, 105–
    106
  See also β-Diketiminates;
    Lanthanides with unusual main
    group ligands
Liposomes, boron neutron capture
  therapy, 316
Lithium. See β-Diketiminates
Lithium amide species,
  trimethylstannyl-trimethylsilyl
  crystal structure, 130f
  dimer, 128
  experimental, 134
  interatomic distances and bond
    angles, 128, 133t
  preparation, 122, 124

M

Magnesium, resistivity of, and
  diboride, 364t
Magnesium-based diamides
  disilylamides, 413t
  possible bonding modes, 414f
  synthesis, 412–415
Main group chemistry
  analogues of carbenes, 2–3
  bonding of dimers of carbene-like
    fragments, 6–7
  boron lone pair orbitals, 12f
  bridging pnictinidene complexes, 9
  carbenoids, 3t
  charged phosphenium complexes,
    11
  cyclic oligomers of arsenidenes and
    phosphinidenes, 3–4
  cyclotetraarsine, 3
  cyclotetraphosphine reacting with
    tetramethyl diarsine, 3–4
  cyclotriphosphine, 3
  dimers of carbine-like fragments, 6
  electron deformation density
    (EDD) study of diphosphenes,
    7–8
  group 13 fragments assembling into
    clusters, 12–14
  molybdenum-phosphorus double
    bond, 8
  paradigm shift, 166–167
  phosphenium ions, 9–10
  phosphididenes and electrons for
    bonding to metal, 8
  phosphinidene production, 5–6
  phosphorus analogues of Wittig
    reagents, 4–5
  singlet and triplet states of four-
    valence-electron group 13
    fragments, 11–12
  triple-decker main group cation,
    14–15, 16
  See also β-Diketiminates;
    Metallacarboranes; N-
    heterocyclic carbenes (NHCs)
Mechanisms
  phosphorus analogues of Wittig
    reagents, 4–5
  proposed, for insertion of sulfur
    into Al–H bonds, 27

silicon multiple bond compounds, 176
Mercury-based disilylamides, synthesis, 416–417
Metal diborides, resistivities, 364t
Metallacarboranes
  carboranyl-thiol-appended half-sandwich titanocene and conversion to halotitanocene, 294–295, 296f
  commercially available, 313
  constrained-geometry metallocenes, 294–295
  crystal structure of half-sandwich chlorogadolinacarborane dimer, 305f
  electron-density map of typical oxo-lanthanacarborane cluster, 303f
  half-sandwich halolanthanacarboranes of carbons adjacent C_2B_4-carborane ligand systems, 302, 304–305
  lanthanacarborane complex with three "carbons apart" carborane ligands, 299f
  lanthanacarboranes with two or three "carbons apart"-carborane cages bonding to Ln(III) metal, 296–298
  nickelaborane, 322–323
  oxide ion-encapsulating tetralanthanide tetrahedron, 300–302
  rotary molecular motor based on nickelacarborane, 322–323
  syntheses of "carbons-apart" lanthanacarborane complexes, 297
  syntheses of half-sandwich halolanthanacarborane complexes, 304
  syntheses of oxo-lanthanacarborane clusters, 301

Metal nanoparticles, poly(methylphenylphosphazene) (PMPP) stabilizing, 339
Metals, resistivities, 364t
Methothionine, sulfur-containing, 424f
Microelectronics. See Boron nanowires
Minerals, hydrated forms of alumina, 21
Models
  ball and spring, of dissociation, 102f
  six-valence electron species, 52–53
Models of biological catalysts, phosphate ester cleavage, 396
Molecular architecture. See Closomers
Molybdenum-phosphorus double bond, phosphinidene complex, 8
Multiple bonds. See Silicon multiple bonds
Muonium (Mu)
  experimental confirmation using, 269–271
  reactions of imidazol-2-ylidenes with, 268–269

N

Nanoelectronics
  fabrication of circuitry, 363–364
  See also Boron nanowires
Nanotechnology
  promise of potential applications, 313
  See also Polyhedral boranes
Nanowires
  computing speed and power, 362–363
  fabrication of nanoelectronic circuitry, 363–364
  ideal components, 364
  room-temperature resistivities of metals and metal diborides, 364t
  See also Boron nanowires

Nerve agents, organophosphate esters, 390, 391–392
NH$_2$ groups, organoaluminum diamide, 29
N-heterocyclic carbenes (NHCs)
 adduct (2) of 1,3-bis-(2,4,6-trimethylphenyl)imidazol-2-ylidene and V(O)Cl$_3$, 256–257
 adducts of Group 4 metal tetrachlorides, 256
 (1,3-bis-(2,4,6-trimethylphenyl)imidazol-2-ylidene)TiCl$_2$(NMe$_2$)$_2$ (6), 260
 complexes of main group halides in high oxidation states, 261–263
 complexes of transition metal halides in high oxidation states, 255–261
 coordination chemistry, 253
 (1,3-diethyl-4,5-dimethylimidazol-2-ylidene)SiCl$_4$ (8), 261–262
 (1,3-diisopropyl-4,5-dimethylimidazol-2-ylidene)SnCl$_2$Ph$_2$ (9), 261–262
 electron configuration, 254f
 first stable carbenes, 254
 hypothetical (1,3-dimethylimidazol-2-ylidene)TiCl$_4$ (4) as model system, 258, 259f
 hypothetical model (1,3-dimethylimidazol-2-ylidene)SiCl$_4$ (10), 263
 hypothetical model (1,3-dimethylimidazol-2-ylidene)SnCl$_2$Ph$_2$ (11), 263, 264f
 infinite stability of, 255
 model complex (1,3-dimethylimidazol-2-ylidene)V(O)Cl$_3$ (3), 257
 model compound (1,3-dimethylimidazol-2-ylidene)TiCl$_2$(NMe$_2$)$_2$ (7), 260, 261f
 N←C→N π-interactions, 254f
 oxo-bridged dinuclear NHC-titanium(IV) chloro species (1), 256
 proposed pathway for dimerization of, 255f
 reaction of, with tetracyanoethylene (TCNE), 273–275
 reactivity with ferrocenium salts, 275
 reactivity with small reagents, 267
 synthesis of ^{13}C labeled, 271
 terminal phosphinidene complexes, 8–9
 triazacyclic, 1,3,4-triphenyl-1,2,4-triazol-5-ylidene coordinated to Re(V) (5), 259
 typical structural motifs for stable carbenes, 253f
 X-ray crystal structure of (1), 258f
 See also Imidazol-2-ylidenes
Nickelacarborane, rotary molecular motor based on, 322–323
Nucleophilic substitution, synthesis of phosphazenes, 335–336

O

One electron oxidation, carbenes, 273–275
Organoaluminum diamide, preparation, 29
Organoaluminum dihydroxide, preparation, 21–23
Organoaluminum diselenol, preparation, 28
Organoaluminum dithiol, preparation, 26–28
Organoaluminum monohydroxide, preparation, 24–26
Organophosphate esters
 chemical warfare and pesticides, 391–392
 structures of nerve gases and pesticides, 392f

uses, 390
See also Phosphate esters
Organotin-sulfur intramolecular interactions
  angles, 430–431
  biocidal activity of organotin compounds, 423–425
  biologically important sulfur-containing residues, 424*f*
  bond distances, angle sums, and closeness to trigonal bipyramid, 430*t*
  bond order (BO) and geometries, 429–430
  compounds containing, 429*t*
  general types of organotin sulfides, 428*f*
  intramolecular organotin 1,1-dithiolate interactions, 425–428
  potential competition, 423*f*
  selected organotin 1,1-dithiolates, 427*t*
  Sn---S and Sn—R_{ax} bond orders, 432
  S–Sn bond order as function of S–Sn intramolecular distance, 433*f*
  S---Sn distances, 431
  structural motifs of organotin 1,1-dithiolates, 426*f*
  transition from tetrahedral to trigonal bipyramidal geometry, 430, 431*f*
Oxidation states. *See* N-heterocyclic carbenes (NHCs); Pnictogens

P

P1 nuclease, phosphate ester cleavage, 394
Paradigm shift, main group chemistry, 166–167
Paraoxon, organophosphate ester, 391–392
Parathion, organophosphate ester, 391–392
Pb. *See* Lead
Pb-nitrogen. *See* Heterocubanes
Perhydroxylation, icosahedral borane derivatives, 315
Permethylation, icosahedral borane derivatives, 314
Persistent radicals. *See* Phosphorus-containing radicals
Pesticides, organophosphate esters, 391–392
Phenyloxenium ion, gas-phase reaction with ethylene, 56
Phenylsilicon cations, attack on pyridine, 59, 61*f*
Phenylsilyl cation
  collision-induced dissociation of unreactive, 61–62
  $HSi:^+$ and benzene complex as unreactive, 60–61
  ion-molecule reactions forming, 62*f*
  π-complex form, 62–63
Phosphadiazonium cation
  coordination complexes of, 281–284
  NMR and structural parameters for complexes of, 282*t*
  P–P coordinate interaction, 284
  solid state structure, 283*f*, 284*f*
  structure, 281
  tetrachloroaluminate, tetrachlorogallate, and heptachlorodigallate salts, 282–283
  See also Coordination chemistry of phosphorus(III)
Phosphastilbene, stable compounds, 6–7
Phosphate esters
  alkaline phosphatase (AP), 393, 394*f*
  attack of P–O–R bond by halide anion, 404*f*

binuclear boron alkoxides and borosiloxides, 398, 399f
binuclear boron halides, 398–400, 401f
binuclear zinc and lanthanide model compounds, 396f
biological systems, 391
chelated boron compounds, 397, 398f
cleavage in biology, 393–396
cleavage using chelated boron compounds, 397–404
cleavage using models of biological catalysts, 396, 397f
dealkylation of trimethyl phosphate with Salen boron chloride compounds, 402t
formation of phosphate coordinated boron cation, 403f
future studies, 405
monoester, diester, and triester, 391f
P1 nuclease, 394
percent dealkylation of various phosphates with Salen boron bromide compounds, 402t
percent dealkylation of various phosphates with Salen phenyl boron bromide compounds, 403t
phosphate dealkylation with boron halide chelates, 400, 404
phosphate triesterase (PTEs), 395–396
phospholipase C enzymes, 395
purple acid phosphatase (PAP), 393–394, 395f
Salen(tBu) ligands, 397, 398f
See also Organophosphate esters
Phosphate triesterase (PTEs), phosphate ester cleavage, 395–396
Phosphaverdazyl radicals
electron nuclear double resonance (ENDOR) experiment, 77f
electron paramagnetic resonance (EPR) characterization, 77–78

preparation, 76
structure and stability, 78
structures, 77
See also Phosphorus-containing radicals
Phosphazenes
alkyl/aryl substituted cyclic and polymeric, 343–344
cyclic, 340–344
derivatization of cyclic, via deprotonation-substitution reactions, 340–341
description, 335
dialkyl, polymer, 336
electrophilic aromatic substitution reaction, 337
isolation of isomers by column chromatography, 343
isomers of calix-4-arene, 341, 342
nucleophilic substitution, 335–336
polymeric, 337–339
polymeric anion intermediate, 338–339
polymerization, 336
poly(methylphenylphosphazene) (PMPP), 337–339
thermal ellipsoid plot of cis-(Me(Ph)P=N)$_3$, 341f
transmission electron microscopy (TEM) of PMPP-gold nanoparticle composite, 339f
Phosphenium cations. See Coordination chemistry of phosphorus(III)
Phosphenium complexes, examples of charged, 11
Phosphenium ions
addition and insertion processes, 53–55
bicyclic cations, 10
insertion reactions, 10–11
quadrupole ion trap mass spectrometry (QITMS), 56
reacting with unsaturated hydrocarbons, 9–10

reactions with 1,4-dienes and with cyclopropanes, 55–56
resembling singlet carbenes, 9
structure, 10f
Phosphine-phosphinidenes, synthesis, 3–4
Phosphines
(silylamino)phosphines, 326–330
(silylanilino)phosphines, 330–333
See also Coordination chemistry of phosphorus(III)
Phosphines, tertiary
gallium trihydride, trihydride/halide, and trihalide complexes, 34–35, 38
See also Gallium
Phosphinidenes
alternative approaches, 5–6
bonding, 8
cyclic oligomers, 3–4
production of free, 5
terminal complexes of N-heterocyclic carbenes, 8–9
Phosphinyl radicals
electron paramagnetic resonance (EPR) characterization, 68–69
preparation, 68
structure and stability, 69–70
See also Phosphorus-containing radicals
Phosphirenylium ion, approach to preparation, 57
Phospholipase C, phosphate ester cleavage, 395
Phosphorenylium ion cyclo-$(CH)_2P^+$
charged cyclopropenylidene analog, 57
energy profiles for reaction with acetylene, 58f
product from PBr^+ and acetylene, 58–59
Phosphorus
calix[5]arene ligands, 249, 250f
calixarene with Ti–O bond, 246, 247f

framework of calix[4]arenes, 238, 240f
insertion of bridging, groups into calix[5]arene, 248
X-ray crystal structures of calix[4]arenes, 242f, 243f
See also Calixarenes; Coordination chemistry of phosphorus(III); Pnictogens
Phosphorus and boron
stable singlet diradical, 83–85
See also Diradicals
Phosphorus-arsenic double bonds, stable compounds, 6–7
Phosphorus-containing radicals
chemical and structural diversity, 67–68
dilithiated tetrakisimidophosphate radicals, 74–76
1,3-diphosphaallyl radical, 72–74
diphosphanyl radical, 70–72
electron paramagnetic resonance (EPR) spectroscopy, 67
history of stable radicals, 66–67
phosphaverdazyl radicals, 76–78
phosphinyl radicals, 68–70
See also Diradicals
Phosphorus-phosphorus double bonds, stable compounds, 6–7
Pnictinidene complexes, bridging, bonding, 9
Pnictogens
canonical forms of Pn^+ cations stabilized by phosphines, 109f
cationic four-coordinate, planar P(I) atom bound to four zirconicenes, 117, 118
cations for +1 oxidation state where Pn=P, 110
compounds containing, in +1 oxidation state, 109
conversion of cations stabilized by PPh_3 to other cations, 111
cyclic dications, 118–119

cyclic triphosphenium cation
  synthesis, 114–115
  early work, 110–112
  electronic structure of cations, 109
  four-, six-, and eight-membered
    rings, 113
  generation of phosphaalkenes of
    type ArHC=PR, 119
  insertion of P^+ into C=C bond, 112
  molecular structure of
    [(dppe)As]$_2$[SnCl$_6$] where
    dppe=cis-1,2-
    bis(diphenylphosphino)ethane,
    115f
  molecular structure of
    [(dppe)P][BPh$_4$], 117F
  molecular structure of [(dppe)P][I],
    116f
  oxidation of acyclic
    "triphosphenium" cations, 111,
    112
  oxidation of diphosphine ligand to
    liberate As–I fragments, 118
  phosphine stabilized
    phosphinidenes, 119
  P(I) cations as possible phospha-
    Wittig reagents, 119
  recent developments, 112–119
  stability of [(dppe)P][BPh$_4$], 117
  stability of iodide salt of
    [(dppe)P]$^+$, 116
  structure identification, 111
  synthesis of As(I) and P(I) cations,
    112–113
  zwitterions containing Pn(I) atoms,
    113
Polyhedral boranes
  carboracycles, 317, 318
  closomers, 318–322
  commercially available boranes,
    carboranes, and
    metallacarboranes, 313
  current applications of boranes and
    carboranes in supramolecular

chemistry and nanoscience,
  316–317
perhydroxylation of icosahedral
  borane derivatives, 315
permethylation of icosahedral
  borane derivatives, 314
rotary molecular motor based on
  nickelacarborane, 322–323
self-assembly of rod structures,
  316–317
stable unilamellar liposomes for
  boron neutron capture therapy,
  316
Polymeric phosphazenes. See
  Phosphazenes
Polymers, silyne and disilyne, 176
Poly(methylphenylphosphazene)
  (PMPP)
  condensation polymerization, 337
  stabilizing metal nanoparticles,
    339
  See also Phosphazenes
Polyphosphazenes
  preparation, 336
  See also Phosphazenes
Potassium. See β-Diketiminates
Purple acid phosphatase (PAP)
  phosphate ester cleavage, 393–394
  proposed mechanism of hydrolysis,
    395f
Pyridines
  attack of phenylsilicon cation Ph–
    S:$^+$ on, 59, 61f
  gallium trihydride,
    trihydride/halide, and trihalide
    complexes of, 38, 41, 44
  See also Gallium

Q

Quadrupole ion trap mass
  spectrometry (QITMS)
  phenyloxenium ions, 56

reactions of PhSi:$^+$ with pyridine and HSi:$^+$ with diethylamine in, 61
six-valence electron charged carbene analogs, 56–57
study of ion-molecule reactions, 56

R

Radicals
bis(disyl)phosphido, 97f, 98f
bis(trimethylsilylamido)(di-isopropylamido)phosphido, 106f
dissociation of X–X molecule to two X, 102f
See also Diradicals; Phosphorus-containing radicals
Reactivity
lanthanacarboranes, 298
stable singlet diradical, 85–88
Rhenium(V). See N-heterocyclic carbenes (NHCs)
Ring structures
molecular gallium sulfides, 44, 47
See also Boron-phosphorus ring and cage chemistry
Rod structures, self-assembly, 316–317

S

Salen compounds. See Boron chelates
SARACEN (structure analysis restrained by ab initio calculations for electron diffraction) method, solving gas-phase structures, 97–98
Sarin gas, organophosphate ester, 391–392
Schiff-base ligands
complex of Zn(II) as unusual main group system, 222–223
conventional "salen", 221f

stabilization of $3d$-$4f$ complexes, 223f
See also Lanthanides with unusual main group ligands
SeH functionality, organoaluminum diselenol, 28
Selenium, reaction with stable singlet diradical, 86
Self-assembly, rod structures of boranes and carboranes, 316–317
Semiconductors
boron nanowires, 365
computing speed and power, 362–363
elemental boron, 364
gate-oxide thickness, 363
nanowire type, 363–364
See also Boron nanowires
Sensors, colorimetric fluoride ion, bidentate boranes as, 211–213
Silanetriyl cations, potential bond forming, 59, 60f
Silanones, Si=O species, 175
5-Silapentafulvalene, synthesis, 171
Silapentatriafulvalene, synthesis, 171
4-Silatriafulvene, synthesis, 170–171
Silatropylium ion
benzene and HSi:$^+$, 59
energy profile for formation, 63f
structure, 63
Silenes
evidence for silene intermediates, 168–169
generation by photolyzing trimethylsilyldiazomethane, 169
isolation of stable, 169–170
Silicon
four- and five-coordinate, calix[4]arenes, 241f
insertion of bridging silyl groups into calix[5]arene, 247f, 248
partial cone and cone conformer of calix[4]arene, 241

use as protecting groups in
  synthesis of 1,2-disubstituted
  calix[4]arenes, 244, 245f
  See also Alkene analogs;
  Calixarenes; N-heterocyclic
  carbenes (NHCs)
Silicon multiple bonds
  absorption maxima and colors of
    matrix-isolated silylenes, 173t
  aromatic compounds with Si atom
    in ring, 171–172
  bond distance of Si=C, 170
  bond length of Si=Si, 174
  calicenes, 171
  dimethylsilylene, 173
  discovery, 167
  disilenes, 172–174
  first bis-silene, 170
  future possibilities, 175–176
  heavier elements, 175t
  isolation of stable silene, 169–170
  reaction mechanisms, 176
  silanones, 175
  5-silapentafulvalene, 171
  silapentatriafulvalene, 171
  4-silatriafulvene, 170–171
  silenes, 168–172
  silylenes, 172–173
  silyne and disilyne polymers, 176
  tetramesityldisilene, 173–174
  triply-bonded silicon compounds,
    175–176
  types of multiply-bonded silicon
    compounds, 175t
(Silylamino)phosphines
  conversion, 330
  Wilburn method for synthesis,
    326–330
(Silylanilino)phosphines
  derivative chemistry, 332–333
  synthesis, 330–332
Silylenes
  absorption maxima and colors of
    matrix-isolated, 173t
  addition and insertion processes, 53

dimethylsilylene, 173
  generation, 172–173
Silyne cation, potential bond forming,
  59, 60f
Silyne polymers, silicon multiple bond
  compounds, 176
Singly bridged isomer, group 14
  element alkene analogs, 186–189
Six-valence electron species
  addition and insertion processes of
    phosphenium ions, 53–55
  charged cyclopropenylidene analog
    phosphorenylium ion cyclo-
    $(CH)_2P^+$, 57
  cyclic phosphirenylium ion, 58–59
  effect of charge on philicity of, 64
  energy profile for exothermic
    reactions of PBr^+ and acetylene,
    57f
  energy profiles for reactions of
    cyclo-$(CH)_2P$:$^+$ and $H_2C=C=P$:$^+$
    with acetylene, 58f
  families of, reactive intermediates,
    54f
  observations by quadrupole ion trap
    mass spectrometry (QITMS), 56
  predicted energy profile for
    addition of $(MeO)_2P^+$ to
    butadiene, 55f
  reaction of PBr^+ with acetylene,
    57–58
  reaction of Ph-O^+ with ethylene,
    56f
  reactions of phosphenium ions with
    1,4-dienes and cyclopropanes,
    55–56
  See also Four-valence electron
    species
Size exclusion chromatography
  (SEC), aggregation and
  dendrimers, 351, 356f
Sn compounds
  carbon dioxide insertion in Sn
    bis(bistrimethylsilylamide), 411
  new silylamine ligands, 418–420

See also Tin (Sn)
Stable radicals. See Diradicals;
   Phosphorus-containing radicals
Stannylstannylene, unsymmetric form,
   182–183
Structure identification, symmetric
   P(I) stabilized cations, 111
Sulfides
   molecular gallium, with ring and
      cage structures, 44, 47
   See also Gallium
Sulfur
   proposed mechanism for insertion
      into Al–H bonds, 27
   See also Organotin-sulfur
      intramolecular interactions
Supersilylenes, reactivity, 59
Supramolecular chemistry,
   applications of boranes and
   carboranes, 316–317

T

T-atom (artificial tetrahedron atom),
   Ga_4S_6 core, 33
Terbium (Tb)
   mixed metal $3d$-$4f$ complexes with
      benzimidazole based ligand,
      232f, 233
   tetra decker complex
      $[Tb_3L_4(H_2O)_2]Cl$, 227, 229f
   unusual Tb_{10} complex, 233–235
   See also Lanthanides with unusual
      main group ligands
Tertiary phosphines
   gallium trihydride,
      trihydride/halide, and trihalide
      complexes, 34–35, 38
   See also Gallium
Tetraalkyldiphosphine, dialkyl
   phosphido radicals, 95
Tetrachloraluminate, salts of
   $[Mes*NP]^+$, 282–283

Tetrachlorogallate, salts of
   $[Mes*NP]^+$, 282–283
Tetracyanoethylene, reaction of N-
   heterocyclic carbenes with, 273–
   275
Tetrakis(disyl)diarsine
   comparison of AsCSi angles in,
      conformers and P analog, 101t
   crystal structure, 99–100
   structures of solid and gaseous, 95–
      100
   thermodynamics of dissociating,
      100–104
   variations of some AsCSi angles in
      crystalline phase, 100t
Tetrakis(disyl)diphosphine
   geometric parameters for
      crystalline, 96t
   geometric parameters for gaseous,
      99t
   intrinsic bond dissociation energies,
      104f
   ligand shape and ligand size, 105–
      106
   radial distribution curve for
      bis(disyl)phosphido radical, 97f
   solving gas-phase structures with
      SARACEN (structure analysis
      restrained by ab initio
      calculations for electron
      diffraction) method, 97–98
   structure in crystalline phase, 96f
   structures of half, and the
      bis(disyl)phosphido radical,
      98f
   structures of solid and gaseous, 95–
      100
Tetralanthanide tetrahedron, oxide
   ion-encapsulating, 300–302
Tetramesityldisilene, synthesis, 173–
   174
Thermodynamics
   ball and spring model of
      dissociation, 102f

comparison of intrinsic bond
dissociation energies for
diphosphines and disilanes,
104f
dissociating diphosphines and
related compounds, 100–104
dissociation energies and intrinsic
bond energies for diphosphine
and disilanes, 105t
energetics of dissociation of
molecule X–X to two X
radicals, 102f
energy of diphosphine molecule,
101
ligand shape and size, 105–106
relationships between energies of
dissociation of sterically
crowded X–X and hypothetical
unstrained molecule, 103f
true dissociation energy, 101
Thin films
chemical vapor deposition, 377
growth by low pressure CVD, 384–385, 388
See also Group 13 chalcogenides
Thiol functionality, organoaluminum
dithiol, 26–28
Tin (Sn)
derivatives of mixed valence
isomer, 182–184
derivatives of singly bridged
isomer, 186–189
derivatives of trans-doubly bridged
alkene analog, 184–186
tin disilylamides preparation and
characterization, 419t
triple-decker cation, 15, 16
See also Alkene analogs; N-heterocyclic carbenes (NHCs);
Organotin-sulfur intramolecular
interactions; Sn compounds
Titanium
resistivity of, and diboride,
364t

See also N-heterocyclic carbenes
(NHCs)
Titanocene, carboranyl-thiol-appended
half-sandwich, 294–295, 296f
Transition metals. See N-heterocyclic
carbenes (NHCs)
Transmission electron microscopy
(TEM),
poly(methylphenylphosphazene)
(PMPP)-gold nanoparticle
composite, 339f
Trimesitylaluminum and -gallium,
hydrolysis by ^{1}H NMR, 25
Trimethyltin hydride, reaction with
stable singlet diradical, 86, 87f
Triphosphenium cations
oxidation of acyclic, 111, 112
synthesis of cyclic, 114–115
Triple bonds, silicon compounds,
175–176
Triple-decker main group cation
dilead, 15, 16
group 13 complexes, 14–15
Tungsten, insertion into calix[5]arene
and control of phosphorus/metal
interaction, 248, 249f
Tungsten-phosphorus bond, terminal
phosphinidene complex, 8

U

Unsaturated hydrocarbons, reaction of
phosphenium ions with, 9–10

V

Vanadium (V). See N-heterocyclic
carbenes (NHCs)
Vinylidenephosphenium ion,
formation, 58
VX gas, organophosphate ester, 391–392

W

Wilburn method
  (silylamino)phosphine synthesis, 326–330
  (silylanilino)phosphine synthesis, 330–332
Wittig reagents
  phosphorus analogues, 4–5
  possible phospha-, 119

Y

Ytterbium (Yb)
  mixed metal $3d$-$4f$ complexes with benzimidazole based ligand, 232f, 233
  See also Lanthanides with unusual main group ligands

Z

Zinc
  complexes from Schiff base ligands with C_4 and C_3 backbones, 226–227
  complexes involving coordination of Zn_2 and $ZnEu_2$ groups, 227, 230f, 231f
  phosphate ester cleavage using binuclear, 296
  tetrameric Zn_2Yb_2 complex, 223, 224f
  trinuclear Zn_3 complex, 223, 224f
  unusual Zn_3CO_8 complex, 221, 222f
  See also Lanthanides with unusual main group ligands
Zinc amides, synthesis of alkyl, 417–418
Zinc-based disilylamides, synthesis, 416–417
Zirconicenes, cationic four-coordinate planar P(I) atom bound to, 117–118
ZnS structures, defect, in gallium(III) sulfide, 33
Zwitterions, containing Pn(I) atoms, 113–114